Balanced Automation Systems

IFIP – The International Federation for Information Processing

IFIP was founded in 1960 under the auspices of UNESCO, following the First World Computer Congress held in Paris the previous year. An umbrella organization for societies working in information processing, IFIP's aim is two-fold: to support information processing within its member countries and to encourage technology transfer to developing nations. As its mission statement clearly states,

> IFIP's mission is to be the leading, truly international, apolitical organization which encourages and assists in the development, exploitation and application of information technology for the benefit of all people.

IFIP is a non-profitmaking organization, run almost solely by 2500 volunteers. It operates through a number of technical committees, which organize events and publications. IFIP's events range from an international congress to local seminars, but the most important are:

- the IFIP World Computer Congress, held every second year;
- open conferences;
- working conferences.

The flagship event is the IFIP World Computer Congress, at which both invited and contributed papers are presented. Contributed papers are rigorously refereed and the rejection rate is high.

As with the Congress, participation in the open conferences is open to all and papers may be invited or submitted. Again, submitted papers are stringently refereed.

The working conferences are structured differently. They are usually run by a working group and attendance is small and by invitation only. Their purpose is to create an atmosphere conducive to innovation and development. Refereeing is less rigorous and papers are subjected to extensive group discussion.

Publications arising from IFIP events vary. The papers presented at the IFIP World Computer Congress and at open conferences are published as conference proceedings, while the results of the working conferences are often published as collections of selected and edited papers.

Any national society whose primary activity is in information may apply to become a full member of IFIP, although full membership is restricted to one society per country. Full members are entitled to vote at the annual General Assembly, National societies preferring a less committed involvement may apply for associate or corresponding membership. Associate members enjoy the same benefits as full members, but without voting rights. Corresponding members are not represented in IFIP bodies. Affiliated membership is open to non-national societies, and individual and honorary membership schemes are also offered.

Balanced Automation Systems

Architectures and design methods

Proceedings of the IEEE/ECLA/IFIP international conference on architectures and design methods for balanced automation systems, 1995

Edited by

Luis Camarinha-Matos
New University of Lisbon
Lisbon, Portugal

and

Hamideh Afsarmanesh
University of Amsterdam
Amsterdam, The Netherlands

Published by Chapman & Hall on behalf of the
International Federation for Information Processing (IFIP)

CHAPMAN & HALL
London · Glasgow · Weinheim · New York · Tokyo · Melbourne · Madras

Published by Chapman & Hall, 2–6 Boundary Row, London SE1 8HN, UK

Chapman & Hall, 2–6 Boundary Row, London SE1 8HN, UK

Blackie Academic & Professional, Wester Cleddens Road, Bishopbriggs, Glasgow G64 2NZ, UK

Chapman & Hall GmbH, Pappelallee 3, 69469 Weinheim, Germany

Chapman & Hall USA, 115 Fifth Avenue, New York, NY 10003, USA

Chapman & Hall Japan, ITP-Japan, Kyowa Building, 3F, 2-2-1 Hirakawacho, Chiyoda-ku, Tokyo 102, Japan

Chapman & Hall Australia, 102 Dodds Street, South Melbourne, Victoria 3205, Australia

Chapman & Hall India, R. Seshadri, 32 Second Main Road, CIT East, Madras 600 035, India

First edition 1995

Printed in Great Britain by TJ Press Ltd, Padstow, Cornwall

ISBN 0 412 72200 3

A catalogue record for this book is available from the British Library

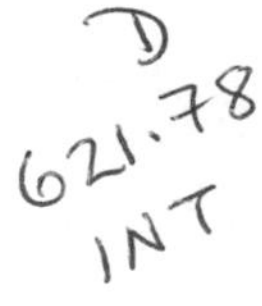

∞ Printed on permanent acid-free text paper, manufactured in accordance with ANSI/NISO Z39.48-1992 and ANSI/NISO Z39.48-1984 (Permanence of Paper).

CONTENTS

Technical Co-Sponsors:

IEEE Robotics and Automation Society
IEEE Systems, Man and Cybernetics Society
IEEE Control Systems Society

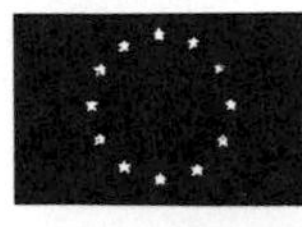

Commission of the European Union

IFIP WG 5.3

SBA - Sociedade Brasileira de Automática

AP3I - Associação de Programas de Integração e Informática Industrial

Introduction
Towards Balanced Automation

The concept. Manufacturing industries worldwide are facing tough challenges as a consequence of the globalization of economy and the openness of the markets. Progress of the economic blocks such as the European Union, NAFTA, and MERCOSUR, and the global agreements such as GATT, in addition to their obvious economic and social consequences, provoke strong paradigm shifts in the way that the manufacturing systems are conceived and operate.
To increase profitability and reduce the manufacturing costs, there is a recent tendency towards establishing partnership links among the involved industries, usually between big industries and the networks of components' suppliers. To benefit from the advances in technology, similar agreements are being established between industries and universities and research institutes. Such an open tele-cooperation network may be identified as an extended enterprise or a virtual enterprise. In fact, the manufacturing process is no more carried out by a single enterprise, rather each enterprise is just a node that adds some value (a step in the manufacturing chain) to the cooperation network of enterprises. The new trends create new scenarios and technological challenges, especially to the Small and Medium size Enterprises (SMEs) that clearly comprise the overwhelming majority of manufacturing enterprises worldwide. Under the classical scenarios, these SMEs would have had big difficulties to access or benefit from the state of the art technology, due to their limited human, financial, and material resources. The virtual enterprise partnerships facilitate the access to new technologies and novel manufacturing methodologies while imposing the application of standards and quality assurance requirements. In contrast to the more traditional hierarchical and rigid departmental structures, the "team work" and groupware are being adapted as a pragmatic approach to integrate contributions from different expertise. Furthermore, as an extension of the concurrent engineering concept, the team work approach is applied to all activities involved in manufacturing, and not only to the engineering areas.

The efforts spent on the implantation of high speed networks (digital highways) and on supporting the multimedia information has opened new opportunities for team work in multi-enterprise / multi-site / multi-platform networks. This new scenario however identifies new requirements in the area of distributed information logistics infrastructure, namely the view integration, view migration, view modification, and view update propagation. Other requirements are identified in the area of information access control, namely access rights to the data, scheduling of access, and the control of data migration. Furthermore, today's market imposes the shorter product life-cycle, the growing demands for customized products, and the higher number of product-variants, that in turn requires greater flexibility / agility from the manufacturing process. Although many of these aspects were anticipated years ago by the academic research, only now they are becoming a reality for the SMEs.
On the other hand, the immense legacy from the past in terms of the manufacturing -equipments, -infrastructures, -processes, -culture and -skills still plays a major role. The social questions raised at present on one hand by the growing unemployment number in the developed countries and on the other hand by the poor economic structure of the developing countries strongly suggest that in order to achieve balanced solutions, the manufacturing automation must be considered in a global context.

The term "balanced automation" as an emerging concept is developed in the framework of the collaborative research activity between the European Union and the Latin America, within the two ECLA (EC-Latin American) projects of CIMIS.net and FlexSys. Balanced automation captures the idea of an appropriate level of technical automation, as opposed to both the pure anthropocentric or the total automation approaches.

Contributing tendencies. Any approach to the design and implantation of a balanced automation system must consider, with an appropriate combination of weights, the results and experiences gained in related recent efforts. The main areas of interest to capture different dimensions of SMEs and their balanced automation can be categorized into the Technocentric-aspects, the Anthropocentric-aspects, and the Economical-aspects.
Every one of these categories contains several areas of research. However in most cases the research in one area is traditionally performed independent of the other areas. As a result, the design of a balanced automation system often encounters mismatch problem and the need to reevaluate and modify Existing research results.
Some of the research that is covered by the three categories are listed below:

-Technocentric aspects cover the Information Modeling and Information Integration, Integrating Infrastructures and facilities, Standards and Reference Models, Intelligent Computer Aided Tools and user-interfaces, Intelligent Supervision Systems, and Robotics;

-Anthropocentric aspects cover the Social Impacts, Skills and Training, Ergonomic Workplaces, Human-Machine Interaction, Team Interaction, and Organization of Responsibilities; and

-Economic aspects cover the cost-effectiveness, quality assurance, both the local and the global organizational aspects, and the important aspects of the strategies for smooth migration and transitions from existing systems to more advanced solutions.

So far, some international projects and also some projects with consortia at the national level have formed forums for melting these efforts. Nevertheless, there is still a prominent need to stress these aspects, their interrelationships, and the development of methodologies for the gradual migration from and/or transformation of existing systems, to replace the more classical approach of "designing completely new systems". Nowadays, opportunities to build new manufacturing systems are tremendously smaller than the needs to transform the legacy scenarios.

Main aspects covered by the book. The topic of balance automation, being a new emerging concept and first "publicized" by the BASYS'95 Conference, is not clearly covered in all papers included in this book. Nevertheless, many of the basic questions are addressed and some preliminary results are presented in various chapters. The main aspects included in this book are:

-Enterprise modeling and organization
-Modeling and design of FMS
-Anthropocentric systems
-Extended enterprises
-Decision support systems in manufacturing
-Balanced flexibility
-CAE/CAD/CAM integration
-Computer aided process planning
-Scheduling systems
-Multiagent systems architecture
-Intelligent supervision systems
-Monitoring and sensors
-Shop floor control

Areas that need more advances. Balanced automation covers a vast area of research. In our opinion, certain areas play a more important role and require additional advanced and/or fundamental research, among which we can name:

-modeling and analysis tools
-team work infrastructure design
-decision support tools
-rapid model prototyping
-migration methodologies
-management of change

Although some of these aspects have been the subject of intense research in the recent past, a new "rethinking" is necessary in terms of balanced automation. For instance, a migration methodology must first produce a realistic and comprehensive image of the existing enterprise, and then use it as a base for the purpose of analysis and diagnosis towards the identification of less balanced parts of the system. On the other hand, manufacturing enterprises must be empowered with the capability to cope with the market dynamism. The ability to manage and represent the effects of market changes within the enterprise provides a strong "mechanism" for continuous improvement and a potential for survival in today's market. These approaches, mostly designed for big industries so far, are still in their infancy and any improvement in the development of either approach sets new requirements and necessary modifications on the development of the others.

Overview of CIMIS.net and FlexSys projects. In 1993, the Commission of the European Union launched some pilot Basic Research Exploratory Collaborative Activities between the European and Latin American academic and research centers. As an exploratory program, the intention was to evaluate the feasibility of such long distance cooperation and to investigate the most effective ways of its implementation. Two of these pilot projects are the CIMIS.net and FlexSys that focus in the area of Computer Integrated Manufacturing.

The CIMIS.net project is focused on research on Distributed Information Systems for Computer Integrated Manufacturing. The CIMIS.net consortium comprises academic and research centers from The Netherlands, Portugal, and Spain from EU, and Brazil and Costa Rica from the Latin America. An aim of the project is to exchange and evaluate experiences in the following areas:

-Integrating Infrastructure for the distributed industrial systems - the paradigm of federated architectures.

-Distributed object-oriented information management systems - distributed access, interoperability, persistency, consistency, and concurrency.

-STEP/Express based modeling of manufacturing applications.

-Experiments on modeling and representation constructs for engineering application - integration of multiple object views with incomplete knowledge, and the temporal aspects, namely support for the versions and revisions of design and history, and the time sequence for the control processes.

-Identification of information management requirements for simultaneous/concurrent engineering, and specification of the supporting functionalities.

The FlexSys project addresses the Design Methods for FMS/FAS Systems in SMEs. This project aims at promoting the collaboration in research, development and evaluation of methods for designing Intelligent Manufacturing Systems. The project involves partners from Portugal and Germany form EU, and Argentina, Brazil and Mexico from the Latin America. One main focus of the research is on the control architecture for SME scenarios. The areas covered include:

- Electronic catalog of CIM components and interactive planning of IMS.
- Manufacturing systems control architectures - distributed approaches.

- Systems monitoring - multisensorial information measurement and interpretation, diagnosis, prognosis, and error recovery.
- Support systems including: transputer-based architectures, distributed object-based systems, simulation of manufacturing systems, etc.
- Distributed factory information and communication systems.

The geographical characteristics of the centers involved in these two projects provide a rich variety of diversified contexts for research on balanced automation; namely, the developed and developing countries, different directions in research and technology applied to the industry, and the diverse cultural and economic situations. The two projects presented a unique opportunity for collaborative research, in spite of the rather limited financial support received from EC. Within this context, through joint research, exchange of scientists, technical meetings and workshops, and the case study missions with some industries, an adequate environment was created for the genesis of the Balanced Automation concept and the initiative of launching the international BASYS Conference.

The concept of Balanced Automation Systems - BASYS is already accepted and applied by research groups outside the CIMIS.net and FlexSys projects. Some of these groups have significantly contributed to this proceedings. We hope that BASYS '95 will be an important milestone in the advanced developments in this area of research.

The Editors

Luis M. Camarinha-Matos
Hamideh Afsarmanesh

Plenary Sessions

PART ONE

Invited Talks

1

The Extended Enterprise - Manufacturing and The Value Chain

Professor J. Browne,
CIMRU,
University College Galway,
Galway,
Ireland.
Tel : +353.91.750414.
Fax: +353.91.562894.
E Mail : Jimmie.Browne@ucg.ie

Abstract

The Extended Enterprise where core product functionalities are provided separately by different companies who come together to provide a customer defined product is made possible by the emerging integration of computing and telecommunications technologies. These technologies facilitate the development of competitive advantage by exploiting linkages in the value chain. If the challenge of CIM (Computer Integrated Manufacturing) was to realize integration within the four walls of the plants, the challenge to manufacturing systems analysts and researchers today is to support inter-enterprise networking across the value chain. Changes in product and process technologies and the emerging pressure for environmentally benign production systems and products further enhance this challenge.

Keywords

Extended Enterprise, Value Chain.

1. LINKAGES IN THE VALUE CHAIN

Writing in 1985 Porter introduced the concept of the Value Chain. In Porters' words "value chain analysis helps the manager to separate the underlying activities a firm performs in designing, producing, marketing and distributing it's products or services". Value adding activities are the building blocks of competitive advantage and the value chain is a set of interdependent activities. Linkages in the value chain define the relationship between the way individual value adding activities are performed. Competitive advantage is also achieved by exploiting linkages, which usually imply efficient information flow and which facilitates the coordination of various value adding activities. For example manufacturing management has long recognised the value of efficient data flow between CAD and CAM systems and technical solutions have been proposed to exploit this linkage.

But of course linkages exist not only within the individual firm's internal value chain but also *between an individual firm's value chain and that of it's suppliers and distribution channels.* Again manufacturing companies have long recognised this and the Kanban ordering and delivery system seeks to exploit the linkage between final assemblers and their components supply chain, to achieve Just In Time delivery.

2. INFORMATION TECHNOLOGY AND THE VALUE CHAIN.

In more recent years the attraction of EDI (Electronic Data Interchange) technology is that it exploits the linkage between the value adding activities of individual firms in the value chain. For example, the ability to electronically share CAD data between specialist suppliers and their customers (the final assemblers of automobiles) in the automotive sector facilitates co-design, and reduces the time to market for new components and ultimately new cars.

Indeed recent developments in information technology and telecommunications are creating new linkages and increasing our ability to exploit old linkages. Keen (1991) used the terms 'Reach" and "Range" to articulate the impact of the new computing and telecommunications technologies on businesses. See Figure 1 below (taken from Browne, Sackett & Wortmann

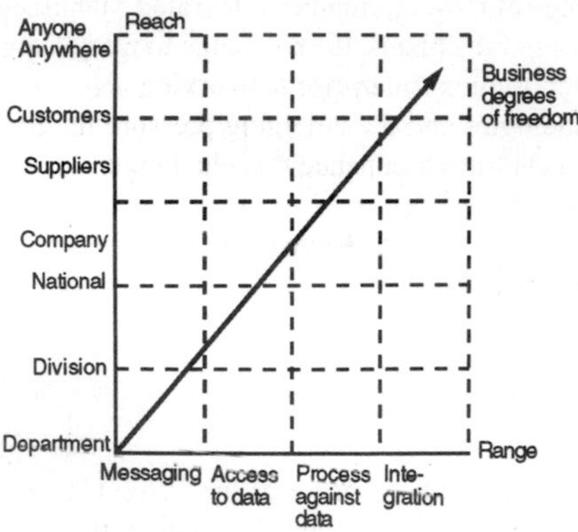

Figure 1 Business and Technology Integration.

1994a). "Reach" refers to the extent that one can interact with other communications nodes - in the ultimate it becomes anyone, anywhere. The recent explosive growth of the INTERNET is testimony to extended reach. "Range" refers to the data and information types that can be supported, ranging from simple electronic messaging between identical computer platforms to complex geometric and product data between hetrogeneous CAD systems on multivendor hardware and software platforms. Clearly the gradual expansion of the reach and range of these technologies provides opportunities to exploit linkages between value adding activities within individual firms or enterprises and indeed between firms and enterprises. But of course the availability of new technology - in this case tele-computing technology - is not sufficient reason to use it. Unless this technology creates competitive advantage it will not find widespread application.

Recent trends towards "focused factories", "core competencies", and the "outsourcing" of non-critical (in a competitive or critical success factor sense) activities provide opportunities for the emerging tele-computing technologies. Furthermore the emerging societal pressure on manufacturers to produce environmentally benign products using environmentally benign processes requires manufacturing firms to look beyond the "four walls of the manufacturing plant". Manufacturing companies must put in place systems which not only manufacture, distribute and service products but also collect at end of life, disassemble, recover or recycle and ultimately safely dispose of products. [See Tipnis 1993]. In effect the responsibility of the manufacturer is <u>extended</u> beyond that of providing and servicing a finished product.

As pointed out by Browne, Sackett and Wortmann (1994b), in the recent past the emphasis was on Computer Integrated Manufacturing and integration within the "four walls of the manufacturing plant". Today the emphasis is changing to include the supply chain to the manufacturing plant, the distribution chain from the plant to the ultimate customers and increasingly the end of life disposal or recycling issue.

The extent to which this change in emphasis has taken place can be seen in the workprogramme for research and technology development in Information Technologies of the 4th Framework Programme of the European Union; the domain 'Integration in Manufacturing' has a subsection entitled 'Logistics in the Virtual Enterprise' whose goal is to achieve an 'information logistics' infrastructure providing the required data at every step of the business process, to underpin the logistics of the supply and the distribution of materials and components'. One of the tasks which has been identified against this objective aims to develop 'innovative IT-based tools to model the interaction between companies forming the extended or virtual enterprise based on a distributed concurrent engineering and co-design approach, and covering all aspects of the product life cycle'. [European Commission, 1994a].

Not surprisingly the 'Industrial and Materials Technologies' workprogramme of the 4th Framework Programme shares a similar perspective. In the introduction to subsection 1.5 entitled 'Human & Organisational Factors in Production Systems', this thinking is clearly set out viz; "... Competitive advantage is increasingly derived from dependency and interdependency between suppliers, component manufacturers, product assembly, distribution, sales and customers. The aim is the integration of new human, business and organisational structures as a key component of the new form of manufacturing systems, involving close **intra-enterprise** and **inter-enterprise** networking to develop customised products in very short lead times, whilst fully utilising human skills" [My emphasis; European Commission 1994b].

3. WHAT IS THE EXTENDED ENTERPRISE?

The Extended Enterprise, as suggested above arises from the convergence of a number of ideas:

1. In many businesses it is no longer possible or indeed desirable to achieve world class capability in all the key functional areas within a single organisation.

2. Manufacturing businesses recognise the value of the "Focused Factory" and are beginning to outsource non core activities.

3. Given 1 and 2, it is now possible to provide core product functions separately by individual enterprises, who come together to provide a market or indeed customer defined value or service. Emerging tele-computing technology facilitates this cooperation or networking of individual enterprises to meet market requirements.

Davidow & Malone (1992) used the term "Virtual Company" to describe such an enterprise - "To the outside observer, it will appear almost edgeless, with permeable and almost continuously changing interfaces among company, suppliers and customers". On a similar vein, the Chief Executive Officer of Intel (Andrew Grove) compared his business to - "...the theatre business in New York, which has an intinerant workforce of actors, directors, writers and technicians as well as experienced financial backers.... By tapping into this network, you can quickly put a production together. It might be a smash hit ... or it might be panned by the critics. Inevitably the number of long running plays is small, but new creative ideas keep bubbling up".

The emergence of the Extended Enterprise would seem to be confirmed by the results of a survey of American manufacturing executives reported by Kim & Miller in 1992. In order to understand changes taking place in the manufacturing environment, the respondents were questioned on the issues of globalisation, competition and cooperation and how they intended to respond to them. Among the programmes they intended to put in place were :

- Change communication patterns and abilities (identified by 32% of respondents).

- Move production and product development closer to markets (identified by 28% of respondents).

- Standardise processes and procedures [including ISO 9000] (identified by 26% of respondents).

- *Develop and expand partnerships and alliances (mentioned by 22% of respondents).*

- *Reengineer the supply chain among suppliers, distribution and sales (identified by 20% of respondents).*

To quote directly from the report :
"The first type of challenge is to design and manage the implementation of physical changes in the location and responsibility of plants, R&D centres, distribution centres and sales offices. These changes represent radical changes in the configuration of supply chains. The second type of challenge *is to change or develop new external relationships, with suppliers, customers, and partners who interact with the supply chain...*

Drucker in his 1990 article in the Harvard Business Review discussed the concept of systems oriented design, "in which the whole of manufacturing is seen as an integrated process that converts materials into goods, that is, into economic satisfactions". He suggested that "a few companies are even beginning to extend the systems concept beyond the plant and into the marketplace ..." As soon as we define manufacturing as a process that converts things into economic satisfactions, it becomes clear that producing does not stop when the product leaves the factory.

Bessant (1994) recognises the issue of inter-firm relationships. He sees a move away "from tight boundaries between firms to blurred boundaries, from 'arms length dealing' to co-operative relationships, from short term to long term relationships, and from confrontational to cooperative relationships/partnerships". Rightly however, Bessant distinguishes the emerging network model of partnerships between firms on the value chain from the vertically integrated firm - "Whereas earlier, more stable environments allowed vertically integrated firms to flourish, exploiting scale economies, the increasingly turbulent and fragmented pattern today requires firms to become focused on distinctive competencies...... The value of networks is that they.... behave as a single large firm with all the implications for flexible response, close contact with customers, manageable scale, innovation, etc."

In an earlier paper [Browne, Sackett and Wortmann 1994b], the extended enterprise was distinguished 'pictorially' as suggested in Figures 2 and 3 overleaf. In the past, the emphasis was an integration inside the four walls of the manufacturing plant (See Figure 2). Today the emphasis has changed to include the supply chain (integration of the supply chain through EDI and JIT) and the integration of manufacturing with distribution planning and control systems [See Figure 3].

In fact it has been argued, and indeed Figures 4 and 5 suggest that the Extended Enterprise represents a logical development of much of the efforts of manufacturing systems specialists over the past ten years or so. Approaches such as MRP, JIT, EDI, WCM, Lean Production, Concurrent Engineering, Supply Chain Management, DRP, Benchmarking and Business Process Re-engineering synthesise into the inter-enterprise networking model and ultimately realise the Extended Enterprise [See Browne, Sackett and Wortmann 1994b].

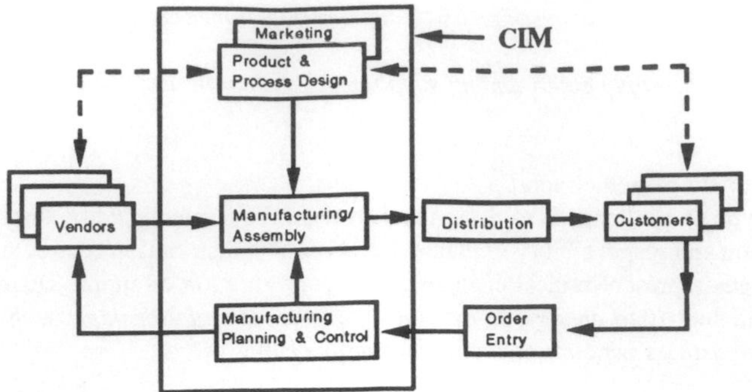

Figure 2 Computer Integrated Manufacturing.

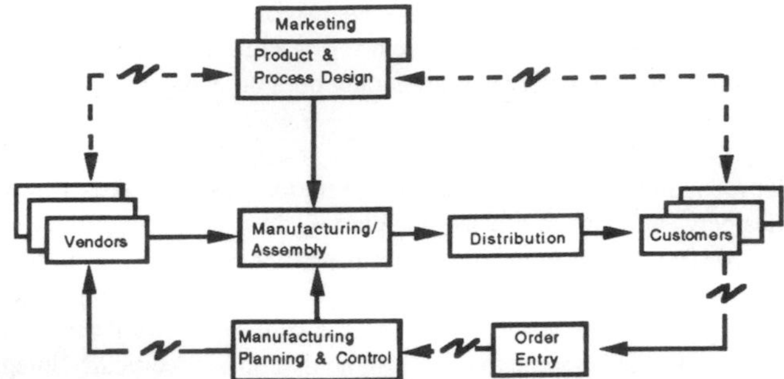

Figure 3 Electronic Data Interchange and The Extended Enterprise.

4. THE EXTENDED ENTERPRISE ~ A RESPONSE TO THE ENVIRONMENTAL CHALLENGE

Today progressive manufacturing companies are developing a total life cycle approach to their products. Some companies are adapting a long term business objective of resource recovery. Resource recovery seeks to maximize the ability to recover products, components and materials and minimize the need for disposal sites through reuse, recycling, reclamation, resale, reconditioning and remanufacturing. In the past responsibility for products ended when the product left the manufacturing plants. Today, the situation has changed radically. The "system boundary", dipicted in Figure 4 overleaf has moved to include post consumer materials management.

Figure 5 overleaf indicates the current balance between reuse/refurbishment and waste disposal; also a more desirable balance for the future.

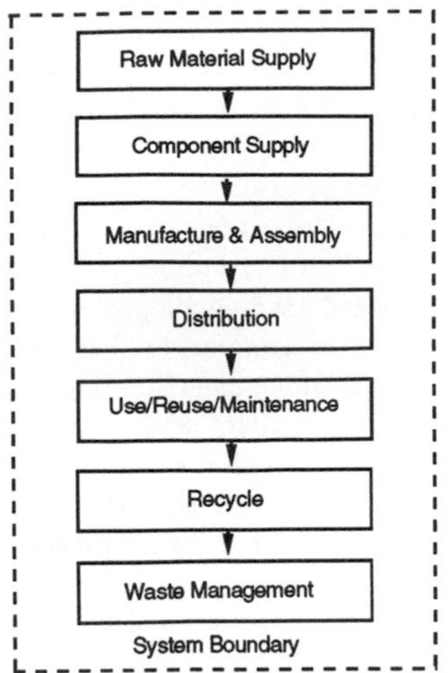

Figure 4 Total Product Life Cycle Perspective in Manufacturing.

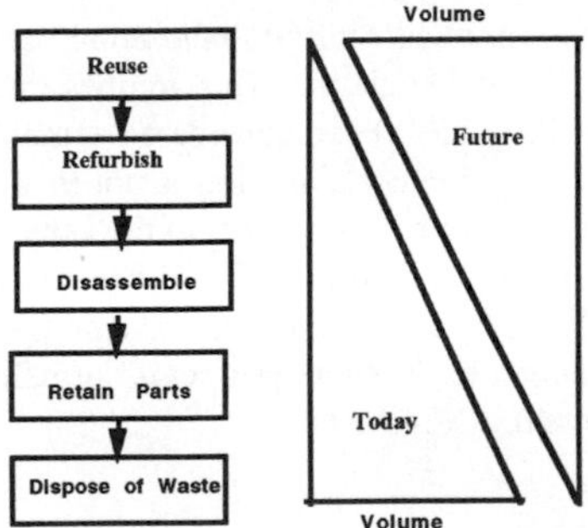

Figure 5 Resource Recovery - Todays Reality and Tomorrows Potential.

Major electronics and telecommunications manufacturing companies are setting up recycling plants to process returned products. Hewlett Packard processes 1200 tons of returned computer equipment in Grenoble each year. Sun claims to earn over 6mECU per annum by reselling old computer parts each year. In Holland, Rank Xerox and Oce disassemble and seek to reuse where possible returned photocopiers. Recently Alcatel, Siemens and Deutsche Telekom have announced a joint activity to recycle 12,000 tons of telephones, facsimile and similar equipment per year.

To indicate the importance of thc resource sustainment issue from a societal and environmental standpoint, we will use the Personal Computer (PC) industry to illustrate the issues. In 1965 the PC did not exist. Today there are estimated to be 140 million PCs in use in the world (1 for every 35 to 40 people). By the year 2010, PCs may well outnumber people. Within five to seven years virtually all current PCs will be discarded. It is estimated that plastics make up approximately 40% of the weight of a PC. The movement towards portable PCs increases the volume of hazardous material used in PCs; for example rechargeable batteries. Computer manufacturers are recognising the trends and are launching ecological programmes. They are beginning to offer 'Green PCs'. The PC manufacturers are beginning to realise that a combination of product takeback, modular design and remanufacturing may offer an environmentally and economically productive path to new product development (RSA, 94, Page 29). The takeback or resource recovery market is of course in it's early stages, and today the volume of consumer products which is taken back is relatively small. The market for secondary products and materials is emerging slowly. However it's importance for the future must not be underestimated.

Resource recovery, in the sense that it involves the original manufacturer taking responsibility for the 'end of life' disposal of his products represents an 'Extended Enterprise' approach.

5. CHALLENGES FOR THE EXTENDED ENTERPRISE.

Clearly the Extended Enterprise involves close collaboration between what were previously autonomous enterprises along the value chain. This implies thrust and the ability to share information which might previously have been considered company confidential. The establishment of such thrust and cooperation/collaboration is not made any easier by the constant changes which are taking place to individual elements of the value chain, and the impact which these changes have on the competitive positions of individual enterprises. Consider the partial value chain presented in Figure 6. Assume for the sake of discussion that the contribution of each element to the valued added of the final product is normalised to 100%. Consider the following assertions in the context of Figure 6.

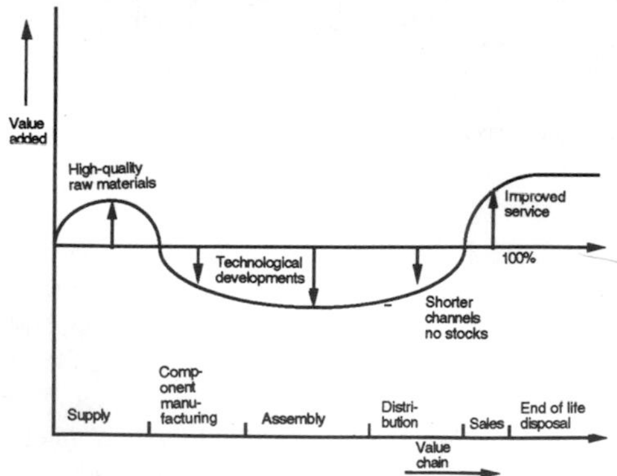

Figure 6

1. New materials such as highly-alloyed steel, composite materials etc. are leading to an increase in value added in material supply. Material costs are rising in the long term, although in the short term the prices are determined by the general economic conditions and the particular situation in a particular industrial sector. However, nearly all metals, oil-based products and minerals are becoming more scarce and therefore more expensive in the long term.

2. The situation with components manufacture is complex. Increased functionality of components for example in electronics, results in expensive but high value added components. The multi-function component eliminates the need for assembly so the value added in assembly is passed back to the component stage.

3. Assembly is likely to move towards lower added cost also, because much functionality is already available in the components.

4. Physical distribution will probably involve lower costs if the interaction between the customer and the manufacturer becomes more intensive.

5. Improved service around the product will increase the added value in sales and after-sales activities.

Another important aspect of component manufacturing is a technological trend towards manufacturing systems where the marginal cost of an additional unit is approaching zero. Miniature products and software products are good examples of this. In fact in general all knowledgeware including for example books, entertainment, CNC code, multimedia software have a marginal cost of almost zero. Furthermore many of today's new products, particularly electronics, computer, consumer electronics and mechatronics products are dominated by embedded software.

The situation depicted in Figure 1 represents in macro terms what is happening in the computer industry where the "valued added" has, over the past five years or so, been pushed *back* into the components side [hence the rise of Intel and the cooperation between IBM, Apple and Motorola to develop the Power PC] and *forward* to the distribution and sales end [and the associated rise of Microsoft on the the one hand and Dell AST, Gateway 2000 on the other]. In effect assembly has become customer order driven, and together with efficient distribution channels has become a source of competitive advantage - customised products in extremely short lead times. In fact the computer industry provide an excellent example of the changing nature of relationships along the value chain, and the need for cooperation between what might ordinarily be considered competitors - in effect IBM "created" Microsoft in order to "sell" their original pc. Now IBM compete with Microsoft with their new "WARP" operating system. Similarly IBM and Apple competed through their respective "pc" and "mac" products. Now they cooperate and share the Powerpc processor.

6. SMES AND THE EXTENDED ENTERPRISE.

SMEs (small and medium sized enterprises) involved in manufacturing seem to fall into two categories :

1. Those who are component manufacturers and who are largely involved in supplying components and assemblies to larger companies and final assemblers. Such SMEs are typically suppliers in sectors such as the automotive sector and the consumer electronics sector.

2. Those who are producers of finished products, such as SMEs tend to be in the capital goods business. For example, the vast majority of machine tool manufacturers in Europe are SMEs as are the majority of producers of moulds and dies.

For category 1 above, component suppliers on the supply chain, the emergence of the Extended Enterprise presents challenges as well as opportunities. Those who are willing (and financially able) to invest in developing appropriate systems e.g. ISO9000 accreditation, EDI links, EDE (Engineering Data Exchange) will benefit from the Extended Enterprise. Those who are weak technically and financially face tremendous difficulties. An indication of the extent of the problems and opportunities presented to supply chain component manufacturers in the automotive sector can be gleaned from a recent article in the Business Section of the newspaper "Independent on Sunday (1995)" headed "Japanese spur suppliers in two speed UK". The article goes on to suggest that "In Britain we are developing a two tier manufacturing sector, divided into those that supply the Japanese and those that don't". It seems that the Japanese car assemblers who have located final assembly plants in the UK are putting tremendous efforts into supplier development - "those companies which get on to Japanese delivery schedules receive a bonus : a series of on site tutorials in Japanese trouble-shooting and production techniques. Those SMEs who adopt best practice through interaction with their Japanese customers are then in a position to bid successfully for work with European car makers.

It is also clear that for SMEs located in peripheral regions or in locations far from their customers (the final assemblers), there are added difficulties. For example within the European Union there are concerns that the emergence of inter-enterprise networking and the Extended Enterprise may lead to a situation where manufacturing enterprises located in peripheral regions of the EU may be placed at a disadvantage. Advanced telecommunications and computing systems (telecomputing) will certainly facilitate information flow within the EU an indeed the global economy, but physical geography will continue to place enterprises located in peripheral regions at a disadvantage in terms of rapid goods flow and logistics. Paradoxically the availability of telecomputing may well accentuate this disadvantage, particularly for traditional products which are expensive to transport. Of course for products which can be transported electronically (e.g. software) no such disadvantage exists and "peripheral" regions (in Europe, Ireland; in global terms India) can profit from the emerging global economy.

CONCLUSIONS.

The Extended Enterprise where individual enterprises based on well defined core competencies cooperate to provide customer driven products is emerging. It is facilitated by the availability of tele-computing technology and is realised as a network of cooperating entities driven by a defined customer requirement. In the past the challenge to manufacturing systems researchers was to support integration within the four walls of the manufacturing plant. Today we must develop systems which support inter-enterprise networking across the value chain.

REFERENCES.

Bessant, J. (1994), "Towards Total Integrated Manufacturing", International Journal of Production Economics, Vol. 34, No.3, Pages 237-251.

Browne, J., Sackett, P.J., Wortmann, J.C. (1994a), "The Extended Enterprise - A Context for Benchmarking", Presented at the IFIP WG5.7 Workshop on Benchmarking - Theory and Practice, Trondheim, Norway, June 16-018.

Browne, J., Sackett, P.J., Wortmann, J.C. (1994b), "Industry Requirements and Associated Research Issues in the Extended Enterprise", Proceedings of the IMSE'94 European Workshop on Integrated Manufacturing Systems Engineering, Grenoble, France, December 12-14, 1994.

Davidow, W.H., Malone, M.S.(1992), "The Virtual Corporation", Harper Business.

Drucker, P.F. (1990), "The Emerging Theory of Manufacturing", Harvard Business Review, May-June 1990, Pages 94-102.

European Commission, (1994a), "Workprogramme for RTD in Information Technologies", Directorate General III Industry, Brussels.

European Commission, (1994b), "Workprogramme in Industrial and Materials Technologies", Directorate General XII, Science Research Development, Brussels.

Independent on Sunday (1995), "Japanese Spur Suppliers in Two-Speed UK", Business Section, Page 3, March 19th.

Keen, P. (1991), "Shaping the Future: Business Design through I.T.", Harvard Business School Press.

Kim, J.S., Miller, J.G., (1992), "Building the Value Factory : A Progress Report for US Manufacturing", Executive Summary of the 1992 US Manufacturing Futures Survey, A Research Report of the Boston University School of Management Manufacturing Roundtable, Boston.

RSA, (1994), Environmental Design Workshops, "Ecodesign in the Telecommunications Industry", Published by RSA, 8 John Adam Street, London WC2N 6EZ.

Tipnis, V.J., 1993, "Evolving Issues in Product Life Cycle Design", Annals of the CIRP, Vol.42, No.4.

BIBLIOGRAPHY.

Anon., (1991), "The Competitive Edge : Research Priorities for US Manufacturing, National Academy Press, Washington D.C.

Berger, S. Dertanzos, M., Lester, R., Solow, R., Thurow, L., (1989), "Toward a New Industrial America", Scientific American, June 1989, Vol.260, No.6, Pages 39-47.

Harman, R., Peterson, L., (1992), "Reinventing the Factory : Productivity Breakthroughs in manufacturing Today", The Free Press, New York.

Furukawa, Y., (1992), "Paradigm Shift in Manufacturing Systems - Forecasting Development Process toward 21st Century from Today's Intelligent CIM", Presentation at the Plenary Session of the 1992 IEEE International Conference on Robotics and Automation, Nice, France.

Nagel, R. Dove, R., (1991), "21st Century Manufacturing Enterprise Strategy - An Industry Led View", Iacocca Institute, Lehigh University.

Gruber, J., Tenenbaum, J.M., Weber, J.C., (1992), "Towards a Knowledge Medium for Collaborative Product Development", Artificial Intelligence in Design, Editor:Gero, J.S., Kluwer Academic Publishers, Pages 413-432.

BIOGRAPHY.

Professor Jim Browne is Director of the CIM Research Unit (CIMRU) at University College Galway in Ireland. He has been engaged as a consultant by companies such as Digital Equipment Corporation, Westinghouse Electric, Apple Computers, De Beers, AT Cross and a number of Irish manufacturing companies, to work in areas such as manufacturing strategy, manufacturing systems simulation and production planning and control. He has also been engaged by the European Union to work on the development of the ESPRIT programme (DGIII) and the BRITE programme (DGXII). He has been engaged in EU research projects with industrial companies such as COMAU in Italy, Renault in France, DEC in Munich, Alcatel in France and a number of European Universities, including the University of Bordeaux, the University of Eindhoven, the University of Berlin and the University of Trondheim.

For the past two years, he has been engaged as a consultant on a project funded jointly by the EU and the State Scientific and Technical Committee of China (SSTCC) - (Peoples Republic of China) to support the National CIMS (Computer Integrated Manufacturing Systems) project of China.

2

Continuous Education in New Manufacturing Systems

M. Tazza,
CEFET-PR.CPGEI
Av. Sete de Setembro, 3165
80-230-901 Curitiba PR Brasil
fax: +55 (041) 224 50 70

Abstract

It will be described a case study in the context of Continuous Education in New Manufacturing Systems. In the case study the term *New Manufacturing Systems* will be restricted to *Flexible Manufacturing Systems* context. The term Continuous Education will be analyzed as one six years experience in teaching and researching activities at post-graduation level at the cefet-pr.cpgei. The education-research program has been developed by adapting the curriculum to the needs of the development of a software tool for the project and analysis of Flexible Manufacturing Systems at workstation level. By *project* it is intended the specification of the behavioral and structural characteristics of an FMS. The term *analysis* means a quantitative description of performance parameters: throughput, sub utilization index, induced wait times and mean value of pallet's population in storage buffers.

Keywords

Flexible Manufacturing Systems, Manufacturing Automation, Computer Integrated Manufacturing, Education in Automated Manufacturing

1 INTRODUCTION

The decade shows an increasing number of publications related with Manufacturing Automation in general and with Computer Integrated Manufacturing (CIM) and Flexible Manufacturing Systems (FMS) in particular. This growth is due to the proposition of some interesting problems to administrators, engineers and researcher community involved with the design, implementation and operation of complex automated manufacturing systems. Such problems are originated from the need of efficient answers to layout organization, group technology, scheduling and control questions. The term Flexible Manufacturing System is a general descritpion of a manufacturing system that is economically renewable when producing medium size number part types at medium sized volumes. Medium size can range from tenths to hundreds for part types and from thousands to tenths of thousands for production volumes. Fig-1 shows the production scope of FMS.

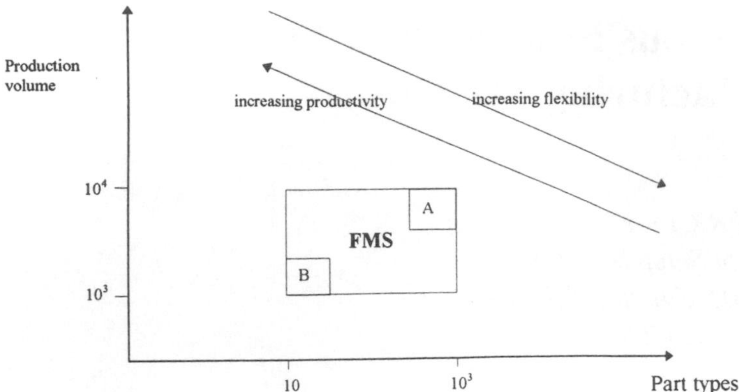

Fig-1: application scope of FMS

It must be noticed that the final values for the interval for volumes and number of part types are strongly related with manufacturing similarities of part types. In other words, a great number of different part types can be produced at high volumes (region A in Fig-1) if there is great similarity in parts process plans. Region B shows the situation of producing a small variety of part types with considerable differences among manufacturing plans.

The major goal of a manufacturing enterprise is to be price/quality competitive in the national and/or international industrial competition. The competitivity is reached by giving a correct solution to the technical-economical problems posed by the efficient implementation and control of large scale automated manufacturing system. The problems reach from controling perturbations in the communication system of the production environment to developing scheduling routines with real time capabilites. Although the application seems to be new, the technical tools for modeling and analysis purposes are based on well know theoretical aspects of computer science. The problems are normally reduced to large size matrix computation, real time monitoring of systems, efficient graph searching, resource allocation, concurrence and synchronization of processes.

We consider that the detailed analisys of production systems design alternatives is of primary importance for efficient final result. The paper describes the basic organization of a software tool (ANALYTICE) developed for the project and analysis of FMSs together with the resulting educational program. Both the tool and a recommended educational program will be detailed. The software is being developed at the Curso de Pos-Graduação em Engenharia Elétrica e Informática Industrial at CEFET-PR. Most of the implementation and some of the specification tasks has been carried on by under-and graduate students, sometimes as lectures tasks, sometimes as M.Sc. dissertation. Section 2 describes the principal organizational and functional characteristics of ANALYTICE. Section 3 describes and discusses the educational program developed for supporting the software tool.

2 THE SOFTWARE TOOL ANALYTICE

The software development process started with a functional specification of the tool. The software should provide functions for the *project* and *analysis* of FMSs. An FMS is defined as a manufacturing system with some degree of flexibility in the manufacturing process. This flexibility is achieved by the use of NC-controlled machine-tools and automated material handling system equipment under the supervision of a control program. In our context, *project* means the specification of the structural and behavioral characteristics of the manufacturing environment. *Analysis* means the numerical description of relevant performance parameters of the manufacturing system when operating under the restrictions imposed by the project. The relevant parameters are those normallly related with performance evaluation models: throughput, induced wait times, index of sub utilization of equipment, mean value of parts and pallets number in storage buffers, work in process, production costs. The restrictions imposed by a specific project represent the system's invariant for this design: number and type of NC-machine-tools and material handling equipment, number of pallets and fixtures, part families, layout, scheduling policies and control programs.

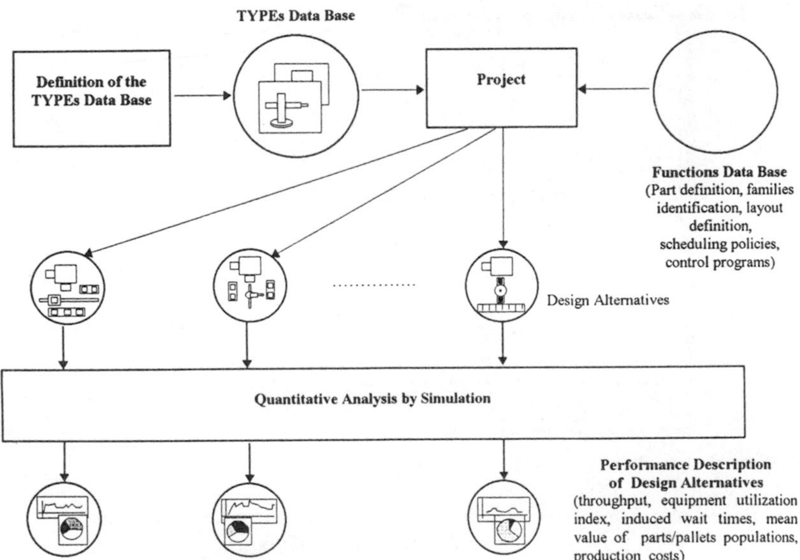

Fig-2: general functional organization ANALYTICE

2.1 General System Organization

The **Definition of TYPEs Data Base** environment in fig-2 is supported by definition and validation modules that allow the structural and behavioral definition and validation of equipment. Starting from a mental model of an equipment or from a description of a real one the user builds an abstraction of the equipment in terms of <process>s, <control signals>,

<responses>, <monitoring variables>, <interaction interface> and <operational parameters>. The *Behavioral Model* of the equipment is represented by a user-defined, syntactically system-directed C-program that establishes the relations among the defined sets of <process>s, <control signals> and <responses>. After consistency verification the system shows a pictorial representation of the result (Fig-3) in an interactive environment.

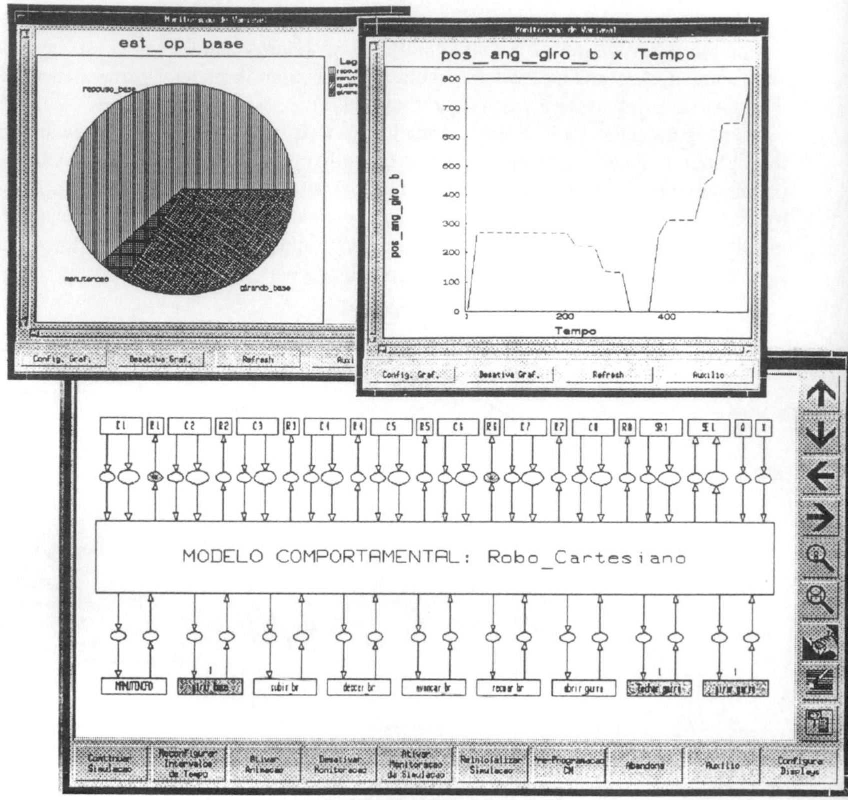

Fig-3: The behavioral model of a robot being validated in ANALYTICE

The user activates <process>es of the equipment model by sending the related <control signal>. The box or the corresponding <process> turns to *on* during at time period computed in terms of the related <operational parameters> (times, speeds, distances). After the computed period the <process> box turns to *off*. The value of those variables defined as <monitoring variables> in the equipment model can be monitored using simulated operation time as independent variable. The graphics of Fig-3 show the value of *operation-state* and *base-position* monitoring variables. A validated behavioral model is called a **CLASS**.

It is possible to attach a geometrical and dynamic model to CLASSes in order to have an animation model of an equipment. Fig-4 shows the definition of the geometric model of the robot CLASS of Fig-3.

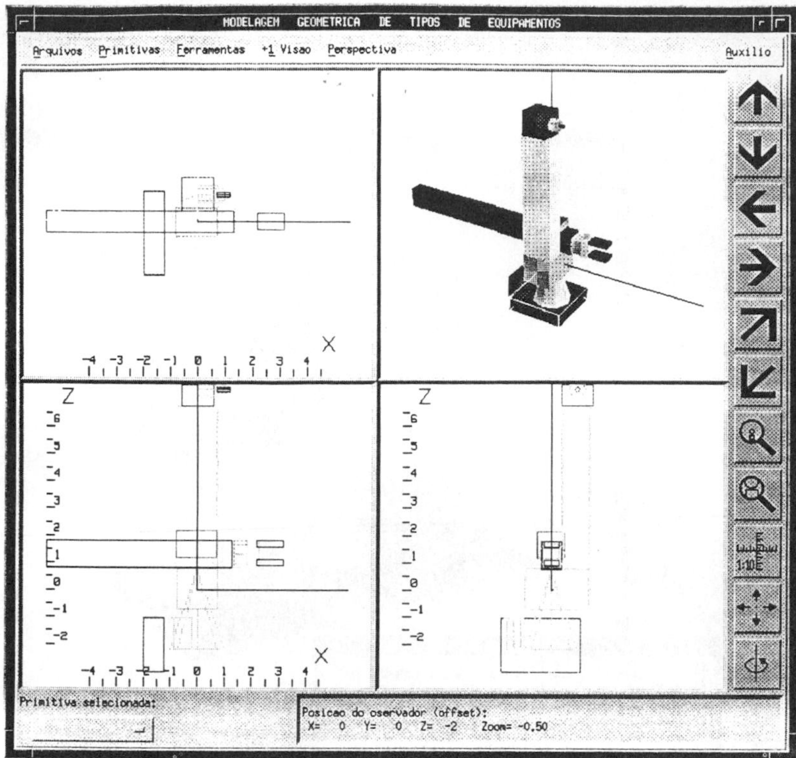

Fig-4: definition of a geometric model for the robot of Fig-3

A geometric model is defined by aggregating 3-D geometric modeling primitives. Each primitive is identified for utilization during cinematic model definition. The cinematic model is based on relations between the set of processes, the set of geometrical components of the geometric model and the set of monitoring variables of the behavioral model. The geometric and cinematic models will be validated by simulation. The set of monitoring variables will be updated by the behavioral model during simulation and the updated values are used at validation time to animate the equipment geometric components. It is possible to define stored program and simulate its execution. A CLASS with validated geometric and cinematic models is called a **TYPE**.

Equipment TYPE's occurrences can be aggregated to define manufacturing workstations. The behavior of workstations can be simulated for performance evaluation purpose. Validated workstation can be stored in the **TYPEs Data Base**. Manufacturing cells result from the aggregation of validated workstation TYPEs together with a material handling equipment recovered from **TYPEs Data Base**. Fig-5 shows a workstation defined in ANALYTICE by this aggregation procedure. The simulation at workstation level is driven by the execution of a control program at workstation level. Programs at this level are basically sequences of control signal to equipment of the workstation.

Fig-5 shows a model of a workstation operating at BOSCH/ Curitiba resulting from the aggregation of equipment The human operator can execute several manufacturing processes, each corresponding to a specific part family.

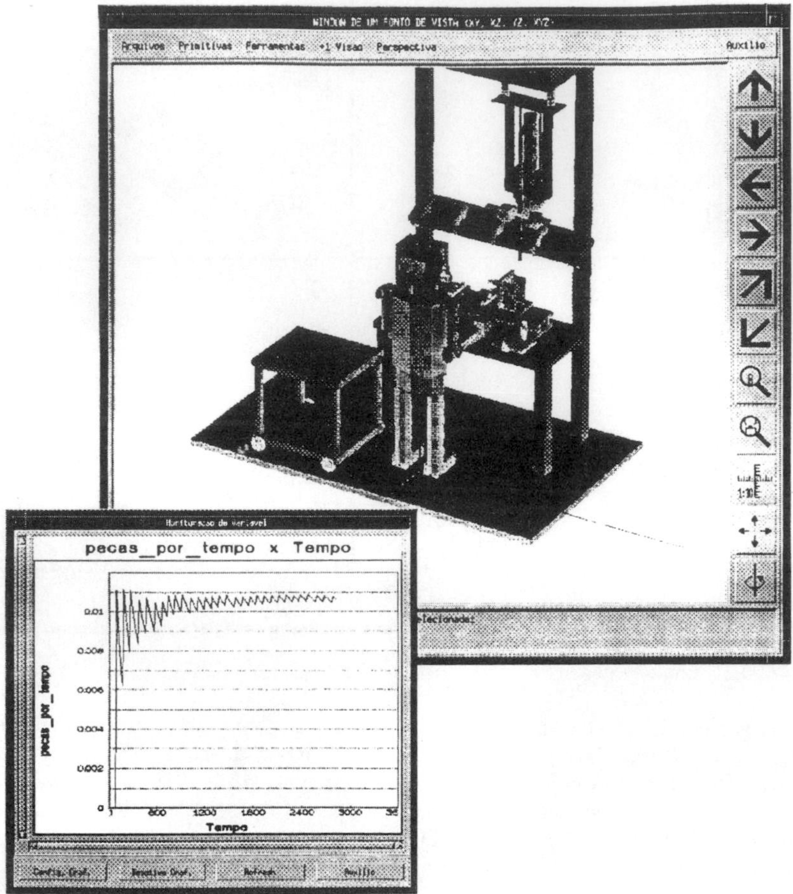

Fig-5: BOSCH manufacturing workstation modeled in ANALYTICE

The graphs shows the production rate versus simulated operation for one of the possible part type.

3 THE EDUCATIONAL PROGRAM

The educational program has been developed together with the software tool. It must be noticed that at the begin of the process no member of the staff has had any idea about FMS.

The first goal was to obtain a clear understanding of the terminology and problems that would appear in FMS context. This understanding was necessary for the proposition of contents of disciplines and of student's work plans. All students at post-graduation level had a formal knowledge on data structures, C programming and UNIX operational system.

3.1 Our Trial and Error Approach

The initial research staft was formed by a research leader with Ph.D. level and systems performance evaluation expertise and three post-graduation students. The first internal documents show a great effort on terminology and basic concept's understanding and in the acquisition of an overall view of the relations among concepts. In order to formalize the ideas, a first publication was proposed to the students, with theme "*The development cycle of FMS*" ([1]). The resulting paper presented a top-down proposal of a macro-tasks sequence for FMS's development cycle: system specification, design, implementation and operation. Each macro-task in the sequence was detailed, identifying and establishing order relations over sub tasks. During this phase of the work terms and concepts *like machine, process and routing flexbility, part types, process planning, part families, group technology, manufacturing cells, pallet types, machine selection, layout definition, scheduling policies, production costs, control part, MHS, AGV, AS/RS, WIP* and *production mix*, had to be understood, functionally related and time ordered. A rapid (about five months) system prototype was developed for identification of future implementation problems.

At the same time the skeletal of a curriculum was establishcd for providing formal education on manufacturing systems in general and FMS in particular. The guidelines for the *curriculum* were the need for mastering specifie topics on manufacturing systems together with specification and analysis techniques. The normal *curriculum* of the course was increased by disciplines on Petri nets ([2]) as formal specification tool, a discipline on performance evaluation of systems using timed Petri nets ([3], [4]). Two additional disciplines on FMS topics, one for generic conceptual and organizational definitions and a second discipline for modeling purpose were offered. Detailed theoretical study of some of the major problems (layout, control and group technology [5], [6], [7]) arising in FMS context were proposed as M.Sc. dissertations. A first altempt for a theoretical approach on scheduling failed.

The specification and implementation of the nucleus of the sortware ANALYTICE brought some insights on the needs of the educational program. The two disciplines on manufacturing systems were reduced to the one related to modeling and analysis of case studies. The students obtained conceptual data over FMS by reading internal documentation for terminological standardization and external technical publications for conceptual familiarization ([8], [9], [10], [11], [12], [13]). The reason for reducing the number of disciplines related with manufacturing systems was the well-known fact that students with formal knowledge are able to read technical papers on group technology, scheduling algorithms and concurrence problems in control of FMS. The Petri Nets discipline was expanded with a review on basic topics of set theory, relations, relations composition, mappings, equivalence relations, equivalence classes and graph theory.

The advances on the development of ANALYTYCE allowed us to propose new dissertation themes. A new attempt to simulate scheduling problems ([14]) was quite successful at theoretical level but failed in producing an environment to validate scheduling strategies. A more specific approach for simulaling FMS control problems ([15]) brought a pretty analysis of control specification languages at workstation and manufacturing cells levels. At that time ANALYTICE helped us to estimate implementation difficulties and development times for control implementation strategies. The major alternatives were Petri nets, Grafcet ([16]), Anteriority tables ([17]) and procedural languages. The decision was the utilization of procedural languages with mechanisms for describing concurrence, sequentialization and synchronization of processes. Four additional M.Sc. dissertations were used to develop the themes of production cost's analysis ([18]), modules for geometric and cinematic modeling of equipment ([19]), modeling and validation of material handling equipment ([20]) and quality control simulation ([21]). Three additional dissertations are at a final stage of development: the modeling of a real case manufacturing workstation ([24], modules for automatic layout definition at manufacturing cell level ([25]) and computation of the number of machines and pallets necessary to achieving gave production rates ([26]).

The educational program for new students was expanded introducing a second optional discipline on the topic of performance evaluation using Markov models. Finally, two new disciplines were offered for both new and older students in partnership with GMD/ Sankt Augustin: Compositional Data Objects ([22]) and High Level Petri Nets ([23]), dealing with formal specification and information modeling techniques.

3.2 Analysis of the Implemented Education Program

We believe that the executed education program gave the students a generalist view of the problems related with the specification and analysis of complex automated manufacturing systems. It is necessary to complement the program with a pratical experience of four to six months period in a real industrial environment. For the development of the next dissertations involving modeling and analysis of manufacturing systems it will be required the acquisition of real data in the industry. A first attempt in this direction is being made ([24]).

For an education program directed to people already working in the industry it seems that a review of formal modeling and analysis techniques is necessary. This review can be executed in parallel with a discussion of topics on new manufacturing systems, working the relations between the formal techniques and the pratical problems. A second step of the education program is based on case studies with the utilization of a manufacturing systems simulator. This two-phases program is depicted in fig-6.

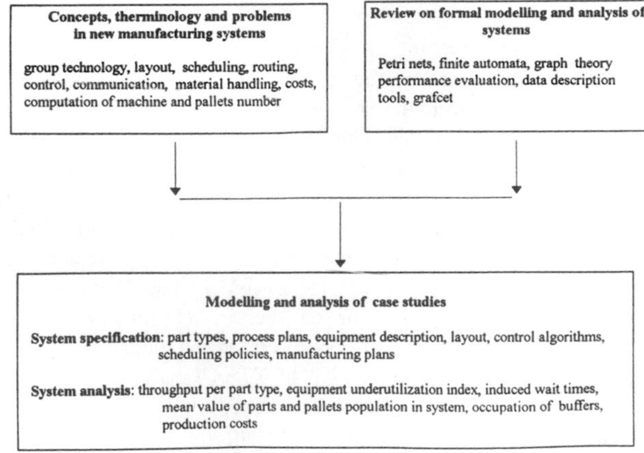

Fig-6: Skeletal of a continuous education program in new manufacturing systems

4 CONCLUSION

It is obvious that the actual education program has been strongly influenced by the software development needs. It must be noticed, however, that the education in formal disciplines together with the utilization of the resulting software for modeling and analysis objectives has brought an automated manufacturing mentality to the students. A program for visiting automated industries and/or dealers of industrial automation equipment in Europe has been partially executed by the students, giving some insight into the real industrial environment. For a modern education program directed to engineers working in a production environment, the utilization of a simulator of manufacturing system is of primary need. Due to the related cost and time it is not possible to develop skill in design and alternative analysis working over real manufacturtng systems. At the other hand, as simulators for complex manufacturing systems are normally complex software packages, a formal understanding of the objects being manipulated and of the resulting performance curves is necessary.

ACKNOWLEDGMENTS

To all my students ([5],[6],[7],[14],[15],[18],[19], [20], [21], [24], [25], [26]) for the entusiasm during the software developmnent,specially to Omar Achraf, Cesar Godoy and Lissandro Bassani that implemented the nucleus of ANALYTICE working with uncomplete specifications.

5 REFERENCES

[1] **Künzle, A., Souzu A. & Stadzisz P.**: *Uma proposta de ciclo de vida para Sistemas Flexíveis de Manufatura*, SUCESU, Anais do XXII Congresso Nacional de Informática, São Paulo, 1989, pp. 736-750.

[2] **Reisig, W.**: *Petri Nets: an introduction*, EATCS Monographs on Theoretical Computer Science, v. 4, Springer Verlag, Berlin, 1985.

[3] **Tazza, M.**: *Análise quantitativa de sistemas*, III EBAI, Curitiba 1988.

[4] **Tazza, M.**: *Quantitative Analysis of a Resource Allocation Problem: A Net Theory Based Proposal*, in Voss et al. (eds.) APN 87, Concurrency and Nets, Springer Verlag, 1987, pp. 511-532.

[5] **Studzisz, P.**: *Especifcação de um ambiente computacional para auxílio ao projeto do arranjo físico de sistemas flexíveis de manufatura.* Dissertação de mestrado, CPGEI / CEFET-PR, Curitiba, ag. 1990.

[6] **Künzle, L. A.**: *Controle de sistemas flexíveis de manufatura - especifcação dos níveis equipamento e estação de trabalho,* Dissertação de mestrado, CPGEI / CEFET-PR, Curiliba, ag. 1990.

[7] **Souza, J. H. F.**: *Tecnologia de Grupo: algoritmos e ferramenta gráfica,* Dissertação de Mestrado, CPGEI / CEFET-PR, Curitiba, ag. 1990.

[8] **Buzacott J. A. & Yao D. D.**: *Flexible Manufacturing Systems: a Rewiew of Analytical Models.* Managment Science, v. 32, n. 7, Jul. 1986, pp. 890-905.

[9] **Kusiak, A.**: *Flexible Manufacturing Systems: a structural approach,* International Journal of Production Research, v. 23, n. 6, 1985 pp. 1057-1073.

[10] **Gallagher,C.C.: & Knight,W.A.**: *Group Technology Methods in Manufacture,* Ellis Horwood Limited, 1st ed. England, 1986.

[11] **Carrie, A.**: *Simullation of Manufacturing Systems,* John Wiley & Sons, Chichester, 1988.

[11] **Valette, R.**: *Nets in Production Systems,* in: Rozemberg. G. (ed). Advanced Course on Petri Nets. Bad Honnef, Springer Verlag, 1987,pp.191-217.

[12] **Browe, J. et al.**: *Classifcation of flexible manufacturing systems,* The FMS Magazine, ap. 1984, pp. 114-117.

[13] **Conway, R. W., Maxwell, W. L. & Müler, L. W.**: *Theory of Scheduling*, Addison-Wesley, Reading, Mass, 1967.

[14] **Sautter F.T.**: *O problema de escalonamento em Sistemas Fleriveis de Manufatura.* Dissertação de Mestrado, CPGEI/CEFET- PR, Curiliba, 1993.

[15] **Tacla C.A.**: *Controle de Sistemas Flexíveis de Manufatura, Especifcação do nível estação.* Dissertação de Mestrado, CPGEI/ CEFET-PR, Curitiba, ag. 1993.

[16] **David, R. & Alla, H.**: *Petri Nets and Grafcet*, Prentice Hall, NewYork, 1992.

[17] **Deschanel, F.**: *Pilotage d'une Cellule Flexible d"Usinage.* These de Doctorat, Universite de Franche-Comte, 1989.

[18] **Gallotta, A.**: *Especifcação e implementação de Módulo de Análise de Custos em FMS.* Dissertação de Mestrado CPGEI / CEFET-PR, Curitiba, mar. 1994.

[19] **Contesini, M.**: *Interface para Projeto Gráfico de Equipamentos de Manufatura,* Dissertação de Mestrado, CPGEI / CEFET-PR, Curitiba, dez. 1994.

[20] **Angonese, V.**: *Modelagem de Equipamentos de Manuseio de Materiais em ANALYTICE,* Dissertação de Mestrado, CPGEI / CEFET PR, Curitiba, nov. 1994.

[21] **Lugo, C. J.**: *Simulação do Controle de Qualidade em Sistemas Flexíveis de manufatura,* Dissertação de Mestrado, CPGEI / CEFET-PR, Curitiba, set. 1994.

[22] **Durchholz, R. & Richter, G.**: *Compositional Data Structures*, The IMC/IMCL Reference Manual, 1. Wiley, 1992.

[23] **Meta Software Corporation**: *Design/CPN,* Internal Functions Programmer's Reference, Version 2.0, Meta Software Corporation 1993.

[24] **Nied, A.**: *Modelagem e Análise de uma Estação de Trabalho em ANALYTICE,* Dissertação de Mestrado, CPGEI/CEFET-PR, Curitiba, conclusão prevista jun. 1995.

[25] **Gallotta, M.P.**: *O problema de dimensionamento em FMS*, Dissertação de Mestrado, CPGEI/CEFET-PR, Curitiba, conclusão prevista para jun. 1995.

[26] **Tsunoda, D.**: *Ambiente para definição automática de arranjo físico em ANALYTICE,* Dissertação de Mestrado, CPGEI/CEFET-PR, Curitiba, conclusão prevista para jun. 1995.

PART TWO

Enterprise Modeling and Organization I

3

Information and command infrastructures for small and medium size enterprises

J. Goossenaerts[a], *A.P. Azcarraga*[b], *C.M. Acebedo*[c] *and D. Bjørner*[a]
[a] *United Nations University, International Institute of Software Technology*
P.O. Box 3058, Macau
[b] *College of Computer Studies, De La Salle University*
2401 Taft Avenue, 1004 Manila, Philippines
[c] *Industrial Engineering Department, De La Salle University*
2401 Taft Avenue, 1004 Manila, Philippines

Abstract

The scope of the problem of designing an information and command infrastructure for Small and Medium Size manufacturing enterprises is explained. The rôle of enterprise and industry modelling during the infrastructure design phase and the rôle of model execution services are summarized. An outline of the MI^{2}CI Philippines project shows an approach to demonstrating and disseminating integrated enterprise and industry modelling know how in support of the operations and projects of SMEs.

Keywords

Infrastructure, enterprise and environment modelling, model execution services, small and medium size enterprises

1 INTRODUCTION

Traditionally, the term infrastructure is used in reference to services in the areas of sanitation, water, power, transportation, irrigation, roads and telecom. Because infrastructure services raise productivity and lower production costs they form an important component in development policies (The World Bank, 1994).

The term *manufacturing industry information and command infrastructure* (abbreviated: MI^{2}CI) is proposed in reference to an information processing and activity control and monitoring system which provides information and control services in support of the operations and projects of manufacturing enterprises. Actors that are anticipated to involve in the design and exploitation of MI^{2}CI services include public-sector industry support organisations, chambers of commerce, trade associations, industry research and development centers, information and communications technology suppliers, telecom network operators and enterprises.

The important rôle of small businesses as instruments of job creation and product and service innovation justifies initiatives to set up information and command infrastructures for them. Meeting the requirements of small businesses requires one to address key issues such as the identification of services and exploitation concepts adapted to SME's needs, the inter-operability of services, access and ease of use, information protection and confidentiality. Addressing this amalgam of issues while avoiding the creation of an amalgam of overlapping and incompatible solutions requires one to construct a global and integrated picture of the problem area prior to the implementation of the infrastructure services. Systematized information about products, business and manufacturing processes, markets, technologies, and regulations is needed.

The concepts of *genericity* and *model life cycle* proposed in the European Prestandard ENV 400.03 (1990) are indispensable in systematizing information for industries prior to the design, implementation and particularization of infrastructure services. The second chapter of this paper proposes three more organizing principles. The *life cycle principle*: the life cycles of products and plants are the prime phenomena-flow based and unavoidable life cycles to engineer, manage and realize. The *interflow principle*: by articulating the interfaces between executing models on the one hand and physical operations and/or projects on shop floor and in engineering offices on the other hand, one can more easily cope with the balance between human work, automation and information technology. The *decomposition principle*: decomposing a manufacturing industry into the generic orthogonal systems plant, team, market and industry allows one to target services and reduce system complexity. After their introduction, the principles are linked to enterprise and industry modelling during the infrastructure design phase, and to the design and execution services.

The MI^{2}CI Philippines Project is the first project in which we try to validate the use of integrated enterprise and industry modelling know how in support of industrial development. It is a joint project of De La Salle University, UNU/IIST and the United Nations Industrial Development Organization (UNIDO). The project assumes that SMEs, to respond to opportunities and changes, and to establish dynamic value-creating partnerships in extended enterprises (Browne *et al.*, 1994), need to deploy advanced information technology within enterprises as well as in the business environment. However, it is also recognized that high degrees of automation are rarely compatible with the socio-economic situation in developing countries, including the high level of unemployment, low investment and poor bankability of SMEs, and the lack of human resources. Therefore, to achieve the target of a balanced deployment of human work, automation and information technology in SMEs, it is necessary for the MI^{2}CI project to take up two work packages: a requirements analysis and an applicability study. The *requirements analysis* of the metal works industry in the Philippines will guide the formulation of a Philippine Business Environment Model and help to formulate an exploitation concept for MI^{2}CI services. The *applicability study* will focus primarily on: (a) the particularization of partial and particular enterprise and industry models from the generic model described by Goossenaerts & Bjørner (1994); (b) the implementation of execution services for such models ; and (c) the implementation of prototype systems for shop floor control and for supply chain operations.

2 PRINCIPLES FOR ORGANIZING MANUFACTURING REQUIREMENTS

The Interflow Orientation

Manufacturing industry and enterprises can be considered as *interflow systems* in which distributed information processing systems (computers and humans) synchronize denoted *computational flow* with *phenomena flow* on shop floors and in engineering offices.

The concept of interflow articulates the distinction between computations and physical activities. It builds on concepts in modelling and simulation. The *interflows* relation between computational flows – executed by a network of information processing entitities – and phenomena flows – on shop floor and in the engineering office – requires interface channels for flow synchronization. The three relations (denoted by the symbols $\boxed{\models}$ (models) , $\boxed{\wr \models}$ (simulates) , and $\boxed{\aleph \models}$ (interflows)) matter for successive steps in the construction of enterprises and industries.

The Life Cycle Orientation

The problem scope for which information and command infrastructures are proposed is – in a bottom-up manner – determined by the life cycles of products and plants. These are the primary – phenomena flow based and unavoidable – cycles to manage, engineer and realize.

(1) The *life cycle of a plant* is concerned with: (a) *Projects* which phase *a plant and its products* into *a market*: the gradual introduction of a new plant and its products into a market (and environment). (b) *Operations*: a plant will respond to *orders* issued by customers, by exploding the orders into *plant programme steps*, by scheduling and carrying out the (production) steps, and by delivering the goods ordered. *Projects* during operations may concern: (i) How to phase in&out *products*, *equipment* and *human resources* into a plant. (ii) How to improve *plant operations.* (c) *Projects* which phase out *a plant and its products*: the gradual withdrawal from use of an old plant and its old products.

(2) The *life cycle of a product* is concerned with: (a) Its *creation* in a (virtual) plant system. (b) Its *usage* by consumers on the market. Usage of the product may include maintenance, repair, upgrading, etc. (c) Its *decomposition* – preferably with recycling or reuse of its composing materials – in a (virtual) plant system.

The Orthogonal Systems of Manufacturing Industry

The focus on (collaborative) labour in the product and plant life cycles and the requirement of orthogonality for the systems to describe these suggest: First, to make, in an enterprise, a distinction between "repetitive" *operations*, to be catered for in a *plant model*; and "one-of-a-kind" *improvements* and *innovations* to be catered for in *projects* carried out by *teams*; Secondly, when extending the viewpoint beyond the boundary of the enterprise to let *extended-enterprise operations* be carried out by so-called *virtual plants*, and to let *extended-enterprise projects* be carried out by *virtual teams.* A virtual enterprise (or extended enterprise (Browne *et al.*, 1994)) comprises a virtual team and a virtual plant; And thirdly, to consider *markets* as the contexts within which enterprises are embedded, and to use the term *industry* to denote this context when considering the projects and virtual projects executed in the market (e.g. industrial policy projects).

The following domains (Table 1) are proposed to organize the objects, processes and issues in industry (see also Goossenaerts & Bjørner (1994)) : (a) A *team* phases in and out a plant or a product in response to *goals* formulated by entrepreneurs or other agents of change. A team refines and enriches the goals into *project programmes* for changing or making a (virtual) plant, it schedules and carries out these programmes, finally, it delivers the new or improved plant, capable of producing goods, with the performance expressed in the goals. (b) A *plant* responds to *orders* issued by customers, by exploding the orders into *plant programme steps*, by scheduling and carrying out the (production) steps, and by delivering the goods ordered. (c) A *virtual plant* comprises several plant systems which synchronize their operations as if they were carried out by a single plant. (d) A *virtual team* comprises several teams which synchronize their engineering work and decisions as if they would carry them out as a team within a single enterprise. (e) A *market* forms the context for the operations of plant systems and virtual plant systems. Market regulations (e.g. such as in Company Law and Tax Law) may affect the operations of plant systems and virtual plant systems. (f) An *industry* forms the context for the projects of (virtual) team systems. Regulations (e.g. such as in Environmental Law, Technical Regulations and international standards) may affect projects.

Table 1 Domains for organizing objects, processes and issues in industry

activities *context*	*operations* "repetitive"	*projects* "one-of-a-kind"
enterprise	plant / production	team / projects
extended enterprise	virtual plant	virtual team
environment	market / trade	industry / (policy) innovation
desirable condition	leanness	agility

The desirable conditions of operations (leanness) and projects (agility) must be realized in different contexts: in the enterprise, in the extended enterprise and in the environment as a whole.

Projects are further classified as follows: (a) *Entrepreneurial Projects* deal with changes of enterprise level programmes during the enterprise life cycle (e.g. the introduction of new products, business process reengineering, investment appraisal). (b) *Engineering Projects* are concerned with product life cycle and production processes for the product. Usually they require innovation. (c) *Quality Improvement Projects* are concerned with product and production process improvements. (d) *Industrial Policy Projects* pertain to the industrial environment as a whole. Virtual plants result from the successful implementation of projects by virtual teams. These require an explicit attention for inter-organisational engineering – in the extended enterprise – and the (standardized) technology infrastructure it requires.

3 INTEGRATING REQUIREMENTS IN REFERENCE MODELS

The use of enterprise models and reference models as tools for organizing and integrating information about enterprise processes is well established. See for instance ENV 40.003 (1990), Scheer (1989), CIMOSA (1993), Spur *et al.* (1994). Goossenaerts & Bjørner (1994) introduce also the concept of industry model. A breakthrough in enterprise modelling for SME depends on progress in industry modelling: many commonalities between (partial) enterprise models for SME have their origin in the market rules (the confluence between enterprise operations and market behavioural rules). The relative weaknesses of SMEs in acquiring know how, capital, technology and human resources justify industrial policy projects for boosting their development. But due to the large number and diverse activities of SMEs it is common that such projects target a market or sector rather than specific firms.

As to the use of models – and advanced computer-based tools– in support of the operations and projects of SMEs it is important to consider the particular reality of SMEs, in determining the kind of activities that should be supported (Bonfatti *et al.*, 1994), and in determining the approaches to investments (Tucker *et al.*, 1994). The ESPRIT RUMS Project (Bonfatti *et al.*, 1994) aims at supporting the following activities: (a) verification and assessment of the current manufacturing organisation; (b) evaluation of technological enhancement hypotheses concerning production resources; (c) definition of realistic manufacturing and purchasing budgets holding uncertainty of forecasts in due consideration; (d) production planning with respect to different time horizons on the basis of a unified model of product and process ; and (e) education and training of human resources involved in the manufacturing process.

Enterprise Modelling and Industry Modelling in the MI^2CI *Project*
In the MI^2CI Philippines Project aspects to be captured in enterprise models and in the business environment or industry model are as follows:
– The partial and particular enterprise models of the metal work industries should capture *internal technological, administrative and human resource facets* and the processes involving them. Internal technological facets are related to engineering design, product engineering design, manufacturing engineering design (CAD, CAE), process planning, materials and parts processing (CAM, CIM), quality assurance, waste avoidance. Internal administrative facets include procurement, inventory, order processing, customer service processing, personnel, budgets, accounts, and cashier. Internal human resource facets include organisation, division of labor, team work, training, quality assurance, etc.
– A Philippine Business Environment Model should capture *external technological, environmental, regulatory, and resource facets* and the services and processes involving them. External technological facets include supplier profiles, tool profiles, transportation nets, sub-contracting, standardization, innovation & technology transfer. Environmental aspects include pollution, environmental impact, material cost. Legal rules and regulations concern reporting obligations, taxes, labour laws, industry association, standardization, quality assurance & product certification, etc. Financial services include stockholders, venture capital banks, government loan agencies, commercial banks, insurance. External human resource facets include workforce profiles (available skills, wages, &c.), training opportunities, professional associations.

Enterprise Modelling
For organizing objects and issues we refer to the orthogonal decomposition. The distinction in the enterprise between repetitive operations and one-of-a-kind projects allows one to split the enterprise models into two components: a plant (interflow) model and a team (interflow) model. These models are related to each other and to phenomena flow. Projects, supported by the team model in execution, transform the plant model and the physical plant. Operations, controlled and supported by the plant model in execution, transform material inputs and produce the products.

Enterprise modelling is concerned with the construction of generic, partial or particular (as defined in ENV 400.03 (1990)) plant interflow models or team interflow models.

The Construction of a Generic Plant Interflow Model.
A construction of a generic plant interflow model is summarized in Table 2. The first three steps deal with statical structural properties of plant systems. The fourth step allows one to define computational flows over *instances* of the templates (cf. types) in a plant structure. Configurations of such instances can simulate plant operations. The fifth step deals with the interfaces between a plant system and plant operations. Some remarks to justify the focus of the five steps in the construction: (**1**) Materials and the work to transform them are described *statically* by means of *part templates* describing units of material and *work templates* describing units of change to material. A *Part-Work Structure* integrates the information in a bill of materials and in process charts and describes the parts and works required for making products. (**2**) Cells and the orders they send and receive incorporate the division of labour and the coordination of results. *Cell Templates* describe cell(s), a cell is a unit capable of sustaining activities/work. Cf. resource model. *Order Templates* describe orders. An order is sent by one cell to another to request the delivery of a part. A *Cell-Order Structure* integrates the properties described in

organization charts and in information flow charts. It is concerned with the division of labour and coordination of results. (**3**) A *Plant Structure* results from joining a part-work structure and a cell-order structure. It integrates the statical properties of a plant. (**4**) *Plant System.* Instances of cell templates and their responses to order instances can simulate the operations at a shop floor. The term *pulse* denotes the response by a cell instance to an order instance it receives. Cf. the order handling process. The term *flow* denotes the integration of pulses. A *plant system* controlled by a model execution system and interfaced to a suitable discrete event generator, can simulate shop floor operations. (**5**) *Plant Interflow Model. Channel templates* describing interface channels are added to the plant system. Interface channels are used to interface a (executed) plant system to a shop floor. They support synchronization through the input and output of particular tasks and signals.

Table 2 The bottom-up construction of a Plant Interflow Model

1	*Part-Work Structure* (*Part Template* $\models$ part(s), *Work Template* $\models$ work(s))
2	*Cell-Order Structure* (*Cell Template* $\models$ cell(s), *Order Template* $\models$ order(s))
3	*Plant Structure* $\models$ plant(s)
4	*Plant System* $\wr \models$ plant operations (*Pulse* $\wr \models$ response to order instance, *Flow* $\wr \models$ operations)
5	*Plant Interflow Model* $\aleph \models$ plant operations (*Channel Template* $\models$ interface channel) *Plant Interflow System, sync(hronize) cycle*

Team Interflow Models

A team working in an (virtual) enterprise aims for improvements or innovations in a (virtual) plant interflow system. In pursuit of a *goal* – the sequence of activities towards its accomplishment has not been secured yet – a team will first – during the design&drafting phase – transform or particularize a plant interflow model (templates, plant structure and plant system consolidate the deliverables during this phase) and next – during the construction phase – construct or upgrade the plant interflow system, by transforming the phenomena flow. In the course of a project, the composition of the team may change: it is adapting it according to changing project activities. In contrast, the cell structure of a traditional plant remains static during operations.

Team interflow models in execution support team activities. A generic team interflow model forms a basis for the implementation of an *Integrated Project Support Environment.*

Industry Modelling

The contemporary market is governed by a large and changing collection of regulations and constraints for the operations of plants, producers and consumers. Coping with the regulations and their changes requires high costs, especially for SMEs. Therefore the SMEs are expected to benefit significantly from more integrated and harmonized industry models, the corresponding market-wide information and command infrastructures, and proper model-driven interfaces between their enterprise infrastructure and the industry infrastructure.

Industry Modelling is concerned with the construction of generic, partial or particular market interflow models or industry interflow models. Such models enhance the understanding, coordination and harmonization of rules and other market information.

A Generic Market Interflow Model

The construction of a generic market (interflow) model parallels the construction of a generic plant (interflow) model. A significant difference is that the cell-order structure is replaced by a person-contract structure. Cells (in plants) communicate with each other in a master-slave relation, client-to-server, whereas legal persons (persons, companies and public bodies) in a market communicate (also) as equals, peer-to-peer. *Contracts* commit two or more legal persons to carry out, during a certain period, a number of exchanges of products or services (including labour, money, etc.). They may express terms of synchronization, obligations, etc., e.g. to incorporate a company a suitable "legal person template" must be selected, this will determine reporting obligations, and – to some extent – the contracts the company can enter into. The legal personality and contracts may also be correlated with the products/services which the enterprise can provide on the market (e.g. banks must be registered in a special commission).

Templates of all legal persons and contracts that exist in a market are joined in a *person-contract structure*. This structure includes person templates for the public bodies that – because of statutory regulations – are involved in the registration and monitoring of (certain) contracts and exchanges. It also includes contract templates such as for incorporation contracts, labour contracts (between citizens and companies), etc. (Goossenaerts & Bjørner, 1994).

Industry Interflow Models

Industry-wide projects such as industrial policy and legislative projects aim at transformations of a market system or sector as a whole. Such transformations may be planned (or designed) in terms of particular market interflow models, prior to their implementation.

Our approach suggests to consolidate the deliverables of industry-wide projects as product, exchange, person and contract templates, (in a) market structure, (in a) market system, (in a) particular market interflow model, and eventually, as (in a) market interflow system.

4 INFORMATION AND COMMAND INFRASTRUCTURE SERVICES

The provision of information and command infrastructure services requires (computer network supported) enterprise model execution services. See CEN/TC310/WG1 (1994a) for a statement of requirements for such services and CEN/TC310/WG1 (1994b) for an evaluation of some related initiatives. As to design services, relevant requirements can be sourced from the literature on computer aided design and project support environments.

In the MI^2CI Project, subsystems of an information and command infrastructure for enterprises and industry are classified in accordance with the domain decomposition in Table 1 (IMES stands for Interflow Model Execution Services and ISDACS stands for Interflow System Design and Construction Services).

An enterprise information and command infrastructure comprises:

Plant IMES: a plant component which supports the repetitive operations of the enterprise (cf. enterprise wide information systems built around enterprise-wide data models (Scheer, 1989)).

Plant ISDACS: a team component which supports the innovations and improvements which the enterprise needs to compete in the market; application packages in a Plant ISDACS may include an *entrepreneuring project coach*, an *engineering project coach* and a *quality project coach*.

An industry information and command infrastructure comprises:

Market IMES: a market component which supports the repetitive operations at the market; application packages in a Market IMES may include *material flow monitors*, *employment monitors*, *(value added) tax collection systems*, *trade and transportation systems*, *an intellectual property monitor*, etc.

Market ISDACS: an industry (team) component which supports the one-of-a-kind (industrial policy) innovations and improvements which the industry needs to sustain its competitiveness in the global market; application packages in a Market ISDACS may include an *industrial policy project coach*.

5 CONCLUSIONS AND FUTURE WORK

The problem domain understanding underlying strategic information and manufacturing technology projects in industrialised countries in combination with achievements in model oriented software development, indicate how we can build information technology applications that may help developing countries to leap forward into the age of sustainable lean/agile supply-based industries. This paper has introduced the concept of information and command infrastructures for SMEs and has explained a framework that can guide the planning of the detailed design and implementation of such infrastructures.

The $\mathsf{MI^2CI}$ Philippines Project will enable us to further test the conceptual and practical validity of the proposed approach. Work in the near future should focus on: (a) the derivation of particular models from generic ones; (b) the elaboration of generic models for markets, teams and industries; (c) the implementation of the generic models; (d) software tool support for the derivation, interfacing, validation, execution and evaluation of particular models; and (e) software tool support for the specification and implementation of interflow models capable of interflowing with enterprise and industry dynamics.

6 REFERENCES

Bonfatti, F., Monari, P.D. and Paganelli, P. (1994) Modelling process by rules: Regularities, alternatives and constraints. In: Vernadat, F., editor, *Proceedings of the European Workshop on Integrated Manufacturing Systems Engineering (IMSE '94)* , INRIA, Grenoble, France, 1994.

Browne, J., Sackett, P. and Wortmann, H. (1994) Industry requirements and associated research issues in the Extended Enterprise. In: Vernadat, F., editor, *Proceedings of the European Workshop on Integrated Manufacturing Systems Engineering (IMSE'94)*, INRIA, Grenoble, France, 1994.

CEN/TC310/WG1 (1994a) CIM systems architecture – enterprise model execution and integration services – statement of requirements. CR1832, CEN/CENELEC, Brussels, Belgium.

CEN/TC310/WG1(1994b) CIM systems architecture – enterprise model execution and integration services – evaluation report. CR1831, CEN/CENELEC, Brussels, Belgium.

CIMOSA (1993) ESPRIT Consortium AMICE, editor. *CIMOSA: Open System Architecture for CIM.* Springer Verlag, Berlin, 2nd, rev. and ext. edition, 1993.

ENV 40 003 (1990) ENV 40 003: Computer integrated manufacturing – systems architecture – framework for enterprise modelling. European prestandard, CEN/CENELEC, Brussels.

Goossenaerts, J. and Bjørner, D. (1994) Generic models for manufacturing industry. Technical report no. 32, UNU/IIST, Macau, December 1994.

Scheer, A.-W. (1989). *Enterprise-Wide Data Modelling - Information Systems in Industry.* Springer-Verlag, Berlin - Heidelberg.

Spur, G., Mertins, K. and Jochem, R. (1994). *Integrated Enterprise Modelling.* Beuth Verlag, Berlin.

Tucker, D.E., Wainwright, C.E.R. and Thethi, A.J.S. (1994) Integrating manufacturing information systems within small and medium enterprises. In: Vernadat,F., editor,*Proceedings of the European Workshop on Integrated Manufacturing Systems Engineering (IMSE'94)*, INRIA, Grenoble, France.

The World Bank (1994) *World Development Report 1994.* Oxford University Press,Oxford,UK.

4

Narrowing the gap of SME for IT innovation with enterprise architectures

C. Meijs
Department of computer science, wageningen agricultural university
Dreijenplein 2, NL-6703 HB Wageningen, Netherlands
e-mail: meijs@rcl.wau.nl

Abstract

Contemporary changes in market structures demands adequate response from Small and Medium sized Enterprises (SME). The rapid evolution of information-technology offers new challenges and exiting new opportunities for innovation. The approach presented in this paper will take the information system architectures of the enterprise as a starting point. It will be extended middle-out to a meta-level dealing with methodology issues and to a project level dealing with involved people, activities and results.

Keywords

information systems architecture, learning, method engineering, reference models

1 INTRODUCTION

It has been continuing difficult for Small and Medium sized Enterprises (SME) to innovate and to gain the benefits from the area of information technology. The introduction of innovative information technology is commonly hampered by:

- Incompatibility of newly introduced systems with the existing computer infrastructure, due to syntax, semantic or pragmatic differences.
- Rapid changes as SME are pressed the to react because the ever-shorter product-life-cycles, while IT projects might have a duration that is longer than expected.
- Lacking of professional IT specialists within the enterprise, with possibly as a consequence that individuals carry their personal wishes through in a craftsman culture.
- Limited resources available for starting innovative IT projects.

A challenge of major importance for the future is to enable SME, to master the growing complexity associated with the effective and efficient use of information technology for their information-related tasks. To our opinion more attention has to be paid to the fitting of information-technology within SME. Therefore we need a clear description of the enterprise architecture, because the concept is surrounded by much confusion and has yet no standard definition.

This paper will focus on:

- the contex of various enterprise architectures and their interactions
- a methodology that emphasizes a reduction of the complexity to analyse and design systems for SME
- illustration of the advocated approach and an evaluation of project results in the dutch agriculture sector.

2 ENTERPRISE ARCHITECTURES

A clear picture of the various architectures and their interactions will improve the planning and implementation of IT projects. We distinghuish information system architectures focussing on the business characteristics, information needs, the technology and on the system applications.

A business architecture establishes a clear understanding of the mission and nature of the enterprise, reflecting the current management and control concepts and philosophy (e.g. TQM or JIT). It provides an overview of production, marketing and the logistics of goods or services. Further, it depicts the units, their employment and related responsibilities.

An information architecture is a personnel and technology independent profile of the major information categories used within an enterprise. It provides a way to relate business functions, data classes, decisions and control.

A systems architecture identifies applications needed to support the information needs of the enterprise, wheter they are computer based such as databases, expert systems, real-time microprocessors and spreadsheets, or non-computer based systems such as libraries, filing cabinets, microfilm, photocopiers or faxes.

The technology architecture consists of the computer infrastructure and digital networks that cooperate to provide support for computing and data communication.

Moreover, interactions among the (parts of) business architecture, information architecture, system architecture and technical architecture takes place as feedbackloops and feed forward loops. Thereover they cannot be carried out sequentially but they must be done in parallel, see figure 1. The planning of information systems should be no longer be separated from the business plan and vica versa.

2.1 Evolving enterprise architectures

Especially the enterprise architectures of SME are strongly influenced by changes in the market structure and the push of new technology.

Value chain

A value chain represents how businesses are linked to each other by the supply of goods and services and may be specialized to a product chain. Negotiations and contracts with suppliers and clients demands for quality of products, processes and information.

Emerging information technology

The processing of information is occupying an increasingly important strategic and economic role. For the introduction of emerging technologies in enterprise architecture, access to a wide range of information is needed.

Balancing with enterprise architectures
The dashed lines between the 4 architectures depicts the interactions. The already mentioned influences of the suppliers and buyers in the value chain and the emerging information technology are also included in the next picture.

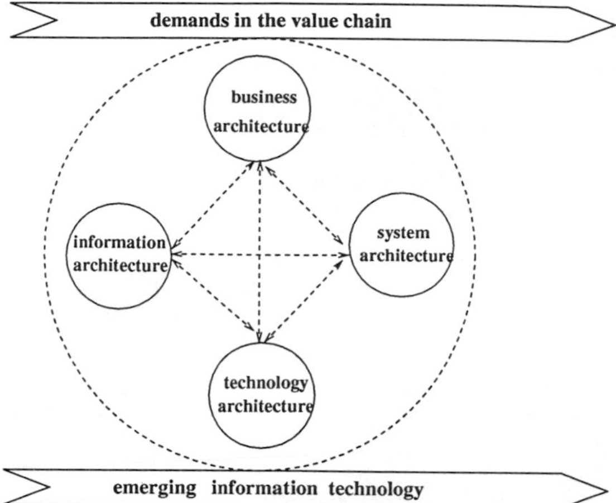

Figure 1 Alignment and balancing the enterprise architectures.

2.2 Reference models for building architectures

For SME the use of reference models appear to advantages. It is often economic impossible to carry out the whole life cycle of information system development, starting with a planning study eventually resulting to implementation and maintenance of tailor-made information systems. Enterprise wide data models are valuable for design [Scheer] (1992).

3 TOWARDS A FLEXIBLE FRAMEWORK

Various frameworks to model information systems have been proposed, among them are those of CRIS [Olle] (1988), CIM/OSA [Esprit] (1989), Zachman and Sowa [Sowa] (1992). A more flexible framework should offer help in the adaption of methods to specific situations in enterprises. Therefore we distinguish three levels in figure 2: the methodology level, the enterprise type level and the project type level.

Some distincions with other frameworks are:

- Both the number and choice of perspectives is open, in [Meijs] (1994) we extended the 2nd level with a chain architecture.
- The number of aspects is not fixed (as in [Sowa] e.g. data, function, time, network, people and motivation). This approach facilitates the decomposition and integration of aspects. Decomposition enables e.g. to split the motivation aspect into goals and constraints. The system archtitecture may contain templates of Abstract Data Types (ADT) that integrates data types and operations.
- The level of detail for describing components related to a perspective and aspect can be refined by hierarchies. Another level of abstraction is introduced by the reference models that creates extra genericity.
- The stages of the system development cycle do not have a dominant position in the framework, like in [Olle], because we want to include evolutionary information systems.

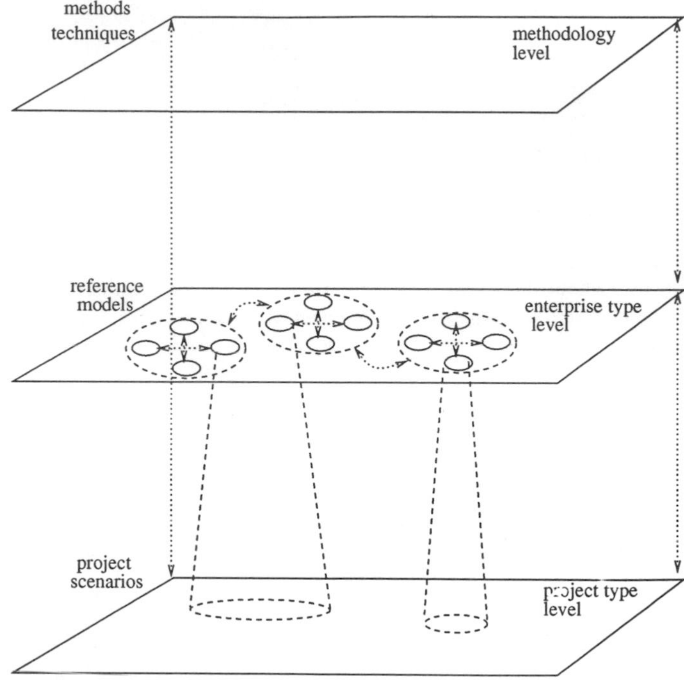

Figure 2 A flexible framework.

In the next paragraph more is said about the methodology level. The use of reference models on the middle level accelerates the planning and implementation of information technology. We will go in more detail for reference information models, that represent generic components for clusters of enterprises. Finally an evaluation of some results at the project level will be presented, using adapted checklists in composing project scenarios. See also for project management [Uijttenbroek].

4 METHODS AN TECHNIQUES

There is a growing awareness that method and techniques cannot be applied unconditionally. Specific characteristics and particular circumstances of focused areas in the value chain of enterprises impact the methodology. Method engineering is considered as the design a methodology with the definition of steps to be conducted. At the 1st level of the framework we may recognize for a methodology a set of steps in the form of a triplet:
< SITUATION, DECISION, ACTION>.
The situation referers to the enterprise(s) type with involved architecture(s), and to the project type level: identified problems, opportunities, involved people and other project constraints. So the description of the situation can be specified and refined to get an adequate understanding of the context for taking decisions about needed actions. Action referes to one or more modelling aspects from the framework. Decisions are related to the intention of a step and may be atomic or compounded.

Decomposition can be decided to reduce complexity. It is then used to guide the modelling process. Integration of multiple specification fragments produced by conflicting basic components (e.g. relevant data may be modelled as a attribute or as entity type) may urge for a refinement with a sub-decision of view integration.

The feature of the framework that enables us to integrate aspects, is useful for putting the next object oriented template belonging to the system architecture into it.

```
Object ORDER
 attributes
   lines : set of (ORDER-LINE)
   creation_date, delivery_date: DATE
   state: {created, delivered, invoiced}
operations
   creation : order lines insertions
   delivery : change order state
   cancellation: delete order lines
constraints
   creation date <= delivery date
   delivery date <= invoice date
event
   pre-condition [quantity.old > replenishment level]
     out of stock
   post condition [quantity.new < = replenishment level]
```

5 REFERENCE INFORMATION MODELS

5.1 Characteristics

Reference models reflecting genericity are constructed for clusters of enterprises in a branch. These models need to be valid and stable for a line of business. Decisions related to control of the enterprise, functions of an enterprise and data classes are important components for the definition of so-called reference information models. Dynamic features may be incorporated by entity life cycle, petri nets, or state transition diagrams.

5.2 Roles of reference information models

- Identification of missing knowledge for learning purposes
  For innovation the involved individual employees of the SME are subjected to a process of change and learning. Effective learning means going through the following phases in a number of cyclic iterations: conceptualization, experimentation, action and reflection [Kolb] (1984).
- Standardization and as a basis for the development of software packages.
  Communication among interest groups is not primarly hampered by technical problems, but by a lack of broadly accepted definitions of products and processes. Different classifications and identification of objects may be prevented by unifying reference models.
  Software industry take some relevant parts of the models as a starting point, e.g. the definitions of entity types and derived indicators. This stimulates standardization of information technology, for instance the electronic data interchange (EDI).
- Selection of software packages.
  An available datamodel may be used to restrict the offered packages to a short list. Functionality of the package is tested by test-set.
- Benchmarking.
  Measuring the quality of business functions and comparing the actual indicators with other enterprises in the line of business, one gets a founded opinion about the performance of the enterprise.

6 EVALUATION AND RESULTS OF PROJECTS

It is estimated that in the Netherlands there are more than 500 000 SME, which may be classified into sectors and finally into one of the 600 branches. There are about 100 000 farms in the dutch agricultural sector, most of these farms employs only a few people. The number of farms belonging to a specific branch ranges from the smallest number for Mushroom cultivation farms (750) to the highest for Dairy farms (36 000).

6.1 Evaluation of projects in the agricultural sector

During the second part of the former decade several reference information models for the agricultural branches as dairy, potplant nursery, poultry and fruit were created in the Netherlands. For each type of farm an reference information model was constructed mainly focussing on the data and process aspect, using the Information Engineering Methodology [Martin]. The construction of the models was a joined effort of the branch organizations, agricultural researchers and information engineers. Our evaluation of projects using these models recommends improvements of the methodology based on the framework of figure 2:

- The branch dimension is tentatively a good top-down criterium for clustering the different types of farms, but there are also branch crossing components on the next level. We need bottom-up techniques to define generic components that might be reused for different branches.
- The hierarchical decomposition of the data flow diagrams restricts the usability and reuse of components in different situational contexts. Related to other disadvantages for the hierarchical approach, especially for the business architectures, as noted in [Peters] (1992), this favours alternatives like object oriented techniques
- The data definitions of the reference models seems to be used most intensive. They are for example important for the introduction of EDI, because it might supports the standardization of exchanging data. However, only little attention was given to some issues about ownership, legization, maintenance of models.
- The decomposition that distinghuises management vs. technical functions results often in isolated systems for admiministrative applications and real-time computers that register data of continuous or discrete processes of the enterprise. Alternatively, we might unify some of the basic elements of both functions by elaborating a bottum-up approach. Interaction analysis improves the adjustments of selected aspects.

6.2 Project types and their scenarios

In co-operation with enterprises IT research projects are conducted by graduated students and supervised by the department of computer science.
- Reverse engineering
A software package for the semi-process industry that was installed at a medium sized enterprise in the food industry. Starting from the systems architecture consisting of several modules for purchasing, production, sales and financial administration, an reverse engineering traject resulted in a data model with Entity Relationship Diagrams.
- A software testing method was elaborated to enable the certification of software for pig farmers. The testing method uses definitions of entity types and ratios that are derived from reference information models for pig farms. The information and system architecture demarcates the boundaries of this project.
- Information resource management and security.
The main activities of this project scenario were: exploration of significant concepts of information resource management and security techniques, interviews with owners of SME, questionnaires to system managers and finally recommendations for auditing purposes of the business and technology architecture. The reference model was constructed after the exploration and validated during the visits to the more than fifty participating enterprises, that were selected across all branches. Confrontation of the concepts from information resource management and security techniques and the inventory of bottle-necks resulted in a handbook for information auditing.

7 CONCLUSIONS AND FURTHER RESEARCH

Reliable and actual information is a key issue for SME. This paper emphasizes in this era of all pervasive applications of information systems and technology the need for alignment of the various architectures. The advocated approach, creates a balance between market needs, via business functions of SME and the emerging information technology. Lessons learned from conducted projects show:
- how to put into practice the building of enterprise architectures, using the business functions and data classes defined in a reference models as a starting point.
- reference information models decrease complexity and cost of projects.
- communication and documentation should be well prepared, attention has to be paid to training and understanding of the involved people.
In the next future new research pojects with empirical results will enrich the methodology according the presented flexible famework.

References

[Esprit] Esprit consortium AMICE (eds) (1989) Open system architecture for CIM, Esprit project 688, Vol 1, Springer Verlag Wien.

[Kolb] D. Kolb (1984) Experimental learning - Experience as the source of learning and development. Prenctice Hall, Englewood Cliffs, New Jersey.

[Martin] J. Martin, C. McClure (1986) Diagramming techniques for analyst and programmers, Prentice Hall, Englewood Cliffs.

[Meijs] C. Meijs , J. Trienekens (1994) Optimization of the value chain using computer aided reference models. In: *Proc of 3rd Ifip wg 7.6 WC on Optmization-based computer-aided modelling and design (eds. J. Doležal, J. Fidler)*, Prague.

[Olle] T. Olle, J. Hagelstein, I Macdonald, C. Rolland, H. Sol F. van Assche and A. Verrijn Stuart (1989) Information systems methodologies, a framework for understanding. Addison-Wesley.

[Peters] T. Peters (1992) Liberation management.

[Sowa] J. Sowa , J. Zachman (1992) Extending and formalizing the framework for information systems architecture. in IBM systems journal, vol 31 no 3.

[Scheer] A. Scheer, A. Hars (1992) Extending data modelling to cover the whole enterprise. in: Communications of the ACM vol. 35 no 9.

[Uijttenbroek] A. Uijttenbroek, P. Anthonisse, C. Meijs, B. Verdoes (1992) Project management principles; Cap Gemini Publishing, Rijswijk.

5

One-Product-Integrated-Manufacturing

G. D. Putnik[a)], *S. Carmo Silva*[b)]
University of Minho, Production Systems Engineering Centre
Azurem, 4800 Guimaraes, Portugal, fax.:+351-53-510268
e-mail: [a)] *putnikgd@eng.uminho.pt,* [b)] *scarmo@ci.uminho.pt*

Abstract

A new organisational frame for manufacturing system integrated over one-product manufacturing is proposed. Integration of a one-product enterprise is realised over a set of independent cells domain. They are connected by the wide-area-network using multimedia telematic technology to provide negotiation in real time. This characteristic provides the possibility to constitute an one-product factory from elements that are localised globally especially for the manufacturing operations which are information based. In this way a concept of a global manufacturing is applied. The life time of a one-product factory/enterprise corresponds at most to a product life time. This organisation form is called One-Product-Integrated-Manufacturing (OPIM) System. The system is characterised by the factory models flowing through the product. OPIM System is a physical implementation of a virtual factory.

Keywords

One-Product-Integrated-Manufacturing, virtual factory, distributed manufacturing system, global manufacturing, factory models flow through product

1 INTRODUCTION

New manufacturing system concepts should provide a start in the XXI century race with qualitative better performances. Characteristics for the new manufacturing system concept could be derived from analysis of the manufacturing systems space representation. Considering this representation, directions of existing manufacturing system concepts improvements start from **Computer Integrated Manufacturing (CIM)** concept, placed in the middle of the manufacturing systems space, and tend to spread in all directions, achieving high flexibility. This strategy leads to a manufacturing system able to:

1. Manufacture from 1 to 1000 products simultaneously,
2. Accommodate lot sizes from 1 to 1 000 000

covering entire space, figure 1. But, to satisfy 1. and 2. efficiently, the manufacturing system must be highly reconfigurable leading to the additional requirement:

3. The new generation manufacturing system should reconfigure for a new product within 1 second, Kim (1990).

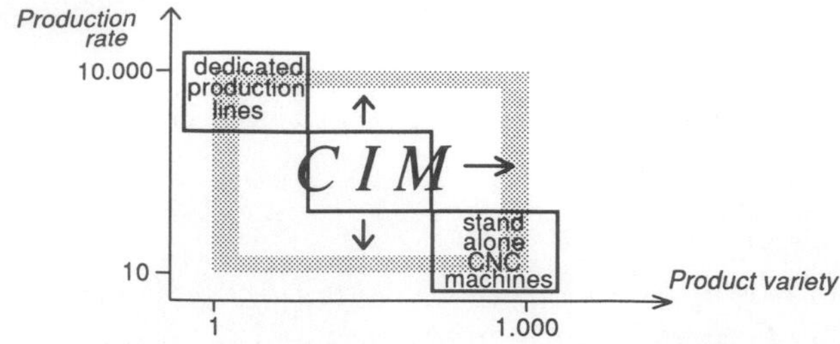

Figure 1 CIM concept, placed in the middle of the manufacturing systems space, tends to spread in all directions, Kim (1990),.

2 A MULTI-PRODUCT-INTEGRATED-MANUFACTURING SYSTEM - THE EXISTING CONCEPT

The strategy of spreading the CIM concept over the entire manufacturing systems space, has some limitations.

First, although systems such as PPC/CAD/CAPP/CAM/FMS/FAS are included in a CIM system this doesn't mean that they are automated. Only aspects of manipulation and memorisation of related objects are automated ("Computer Aided" attribute) but the design functions are still in the human domain. This is a very serious limitation in achieving the target manufacturing system of the future. It is accepted that the aspect of design automation is solvable by introducing artificial intelligence concept in CIM systems. In this way we come about to the concept of **Intelligent Manufacturing Systems (IMS)**.

Second, in CIM and IMS concepts the hardware and software flexibility is an important issue, but study on organisational structures is not emphasised. However, new studies on organisational structures present models which we understand to accommodate concepts like CIM and IMS.

A list of manufacturing system concepts including organisational aspects is given bellow.

- **One-of-a-Kind Production (OKP) Systems** , Rolstadas (1991), Mertins et al. (1992), Vepsalainen (1990).
- **Global manufacturing** , Mitsuishi et al. (1992), Lalande (1992), Carlson et al. (1992), O'connor et al. (1990), Horgan (1990).
- **The next generation of FMS systems based on metamorphic machines** , Ito (1992), Yoshikawa (1984).
- **A fractal factory and Holonic manufacturing system** , Warnecke (1994), Winkler (1994),
- **A virtual factory** , Kim (1990), Staffend (1992).

All these sophisticated, highly performing models, are based on the paradigm that a manufacturing system processes products.

We will call this paradigm a **Multi Product Manufacturing System**, figure 2.

The figure emphasises the fact that multi-product integrated manufacturing system is characterised by the **flow of products** through it.

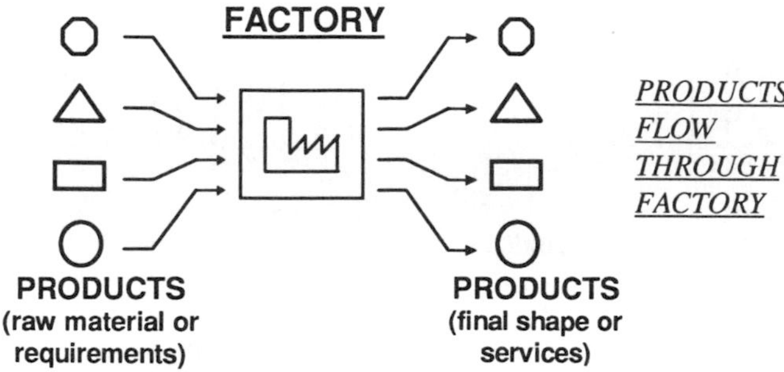

Figure 2 Multi-Product Manufacturing : the *flow of products* through a factory.

All models mentioned above, are **always based on the existing hardware structure of one company**. But, using the system for a variety of products means loosing some technical efficiency because it is clear that each particular machine can be used with maximum efficiency just for one type of products which employs the maximum capability of the machine, figure 3a.

The conclusion is that, in any case, with a fixed hardware structure a company cannot ever achieve maximum efficiency. This phenomena is the main reason for introduction of "metamorphic machines" capable to adapt to any particular product. It is not clear if and how could metamorphic machine design principles be applied to organisational design.

3 ONE-PRODUCT-INTEGRATED-MANUFACTURING SYSTEM (OPIM) - THE FUTURE CONCEPT

Multi-product-manufacturing systems are conceived for a variety of products. This means loosing some technical efficiency of machines when compared to a dedicated system. Full performance facility is possible through a manufacturing system for a single product, figure 3a. Thus, we could think of establishing a factory/enterprise just for a particular product, and for a number of products we would need at least equal number of factories/enterprises, figure 3b.

3.1 Designing / integration of the OPIM system

Under the One-Product-Integrated-Manufacturing (OPIM) concept a new manufacturing system structure is conceived for every new product. The design and manufacturing processes for a product should be decomposed in a sequence of particular tasks. The most suitable resources for each task should be selected. Searching for the resources for task realisation means selecting the structure elements.

The domain for resources selection is the set of all cells, Putnik (1995), (machine tools, transportation devices, computers, manufacturing cells) which can realise the necessary

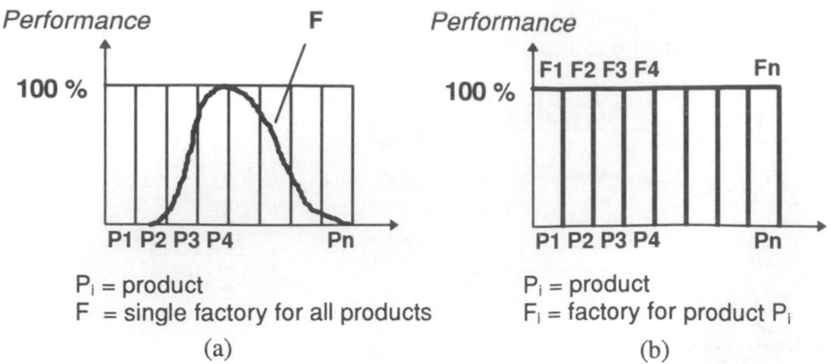

Figure 3 Facility performance for (a) multi-product manufacturing and (b) OPIM system

manufacturing functions, connected by public communications network, using multimedia telematic technologies.. To provide optimal choice the domain, the "market" of cells, candidates, should be as large as possible. If the domain is limited to the set of machines within an existing factory then the concept will be reduced to a One-of-a-Kind production system or to a virtual factory.

The complexity of OPIM system design can be illustrated as follows. Let the domain for OPIM system design be represented as a connected graph (OPIM systems building blocks, cells, are connected by public communication network). If there exist a manufacturing system structure which can be defined over the available domain it means that the design task can be reduced to a task of finding a subgraph, of the domain graph, which satisfies given requirements for manufacturing. This task is hard, belongs to NP class, because a set of N manufacturing system elements, defines 2^N manufacturing system elements subsets, the power set, for building more complex structures. This means that the OPIM system design must be based on AI based technologies for decision making. The lattice of the refereed power set is shown in the figure 4. as a OPIM system design/integration domain.

The design process should be performed automatically to achieve *a real time reconfiguration* and a good solution. All elements of domain set and all new designed elements/cells should have a formal representation (Putnik (1995)) in a common knowledge base. The OPIM system design process is a process of defining an interpretation, instance, of a general model of a manufacturing system. The design process is realised through negotiation between the leading company, which initiates the OPIM process, and the number of cells, candidates, for realisation of manufacturing tasks, including design, planning and control and manufacturing. During the design process there is a number of virtual factories which forms the solution space. Based on a defined performance criteria, the initiator selects the optimal factory structure. We could say that we have a flow of virtual factories through the product.

We will call this paradigm a **One-Product-Integrated-Manufacturing System** concept. The OPIM system is characterised by the **flow of a factories** through a product, figure 5. This is the **inverse model** of the multi-product-manufacturing system.

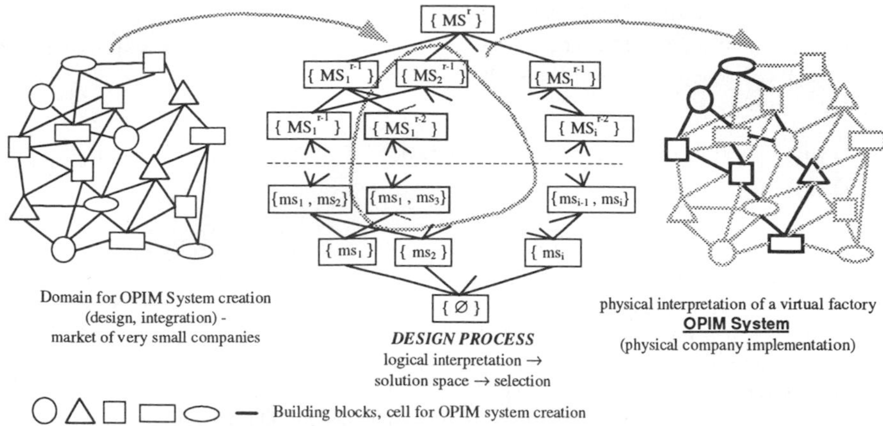

Figure 4 The lattice of the manufacturing system elements power set as a OPIM system design / integration domain.

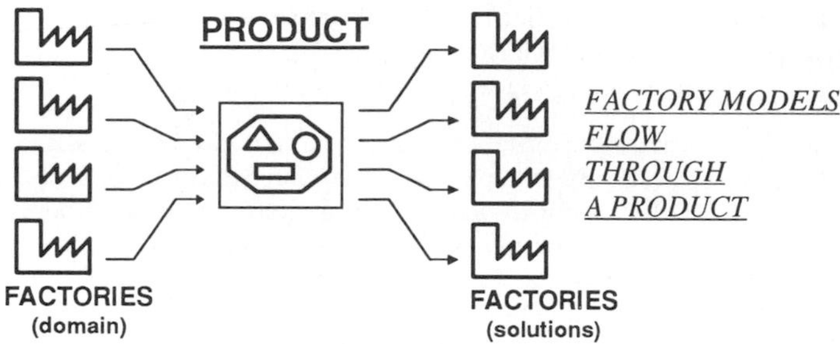

Figure 5 OPIM system : the *flow of factories* through a product.

3.2 Some characteristics of the OPIM system

A list of some characteristics of the OPIM organisation form is given :

- The optimal structure is built up from primitive cell, single machines. Each cell is specialised in one type of service (design, planning, management and manufacturing) for one type of product. The resulting physical OPIM system should be the best performing system for the product under the available manufacturing cell market.
- All design, planning and management functions are information based and independent of distances between the processing units. The processing units can be located at any point of the world if they are connected by the communication network. This is a pure application of a global manufacturing concept.
- The integrated OPIM system is essentially distributed manufacturing system.

- OPIM System is a physical implementation of a virtual factory.
- The flexibility of the OPIM concept is in the design stage. The OPIM system itself is not flexible. It is a system dedicated to one specific product.

4 OPIM SYSTEM IN A SPACE OF MANUFACTURING SYSTEMS MODELS

Using a formal, rigorous definition of manufacturing it is possible to specify a manufacturing systems definition space. The richer the space is, the better the chances are for qualitative development of manufacturing systems. Through specification of the manufacturing systems definition space it is possible to build a rigorous taxonomy of manufacturing systems/models. The manufacturing systems space, defined in chapter 1, figure 1, which gives emphasis to flexibility, should be enlarged by new aspects.

The manufacturing systems definition space will be called **"the manufacturing systems reference model"**. A basic reference model can be defined in a three-dimensional space:

1. **degree of integration** (structure and functionality). Integration as a key issue for developing/designing the "factory-of-the-future" model. The concept of CIM is a manufacturing system concept defined at a high degree of Computer-Aided integration.
2. **decision making** . Intelligence , with a decision making technologies as a key issue.
3. **organisation** . Organisational aspects of the target manufacturing system.

In this reference model each of the manufacturing system concepts mentioned above, section 2. and also OPIM system, are defined by triples, points in a three-dimensional space, figure 6.

Integration in the OPIM system is defined over a globally distributed domain. The high degree of integration is based primarily on WAN technologies. This concept is expressed through the concept of *'Global Manufacturing'* with the additional attribute *'Integrated'*. The OPIM system implementation and utilisation is design intensive (belonging to hard problems - NP class) for which we need in any case nondeterministic algorithms, that is, *intelligence*, for achieving good solutions. Finally, the OPIM system itself is a *particular organisational frame*. Thus,

One-Product-Integrated-Manufacturing System is defined by a triple:

(*'Global Integrated Manufacturing', 'Intelligent Systems', 'OPIM Organisation Framework'*)

5 CONCLUSION

Users of OPIM system are any size company, very small, SME and large companies, or any person which is self-employed offering and requiring teleservices on manufacturing activities.

- Companies can benefit from OPIM in two ways : 1) by strengthening competitiveness through access to competitive resources, services and knowledge (which is a high value-

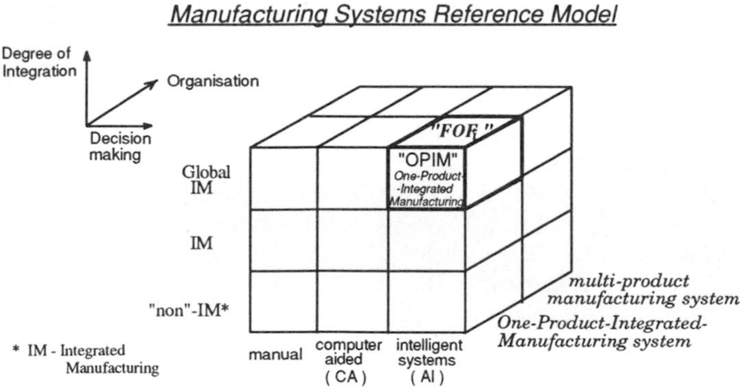

Figure 6 OPIM system in a space of manufacturing system models.

aided resource) provided by teleservices and 2) by dynamic business process re-engineering leading to new products, services and business supported by a networked underlying universe of resource entities. Further advantages can also be envisaged, for large companies, through decentralisation and business distribution through competitive locations.

- Self-employed people can benefit from OPIM through integration and co-operation with SME and large companies for competitive development of complex processes, products, services and business. Competitive base, in this case, is, for example, avoidance of travelling and transportation cost, use of free and low-cost facilities (own home) and independence of distance.
- OPIM can also offer lower prices, high quality of a products and competitiveness, and high flexibility for creating manufacturing systems through large number of small companies which can include one machine and one self-employed person.

A number of researching issues are imposed. We list some :

- Formal specification and design of the OPIM system, the intelligence for OPIM, transformation of information Company to Knowledge Company, communication based on multimedia telematic technology and related problems, task-resources allocation management of the OPIM system, distributed and real-time CAD/CAPP/CAM/PPC applications, economy of the OPIM system, legal aspect of the OPIM system and intellectual property issues.

Other topics related with OPIM model are, for example, “architectures for systems integration, federated architectures, Computer Integrated Business”, integration of self-employment in global economy “independently of their geographical location”.

6 REFERENCES

Carlson G., Parks R. (1992) EDI: Effectively Linking Customers and Suppliers, *Proceedings* AUTOFACT '92, Detroit, pp 31-37 - 31-43:

Horgan J. (1990) Engineering Design Service in an Enterprise, *Proceedings* AUTOFACT '90, Detroit, pp 26-31 - 26-42.

Ito Y. (1992) Amalgamation of human intelligence with highly automatised systems - An approach to manufacturing structure for the 21st century in Japan, in *Human Aspects in Computer Integrated Manufacturing* (eds. Olling G.J., Kimura F.), Elsevier Scientific Publ.

Kim S. H. (1990) Designing Intelligence, Oxford University Press.

Lalande M. (1992) Communication of Engineering Data in a Global Environment, *Proceedings* AUTOFACT '92, Detroit, pp 28-1 - 28-8.

Mertins K., Albrecht R., Steinberger V. (1992) Business process-oriented order control: An integrated approach for distributed systems, in *Human Aspects in Computer Integrated Manufacturing* (eds. Olling G.J., Kimura F.), Elsevier Scientific Publ.

Mitsuishi M., Nagao T., Hatamura Y., Kramer B., Warisawa S. (1992) A Manufacturing system for the Global Age, in *Human Aspects in Computer Integrated Manufacturing* (eds. Olling G.J., Kimura F.), Elsevier Scientific Publ.

O'connor F. A., Diesslin R., Lamoureux J. (1990) Information Systems Strategy: A Case Study Within a Mature Industry, *Proceedings* AUTOFACT '90, Detroit, pp 26-1 - 26-23.

Putnik G. D., Carmo Silva S. (1995) Knowledge Integrated Manufacturing Systems, *Proceedings* 1st World Congress on Intelligent Manufacturing Processes and Systems, Mayaguez, pp 688-698.

Rolstadas A. (1991) CIM and one of a kind production, in *Computer Applications in Production and Engineering: Integration Aspects* (eds. Doumeingts G., Browne J., Tomljanovich M.) Elsevier Science Publ.

Staffend G. S. (1992) Making the Virtual Factory a Reality, *Proceedings* AUTOFACT '92, Detroit, pp 15-9 - 15-16.

Vepsalainen A. (1990) Systems That Turn Manufacturing into Service, in *Advances in Production Management Systems* (ed. Eloranta E.), Elsevier Scientific Publ.

Warnecke H. J. (1994) Future Competitiveness of European Industries through Applied Research, Workshop *Production Management Optimisation in Real Time*, Technotron S.A, Lisboa.

Winkler M., Mey M. (1994) Holonic Manufacturing Systems. *European Production Engineering*, **18**, No 3-4, pp 10-12

Yoshikawa H. (1984) Flexible Manufacturing Systems in Japan, *Proceedings* IFAC 9th World Congress, Budapest, pp 34-42.

7 BIOGRAPHIES

Dr. Goran D. Putnik received his MSc and Ph.D. from Belgrade University, both in domain of Intelligent Manufacturing Systems. Dr. Putnik's current position is assistant professor in the Department of Production and Systems, University of Minho, Portugal, for subjects CAD/CAPP, Intelligent Manufacturing Systems and Design Theory. His interests are machine learning and manufacturing system design theory and implementations.

Dr. Silvio do Carmo Silva received his MSc from University of Wales and PhD from Loughborough University of Technology, both in Great Britain, both in the domain of production systems management and design. Dr. Carmo Silva's current position is assistant professor in the Department of Production and Systems, University of Minho, Portugal, for subjects Production Management and Computer Integrated Manufacturing Systems. His interests are manufacturing systems design and operation.

PART THREE

Intelligent Supervision Systems

6

Transition Enabling Petri Nets to Supervisory Control Theory

G.C. Barroso[†], *A.M.N. Lima*[‡] *and A. Perkusich*[‡]
[†]*DF/UFC*, [‡]*DEE/UFPB*
Av. Aprígio Veloso, 882 - 58109-970 - Campina Grande - PB - Brazil
FAX +55 83 333 2480 - e-mail: {gcb,marcus,perkusic}@dee.ufpb.br

Abstract

This paper presents a Petri Net (PN) model conceived to be employed with the Supervisory Control Theory (SCT). The supervisor synthesis is obtained by processing both the system and specification models through the proposed algorithms. These algorithms make possible to obtain the controller of a discrete event system based on a given specification. Its simplicity and efficiency are demonstrated using a typical manufacturing cell problem.

Keywords

Supervisory control, discrete event systems, Petri nets, manufacturing systems.

1 INTRODUCTION

Over the last years the use of the *Supervisory Control Theory* (SCT) in the design of controllers for Discrete Event Systems (DES), has rapidly increased. Among other formalisms, Petri nets (PN) (Murata, 1989) have been widely used as a powerful tool in the analysis and design of discrete event systems (Giua and DiCesare, 1994; Sreenivas, 1993). In general, the controller synthesis is achieved by changing the model structure. In most cases the structure modification does not follow any standard procedure.

This paper investigates a new approach to the supervisors synthesis, using the SCT and an extension of a class of Petri nets, called *Petri Nets with Transition Enabling Functions* (*PNTEF*)(Papelis and Casavant, 1992). Two new algorithms are introduced in order to synthesize the controller of a DES.

The present paper is organized as follows: In Section 2 the the basic definition of the SCT is presented. In Section 3 *PNTEF*s are defined. In Section 4 the proposed algorithms to synthesize the discrete event controller are outlined. In Sections 5 and 6 two examples are discussed to exemplify the proposed approach. And finally in Section 7 the main conclusion of the paper are presented.

2 GENERAL DEFINITIONS

2.1 Supervisory Control Theory

In the SCT (Ramadge and Wonham, 1989), the system behavior is represented by a 5-tuple $G = (\sum, Q, \delta, q_0, Q_m)$, called generator. The symbol $\sum$ represent the set of event labels, or event alphabet; Q is a set of states; $\delta : \sum \times Q \rightarrow Q$ is the transition function defined at each $q \in Q$, so that $\delta(\sigma, q)$ is defined only to some subset $\sum(q) \subset \sum$ that depends on q. The marked states Q_m represent the end of an event sequence, and also represent the execution of a task by the system. The set of all *strings* formed by any number of symbols from $\sum$, including the empty symbol ϵ, is denoted by $\sum^*$. The transition function δ is extended to process strings from $\sum^*$.

A state q is said to be *accessible* iff $\exists s \in \sum^* \mid \delta(s, q_0) = q$. A state q is said to be *coaccesible* iff $\exists s \in \sum^* \mid \delta(s,q) \in Q_m$. G is *trim* iff it is accessible and coaccessible.

Each generator G has two associated languages: $L(G)$ is the language generated by G and $L_m(G)$ is the language marked by G. These languages are sets of *words* defined on $\sum$. The language $L(G)$ represents the physically possible behavior of the system, while $L_m(G)$ represents the tasks it is able to complete.

To control a DES it is necessary to admit that some events may be disabled when desired. To model such control action, the $\sum$ set is partitioned into: i) $\sum_c$ the *controllable* event set and ii) $\sum_u$ the *uncontrollable* event set, such that $\sum = \sum_c \cup \sum_u$ and $\sum_c \cap \sum_u = \emptyset$. All the events in $\sum_c$ may be disabled at any time, while those in $\sum_u$ suffer no influence of an external control action.

A *control input* for G consists of a subset $\Gamma \subseteq \sum$, satisfying $\sum_u \subseteq \Gamma$. If $\sigma \in \Gamma$, then σ is enabled by Γ, otherwise, σ is disabled. The condition $\sum_u \subseteq \Gamma$ means that the events in $\sum_u$ are always enabled.

Let $\Gamma \subseteq 2^{\Sigma}$ denote the set of control inputs. A DES represented by G, with a set of control inputs Γ is called a *controlled* DES (CDES). For convenience, one refers to a CDES by its underlying generator G.

Controlling a CDES G, consists of generating a sequence of elements γ, γ', γ'', ... in Γ, in response to the previously observed events generated by G. Such a controller will be called a *supervisor*.

A supervisor is a map $f : L \rightarrow \Gamma$, specifying for each possible string of generated events w, the control input $f(w)$ to be applied.

A supervisor is represented by an automaton and an output map, i.e.: $S = (\sum, X, \xi, x_0)$ is an automaton and $\psi : X \rightarrow \Gamma$ is the output map.

One says that the pair (S, Ψ) realizes the supervisor f if for each $w \in L(G/f)$, $\Psi(\xi(w, x_0)) = f(w)$, where $L(G/f)$ represents the closed behavior of the composed system G supervised by f.

The basic supervisory control problem may be stated as follows: Given a DES G with open-loop behavior given by L, what closed-loop behavior $K \subseteq L$ can be achieved by supervision? To solve this problem, it is necessary to define the concept of *controllability*.

Given two arbitrary languages L, $K \subseteq \sum^*$ and an alphabet $\sum = \sum_c \cup \sum_u$, one says that K is *L-closed* iff $K = \bar{K} \cap L$, and K is *L-controllable* iff: $\bar{K} \sum_u \cap L \subseteq \bar{K}$, where $\bar{K}$ is a prefix of K, i.e., the set of all prefixes of the strings in K.

Given a generator G, the language $K \subseteq L_m(G)$ models the desired specification. The language K represents the tasks to be executed under supervision. Then, the goal is to find a proper supervisor S to G so that the closed-loop system satisfies the condition $L(S/G) = K$.

There exists a *proper* supervisor so that $L(S/G) = K$ iff K is $L_m(G)$*-closed* and $L(G)$*-controllable* (Ramadge and Wonham, 1989). When these conditions are not satisfied, it is always possible to meet a *supremal controllable sublanguage* $K^\dagger$ such that $K^\dagger \subseteq K$. In the case of finite state generators, $K^\dagger$ is always computable (Ramadge and Wonham, 1989).

3 PNTEF

A Petri net with transition enabling functions is a 4-tuple $PNTEF = (N, l, M_0, \Phi)$, where: $N = (P, T, I, O)$ is a structure of a place/transition Petri net, where: $P = \{p_1, ..., p_n\}$ is a finite set of places, represented by circles; $T = \{t_1, ..., t_m\}$ is a finite set of transitions represented by bars or boxes; $P \cap T = \oslash$ and $P \cup T \neq \oslash$. $I : P \times T \to \aleph$ is the input function that specifies the arcs directed from places to transitions ($\aleph = \{0, 1, 2, ...\}$); $O : T \times P \to \aleph$ is the output function that specifies the arcs directed from transitions to places; $l : T \to \sum$ is a function that labels the transitions with symbols of the alphabet $\sum$; M_0 is the initial marking; $\Phi = (\varphi_1, ..., \varphi_m)$ is a set of logical expressions associated with the transitions.

A transition t_j is enabled in a *PNTEF*, given a marking M', if: $M'(p_i) \geq I(p_i, t_j), \forall p_i \in P$; the logical expression φ_j associated to t_j is truth; the firing of t_j, enabled by M', generates a new marking M'', so defined: $M''(p_i) = M'(p_i) + O(t_j, p_i) - I(t_j, p_i)$

4 SUPERVISOR SYNTHESIS ALGORITHMS

In this paper the synthesis of the supervisor is obtained executing the two algorithms presented in this section. The first one is a modified version of the reachability tree algorithm presented by Murata (Murata, 1989). The second algorithm is derived from the one proposed by Ziller (Ziller and Cury, 1994).

4.1 Modified Reachabillity Tree

For modeling physical systems, it is natural to consider an upper bound for the number of tokens that each place can hold. A PN with this constraint is called *finite capacity Petri net*, and not all the markings are present in its *reachabillity tree* (Murata, 1989). Some information is lost when one uses the symbol w to replace the markings where the number of tokens tends to grow infinitely.

Algorithm 1 *Modified Reachabillity Tree Algorithm* for finite capacity PNs.
Begin

1. Label the initial marking M_0 as the *root* and tag it *new*;
2. While new markings exist do:
   (a) Select a new marking M;
   (b) If M is identical to a marking that already exists, tag it *old*;
   (c) If no transition is enabled at M, tag it as *dead*;
   (d) While enabled transitions exist at M, do the following for each enabled transition:
      i. Obtain the marking M' that results from firing t at M;
      ii. If there exists a marking M'' such that the capacity of a place is exceeded, then replace $M'(p)$ by w for each p such that its capacity is exceeded;
      iii. Introduce M' as a node, draw an arc with label t from M to M' and tag M' *not-permitted* if the capacity of a place is exceeded, otherwise tag it *new*.

End

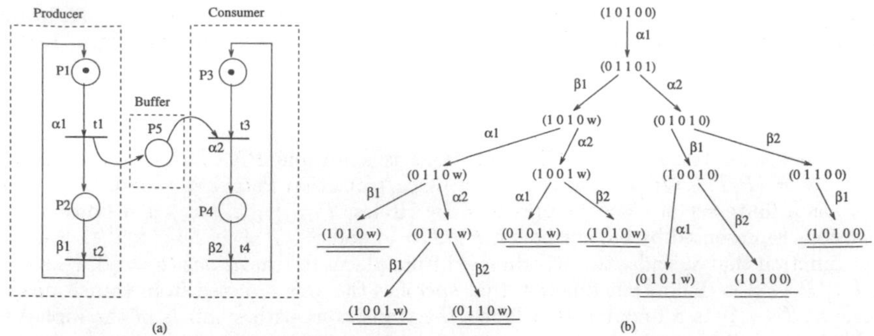

Figure 1 (a) PN model; (b) Reachabillity tree of a producer/consumer system

4.2 Algorithm to Construct a Generator for *supC(L)*

Different algorithms have been proposed to compute *supC(L)*. Generally these algorithms require the computation of the *trim* components of *G*. If the system and the specification are modeled by finite automata, then the *supC(L)* exists (Ramadge and Wonham, 1989). A new algorithm to find the *trim* component of a system being modeled by a finite capacity PN and then compute the *supC(L)*, is introduced. To compute *supC(L)* (item 5 of the algorithm) it is applied the algorithm presented by Ziller (Ziller and Cury, 1994).

Algorithm 2 *Algorithm to construct a generator for supC(L) from a PN generator:*
Begin

1. Create a dynamic list of states, *block_list,* and initialize it with the blocking states (blocking markings);
2. Add to the list the states that the only output event (transition) is input of a blocking state;
3. Add to the list the states that have, at least one output transition, labeled by an uncontrollable event, that is an input transition of a blocking state;
4. Create a dynamic list of states and events, *danger_list,* and initialize it with the *block_list* ancestors (states) together with the events that link *block_list* states to *danger_list* states, since the ancestor does not belong to the *block_list*;
5. Given the desired system specification, compute the supremal controllable sublanguage *supC(L)*;
6. Add to the *danger_list* the states and their respective output events that will be disabled to execute the *supC(L)* since these states do not belong to the *danger_list*.

End

5 PRODUCER/CONSUMER EXAMPLE

A typical producer/consumer system may be modeled by the PN of Figure 1(a). To execute a given task, it is necessary to incorporate some control action to this model. To construct the reachabillity tree of the PN generator *G*, and to derive the generator of the maximal possible specification $K^{\dagger}$, the Algorithms 1 and 2, are executed sequentially. To

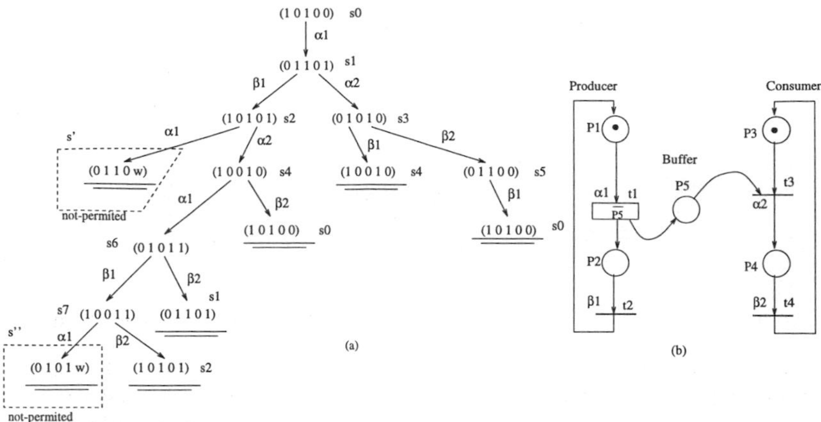

Figure 2 (a) Modified reachabillity tree of the producer/consumer PN model; (b) PNTEF supervisor for the producer/consumer system

implement the supervisor a *PNTEF* with the same structure of the PN model of G, is employed.

Considering a finite capacity *buffer* modeled by the place p_5, it is necessary to limit the number of tokens in that place. Observing the reachabillity tree, Figure 1(b), one can see that the number of parts produced and deposited in the *buffer* may become too big before one of them be consumed.

Suppose that p_5 in Figure 1(a), has capacity equal to one. Executing the Algorithm 1, the new reachabillity tree sketched in Figure 2(a), is obtained.

Notice that executing the reachabillity tree algorithm, the system state information is lost if the sequence $(\alpha_1\beta_1)^*$ is fired from the *root* (marking M_0). Moreover, executing the Algorithm 1, all the states where the net capacity is not exceeded, are enumerated. Besides that, one obtains information about the strings on the path from the *root* to a *not-permitted* state. Figure 2(a) shows all the physically possible strings for a system whose *buffer* has capacity equal to one, that is, states marked with a dashed line boxes should not be reachable. Consider the following specification: the initial state must be always reached.

To control the system, let $\sum = \{\alpha_1, \alpha_2, \beta_1, \beta_2\}$ where $\sum_c = \{\alpha_1, \alpha_2\}$ and $\sum_u = \{\beta_1, \beta_2\}$. In this case, it is necessary to disable α_1 in markings $M_2 = [10101]^T$ and $M_7 = [10010]^T$ to achieve the desired specification.

Given the PN model of the plant G, and applying Algorithm 2, either the automaton that generates the language representing the desired specification may be obtained, or then its maximum possible approximation.

Applying Algorithm 2, first, the *trim* component of the plant G is computed, and then, the generator of the least restrictive language for the desired specification. Notice that the *danger_list* has all the states (markings) and their respective events to be disabled. The maximum controllable sublanguage is generated by the automaton derived from the reachability tree of the Figure 2(a). The reachabillity tree of this automaton is presented in Figure 2(a). The markings in the dashed line boxes have to be excluded.

The supervisor may be implemented by the *PNTEF* of the Figure 2(b), i.e.: $N = (P, T, I, O)$ is a PN structure, identical to the PN structure of G; l (See Section 3) repre-

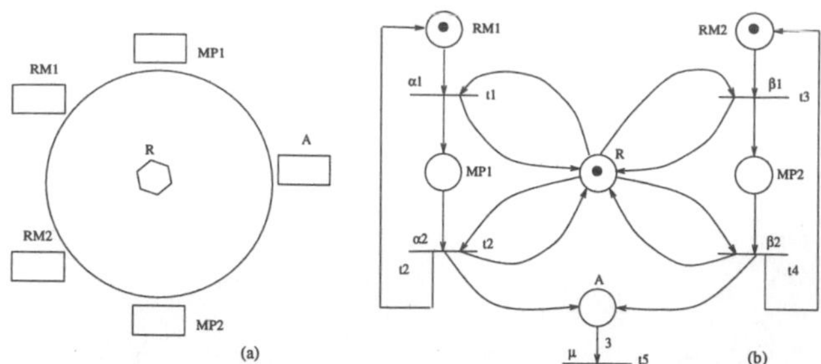

Figure 3 (a) Example of a manufacturing cell; (b) PN model of a manufacturing cell

sents the following mapping: $t_1 \rightarrow \alpha_1$; $t_2 \rightarrow \beta_1$; $t_3 \rightarrow \alpha_2$; $t_4 \rightarrow \beta_2$; $M_0 = [10100]^T$ is the initial marking; $\varphi_1 = \bar{p}_5$ while $\varphi_2 = \varphi_3 = \varphi_4 = 1$, because there are no restriction to the firing of transitions t_2, t_3 and t_4, since when they are enabled in some marking.

Considering that $t_1(\alpha_1)$ may not fire in two markings, only the logical expression $\bar{p}_5$ associated with t_1 is shown in Figure 2(b). This means that t_1 will not be always enabled when the system is in a state that p_5 hold a token. Since the other transitions have no additional firing restrictions, the logical expressions associated to them are $\varphi_2 = \varphi_3 = \varphi_4 = 1$, and are implicitly considered in Figure 2(b).

The logical expression $\bar{p}_5$ may be substituted by an inhibitor arc, and the net transformed in another net by the *complementary-place transformation* (Murata, 1989).

The *PNTEF* supervisor may be then executed in parallel with the plant, obeying to the transition firing rule.

Notice that specifying another task for the system, e.g., $K = (\alpha_1\beta_1\alpha_2\beta_2)^*$, it is enough to run steps 5 and 6 of the Algorithm 2. Therefore, the new states and their respective output events to be disabled are determined.

6 A MANUFACTURING CELL

Considering a manufacturing cell, as shown in Figure 3(a), it is possible to identify two raw-material deposits RM1 and RM2; two processing centers MP1 and MP2; and an assembling center A. A robot R moves raw-material from deposits RM1 and RM2 to the processing centers MP1 e MP2, respectively, and from these centers to the assembling center A. Figure 3(b) shows the PN model for this cell, where each three assembled parts constitute an item of this cell. Items are then transported to another cell. Exit of an item, from A, is modeled by transition t_5 (event μ) and is accomplished by an external agent. The maximum number of parts that the assembling center A supports is three. These parts are enough to assemble an item, i.e., these are the upper bound in the number of tokens that place A may hold in any marking of the net.

Let $\sum = \{\alpha_1, \alpha_2, \beta_1, \beta_2, \mu\}$, where $\sum_c = \{\alpha_2, \beta_2\}$ and $\sum_u = \{\alpha_1, \beta_1, \mu\}$. After executing Algorithm 1, the modified reachabillity tree for the PN model of the manufacturing cell, is obtained. This tree is shown in Figure 4.

The automaton, shown in Figure 5, represents a specification to obtain a given item. An item is assembled with two parts processed by MP1 and one processed by MP2. It is necessary to assemble first parts processed by MP1.

Executing the Algorithm 2, the generator of the *supC(L)* is obtained (In this case, the same automaton of the specification). See Figure 5. The Algorithm provides the

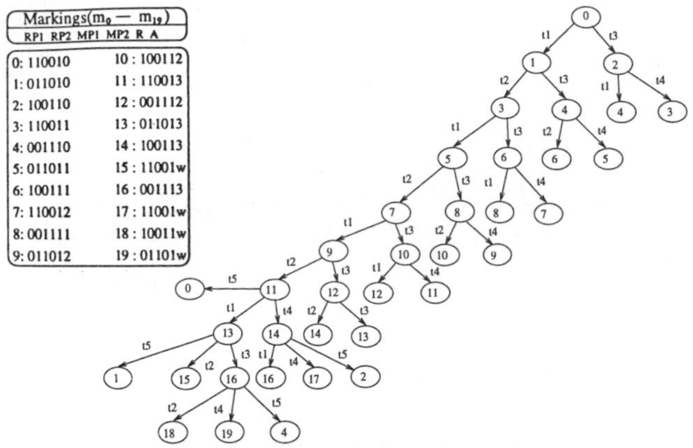

Figure 4 Reachabillity tree for the system with capacity three

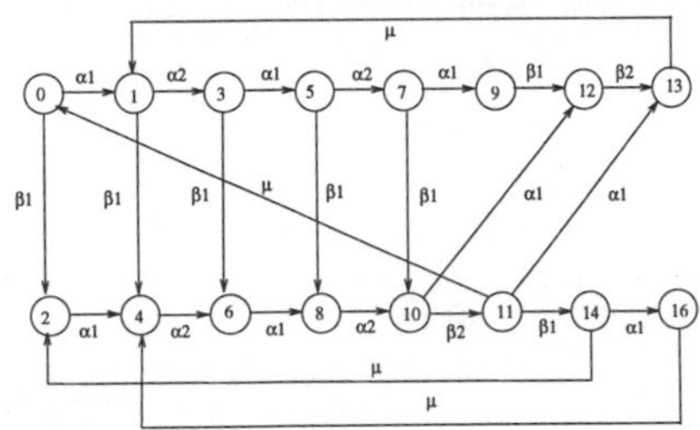

Figure 5 Automaton generator of the specification

danger_list, i.e., the list of states and respective output events to be disabled in that states or markings. See Figure 6(a). The supervisor is then shown in Figure 6(b). Notice that, given the *danger_list*, Figure 6(a), $t_2(\alpha_2)$ may fire only when the number of tokens in the place A is less or equal to 1 ($A \leq 1$). In the same way, $t_4(\beta_2)$ may fire only when the place A holds a number of tokens equal to 2 ($A = 2$). In any other situation, these transitions will be disabled, as shown by the respective associated logical expressions.

7 CONCLUSIONS

This paper introduced an extended Petri net model with transition enabling functions. This extension was used in modeling and synthesis of supervisors. A modified reachabillity tree algorithm was also introduced. This algorithm allows obtaining all the markings in a not modified finite capacity PN model. Finally, an algorithm to the construction of a $supC(L)$ generator was introduced. This algorithm is executed based on the modified reachabillity tree of the system PN model, and enumerates all the states and respective output events to be disabled.

Using the *PNTEF* model and the enumerated algorithms, it was proposed a systematic approach to solve the supervisory synthesis problem, with no modification in the structure of the net that models the system. On the other hand, this approach do not solve the

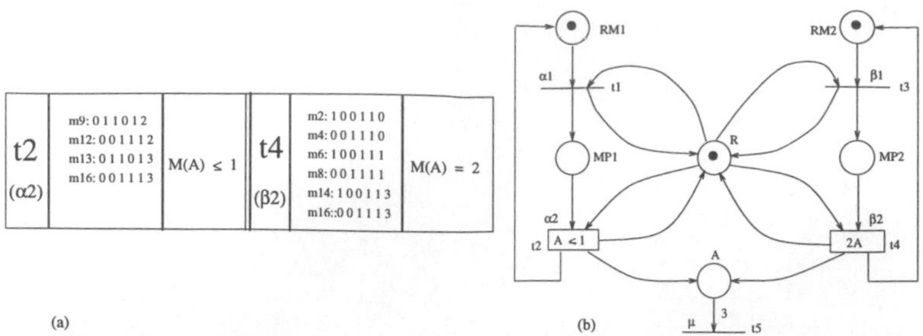

t2 (α2)	m9: 0 1 1 0 1 2 m12: 0 0 1 1 1 2 m13: 0 1 1 0 1 3 m16: 0 0 1 1 1 3	M(A) ≤ 1	t4 (β2)	m2: 1 0 0 1 1 0 m4: 0 0 1 1 1 0 m6: 1 0 0 1 1 1 m8: 0 0 1 1 1 1 m14: 1 0 0 1 1 3 m16::0 0 1 1 1 3	M(A) = 2

Figure 6 (a)Table of markings and their respective events to be disabled; (b) PNTEF supervisor of the manufacturing cell

classic state explosion problem of DESs, once the modified reachabillity tree has been employed. To solve this problem one possible solution is to employ a modular approach for the analysis and synthesis of the controller. Actually the use of *G-Nets* (Perkusich, 1994) as a modular formalism is under investigation.

8 REFERENCES

Giua, A. and F. DiCesare (1994). Blocking and controllability of petri nets in supervisory control. *IEEE Transactions on Automatic Control* **39**(4), 818–823.

Murata, T. (1989). Petri net: properties, analysis and applications. In: *Proceedings of the IEEE - vol. 77, num. 4.* pp. 541–580.

Papelis, Y.E. and T.L. Casavant (1992). Specification and analysis of parallel/distributed software and systems by petri nets with transition enabling functions. *IEEE Transactions on Software Engineering* **18**(3), 252–261.

Perkusich, A. (1994). Analise de Sistemas Complexos Baseada na Decomposiç ao de Sistemas de G-Nets. PhD thesis. Universidade Federal da Paraíba - Campus II. Campina Grande, Paraíba.

Ramadge, P. J. G. and W. M. Wonham (1989). The control of discrete event systems. *Proceedings of the IEEE* **77**(1), 81–97.

Sreenivas, R. S. (1993). A note on deciding the controllability of a language k with respect to a language l. *IEEE Transactions on Automatic Control* **38**(4), 658–662.

Ziller, R. M. and J. E. R. Cury (1994). On the supremal l-controllable sublanguage of a non prefix-closed language. In: *Anais do 10. Congresso Brasileiro de Automática, Rio de Janeiro.* pp. 260–265.

Giovanni C. Barroso received the B.S. degree in EE from UFC, Fortaleza, CE, Brazil, in 1982, the M.Sc. degree in EE in PUC/RJ, Rio de Janeiro, RJ, Brazil in 1987. He is a faculty member of DF/UFC, since 1989. He is a Ph.D. candidade at UFPB, Campina Grande, PB, Brazil.

Antonio M.N. Lima received the B.S. and M.S. degree in EE from UFPB, Campina Grande, PB, Brazil in 1982 and 1985 respectively, and doctaral degree in EE from INPT, Toulose, France, in 1989. He is a faculty member of DEE/UFPB since 1983.

Angelo Perkusich received the B.S. degree in EE, from UNESP, Barretos, SP, Brazil, in 1982, the M.S. and doctoral degree in EE from, UFPB, Campina Grande, PB, Brazil in 1986 and 1994 respectively. He is a faculty member of DEE/UFPB since 1991.

7

Planning, Training and Learning in Supervision of Flexible Assembly Systems*

L. Seabra Lopes and L.M. Camarinha-Matos
Departamento de Engenharia Electrotécnica, Universidade Nova de Lisboa, Quinta da Torre, 2825 Monte Caparica, Portugal.
Fax: +351-1-2957786. E-mail: {lsl, cam}@uninova.pt.

Abstract

In the context of balanced automation systems, a generic architecture for evolutive supervision of robotized assembly tasks is presented. This architecture provides, at various abstraction levels, functions for dispatching actions, execution monitoring, and diagnosing and recovering from failures. A planning strategy and domain knowledge for nominal plan execution and error recovery is described. Through the use of machine learning techniques, the supervision architecture will be given capabilities to improve its performance over time. The participation of humans in the training and supervision activities is considered essential. The combination of human interactivity with automatic aspects (planning, learning, ..) is discussed.

Keywords

Robotized Assembly, Action Sequence Planning, Monitoring, Failure Diagnosis, Machine Learning, Programming by Demonstration.

1 INTRODUCTION

Manufacturing companies throughout the world are, today, receiving pressures from a variety of sources. The globalization of the economy is leading to the production, within short time frames, of high quality and competitively priced products. Another factor is the increasing complexity of products, resulting from their complex functional definition, from a higher number of product varieties and details, from the need to prevent negative impact in the environment, etc. Even small changes in a product may lead to extensive engineering activities, with a number of improvement cycles. On the other hand, independently of the product end users, manufacturing companies make up a global chain of producers/consumers. A company alone does not have control over the complete cycle of making a finished product. Rather, it must be prepared to handle errors and defects in the partial products, received from suppliers.

---

* This work has been funded in part by the European Community (Esprit projects B-Learn and FlexSys) and JNICT (project CIM-CASE and a Ph.D. scholarship).

A great number of companies are therefore undergoing massive transformations. From the perspective of the product development lifecycle, the approach of Concurrent Engineering aims at reducing its duration by integrating all the related functional inputs and needed expertise as well as facilitating the flow of information among concurrent activities. Other concepts are being applied in industrial practice. For instance, leanness (Lean Manufacturing) is concerned with reducing wasteful activities, unnecessary inventory, long lead times, etc. More ambitious, the concept of Agile Manufacturing is aimed at enabling the production of highly customized products, when and where the client needs them. This certainly involves integrating concurrent activities and implementing leanness, but also redefining the relationship between the enterprise components and the customer.

In terms of the manufacturing resources, many companies, whose production lines are oriented towards a single product, or a small number of varieties, are now struggling with the need to support a larger number of product varieties, often to be produced in small amounts. This is not a trivial problem to solve, since in traditional automation systems, the setup time required to change over to a different product variety is orders of magnitude greater than the cycle time. Usually, production lines are composed of specially designed machines, integrated by materials handling systems. The manufacturing environment is very structured.

The ability to produce highly customized products, in order to satisfy market niches, is a competitive advantage that companies are trying to possess. This requires the introduction of new features in automation systems — flexibility, adaptability, versatility — relaxing cell structuration constraints and leading to the concept of Flexible Manufacturing Systems (FMS). To accommodate the wide range of product-processing techniques in an FMS more versatile machines must be included: easily integratable CNC machines, multi-operation devices, robots with multiple tools / end-effectors, modular fixtures and feeders. A rich sensorial environment and advanced communication infrastructures (and related protocols: MAP/MMS, fieldbus, ...) are other important hardware requirements to build up an FMS.

The efficiency and economical success of flexible manufacturing systems depends, however, on the capacity to handle unforeseen events, which occur in a great number, due to the reduction in system structuration constraints. The complexity of flexible manufacturing processes makes this task difficult to perform by humans. Therefore, in manufacturing systems, flexibility and autonomy are tightly related concepts. The introduction of "intelligent" functionalities, at programming level, is vital to achieving flexibility and autonomy: interactive programming / planning, in order to reduce set up costs and assure a rapid response to product innovation, process changes, or shifts in demand; sensorial perception and status identification; on-line decision making capabilities, such as monitoring, diagnosis and error recovery, to cope with the non-deterministic behavior of less structured cells; information integration, viewing the assembly system as part of a more general CIM system.

Of course, it will not be possible to achieve full system autonomy, since it is very difficult to anticipate all possible events and failures at programming time. To be successful, even the most advanced technologies depend on adequate consideration of the human factor.

The work being conducted in our labs, towards an architecture for supervision of assembly and manipulation tasks, considers, at different planning levels, functions for dispatching actions, monitoring their execution, and diagnosing and recovering from failures. The planning approach is initially presented, identifying the interactive aspects and describing the automatic ones. The experimental setup is described with emphasis on the integration effort under a perspective of migration from legacy systems

Our main concern is the acquisition of knowledge about the tasks and the environment to support monitoring, diagnosis and error recovery. A training methodology was devised, in which a human tutor can provide the system with apriori knowledge, with examples of desired behavior and examples of concepts to be learned (programming by human demonstration). Training is considered both at initial system programming and when unanticipated situations arise. Based on the demonstrated examples and using machine learning techniques, the system will incrementally build the needed models.

2 PLANNING AND SUPERVISION

2.1 Planning Framework

Planning in manufacturing and assembly is typically done in a hierarchical fashion, starting in long-term planning activities, at the upper level, and detailing until the necessary executable plans are generated. At the lower planning levels, product and process oriented planning, scheduling at the shop floor and detailed execution planning are considered. Product oriented planning determines a feasible assembly precedence graph taking into account the constraints derived from the product model (bill of materials, geometric model, tolerances model, materials model, etc.). Process oriented planning, taking into account feasibility conditions from the technologic point of view, generates additional constraints, more concerned with the resources (robots, end-effectors, feeders, fixtures, etc.) that must be used.

Several attempts to adapt generic planners (developed in the AI community and having in mind toy problems as "the blocks world"), to realistic robotics tasks have been made. Other approaches are strongly geometric-reasoning-based but require heavy processing procedures and achieved results still present some limitations. On the other side, the most adequate spatial-related solutions are not completely justified by pure geometric reasoning but depend on other technological constraints.

We feel that the two phases of product oriented planning and process oriented planning can more easily be performed in interaction with the humans. Various interactive planning systems (Computer Aided Process Planning — CAPP) have been developed in recent years (Furth, 1988) to help the human expert in generating appropriate process plans. Most of these systems follow an interactive generative approach, in the line of a decision support system, helping the production engineer in the generation of a feasible sequence of assembly steps. The process plan typically includes information on abstract operations to be performed and respective precedences, goal positions, mating referentials, approaching directions, possible part grasping zones, part stable poses, types of resources that can be used in each operation, etc.

Finally, when all the information about an assembly task is gathered and a specific cell is selected, the high-level specification contained in the process plan is instanciated. It is now the moment to plan the execution, i.e., to determine all needed actions, their characteristics and parameters and their optimal sequence.

2.2 Planning Strategy and Representation

For nominal planning, and also for failure recovery, a planner , in the AI sense of the term, was developed. This planner, implemented in Prolog, uses a domain independent planning strategy, but takes into account domain knowledge provided in a pre-defined format. The planning strategy is, basically, a best-first forward search procedure. From the initial state of the world, new states are generated, by applying operators of the considered domain, until the goal state is reached. In each step, a set of legal operator instantiations is determined, and evaluated according to domain dependent heuristics. The operator with highest score is selected for continuing the search. The way the planner uses the provided domain knowledge to select operators makes it a non-linear planner, since it can handle interactions between goals.

The domain specification includes: definition of operators; templates of facts about the world; typical precedences between facts in the goal state; and measures of contribution of facts asserted in a given phase to goal facts. An operator is defined as a Prolog clause having the following format (see example in Fig. 1):

```
operator(Op,Info,Keep,Del,Add).
```

Where: `Op` — template of the operator, specifying name and parameters.
`Info` — list of facts about the world.
`Keep` — pre-conditions not removed by execution of the operator.
`Del` — pre-conditions removed by execution of the operator.
`Add` — conditions that become true as consequence of executing the operator.

In order to handle non-linear planning problems, this planner can take into account known precedence relations between facts in the goal state. The goals not preceded by any other goals in the goal state are selected to be solved first. When each of them is solved, some new goals will be considered by the planner, and so on. In the blocks world, the following precedence relation would be enough: `typical_precedence(on(B,C),on(A,B))`.

In each phase of the planning process, the pre-conditions of all operators will be matched to the current world state. The result is a set of legal operator instanciations whose usefulness, concerning the solution of the goals currently being considered, is evaluated. This evaluation is made taking into account the contributions of each of the added facts to each of the goals currently being considered. This evaluation is, again, domain dependent, and is made based on Prolog rules of the following form:

```
contrib_to_goal(CS,F,G,Score):- <Rule Body>.
```

Basically, the rule says that in a given state `CS`, an operation that asserts the fact `F` will contribute to goal `G` in a way measured by the returned `Score`.

2.3. The Assembly Domain Knowledge

In Assembly, domain knowledge is much more complex than in the blocks world. At the center of the problem are the graphs of precedences for assembly and for disassembly, the graph of connections between components, and of course geometrical data. A good human interface is necessary to acquire all this information. Graphical simulation can be used to acquire information on positioning, grasping, trajectory skeletons, etc. For instance, the needed referentials (grasp and approach positions, mating referentials, etc.) can easily be "suggested" by the human expert by pointing on a screen. The evaluation of the precise values of such referentials can then be done by an automatic function. This leads to a "light" procedure, implementable with available technology (see, for instance, systems like ROBCAD or IGRIP).

The planning strategy, described above, is used to determine sequences of actions, at a sufficiently high level of abstraction. Still, the planner must take into account the positions of parts in feeders and pallets, the mating positions, the mating precedences, the tools for grasping and mating, the needed elementary skills, etc. Examples of an operator, a precedence rule and a contribution rule are presented in Fig. 1. Besides the presented `assemble` operator, the planner can make use of a `disassemble` operator, especially useful in correcting assembly errors, and various operators for picking and placing parts in/from fixtures, feeders or the free workspace, for fetching and storing tools, for feeding parts and pallets, etc.

It is not possible to completely describe here a realistic planning example. Just for illustration, we mention an experiment with the Cranfield Benchmark, a well known laboratory product used for testing in the assembly domain. It is a pendulum composed of seventeen parts: two side plates, four

```
operator(
  assemble_component
      (R,T,Obj,Comp,Part,Prod,Fix,Cp,Geom),
  [% Info:
      object_type(Obj,Part),
      part_tool(Part,T),
      component_contacts(Comp,LComp),
      mate(Prod,Comp,Part,Geom,Prec,Succ)],
  [% Keep-PC:
      current_tool(R,T),
      not(assembled(Comp,Prod,Fix)),
      fixture_with_product(Fix,Prod),
      not(robot_arm_breakdown(R)),
      not(tool_breakdown(T)),
      not(defective(Obj)),
      all(C,Prec,assembled(C,Prod,Fix)),
      all(C,LComp,
         [ assembled(C,Prod,Fix)),
           represented_by(C,X)]
           -> [not(defective(X))]) ],
  [% Del-PC:
      current_arm_position(R,Cp),
      object_in_robot(Obj,R) ],
  [% Add-C:
      assembled(Comp,Prod,Fix),
      represented_by(Comp,Obj),
      robot_free(R),
      current_arm_position(R,Dp) ]
).

typical_precedence(
        assembled(C1,P,F),
        assembled(C2,P,F))
:- clause(mate(P,C2,_,_,Prec,_),true),
   member(C1,Prec).

contrib_to_goal(CS,
  pallet_available(Pal,_),
  assembled(C,Pd,_),
  3) :- member(part_in_pallet(Ob,Pal),CS),
        member(object_type(Ob,Part),CS),
        member(mate(Pd,C,Part,_,LP,_),CS),
        check_prec(CS,C,Pd,LP).
```

Figure 1 Assembly Planning Knowledge.

spacer pegs, a shaft, a lever, a cross bar and eight locking pins. Except for the locking pins, all the other needed mate operations are stack operations, the most common in industrial practice. All mate operations require some degree of compliance. Compliance, however is not handled at the level of action sequence planning. In our experimental setup we have special purpose feeders (which limits cell's flexibility) for side plates and cross bars. The locking pins and the spacer pegs are fed to the system in a pallet and the lever and the shaft in another pallet. Three different tools must be used. Running the planner with the incorporated domain knowledge on this problem produces an optimal plan with 53 operations without any backtracking, which may be considered a good result.

Each of the operations in the generated plan must be expanded, in a more or less deterministic way, into a sequence of resource-level operators, like `move`, `approach`, `transfer`, `peg_in_hole` or `grasp`. This involves lower levels of planning, more concerned, for instance, with trajectories or compliance. A skill acquisition approach to compliance, using learning techniques, seems promising (Kaiser, *et al.*, 1994). However, as we are more concerned with execution supervision, we prefer to simplify the planning functionalities of the lower levels in order to be able to realize a working prototype.

2.4 Supervision Functions

Two plan levels were already mentioned. A hierarchical specification of the mate precedence graph or the learning of macro-operations may originate additional plan levels. A hierarchical plan is an advantage for supervision, since it provides different contexts for error detection and recovery. At the lower levels, error recovery will tend to consist of simple reflexive actions. At the upper levels error recovery will require more extensive diagnostic reasoning and planning. The architecture of an Intelligent Execution Supervisor should reflect the hierarchical structure of the plans. For each plan level, the main functions are (Camarinha-Matos, *et al.*, 1994):

Dispatching and Global Coordination — The global coordination activities performed by a high level controller include: dispatching actions to the executing agents; synchronization of activities among agents and with external events; world model update; information exchange.

Monitoring of Assembly Plans — The monitoring function is used to detect non-nominal feedback in the system during plan execution.

Failure Diagnosis — This diagnosis function will firstly check if there really is a failure (failure confirmation) and update the internal model. Then this function will try to classify and explain the failure. At each execution level, different levels of explanation for a detected failure may be generated, depending on the amount of information available. For example, a gross diagnostic can be "pick fail". A more detailed diagnostic could be "pick fail due to object sliding". The least detailed explanation would be "deviation detected".

Failure Recovery — At each supervision level, the recovery function is called when the diagnosis function confirmed a failure and found an explanation. The recovery function will try to determine a recovery strategy to bring the execution to a nominal state. One basic question is how to build recovery strategies? Since the detected error is some unexpected (abnormal) event and, therefore, the nominal plan is not to be altered.

3 THE TRAINING METHODOLOGY

One important aspect is the acquisition of knowledge to support the supervision functions. In real execution, a feature extraction function is permanently acquiring monitoring features from the raw sensor data (namely from force & torque data and from discrete sensor data). These features are used to guide the control function of the operation being executed as well as to detect deviations between the actual behavior and the nominal operation behavior. For example, let's consider that, during the execution of a `Transfer` operation, in which the robot carries a part to be assembled, an object, unexpectedly originating in the environment, collides with the gripper causing the part to move without falling. The first diagram, included in Fig. 2, shows the perceived sensor data during actual execution. The second diagram shows a qualitative model of the operation. The third diagram shows a qualitative interpretation of the raw sensor data in terms of the features used in the operation model. Since a deviation is detected, the diagnosis function is called to verify if an execution failure occurred and, in that case, determine a failure classification and explanation. For this function, additional features must be extracted. Diagnosis is a decision procedure that needs a model of the task, the system and the environment. The final step, based on the failure characterization, is recovery planning.

The problem of building the knowledge base, and in particular the models that the monitoring, diagnosis and recovery functions need, is not easily solved. Even the best domain expert will have difficulty in specifying the necessary mappings between the available sensors on one side and the monitoring conditions, failure classifications, failure explanations and recovery strategies on the other. Also, a few less common errors will be forgotten. Known prototype systems show limited domain knowledge, as they are intended mainly for exemplification and not to be used as robust solutions in the real world.

In a very initial phase, the models of the elementary operations of the system must be built, considering both sensing and action aspects. Some operation models can be easily hand-coded. For instance, part feeding or robot motion along a given path are operations for which the control function is relatively easy to specify. Ensuring the nominal conditions for these operations is also simple. For instance, a few binary sensors are enough to identify the status of a part feeder. To hand-code an evaluation function based on these sensors is a trivial task.

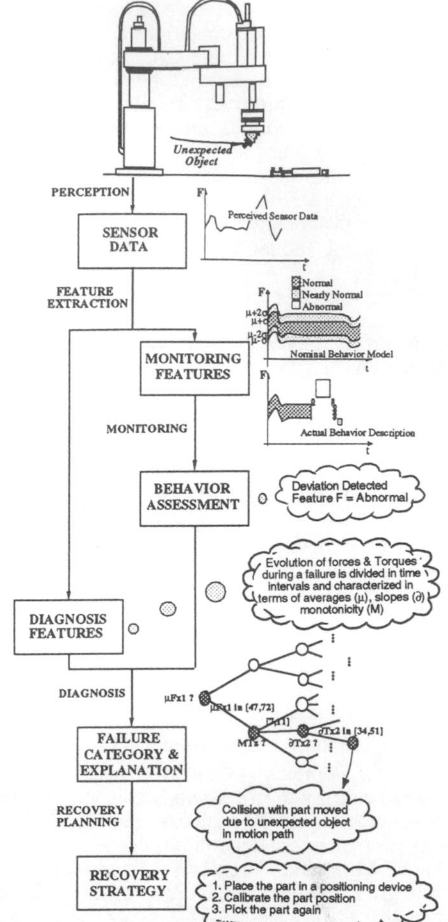

Figure 2 Example of the Error Detection and Recovery Cycle.

For complex operations, that don't have a well defined model, the control functions are very costly to program in the classical way. An example is compliant motion in robotized assembly: the variations in size, weight, friction coefficient, etc., of the parts involved are so wide that a clear generic model of the needed compliance does not exist. In the same way, evaluating the behavior of the system during execution of a given operation can be much more complex than assessing the status of a feeder based on binary sensors. The paradigm of Programming by Human Demonstration (in this case Robot Programming by Demonstration (McCarregher, 1994; Kaiser *et al.*, 1994)) seems indicated to overcome this type of difficulties. According to this paradigm, complex systems are programmed by showing particular examples of their desired behavior. In our approach, interaction with the human, seen as a tutor, is fundamental. Functions for training and learning are included in the supervisor architecture (Fig. 3).

An adequate user interface facilitates transfer of the human's knowledge to the system. This knowledge can be coherent and complete or can be empirical, based on real examples of complex situations which have no known model. For instance, to teach the robot how to insert a peg into a hole (a compliant operation), the tutor can guide the robot arm to reach that goal, while collecting both force & torque data and actuated velocities. Then the best tutoring experiments (according to some criteria, for instance the minimization of insertion time) are selected to serve as examples that a neural network or a fuzzy controller use to learn the control function of the operation (application of learning in compliant motion are being investigated by our partners in the B-LEARN project (Nuttin, *et al.*, 1994)).

In which concerns the definition and characterization of elementary operations, our research focus has been centered in execution evaluation, including monitoring (detection of failures) and diagnosis (classification and explanation). For instance, the knowledge required for monitoring of motion operations was obtained by training in the following way: traces of all testable sensors were collected during several runs of each operation; for each continuous feature, the typical behavior of the attribute during execution of the operation was calculated as being the region between the average minus standard deviation behavior and the average plus standard deviation behavior.

The system is also trained to identify/classify execution failures. The result is the creation and refinement of a qualitative failure model, composed of failure descriptions and taxonomic and causal relations between them. For training, several external exceptions were manually provoked and the effects as well as the corresponding traces of sensor values recorded. Examples of provoked external exceptions are: unexpected objects in robot arm motion path; misplaced or missing parts; defective parts or assemblies; misplaced or missing tools. For some of the resulting execution failures, it was easy to hand-code classification rules. For others, classification knowledge was generated by inductive algorithms. When flat classifications were generated, the leave-one-out error rate was too high. Taxonomies of failure classifications, generated by SKIL, presented a much better degree of accuracy, but still don't ensure full reliability of the system. Further evaluation of learning techniques is still necessary.

Finally, recovery strategies were programmed for the most common errors. A good user interface should facilitate also the specification of recovery strategies. Feasible possibilities, that we didn't consider, are graphical simulation and virtual reality. In the current prototype recovery strategies are entered in textual form.

In any case, the main idea is that programming the supervision system includes both traditional programming (implanting in the system monitoring, diagnosis and recovery rules, known *a priori*) and programming by demonstration (examplar-based approach).

It is important to distinguish between training of elementary operations and specific task training (Fig. 3). In the first case, the implanted knowledge and the provided examples are concerned with characterizing the basic primitives of the system (elementary operations, those used in task planning) and their related skills. Only after this phase, the system should start training and then executing specific tasks.

In real execution, when a failure is identified, and there is no known recovery procedure for the situation, recovery planning is attempted. If a plan is found and, when applied, brings the

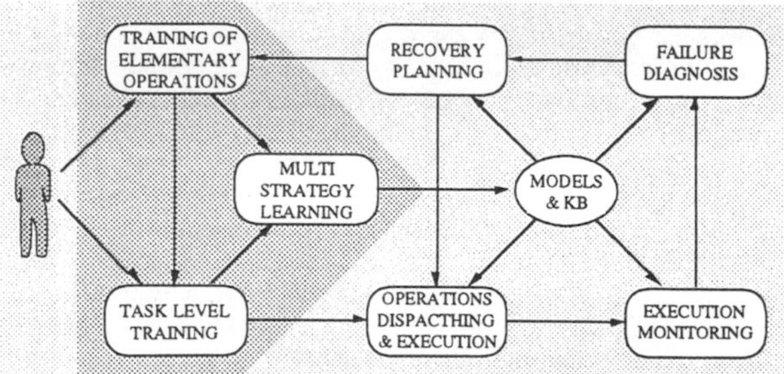

Figure 3 Training and Supervision Architecture

system back to nominal state, this plan will be generalized and stored for future use. If it is not possible to recover automatically, help from a human expert is requested. Eventually, the solution provided by the human is also generalized and stored. This kind of incremental programming and generalization may also be viewed as training. At the level of elementary operations, execution evaluation should contribute to the tuning of the related skills (Kaiser, *et al.*, 1994; Nuttin, *et al.*, 1994).

This approach allows for progressive independence of the human operator, who will not need to be permanently supervising the assembly process. On the other hand, as the system relies more on continuous self-evaluation of its behavior than on very accurate models, the requirements on operator specialization are relaxed. The main limitation of the proposed approach is the cost of training in the initial phase: generating a sufficient number of examples of each learning concept, mainly when physical devices are involved, is a **tedious task**.

Training and execution form a long-term learning cycle during which the models of the operations, the manipulated objects and the environment are built and refined. Experience gained in making a certain product is useful for making other products. In this way, training a specific task is simplified, especially from the humans perspective. Long-term learning, however, poses many problems concerning knowledge representation, an issue with big impact in the reasoning processes.

4 EXPERIMENTAL DEVELOPMENTS

4.1 Migration from Legacy Systems

Migration from legacy systems is one of the most challenging aspects of manufacturing systems. Existing devices and systems are often too valuable to be thrown away. However, their integration in more advanced supervision architectures poses many problems, some of which we have faced into installing our experimental setup.

The setup being used in the experiments is a robotic assembly cell, composed of one SCARA robot, three robot grippers, magazine and corresponding tool exchange mechanism, two special purpose feeders, one fixture and sensing devices. Needless to say, the robot control language is not suited to write intelligent control software. Moreover, it does not give information about errors and does not provide guarded movements.

As feedback information sources for our supervision system, the following discrete information sensors were integrated in the cell: a) in each gripper, to detect if it is open, closed or clamping; b) in feeders, to detect part presence, part stock existence and feeder problems; c) in the wrist of the robot, to find out which tool is attached, if any; d) in each tool place to detect tool presence; e) in the fixture, to detect if the jig which will hold the assembly is present. The most frequent execution failures are anticipated to be those in which the robot arm is involved, including collisions, obstructions and handling failures. Therefore, a force and torque sensor, which seems a good candidate to give information about those failures, was also included.

A major problem found is that the robot controller does not provide guarded movements. Since communication via Teach Pendant (TP) is quite fast, the solution was to decompose each motion command into a series of incremental steps, executed sequentially via TP, until some condition is verified. A TP Emulator process was used to achieve this goal.

As it is not easy to make acquisition of large quantities of sensorial data in Unix workstations, and, on the other side, concurrency in Unix affects the sensor sampling rate, a program called Low-level Monitor is run in a dedicated PC, where it is quite simple and cheap to implant a data I/O board. This program will check conditions during the execution of actions, as specified by the Intelligent Supervisor, and is able to answer questions about the state of the system during the diagnosis phase. Communication between the Low-level Monitor and the Intelligent Supervisor, running in Unix, is accomplished via an RS232C line.

To integrate and make available the services of sensors and actuators, a Cell Front-End was developed in the Unix environment. This front-end is composed of a set of server processes which interface clients (namely the intelligent supervisor) to the cell resources. An Operational Server process makes the interface with actuators. A Monitor process makes the interface with sensors via de Low-Level Monitor PC. In general, migration from legacy systems involves integrating existing device controllers in a more flexible programming environment, usually UNIX. Under a client-server approach, the services of the physical devices are mirrored by (possibly a hierarchy of) Unix processes.

4.2 Learning of Hierarchical Diagnostic Knowledge

As emphasized in section §3, the difficulty in hand-coding the models that the supervision functions need, raises the question of how to build such models automatically. The classification phase of the diagnosis task can be performed based on knowledge generated by induction. Several inductive learning algorithms and systems (e.g. , ID3 (Quinlan, 1986), (Hirota, *et al.*, 1986), Smart+ (Botta, Giordana, 1994), CONDIS (Seabra Lopes, Camarinha-Matos, 1995)) were applied to the assembly domain (Camarinha-Matos, *et al.*, 1994; Seabra Lopes, Camarinha-Matos, 1995). These techniques are only able to learn uni-dimensional concept descriptions: the resulting knowledge is only able to assign classes to objects from a given domain. In the assembly domain, for example, these algorithms and systems cannot handle the problem of discriminating collisions from obstructions and normal situations, handling simultaneously the problem of discriminating between different types of collisions.

Having as motivation the automatic construction of the models required for the Assembly Supervisor, the idea of generating a concept hierarchy became more attractive. A new algorithm, SKIL (structured knowledge generated by inductive learning), was developed to perform that task (Seabra Lopes, Camarinha-Matos, 1995). The concepts in the hierarchy learned by SKIL are characterized by a set of symbolic classification attributes. At the lower levels of the hierarchy, concepts are described in more detail, i.e., more attribute values are specified. Moreover, in detailing or refining a concept, in which attributes take certain values, it may make sense to calculate other attributes. Therefore, the user should provide a set of attribute enabling statements of the form (A_i, A_{ij}, EA_{ij}), meaning that when the value of A_i is determined to be A_{ij}, then attributes in EA_{ij} should be included in the set of attributes to be considered in the continuation of the induction process. For example, when learning the behavior of a `Transfer` operation, if a collision is found, it may make sense to determine some characteristics of the colliding object, like size, hardness and weight. This could be expressed by the following attribute enabling triple:

```
(failure_type, collision, {obj_size, obj_hardness, obj_weight})
```

The training data is characterized by a set of discrimination features, which can be numerical or symbolic. Each example in the training set is composed of a list of attribute-value pairs followed by a vector of feature values.

To illustrate an application of this algorithm, consider the macro-operation «Pick and Place» of a part and three of the basic primitives involved: a) approach to grasp position (`Approach-Grasp`); b) `Transfer` (of part); and c) approach to the final position (`Approach-Ungrasp`). During the training phase, each of the selected operations was executed many times and several external exceptions were simulated. In this case, the goal of applying SKIL is to learn a taxonomy of failure descriptions. The classification attributes and enabling triples used in the `Approach-Ungrasp` primitive were as specified in Fig. 5. The training set was composed of 117 examples. After running SKIL on the domain specification and training set, a decision tree was obtained having 71 nodes. The concept hierarchy contained in the tree has 59 nodes, being 10 of them internal nodes and 49 terminal nodes. Examples of the corresponding rules are:

```
∀x: ( Fz1(x,[7,21[) & Fx1(x,[-4.5,1[) & Dx1(x,[-0.5,0.5[) )
     => ( behavior(x,normal) & part_status(x,ok) )

∀x: ( Fz1(x,[-995,7[) & Dz2(x,[-542,-51[) & Fx3(x,[-464,-13[) )
     => ( behavior(x,failure) & part_status(x,moved) &
          failure_type(x,obstruction) )
```

Performing the leave-one-out test with the same data and algorithm, the resulting average error rate is 15%. It should be noted that, using a traditional inductive learning algorithm (such as ID3 (Quinlan, 1986)) and "flat" concept descriptions, equivalent to all combinations of values of the classification attributes, a much higher error rate is obtained.

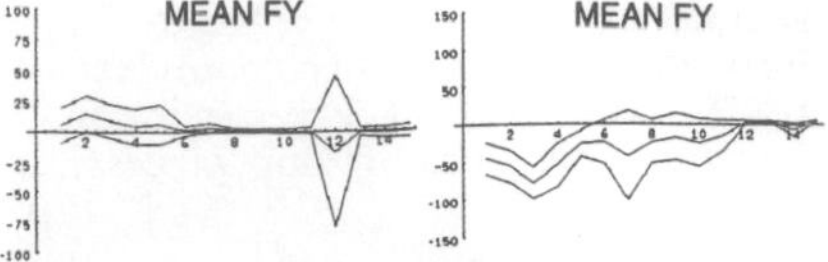

a) Collision in part & part lost b) Front collision and part moved

Figure 4 Typical force behavior in two types of failures in Approach-Ungrasp.

Attribute	Attribute Values
behavior	{ normal, failure }
part_status	{ ok, moved, lost }
failure_type	{ collision, obstruction }
collision_type	{ part, tool, front }

a) Attributes

Attribute	Attribute Value	Enabled Attrib
behavior	failure	{ failure_type }
failure_type	collision	{ collision_type }

b) Enabling triples

Figure 5 The Approach-Ungrasp problem specification.

When the user wants to get more and more information about a failure situation, the number of classification attributes and their values increases. If these attribute values are to be combined to produce "flat" classifications or labels, the number of labels increases exponentially, and the problem becomes intractable. This is the case of the information collected during failures of `Approach-Grasp` and `Transfer` (respectively 88 and 47 examples):

behavior — generic information about the operation behavior; can be normal, collision, front collision or obstruction; what will be learned is, in fact, a model of the behavior (either normal or abnormal) of the system when performing these operations.
body — what was involved in the failure, e.g., the *part*, the *tool*, the *fingers* (*left*, *right* or *both* fingers).
region — region of body that was affected, e.g., *front*, *left*, *right* or *back* side, *bottom*, ...
object size — size of object causing failure: *small*, *large*.
object hardness — can be *soft* or *hard*.
object weight — can be *low* or *high*.

For the `Approach-Grasp` primitive, the domain specification, provided to SKIL, included 10 classification attributes, 28 attribute values and 8 enabling triples. A decision tree and concept hierarchy with 93 nodes was generated. The error rate (30%) is much higher than in the previous problem when SKIL was also applied (15%). This is understandable since the target concept is much more complex and a smaller training set was provided (only 88 examples). For the `Transfer` problem, in which only 47 examples were collected, the taxonomy generated by SKIL has an error rate of 34%. We confirm what was expected in this type of domain: as a general trend, when the number of occurrences of each attribute value in the training set increases, the corresponding error rate decreases.

Training the system to understand the meaning of sensor values and learning a qualitative and hierarchical model of the behavior of each operation is a crucial step in diagnosis. Programming such model would be a nearly impossible task.

Since the human defines the "words" (attribute names and values) used in the model, the human is capable of understanding the more or less detailed description that the model provides for each situation. It is then easier to hand-code explanations for the situations described in the model. The explanation that must be obtained for the given execution failure includes, not only the ultimate cause (an external exception or system fault), but also the determination of the new state of the system. The failure explanation rules that we are using have the following format:

```
explanation(Op,FD,C,Del,Add) :- ... .
```

The first argument, `Op`, is the elementary operation during which the failure occurred or, in, general the execution context. The second argument, `FD`, is a description of the execution failure, obtained from the sensor data using the taxonomic model. Then, `C` is the ultimate cause of the failure. Knowing the exact cause can be irrelevant. What is important is to identify the state of the system after the failure. The arguments `Del` and `Add` specify what facts must be deleted from the state description and what facts must be added.

For example:

```
explanation(
    get_tool(R,T1,TP1),
    [ [failure_type,wrong_tool],[current_tool,T2] ],
    unknown,
    [ current_tool(R,T1),tool_in_magazine(T2,TP2)],
    [current_tool(R,T2),Fact]
) :- ( toolplace(TPx), Fact = tool_in_magazine(T1,TPx);
       Fact = tool_lost(T1) ).
```

Diagnostic reasoning and causality have been studied for some time, and tested frequently in domains like electronic circuits, but there is no unified theory for these matters. Moreover, approaches like the one presented in (de Kleer, 1990) structure the problem considering that the main goal is to determine the faulty components in a system. However, in the assembly domain, not only system faults, but also external exceptions can be causes of off-nominal feedback. The described representation, that we are using, seems to provide useful results, although additional evaluation and refinement are needed.

4.3. Recovery Planning

Finally, when an explanation for an error is obtained, recovery planning is attempted. However, the most common case, probably, is that, not only one, but several possible explanations are found. The available sensor information is not enough to assert, with significant certainty, which is the state of the system. For instance, in the "wrong tool" example, presented above, the explanation rule provides several explanations, namely that the needed tool is in some other toolplace, TPx (and this represents already several possibilities), or lost. In this case, the recovery planner has to assume one of the explanations, generate a recovery plan and use it. If some problem arises, probably the assumption was not correct and some other explanation must be considered.

To be noted is that the initial plan is generated for the nominal conditions. The representation of the operators contains several assumptions. For instance, `not(defective(Obj))` is one of the pre-conditions of the `assemble` operator. Imagine that, in executing the assembly plan of the Cranfield Benchmark, mating a spacer peg with the side plate fails. Various explanations are possible: the spacer peg might be defective, the side plate might be defective, some unexpected object originating in the environment was obstructing the mate operation, etc. Unless some sophisticated sensorial feedback is available, the robot will have to recover based on assumptions or beliefs. The recovery actions will be, simultaneously, verification actions.

One aspect that will be address in the near future is the application of learning techniques in error recovery. A learning feedback loop will have to be implemented (Evans, Lee, 1994). In a first step, when an error, for which no recovery strategy is known, is detected, recovery planning is attempted. If recovery planning fails to generate a plan, eventually the human operator will provide one. In a second step, if the obtained recovery strategy is successful, it will be generalized and archived for future use. This will be attempted in the next phase.

5. CONCLUSIONS AND FUTURE WORK

In the context of balanced automation systems, an evolutive execution supervision architecture was presented. Flexibility implies increasing the on-line decision making capabilities, for which dispatching, monitoring, diagnosis and error recovery functionalites have been devised. The lack of comprehensive monitoring and diagnosis knowledge in the assembly domain points out to the use of machine learning techniques, leading to an evolutive architecture. The participation of humans in the training and supervision activities is considered fundamental in order to achieve flexibility in assembly systems. The general approach is, therefore, to collect examples of normal and abnormal behavior of each operation or operation-type/operator and generate a behavior model that the diagnosis function will use to verify the existence of failures, to classify and explain them and to update the world model.

Concerning the failure identification/classification problem, the application of the algorithm SKIL, that generates concept hierarchies with a higher degree of accuracy, provided interesting results. Further research will focus on efficient generation of examples, feature construction and selection and long-term learning. Developments in planning for nominal execution and for error recovery were presented. The used planning strategy is domain independent, non-linear, forward chaining and depth/best-first. The representation of the assembly domain knowledge was presented. Future research in this topic includes the learning aspects in recovery planning.

6 REFERENCES

Botta, M. and Giordana, A. (1993) SMART+: A Multi-Strategy Learning Tool, *Proc. Int'l Joint Conference on Artificial Intelligence (IJCAI-93)*, pp. 937-943.

Camarinha-Matos, L.M., Seabra Lopes, L. and Barata, J. (1994) Execution Monitoring in Assembly with Learning Capabilities, *Proc. of the 1994 IEEE Int'l Conf. on Robotics and Automation*, San Diego.

de Kleer, J., Mackworth, A. K., and Reiter, R. (1990) Characterizing Diagnoses, *The Eighth National Conference on Artificial Intelligence (AAAI-90),* Boston, 324-330.

Evans, E.Z. and Lee, C. S. G. (1994) Automatic Generation of Error Recovery Knowledge Through Learned Reactivity, *Proc. of the 1994 IEEE Int'l Conf. on Robotics and Automation*, San Diego.

Furth, B. (1988) *Automated Process Planning*, NATO Advanced Study Institute on CIM: Current Status and Challanges, Springer-Verlag, NATO ASI Series.

Hirota, K., Arai, Y. and Hachisu, S. (1986) Moving Mark Recognition and Moving Object Manipulation in Fuzzy Controlled Robot, *Control-Theory and Advanced Technology*, vol. 2, no. 3, pp. 399-418.

Kaiser, M; Giordana, A. and Nuttin, M (1994) Integrated Acquisition, execution, evaluation and Tunning of Elementary Operations for Intelligent Robots, *Proc. of the IFAC Symp. on Artificial Intelligence in Real-Time Control*, Valencia, Spain.

Quinlan, J. R. (1986) Induction of Decision Trees, *Machine Learning* , 1, pp. 81-106.

Seabra Lopes, L. and Camarinha-Matos, L.M. (1994) Learning to Diagnose Failures of Assembly Tasks, *Proc. of the IFAC Symp. on Artificial Intelligence in Real-Time Control*, Valencia, Spain, October 1994.

Seabra Lopes, L. and Camarinha-Matos, L.M. (1995) Inductive Generation of Diagnostics Knowledge for Autonomous Assembly, to appear in *Proc. 1995 IEEE Int'l Conf. on Robotics & Automation*, Nagoya, Japan.

McCarragher, B.J. (1994) Force Sensing from Human Demonstration using a Hybrid Dynamic Model and Qualitative Reasoning, *Proc. of the 1994 IEEE Int'l Conf. on Robotics and Automation*, San Diego.

Nuttin, M., Giordana, A., Kaiser, M. and Suárez, R. (1994) — *Machine Learning Applications in Compliant Motion*, Esprit BRA 7274 B-LEARN II, deliverable 203.

7 BIOGRAPHIES

Luís Seabra Lopes is a doctoral candidate in the Electrical Engineering department of the New University of Lisbon (UNL), with a grant from JNICT, the portuguese research board. He graduated from the Computer Science department of UNL in 1990, and then became member of the Intelligent Robotics Group of the same university. He is participating in the european ESPRIT project B-LEARN, one of the first projects to investigate applications of Machine Learning to Robotics. At the moment his main interests are Robotics, Machine Learning and Intelligent Supervision.

Luís M. Camarinha-Matos is professor at the New University of Lisbon. He graduated in Computer Engineering from the Computer Science department of UNL in 1979. He then became assistant professor at the same department and participated in several european research projects. He received the Ph.D. degree in Computer Science, topic Robotics and CIM, from the same university in 1989. He was co-founder of the Electrical Engineering department of UNL were he is currently auxiliar professor (eq. associate professor). He also leads the group of Robotic Systems and CIM of Uninova. His main interests are: CIM systems integration, Intelligent Manufacturing Systems, Machine Learning in Robotics.

8

Automated Planning for restarting Batch Procedure in the Wake of an Incident (Single Product)

Djamel Rouis
URA-CNRS 817 Departement de genie Informatique. UTC,BP 649; 60206 Compiegne Cedex-France;drouis@hds.univ-compigne.fr;+33 44 23 44 77

Abstract

The object of this research is to study the automated planning of batch procedure (batch sequence) which is allowed to restart the procedure that the execution has been interrupted by an incident. The proposed algorithm can replan the sequence of phases if an interruption happened in the plant for two cases illustrated. The first case, when the plant must satisfy the previous fabrication order before satisfying the present. The second one, when the plant work with two fabrication orders, knowing that, the previous fabrication order has been realized. We proposed also the global architecture for automating the restarts of the sequence of phases. In assumption, the incident that provoked the stoppage in the current cycle of production, identified and that causes suppressed. We do not take into consideration the diagnostic of the failure.

Keywords
Artificial Intelligence, Automation, Planning, Manufacturing.

1 INTRODUCTION

There is no point view that deals with the automated planning for restarting batch sequence in the wake of an incident. It exists in Cott and Maccietto (1989) an approach that deals with the implementation of computer-based system that integrates planning and plant control of batch chemical plant. The author gives a "flash" that is only a general planning control level structure. Our approach is a continuation of the approach that is deal it in Rouis (1994). The high level of the global architecture in that approach is not automated yet. Now, we try to solve this problem for giving a solution that concerns two cases for single product. The first one, when the plant must satisfy the previous fabrication order before satisfying the present. The second one, when the plant work with two fabrications orders. Our paper is organized as following :

1-Problem formulation. 2-Planning. 3-Illustration.4-Architecture.5. Consequences.
In the literature, we found a fabrication order as file that indicates what must do when and how to manufacture a product in details. This file contains essentially a copy of fabrication game as a conforms size of the executed lot. We find in this game a set of parameter such as: -The time unit, the fixed time of adjustment of machines and the list of operations that are no labelled in the order of execution.- There are also in this game the periods like a start and finish dates of each operations - The indication of the priority for executing the necessary plan - The data base about the insurance quality controled by customer at the delivery Bennassy (1990).
In our approach, It is a list not necessarily ordered for obtaining the desired product. This fabrication orders contain the all information that is necessary for the standard batch procedure and that is previously defined (single product).Ex:FO(k)= (prQ(p1), prQ(p2),prQ(p3))), knowing that "prQ(p1)" is the product p1 with a quantity Q.

A *production plan* P(k) is a fabrication order must contain a list that is not necessary ordered to obtain a desired product. It must contain also the standard procedure for each form of product. A production plan contains severales phases that may contain the *operations* that are reached by an incident and executed operations. We define also a *phase* like a set of operations that may be executed in parallel for obtaining one or several products.

Phase k1 =[op1, op2,op3,..opn]

2 PROBLEM FORMULATION

The envisaged problem is concerned with the automated planning of batch procedure (batch sequence) which is allowed to restart the procedure that the execution has been interrupted by an incident.

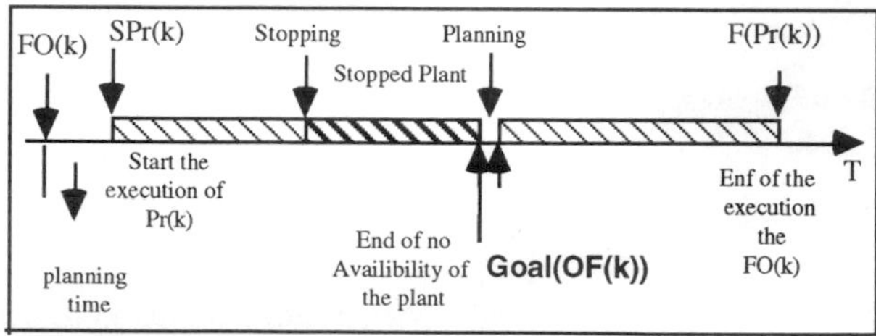

Figure 1 The process of planning is activated only during of no running periods of the plant

FO (k): It's a k 'ieme **F**abrication **O**rder.
Pr(k): It's the *"normal"* batch Procedure about the execution of FO(k).

When the stoppage is happened during the execution of Pr(k), The execution renewal can make obligatory certain phases of execution that is no presented in Pr(k). The process of planning can be activated only during of no running periods of the plant because it is used a description of initial state of the plant that must be true so long as the plan not be obtained as shown in figure (1). The process of planning is for object: -- To finds a set of phases among a set of S feasible phases from the description of state of the plant at the date" *End of availability"*. -- To ordered totally or partially these phases to produce a batch procedure that allows to restart the plant and to obtain a final product (complete realization of FO(k)).

Case 1

In this case, the plant must satisfied the previous fabrication order before the satisfaction the present as shown in figure (2).

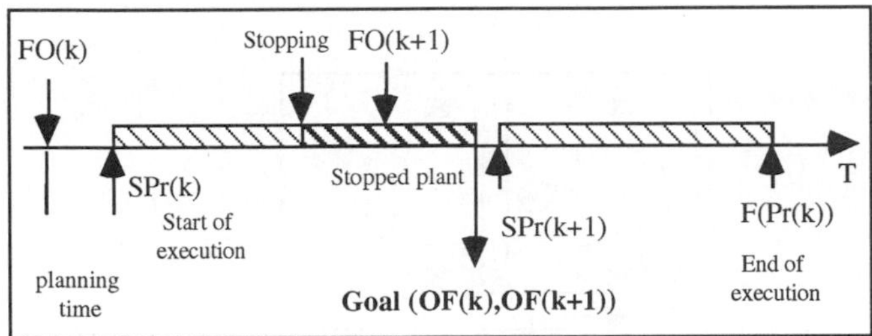

Figure 2 Satisfiying the previous FO before the present FO

Case 2

In this case, the fabrication order **FO(k)** has been realized. The plant now work with two fabrication orders **FO(k+1), FO(k+2)** as shown in figure (3).

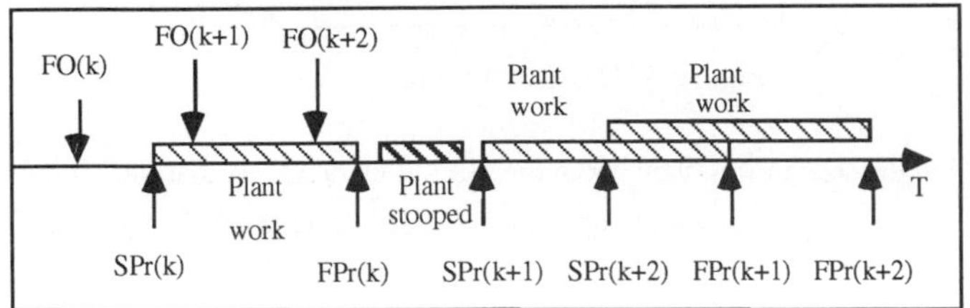

Figure 3 FO(k) is realised and the plant works with FO(K+1) and FO(K+2)

3 ILLUSTRATION

3.1. Description of the Process

There is two reactifs (A) and (B) of row materials that are weighted into two different tanks (D1) and (D2). The reactifs are transferred into chemical reactor (R3) (see fig (4)) preliminary charged with a quantity of solvent. The reactor is heated until the reaction is happened. His temperature is maintained during prescribe interval of time mentioned in the batch procedure. When the reaction is happened, the reactor is cooled and the product is transferred to storage tank. We have introduce the valve V12,V22,V32 as the valve that allowed for purging tanks D1,D2,R3 respectively. The valves are used to recover the reactif after an incident and during weight.*(Ex, V10 still opened and the quantity of (A) is higher than desired quantity)*, or the reactor cannot work normally.

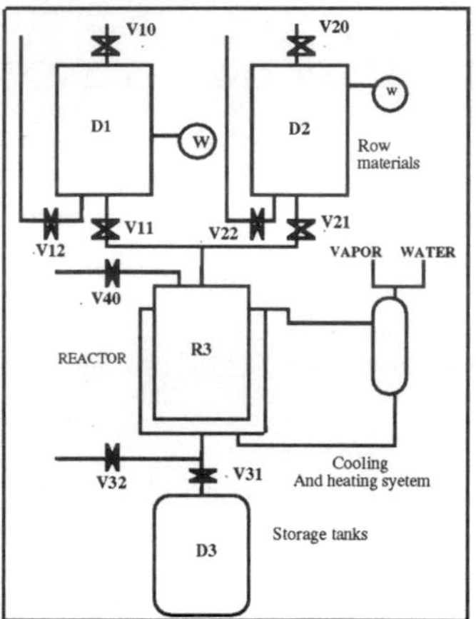

Figure 4 The Batch process

For determening the state of this plant about the cases illustrated , we start by giving a description of the set of phases before applying our search strategy as shown in figure (5).

Description of the set of phases

We can note the predicate first order by:

Tank (D1,< Pr(Qa),Tad>) => The tank D1 contains a liquid product (monomer) with a desired quantity Qad at the desired temperature Tad

Case 1

We precede a separate description for each batch (i) and batch (i+1) procedures:

A-Event (i+1) (The present fabrication order)

A-1-**Pooring-L Phase** Phase (1) (i+1): {Dosage Pr_a ; Dosage Pr_b }

Pre: Tank (D1,<Pr(o)$_{i+1}$, >) $\wedge$ Tank (D2,<Pr(o)$_{i+1}$, >)
Del: Pre
Add: Tank (D1,<Pr(Qa)$_{i+1}$, Ta >) $\wedge$ Tank (D2,<Pr(Qb)$_{i+1}$,Tb >)

<u>with:</u> Tank (D1,<Pr(o)$_{i+1}$, >)

The tank D1 contain no row material (A). The temperature is not mentioned.

We note also: (Qa)<(Qad) ; (Ta)<(Tad).

The quantity Qa is lower than desired quantity, the same as the temperature.

A-2- **Purge Phase** phase (8) (i+1): {Purge Tank_a ; Purge Tank_b ; Purge_Reactor}

Pre: Tank (D1,<Pr(Qa)$_{i+1}$,Ta>) $\wedge$ Tank (D2,<Pr(Qb)$_{i+1}$,Tb>) $\wedge$Reactor(R3,< Pr(Qd)i,Td>)
Del: Pre
Add: Tank (D1,<Pr(Qad)$_{i+1}$,Tad>) $\wedge$Tank (D2,<Pr(Qbd)$_{i+1}$,Tbd>) $\wedge$Reactor(R3,< Pr(o)i,Td>)

A-3-**Feeding Solvent Phase** Phase (3)(i+1) :{Feeding _solvent }

Pre: Reactor (R3,<Pr(o)$_{i+1}$, >) $\wedge$ Reactor (R3,Pr(Qmd)$_{i+1}$,Tmd>)
Del: Reactor (R3,<Pr(Qmd)$_{i+1}$,Tmd>)
Add: Reactor (R3,<Pr(Qsd)$_{i+1}$,Tsd>)

A-4- **Feeding Mixture Phase** Phase (2)(i+1) :{ feeding _Mixture }

Pre: Tank (D1,<Pr(Qad)$_{i+1}$,Tad>) $\wedge$ Tank (D2,<Pr(Qbd)$_{i+1}$,Tbd>)
Del: Pre
Add: Reactor (R3,<Pr(Qmd)$_{i+1}$,Tmd>)

A-5-**Heating reactor Phase** Phase (4)(i+1) :{ Heating_reactor}

Pre: Carry_Reactor (R3,<Tam,Patm>)$_{i+1}$ $\wedge$Reactor (R3,<Pr(Qsd)$_{i+1}$,Tsd>)
Del: Reactor (R3,<Pr(Qsd)$_{i+1}$,Tsd>)
Add: Carry_Reactor (R3,<Tex,Pm>)$_{i+1}$

A-6-**Cooling reactor Phase** Phase (5)(i+1) :{ Cooling_reactor }

Pre: Carry_Reactor (R3,<Tex,Pm>)$_{i+1}$
Del: Pre
Add: Carry_Reactor (R3,<Tam,Patm>)$_{i+1}$

A-7-**Purge reactor Phase** Phase (6)(i+1) : {Purge_reactor }

Pre: Reactor (R3,<Pr(Qmd)$_{i+1}$,Tmd>) $\wedge$Carry_Reactor (R3,<Tam,Patm>)$_{i+1}$
Del: Carry_Reactor (R3,<Tam,Patm>)$_{i+1}$

Add: Reactor (R3,<Pr(Qcd)$_{i+1}$,Tcd>)

A-8-**Storage product Phase** Phase (7)(i+1) :{ Storage_product}

Pre: Reactor (R3,<Pr(Qcd)$_{i+1}$,Tcd)
Del : Pre
Add: Tank (D3,<Pr(Qcd)$_{i+1}$,Tcd)

We must satisfy the goal G_1(i+1) for the fabrication order FO(i+1).

Description of the Set of Phases for the Event (i)

B. Event (i) (the Previous fabrication order)

We suppose it not exits no row materials in tanks (D1),(D2) of the batch procedure (i). The reactor is founded at the heating phase at the same procedure. The product into the reactor is adulterated.

B-1-**Heating reactor Phase** Phase (4)(i) :{Heating_reactor }

Pre: Carry_Reactor (R3,<Tam,Patm>)$_i$
Del: Pre
Add: Carry_Reactor (R3,<Tex,Pm>)$_i$

B-2-**Cooling reactor Phase** Phase (5)(i) :{ Cooling_reactor }

Pre: Carry_Reactor (R3,<Tex,Pm>)$_i$
Del: Pre
Add: Carry_Reactor (R3,<Tam,Patm>)$_i$

B-3-**Purge reactor Phase** Phase (6)(i) :{ Purge_reactor }

Pre: Reactor (R3,<Pr(Qmd)$_i$,Tmd>) $\wedge$Reactor (R3,<Tam,Patm>)$_i$
Del: Carry_reactor (R3,<Tam,Patm>)$_i$
Add: Reactor (R3,<Pr(Qcd)$_i$,Tcd>)

B-4-**Storage Product Phase** Phase (7)(i) :{ Storage_product}

Pre: Reactor (R3,<Pr(Qcd)$_i$,Tcd)
Del: Pre
Add: Tank (D3,<Pr(Qcd)$_i$,Tcd)

B-5-**Purge Reactor Phase** Phase (8)(i) :{ purge_reactor }

Pre: Reactor (R3,<Pr(Qw)$_i$,Tw>)
Del: Pre
Add: Reactor (R3,<Pr(o)$_i$, >) With W : Waste

3.2 Description the States of the Plant

The initial states (So) of the plant:

S0 ={Tank (D1,<Pr(o)$_{i+1}$, >) $\wedge$Tank (D2,<Pr(o)$_{i+1}$, >); Reactor (R3,<Pr(Qw)$_i$,Tw>) $\wedge$Carry_Reactor (R3,<Tam,Patm>)$_i$}

We apply a forward search strategy . The goal for achieving is G_T =(G_1(i+1) ; G_2(i))

On the other words we have:

G_T={Tank (D3,<Pr(Qcd)$_{i+1}$,Tcd>) ; Tank (D3,<Pr(Qcd)$_i$,Tcd>)}

Since the goal is not include in that situation we select the applied phases and that might be executed in parallel:

Ph(1) : {Phase (1)(i+1) ; Phase (4) (i) }

Because the precondition of these two phases are include in the previous State So

Pre (Phase (1) (i+1)) $\cap$ Add (Phase (4) (i) = ø

These two phases are applied simultaneously because there are used a different equipment. Since the event (i) is lower than the event (i+1) that means the event (i) must be realized before the event (i+1).

u Del (Ph(1)) ={ Tank (D1,<Pr(o)$_{i+1}$, >) $\wedge$ Tank (D2,<Pr(o)$_{i+1}$, >) ; Carry_Reactor (R3,<Tam,Patm>)$_i$}

u Add (Ph(1))={Tank (D1,<Pr(Qa)$_{i+1}$, Ta >) $\wedge$Tank (D2,<Pr(Qb)$_{i+1}$, Tb >) ; Carry_Reactor (R3,<Tatm,Patm>)$_i$}

The state (S1)

S1 ={ Tank (D1,<Pr(Qa)$_{i+1}$, Ta >) $\wedge$Tank (D2,<Pr(Qb)$_{i+1}$, Tb >) ; Carry_Reactor (R3,<Tatm,Patm>)$_i$ $\wedge$Reactor (R3,<Pr(Qw)$_i$,Tw>) }

Since the goals G_T is not include in the situation S1, we select the applied phases:

Ph(2) : {Phase (8)(i+1) ; Phase (5) (i), Phase (8)(i) }

u Del(Ph(2)) ={ Tank (D1,<Pr(Qa)$_{i+1}$, Ta >) $\wedge$ Tank (D2,<Pr(Qb)$_{i+1}$, Tb >) ; Carry_Reactor (R3,<Tatm,Patm>)$_{(i)}$ $\wedge$Reactor (R3,<Pr(Qw)$_i$,Tw>) }

uAdd(Ph(2))={Tank (D1,<Pr(Qad)$_{i+1}$, Tad>) $\wedge$Tank (D2,<Pr(Qbd)$_{i+1}$, Tbd >) ; Carry_Reactor (R3,<Tatm,Patm>)$_{(i)}$ $\wedge$Reactor (R3,<Pr(o)$_i$, >) }

The state S2

$$S2=\{\ \text{Tank}\ (D1,<Pr(Qad)_{i+1}\ ,\ Tad>)\ \wedge\ \text{Tank}\ (D2,<Pr(Qbd)_{i+1},\ Tbd>)\ ;$$
$$\text{Carry\_Reactor}\ (R3,<Tam,Patm>)_{(i)}\ \wedge \text{Reactor}\ (R3,<Pr(o)_i,Td>)\ \}$$

Since the goals G_T is not include in the situation S2, we select another times the applied phases:

$$Ph(3):\ \{\text{Phase}\ (2)(i+1)\}$$

$$u\ Del(Ph(3))=\{\text{Tank}\ (D1,<Pr(Qad)_{i+1}\ ,\ Tad>)\ \wedge \text{Tank}\ (D2,<Pr(Qbd)_{i+1},\ Tbd>)\ ;$$
$$\text{Carry\_Reactor}\ (R3,<Tam,Patm>)_{(i)}\}$$

$$uAdd(Ph(3))=\{\text{Reactor}\ (R3,<Pr(Qmd)_{i+1},\ Tmd>);\ \text{Reactor}\ (R3,<Pr(o)_i,>)\}$$

Determining the New State of the Plant

$$S3=(S2-\ U\ Del(Ph(3)))+U\ Del\ (Ph(3))$$

$$S3=\ \{\text{Reactor}\ (R3,<Pr(Qmd)_{i+1},\ Tmd>);\ \text{Reactor}\ (R3,<Pr(Qcd)_i\ ,Tcd>)\ \wedge \text{Reactor}$$
$$(R3,<Pr(o)_i,\ >)\ \}$$

Since the goal G_T is not include in the situation S3, we select the applied phases that can be executed in parallel

$$Ph(4):\ \{\text{Phase}\ (3)(i+1)\}$$

$$UDel(Ph(4))=\{\text{Reactor}\ (R3,<Pr(Qmd)_{i+1},\ Tmd>)\ ;\ \text{Reactor}\ (R3,<Pr(Qcd\ )_i,Tcd>)$$
$$\wedge \text{Reactor}\ (R3,<Pr(o)_i,>)$$

$$UAdd(Ph(4))=\ \{\text{Reactor}\ (R3,<Pr(QSd\ )_{i+1},\ TSd>);\ \text{Tank}\ (D3,<Pr(o)_i\ ,\ >)$$

The state S4 of the plant

$$S4=\ \{\text{Reactor}\ (R3,<Pr(QSd)_{i+1},\ TSd>);\ \text{Tank}\ (D3,<Pr(o)_i,\ >)\}$$

We can see the goal G2(i) is reached : Tank $(D3,<Pr(o)_i,\ >)$

On the other hands G1(i+1) is not reached yet. We are satisfied the previous fabrication order FO(k).Since the goal G1(i+1) is not reached we select another times the only phases for batch procedure (i+1).

$$Ph(5):\ \{\ \text{Phase}\ (4)\ (i+1)\ \}$$

$$UDel(Ph(5))\ =\{\ \text{Reactor}\ (R3,<Pr(QSd)_{i+1}\ ,\ TSd>)\}$$

$$UAdd(Ph(5))=\{\text{Carry\_Reactor}\ (R3,<Tex,Pm>)_{i+1}\ \}$$

Determining the state S5

$$S5=\{\ \text{Carry\_Reactor}\ (R3,<Tex,Pm>)_{\ i+1}\ ;\ \text{Tank}\ (D3,<Pr(o)_i\ ,\ >)\}$$

The goal G1(i+1) is not reached. That means is not include in the state S5.The applied phases are:

$$Ph(6): \{Phase\ (5)(i+1)\}$$

UDel(Ph(6))={ Carry_Reactor (R3,<Tex,Pm>)$_{i+1}$}

UAdd(Ph(6))={ Carry_Reactor (R3,<Tam,Patm>)$_{i+1}$}

The state S6 of the plant

S6= { Carry_Reactor (R3,<Tam,Patm>) $_{i+1}$; Tank (D3,<Pr(o)$_i$, >)}

The goal G1 (i) is not reached. The selected phase is (Ph(7)) whitch is Purge_reactor $_{(i+1)}$

UDel(Ph(7))= { Carry_Reactor (R3,<Tam,Patm>) $_{i+1}$ }

UAdd(Ph(7))= { Reactor (R3,<Pr(Qcd)$_{i+1}$, Tcd >)}

The state S7 of the plant

S7= { Reactor (R3,<Pr(Qcd)$_{i+1}$, Tcd >) ; Tank (D3,<Pr(o)$_i$, >) }

Another times the goal G1(i+1) is not reached yet. The last selected phase is Storage_product$_{(i+1)}$

UDel (Ph(8)) = { Reactor (R3,<Pr(Qcd)$_{i+1}$, Tcd >)}

UAdd (Ph(8)) = { Tank (D3,<Pr(Qcd)$_{i+1}$, Tcd >)}

The state S8 of the plant

S8= { Tank (D3,<Pr(Qcd)$_{i+1}$, Tcd >) ; Tank (D3,<Pr(o)$_i$, >)}

Finally, the goal G1(i+1) is reached in the state S8. In the other hand , the goal G2(i) is reached in the state S4. Then, there are a satisfaction of the fabrication order FO(i) before a satisfaction of the fabrication order FO(i+1).

The Plan of the case 1

Initiale State ----> Poor(A) $_{i+1}$ / Poor(B) $_{i+1}$ / Heating_ R$_{(i)}$ ----->S1-----> Purge(A) $_{i+1}$ / purge(B) $_{i+1}$ / Cooling R$_{(i)}$ --->S2 ---->Purge_reactor $_{(i)}$ / Feeding_mixture $_{(i+1)}$

S3 -----> storage_product $_{(i)}$ / Feeding_solvent $_{(i+1)}$ -----> S4 $G_{2(i)}$ ----> Heating_reactor $_{(i+1)}$ -----> S5 ---->

Cooling_reactor$_{(i+1)}$ --->S6 -->Purge_reactor$_{(i+1)}$ --->S7-->storage_product$_{(i+1)}$ ---->S8 $G_{1(i+1)}$

with: G1 $_{(i+1)}$ = Tank (D3,<Pr(Qcd)$_{i+1}$, Tcd >)
} -->G_T ={ G1(i+1) ; G2(i)}
G2 $_{(i)}$= Tank (D3,<Pr(Qcd)$_i$, Tcd >)

Case 2

In this case, we can suppose the fabrication order as the events that come simultaneously but heir time of execution are different.

Fabrication order FO(k+1) ={X'$_1$ (quantity) Qa , X'$_2$(quantity) Qb ; for manufacturing the single product (Qc) $_{k+1}$, knowing that X'$_1$,X'$_2$ is members of N}

Fabrication order FO(k+2) ={X"1 (quantity) Qa, X"2(quantity) Qb ; for manufacturing the single product (Qc) $_{k+2}$, Knowing that X"$_1$,X"$_2$ is members of N }

We consider for the batch procedure (i+1), where Pr(k+1) represents the execution of the fabrication order FO(k+1) at the time t $_1$ and Pr(k+2) also represents the execution of the fabrication order FO(k+2) at the time t $_2$. We take always the fabrication order as the events.

FO(k+1) is equivalent at the event (k+1)

FO(k+2) is equivalent at event (k+2).

With : T FO(k+1) < T FO(k+2) => event (k+1) < event (k+2)
T : the execution time of batch procedure.

Since these two fabrication orders FO(k+1),FO(k+2) arrived when the plant work, the Purge phase is not applied in this case. It is useless to do the description for the set of phases Each procedure is expressed separately as expressed in the case n°2. It can change only the indices (i) to (K+2) and (i+1) to (k+1). We can eliminate the operators Purge_Tank (A) et Purge_Tank (B).We displaced feeding_mixture between the states (S1-S2), feeding_solvent between the states (S2-S3) and heating_reactor between the states (S3-S4).

Then the goal is reached $G_{1(k+1)}$ of the batch procedure Pr(k+1) in the state S4, but $G_{2(K+2)}$ of the batch procedure Pr(k+2) is reached in the state S7.

Knowing that, the goal $G_{O(K)}$ is reached at the state S0 of the plant. That means, the final product is found at the storage phase

The Plan of the case 2

$G_{O(k)}$ Poor(A) $_{k+2}$ Feeding mixture (a),(b)$_{(k+2)}$
Initial state ---->Poor(B) $_{k+2}$ -----> S1-----> --->S2 --->
Heating_reactor$_{(k+1)}$ Cooling_reactor $_{(k+1)}$

Feeding solvent $_{(k+2)}$ Storage_product $_{(k+1)}$ $G_{1(k+1)}$
Purge_reactor $_{(k+1)}$ -----> S3 ----> Heating_reactor $_{(k+2)}$ -----> S4 ---->

Cooling_reactor $_{(k+2)}$ -->S5 -->Purging_reactor $_{(k+2)}$ -->S6-->storage_product$_{(k+2)}$ $G_{2(k+2)}$
------> S7

With:

$$G_1\ (k+1) = \text{Tank}\ (D3, <Pr(Qcd)_{k+1}, Tcd >) \quad G_2\ (k+2) = \text{Tank}\ (D3, <Pr(Qcd)_{k+1}, Tcd >) \quad \} \rightarrow G_T = \{ G_{1(k+1)} ; G_{2(k+2)} \}$$

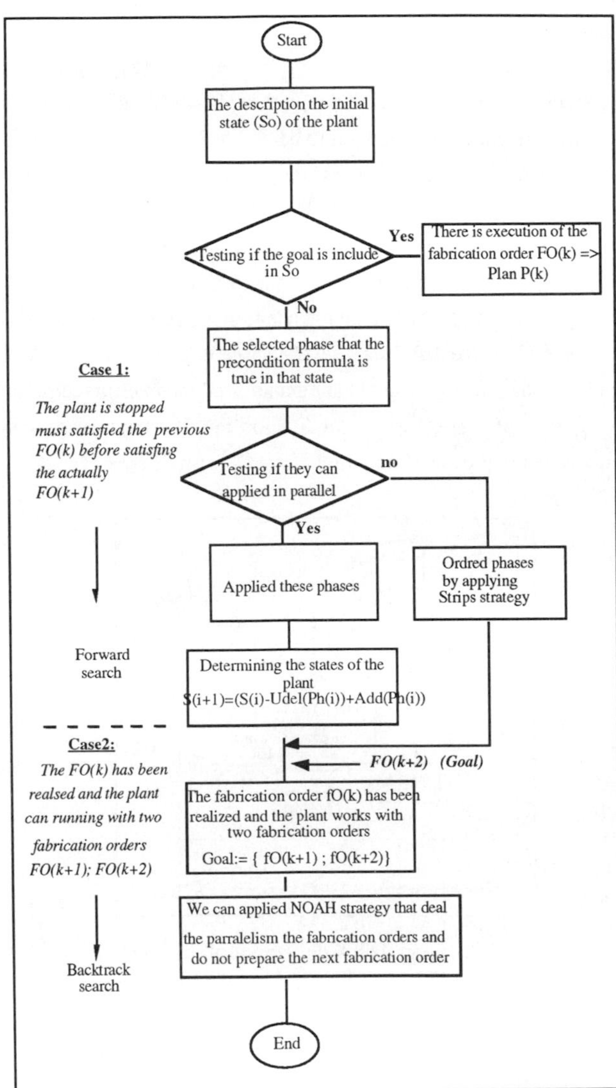

Figure 5 The Search Strategy

4 ENVISAGED GLOBAL ARCHITECTURE

Our planning system is composed of the following elements: Planning, algorithm, and supplementary elements like: real time data base, simulator.
The planning can take into account the market demands for product, production time and energy resources .. etc. This activity is relative of the long term (days) period in outlook. It takes only the general capacity of global of the process. From this information, it can generate the production plan (reference) for sending it to batch level control system. For generating the production plan, the planning needs a real time data base that is necessary of the set of phases and goals to reached. This is the fabrication order. The planning needs also the initial condition of the process. The role of batch level control system (BLCS) is charged to execute the production plan established by planning in versus the information received by the lower level that is illustrated in Rouis (1994). When the interruption happens in the batch plant, the planning system receives by feedback the state of the plant. The real time data base can stock the initial condition of the process. The planning may send the interrupted production plan for replanning the sequence of phases by our search algorithm. It is convenient to simulate and animate the sequence of phases by simulator.

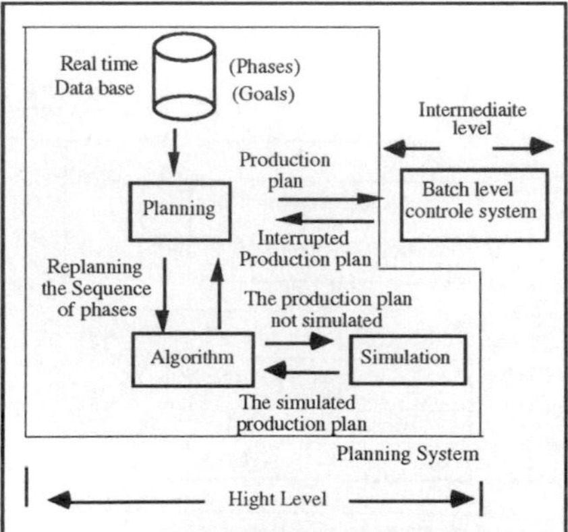

Figure 6 The global Architecture of the High level

5 CONSEQUENCES

For the single product, the batch procedures are practically the same. Then, It is useless to replan. We must change only the quantities of the row material or product. That is mean

Pr(k+1) is practically equal at Pr(K+n) by taking into an account the stoppage of the plant during the incident was happened.-It is impossible to set to work the plant that is produced a single product with the satisfaction of two fabrication orders at the same time like FO(k+1) and FO(k+2).- Before replanning, the fabrication order cannot be taken into consideration when the present fabrication order is complete.
In the future work, we try to give a solution of a define lack of integration between the planning and control activities the same as evoked by Cott and Maccietto (1989).

6 REFERENCES

BENASSY (1990) "Gestion de Production " Hermes , Mars 90.

COTT and MACCIETTO (1989) An Integreted approach to computer -aided operation of batch chemical plants.Compt chem Eng vol 13 n°11/12 pp 1263-1271, 1989.

FIKES and NILLSON (1971) " STRIPS : A new approach to the application of theorem proving to the problem solving " . Artificial intelligence , vol 2 pp 189-208, 1971.

FUSILLO and POWER (1987) "A synthesis method for chemical plant operating procedure " Computer Chemical Engineering Vol 12,N° 9/10 , PP 1023-1034, 1988.

IVANOV .V.A , V.V.KAFAROV , V.L.KAFAROV and A.A. REZNICHENKO (1980 b) Design principles for chemical production startup algorithm . Automat .Remote Control 41, 1023,1032(1980 b)

FUSILLO and POWER (1988) " Computer-aided planning of purge operations " Aiche Journal Vol 34, N°4 PP 558-566,1988 .

IVANOV.V.A ,al (1980 a) On algorithmization of the startup of chemical production . Engng Cybernetics 18,104-110- (1980 a) .

KINOSHITA.A, T.UMEDA,E. O'SHIMA (1981). An algorithm for synthsis of operational sequences of chemical processing plants . In Computerized Control and operation of chemical plants, Vienna, Autria, Ostereichisher chemiker , Vienna, Autria, (1981).

RIVA .J.R, D.F.RUDD (1974) . Synthesis of failure-safe operations . Aiche Jl 20, 320,325 -(1974).

D. ROUIS (1994) Automation of batch process restarts in the wake of the incident. In proceeding of the Asian Control conference (ASCC) ; Vol 2, P765 ,July 27-30 (1994) TOKYO -JAPAN

LAKSHMANAN ,G. STEPHANOPOULOS (1988) . Synthesis of operating procedures for complete chemical plant-II-. A Nonlineair planning methodology. Computer Chem Engng Vol 12. No 9/ 10 PP 1003-1021.

[LAZARO 89] "A comprehensive approach to production planning in multipurpose batch plant " Computer chemical engineering.

LOSCO (1987) " Computer control application in a batch polymerisation plant " Pocc .CEF 87,Ciardini naxo,Italy 87.

PART FOUR

Modeling and Design of FMS I

9

An Object-Oriented Approach to the Design of Flexible Manufacturing Systems *

J. Reinaldo Silva †
Computer System Group, University of Waterloo, Waterloo, Canada
e-mail: reinaldo@csg.uwaterloo.ca

H. Afsarmanesh
Computer System Department, University of Amsterdam, The Netherlands
email: hamideh@fwi.uva.nl

D.D. Cowan
Computer System Group, University of Waterloo, Waterloo, Canada
email: dcowan@csg.uwaterloo.ca

C.J P. Lucena
Computer Science Departement, PUC-Rio de Janeiro, Brazil
email: lucena@inf.puc-rio.br

Abstract

In this paper a hybrid top-down/bottom-up method that can be viewed as an extension of the traditional dynamic modeling technique using Petri Nets and Parametric Design is presented as an approach to the design of Flexible Manufacturing Systems. The resulting method supports a clear separation of functionality among the design objects by using the ADV/ADO object-oriented design framework. Thus, the designs as well as the general functional models can be reused. Comparing the method described in the paper with the object-oriented architecture introduced and employed in the PEER object-oriented database system suggests an implementation approach which can support the object clustering properties of ADV's and ADO's.

Keywords

OO-design, Flexible Manufacturing Systems, Petri Nets, Abstract Data Views, design reusability

1 INTRODUCTION

Flexible manufacturing systems often have two conflicting characteristics: a clear physical model given by a set of production processes and machines, and a logical or abstract model usually related to a production plan and accompanying control algorithms. Both models are

*This work is part of the research activities of the Esprit/ECLA Flexsys 76101 and Cimis.net 76102.

†Partially supported by FAPESP. On leave from University of São Paulo, Brazil.

necessary to describe the complete system and yet the design methods related to the two models are inherently incompatible. Such a substantial difference in design approaches could easily lead to errors at the early stages in the lifecycle of the entire artifact‡, because the products of the design at this stage are mismatched and relationships are not always clear. Errors made in these initial phases are very expensive to remove since they are often found late in the development process as the integrated flexible manufacturing system is realized.

Development of a method for design of logical or abstract objects such as software that would simplify conversion to prototypes might alleviate this design issue. Then the logical products of the design could be incorporated into the physical model and the incompatibilities could be observed and corrected. Several attempts have been made to find such a method, most of them associated with software development. Typical examples are rapid prototyping, bottom-up (Sommervile, 1992), outside in (Marca, 1988) development and object-oriented design [Booch 91]. Development of such a method might also make design formalisms and tools more attractive to the practicing design engineer because of the ability to produce a prototype that would make the benefits of formal approaches more concrete, thus bridging the gap between new achievements in Design Theory and practical applications in Engineering Design.

In this paper we describe a hybrid design method based on providing abstractions of both the logical components such as the software, and the physical components of the flexible manufacturing system. The abstractions represent the behavior and functionality of both types of components through their interfaces (Cowan, 1995). By using suitable constructors to combine the components and appropriate information hiding mechanisms we can reuse both types of components and hence use previously designed elementary components. The external behavior of each component could be validated at any point in the development process using its dynamic model expressed in PFS/MFG.

External behaviors or interfaces and their corresponding object models are represented here by Abstract Data View /Abstract Data Object (ADV/ADO) pairs (Cowan, 1993a)[Alenc 94](Cowan, 1995) where ADVs are extensions to the object model to support the specification of interfaces. The design approach will be demonstrated using an example of a discrete control algorithm for shop floor control (Bruno, 1986). Finally we describe an implementation of these ideas based on a federated architecture using the PEER database (Afsarmanesh 1993)(Tuijnman, 1993)(Afsarmanesh 94).

2 DESIGN OF FLEXIBLE MANUFACTURING SYSTEMS

A Flexible Manufacturing System (FMS) is a cooperating set of process machines (usually numerically controlled) connected by an automated transportation system (Tempelmeier, 1993). We add to this definition by allowing the FMS to have local storage facilities and local computers dedicated to information handling and control. These added components are not necessary to justify our technique, but were introduced in order to test our approwith a complex system.

A version of the life cycle for an FMS (Tempelmeier, 1993) is presented in Figure 1, where the two main activities of integrated production planning and model optimization are highlighted. Most of the current design methods focus on these activities since they are critical factors

‡The term artifact is used to represent the goal of the design process, and could refer to a physical object such as a mechanical part or a logical object such as a software system or a control algorithm. or some combination of these two categories.

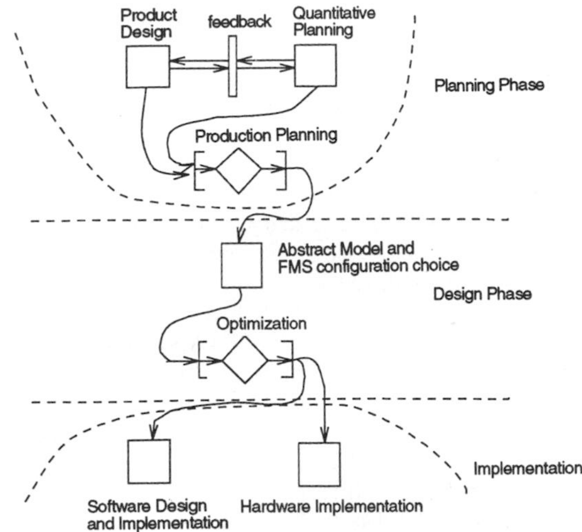

Figure 1 Life cycle of an FMS in PFS/MFG representation

in producing a good design. However, modules such as Product Design, Software Design (especially associated with cell and factory control) and Modelling could make significant contributions to the overall result if a more flexible design approach is used.

We concentrate on an approach that reinforces abstraction in the modelling and evaluation of configuration options, and that could provide the basis for further optimization as well as requirement and production plan validation. The representations we use are intended to support software design and specification. We use a general integration plan as input and rapidly generate abstract models of the FMS configurations that could be refined further through the proper choice of machines and layout. The advantage of this approach is that consistent abstract models can be shared by all those professionals with different backgrounds participating in the design thus, making it easier to reinforce concurrence at each design phase. In addition, when changes and adaptations are required, a common occurrence in today's dynamic marketing environment, the appropriate abstraction will be available.

Another important aspect of FMS design is the manner in which the proposed approach handles complexity. Complexity depends on the size of the system (the number of machines) and of the degree of flexibility, where flexibility is a function of the number of interrelation operations. Current definitions of FMS (Ranta, 1990) range from small (2-4 machines) to complex systems (15-30 machines). Complexity can vary with the size of FMS in several ways. For instance, if a set of numerically controlled machines do the same set of operations with a mix of products, the workload balance will be simpler than if the set of operations is different and the manufacturing of each type of product in the mix has to be put in a sequence in addition to balancing the overall workload. Thus, modeling is very important for small systems, since small and medium-size enterprises usually attempt to obtain more and varied production from small FMSs.

Reuse of old models stored and retrieved from a database can substantially reduce the cost of redesign and adaptations. Our design approach is reuse-oriented in that components and their relationships are specified solely by their interfaces.

2.1 Petri Net and PFS/MFG Modeling

Petri Net methods have been used extensively to model FMS (Silva M., 1993)[Proth 93](DiCesare, 1993)(Proth, 1993a). Such methods provide formal models for FMS, and are able to handle the complexity inherent to features such as dependencies between cell operations, parallelism, concurrence of tools and/or material. However, Petri Net models (principally those based on Place/Transition Nets [Reisig 89] do not adequately support reusability or abstraction.

Recently, there has been more research in high-level Petri Nets with the objective of introducing abstraction into net-oriented design methods. PFS/MFG (Program Flow Schema/Mark Flow Graph) is one type of high-level net 1988 [Miyagi 88] specifically for the design of discrete manufacturing systems. The idea is to add abstraction to the modeling power of Petri Nets (Condition/Event) and produce less complex graph models for qualitative and quantitative analysis. The structure of a specific PFS/MFG representation can be translated into Dynamic Logic (Silva J., 1992), and then artificial intelligence tools can be used to assist with the modeling and analysis of target systems (Lucena, 1989).

A revision of the conventional PFS/MFG representation that was introduced in (Silva J., 1992a) allows the simulation of a Petri Net model at any level of abstraction, thus, making these formalism more suitable for the top-down design of discrete production systems. The revised formalism was used in a cognitive design model [Takeda 90] (Tomiyama, 1992) that was applied to FMS in (Silva J., 1994). The addition of modularity, information hiding and separation of concerns indicated that a hybrid approach to design may be possible.

3 THE ADV/ADO MODEL

Some of the characteristics of the hybrid-design approach presented in this paper such as, abstraction, information hiding, nesting, and polymorphism can be found in object-oriented design [Booch 91], an approach to design that is primarily bottom-up. However, the majority of the practical design problems in engineering, particularly in the design of FMS, have a functional flavor, and are more closely related to top-down methods. We do not intend to debate which method is more appropriate, rather we wish to combine and integrate the two approaches to obtain a good design solution.

Thus, we first tried to find a framework where functionality could be implicitly or explicitly applied depending on the specific requirements of a particular design phase. Such a feature could be used to combine existing detailed elements with those for which only the general behaviour is known. In other words, we could combine existing elements (reusable blocks) with abstract descriptions. The ADV/ADO (Cowan, 1993)(Cowan, 1993a) which was originally created to allow a clear separation of the interface including the user interface from the application component in software design is used in this hybrid-design approach.

An Abstract Data Object (ADO) has a static description/model of the artifact and methods (behaviors) that can query or change its internal state. ADOs can be combined through operators for composition (nesting) and aggregation (Alencar, 1994) to build complex elements.

The interaction between ADOs, or between an ADO and an external medium such as a user or network, is through an interface object called an Abstract Data View (ADV). ADVs are in fact extended ADOs that handle input and output events or the exchange of messages and data among existing ADOs. An ADV specifies the external behavior or functionality of a model because an ADV specifies an interface to an ADO. For instance, a numerically controlled (NC) machine with a local magazine can be represented by an ADO where its interaction with the outside world is an enclosing ADV that accepts a program of operations and returns an error message, a progress report of the process, or an acknowledgement indicating that the operation was completed. Both the ADV and ADO are different objects which have their own behavior (methods). Interactions between them can be represented by another ADV which is responsible for the coupling or aggregated behavior, as might happen during the downloading of an APT program and the loading of the NC machine.

This simple example illustrates how to represent the strong separation between the pre-conditions to evoke the machine services and the process. The example could be extended to a manufacturing center, or a Flexible Manufacturing Cell and its integration in the overall FMS.

Top-down design is supported since ADVs provide a functional design interface connected to abstract models of subsystems that could be developed later. Instead of specifying a specific DNC machine or setting general features such as the number of axes or the set of tools in our simple example, we could perform a design with a statistical model of success performing operations or a lower bound time on machine operations as qualitative criteria. Such qualitative rules can guide the modeled behavior encapsulated in ADV's and could be developed later in the ADO model.

Two types of consistency relations between ADVs and ADOs are defined in [Cowan 94a]: horizontal and vertical consistency. A vertical consistency relation is defined between an ADV and its owner ADO as:

$$\mathcal{R} = \{(x, y) | x \in S \wedge y \in P \wedge ADO_j \vdash (x \rightarrow y)\} \quad (1)$$

where S and P are respectively the set of inputs and outputs of the ADV. Thus, any transaction with the outside world must be valid in the ADO model.

The same ADO or subsystem can interact with the external world in several different ways (Tomiyama, 1992), and they must all be consistent with each other. This form of consistency is called horizontal consistency in (Cowan, 1993a). We rephrase horizontal consistency in our cognitive model of design with the expression:

If $\mathcal{A}^j$ is the set of ADV's for the ADO j and

$$\exists \mathcal{A}^j{}_i.(x, y) \in \mathcal{R}_{ij} \Rightarrow [\exists \mathcal{A}^j{}_k.(x, z) \in \mathcal{R}_k j \Rightarrow z = y] \quad (2)$$

that is, ADV's cannot contradict each other.

The concept of consistency ensures the correct integration of concurrent designs or existing components such as the software and hardware implementation in Figure 1 and provides a strong foundation for the reuse process.

Two critical operations in the reuse process are: the search for a reusable component, and the adaptation of suitable candidates in the overall design. The efficiency of the first operation depends upon reducing the search space by providing in advance expected characteristics of the candidates or by performing the search using very short descriptions or metaphors for the components. We believe that in engineering and especially in the design of FMS, synthetic functional models described by ADVs would provide a good search space for reuse if vertical consistency with the ADOs is guaranteed. The importance of consistency to this reusability was

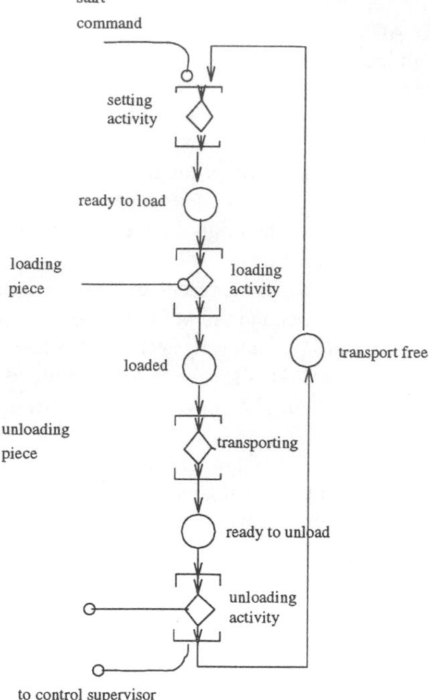

Figure 2 Model definition of a transportation system which takes one piece at a time

discussed in [SilvaJ 92]. The integration of reusable components can be guided by functionality and the maintenance of horizontal consistency.

More details about the concept of ADV and ADO can be found in [Cowan 94a] and in the related bibliography mentioned there. In the next section we will describe a short example to illustrate the basic concepts.

4 A SMALL EXAMPLE

In this section we revisit a short example proposed by (Bruno, 1986) to show how the hybrid method could be applied to the design of FMS. This example will be recast using the ADV/ADO approach with PFS/MFG used as a modelling tool.

Initially, we will try to identify the objects of the system [Booch 91]. In the area of FMS and Computer Integrated Manufacturing (CIM), composable objects can be easily identified. Some of the basic FMS constituents are: a transportation system, a NC machine, local storage and a local controller. The composed system is a flexible cell totally automated and controlled by signals which are sent from a central station.

A conventional Petri-Net approach (Silva M., 1993)[Proth 93] would model each one of these objects with decision-free nets, that is, nets without any kind of conflict, leaving all decisions to

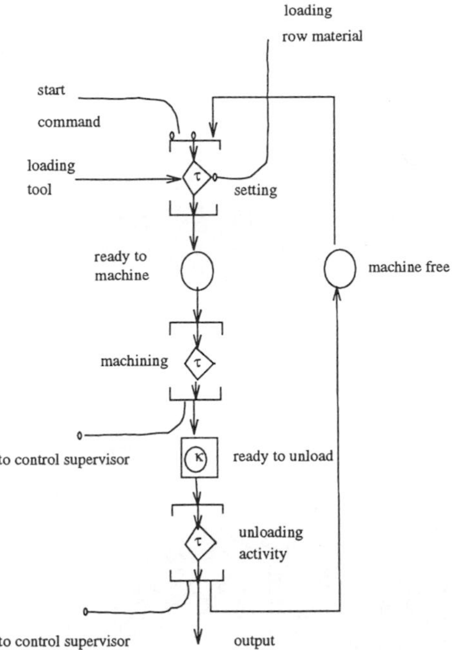

Figure 3 Model definition of a machine process

the control module. The internal behaviour of each object would be modeled in order to identify the control entries and outputs. For the current example, we just rewrite the abstract models presented by (Bruno, 1986) in PFS/MFG.

In a first approach, the transport facilities would be modeled as a system that processes one item at a time (an AGV). The whole operation can be started by a control sign (all remaining tasks such as adjusting positions and tracing a trajectory are performed locally). Figure 2 shows a PFS/MFG representation of the transport system.

Notice that Figure 2 is a decision-free net with external interactions with the control system represented by the start control signal and the signal sent to the supervisory system saying that the transportation process is finished. From the point of view of generating control algorithms and software, we could say that the interactions with the control system represent the principal control flow, to distinguish them from the (secondary) flow of items, such as the exchange of a piece with the outside (load and unload operations). Such a distinction is a key point in the modelling of discrete systems (Proth, 1993)[DiCesa 93][Miyagi 88].

The PFS/MFG high level representation of an NC machine is illustrated in Figure 3, and the storage system is represented abstractly in the diagram of Figure 4. We assume in this example that a complete design for these three elements can be reused taking the abstract model presented in Figures 2, 3 and 4 as a "target model" (Silva J., 1992). Our focus will be in the design of the discrete controller for the FMC.

In the revised presentation of PFS/MFG (Silva J., 1992), a flow relation is defined as an

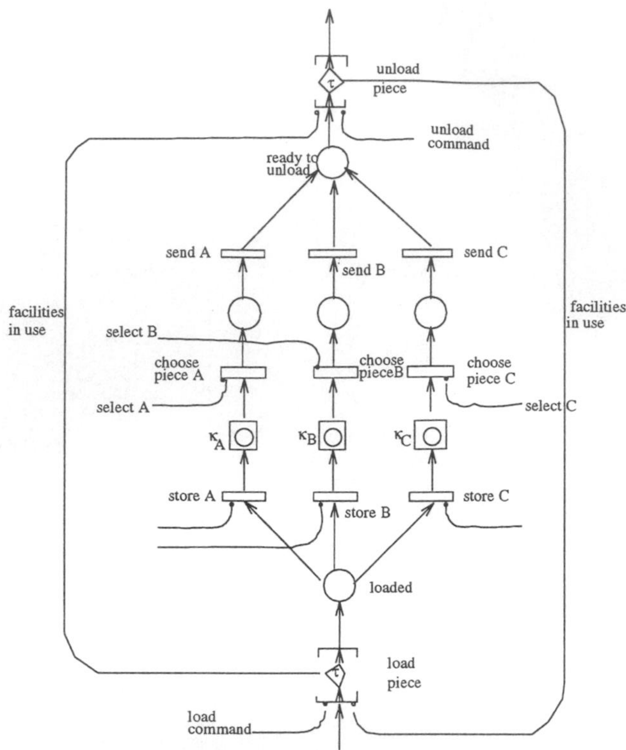

Figure 4 Model definition of the storage system.

object instance of a more generic class: the general relation between boxes (passive elements) and activities (active elements). Boxes are represented by B and each box has an attribute "kind" to indicate whether the box represents a storage, assembly, or distribution element[Miyagi 88]. The activities are classes with at least one attribute to specify the estimated time spent answering a call or finishing a process operation (if t=0 the activity will collapse into the representation of an instantaneous event of the conventional C/E Petri Net). The relationship between boxes and activities is given by:

$$\mathcal{F} = (B \times A) \cup (A \times B) \tag{3}$$

Each pair in this relation is a class, whose subclasses are *gates* and *flows*. Gates stand for non-structured relations and could be instantiated by external or internal gates. External gates represent control signs or calls for external pieces, information, or material, such as the *start command* in Figure 2 (a control sign), the *loading/unloading piece* command in Figure 2 (a call for an external piece), and the *to control supervisor* in Figure 3 (information about the process operation). An example of an internal gate is the *facilities in use* connection in Figure 4.

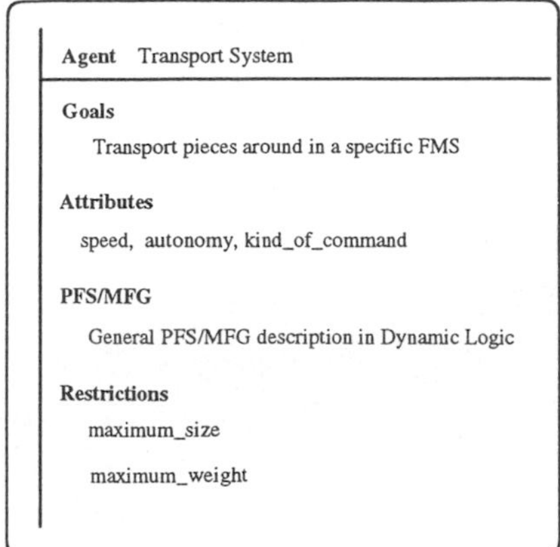

Figure 5 Informal definition of the ADO for the Transportation System.

Introducing this internal gate relation synchronizes the load/unload operation in the storage system, a requirement motivated by a need to share a manipulator robot[§].

Activities represent actions which are accessible only if their pre- and post-conditions are activated and deactivated, respectively. Activities can also encapsulate a cluster of other activities and conditions. For instance, the setting activity in Figure 2 can be refined as a subnet (also without conflicts) composed from two other activities, the loading of a piece (from outside) and the loading of a machine tool (from an internal magazine). Of course, each one of these aggregated activities can be further refined.

As mentioned earlier, the elements such as the transportation system, machine (a numerical control process) and storage are basic objects in the domain of Factory Automation and can be reused through a "standard" object-oriented technique or a combination of these methods and search techniques based on analogy or metaphors (Silva J., 1992). These reusable components can be integrated using a top-down design of the control system, which is depicted later in this paper.

The models of reusable components can be encapsulated as ADOs, together with some short documentation and other attributes (including a detailed model expressed in PFS/MFG). The representation of the ADOs use the techniques described in [Fields 93]. For instance, the ADO *Transport System* is shown in Figure 5.

An important feature of our design representation is that we could analyse the target system properties and/or simulate its behaviour at any level of abstraction by using the PFS/MFG internal model, which is a valuable validation mechanism for control engineering. One function of the ADVs is to specify the proper pre- and post-conditions for the internal activities encapsulated in an ADO. For instance, in the Figure 6 we have a very abstract ADV, representing the major functionality of the transport system which is to move pieces. As it is shown in Figure 2 the operation of a successful transport system depends on external signs and interaction with the control station. The ADV that represents the functionality of the transport system combines its

[§]The introduction of this requirement here suggests that "ad hoc" requirements or economic constraints could be included in the design method.

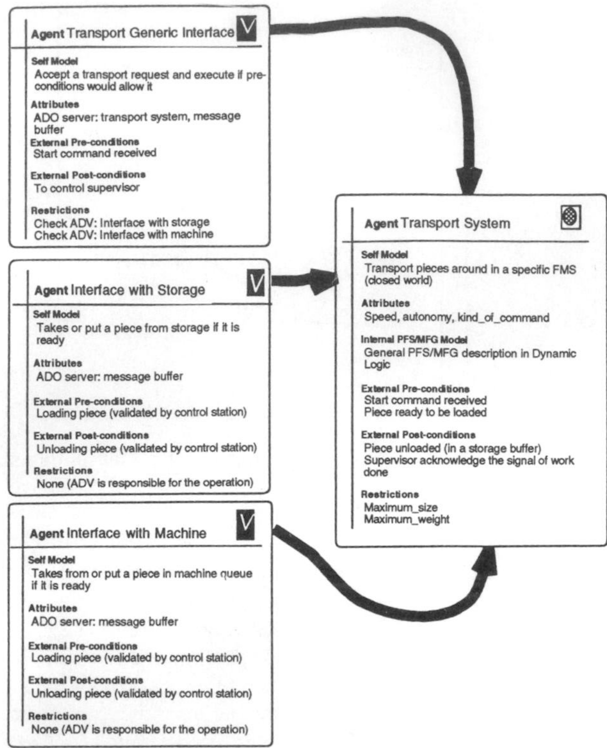

Figure 6 Cluster of behavior and model for a Transport System.

features with features of other ADVs to model the interaction with the control station. Hence, a request to transport a piece from one location to another would depend on the state of the condition $freeAGV$ (see Figure 2), which is based on the current state of the ADO model of the AGV, and on the interaction with the ADVs that represent the interface between the AGV and the control station, the supervisor, and the supplier/consumer of pieces.

If we build all the basic elements as encapsulated ADOs and ADVs, the construction of the control software requires two substantial steps:

i) build all ADV links between the element objects (storage, machine and transport system) and the control station ADO;
ii) support all links with discrete control modules and algorithms.

Figure 7 shows a schema of the ADV-links between the control station and the other elements already described. This is the core specification to model a centralized controller which is the most common solution for a control problem where all decisions are left to the controller. However, it should be noticed that a similar framework based on the ADV/ADO approach would work for other control techniques, where local controllers would have a more complex role in the process or a hybrid solution of clusters with more intelligent local controllers and pure server mechanisms.

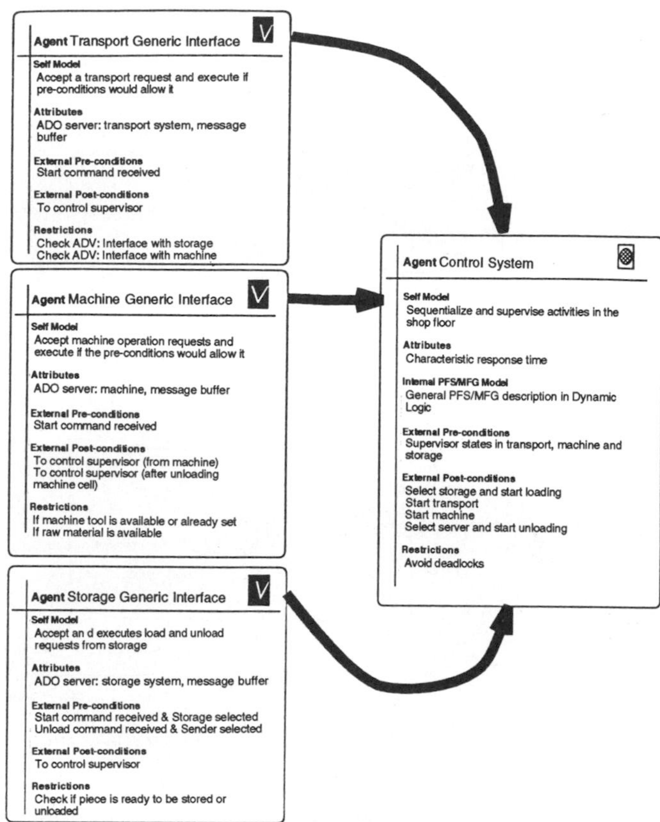

Figure 7 Cluster of ADV's connections with the Control System.

For instance, suppose that the transportation system receives a signal to start an operation¶. However, the loading process depends on the transportation system to recognizing in real time if the piece is ready to be removed. We are assuming it could be done by sensor signals whose interpretation would lead to a decision without any external intervention. That would make the system more reliable, even if in our case study the controller needs to ensure that the correct piece was selected or stored.

Supporting a combination of centralized and distributed systems is very important in the design of modern factory automation where legacy systems (relying on centralized control) are merged with modern autonomous processes.

Another interesting application of hybrid (centralized/distributed) systems is to the information-control problem, that is, to place information systems that supply information to control decisions at different abstraction levels. (Kagohara, 1994) shows a model based on layers, composed of production planning followed by a product design (CAD and CAPP) and finally a shop-floor control-layer. A system to generate coordinated control plans can also be designed using ADVs and ADOs as in the current example.

¶Since there is only one storage system and one machine, a simple argument would be enough to denote an operation from the storage to the machine or from the machine to the storage.

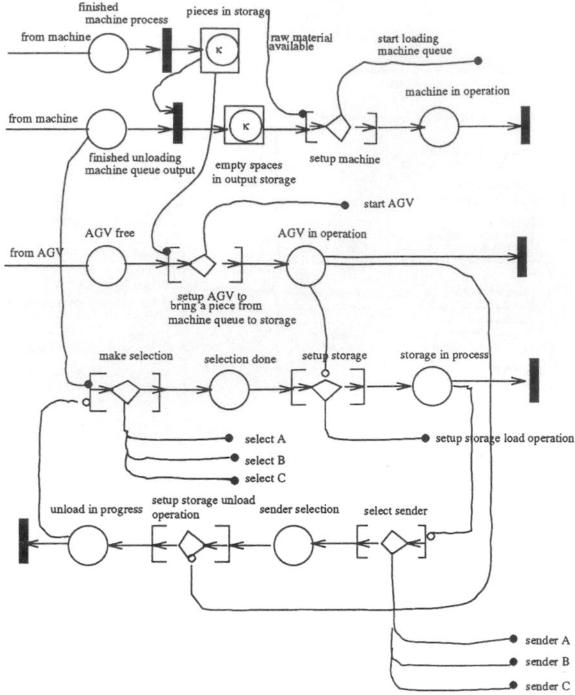

Figure 8 High level modelling of the discrete controller.

Figure 7 shows the abstract model of the control algorithm, and Figure 8 shows details of its internal PFS/MFG net, where the output events represent that control signs sent to the operator display or supervisory system.

In the next section we show the connection of the methodology with the information-sharing mechanism of the PEER database.

5 PEER IMPLEMENTATION APPROACH

As described earlier in the paper the model description and definition chosen for the ADV/ADO is closely related to the object-oriented approach of the PEER federated database management system. In this section, a brief description of the PEER system will be presented, and then by applying PEER to the example presented in Section 4, an implementation approach for the ADV/ADO model of the shop floor environment is addressed.

5.1 PEER database architecture

PEER (Afsarmanesh, 1993)(Tuijnman, 1993)(Afsarmanesh, 1994) is a federated object-oriented database management system designed and developed at the University of Amsterdam. A federated PEER network consists of a loose federation of autonomous, heterogeneous distributed database systems. In the remainder of this section, we give an overview of several concepts developed in PEER to support the rich and complex CIM application area. PEER supports autonomous cooperating agents which share and exchange information. This is achieved by a sophisticated schema derivation/integration mechanism, which supports importing remote information and restructuring and integrating it with local information (Afsarmanesh, 1993). The sharing of information among team members in PEER is negotiated to preserve the referential integrity (Tuijnman, 1993). PEER also offers support for object clusters shared in the network. An object cluster represents an entity that groups together a set of objects that are interrelated through a directed acyclic graph hierarchy. The representation, identification and boundaries of object clusters, and a linearization mechanism to transform object clusters into a linear format is fully described in (Tuijnman, 1993).

The object-oriented data model and the language of PEER is primarily based on 3DIS (Afsarmanesh, 1989). However, this model has been extensively extended to support more semantics, to represent specific concepts and entities related to manufacturing, and to support the kernel structure for the distributed architecture of PEER. The PEER data model supports the fundamental abstractions of instantiation, generalization, and aggregation. Any identifiable piece of information (both data and meta-data) is uniformly represented as an object. PEER offers a number of facilities that are useful in Concurrent Engineering for CIM environments and are briefly described in this section. The nucleus of a PEER system in a CIM cooperation network is the PEER model and language, and a layer of modeling constructs and operations that are defined specifically to represent the generic abstract data types used in engineering and industrial manufacturing. The distributed object management of PEER is handled by a layer on top of the nucleus, that supports the distributed schema management and object sharing. For the exchange of object clusters as entities between PEER agents and between a PEER agent and an application program, a linear representation is generated. The last layer of the PEER architecture is the interface that supports the access to and communication with users and application programs at the agent level.

Information management in a network of agents is supported by PEER through distributed schema management including integration and derivation of local information and information available from other PEER agents. The semantic interrelationships (loose or tight integration) among the data and knowledge of different agents are established systematically and incrementally. Several schemas coexist in every PEER agent; namely, the local, export, import, and integrated schemas. The local schema LOC in each PEER agent specifies the type structure of the information stored locally at that agent. Part of the local information can be made available to other PEER agents, by specifying one or more export schemas (EXPs), that defines a view on the local information. An export schema in PEER can restrict the exported local information available to other PEER agents. Other PEER agents can acquire these export schemas and designate them as import schemas (IMPs), thus, making the information of other PEER agents available locally. The integrated schema (INT) defines a single uniform type structure on the information and specifies the derivation and integration of the local and imported information. Since the integrated schema is local to a site, different PEER agents may establish different correspondences between their schema and other sites' schemas, thus there is no single global schema for the network.

The implementation of the PEER federated system is written in the C programming language, and runs on a network of SUN workstations. This implementation supports a distributed multinode environment and includes two tools, a schema manipulation tool (SMT) and a database browsing tool (DBT). PEER tools are developed using X-windows on SUN workstations. PEER has been implemented at the Computer Systems group of the University of Amsterdam. PEER has been used in the ESPRIT ARCHON project No. 2256 (designing an architecture for the cooperation of expert systems in a multi-agent system), and the ESPRIT CIM-PLATO project No. 2202 (supporting the coexistence of diverse CIM tools in a planning tool box), and is currently being applied to the ESPRIT CIMIS.net basic research project ECLA 004:76102 (focusing on distributed information systems for CIM), and the DIGIS project (the integration of genomic information systems).

5.2 Application of PEER to the example

In this section we briefly describe the implementation of the small example defined in Section 4, using the PEER architecure. For every agent defined for the environment, such as those represented in Figures 6, 7, and 8, a corresponding PEER agent will be defined. The LOC schema for each agent will contain all the information that is locally stored in that agent. Therefore, the "Self Model", "Attributes" and "restrictions" are represented in this schema. While the "Attributes" will be represented in PEER by static PEER objects, the "Self Model" and "restrictions" will be represented by dynamic PEER objects. Dynamic objects or behavioral objects in PEER consists of three categories. One category defines the "constraint evaluators". The "restrictions" defined for agents in Figures 6, 7 and 8 fall into the category of constraint evaluators. Constraint evaluator objects have an associated executable piece of code (method). This code will run after any relevant modification to the data of this agent. Constraint evaluators will check for the consistency of the database state. If the data is modified in a way that violates the restriction rules defined for the agent, then the database is in an inconsistent state, and the modification will not be accepted and must be redone. The "Self Model" defined for the agents in the example consists of the routines that must be executed when the information about the "External-Pre-conditions" are satisfied. The routines performing the self model produce some new results that must be stored within the agent and produce the information about the "External-Post-conditions". Using PEER the "Self model" of an agent will be represented as a dynamic object of the category "storage transaction". Storage transactions are long routines that can be executed and will produce some data to be stored in the agent.

Every agent defines a number of EXP schemas derived from the LOC schema that supports the sharing and exchange of information among the agents and consequently provides the means for cooperation among distinct agents. The EXP schemas defined by one agent represent the part of LOC information that this agent will share with other agents. Another Agent can import an EXP schema (called IMP schema there) and then integrate it with its LOC schema to create its integrated view (INT schema) of all the information that it needs to access. The information represented as "External Post-condition" in an agent will be included in an EXP schema so that other agents can access. Therefore, other agents can access these post-conditions to verify their pre-conditions. Namely, another agent's (A2) external post-condition that is included in A2's export schema will be imported by this agent (A1) as its imported (IMP) schema to become A1's external pre-condition. Every agent will create its own INT schema, through integrating its LOC schema with its IMP schemas. Thus, an agent through its INT schema has access to all the information it needs to check for its pre-conditions and to run its self model.

A PEER implementation of the ADV/ADO system is planned. This implementation will follow the guidelines described above for the definition of the agents involved and their interconnections.

6 CONCLUSION

In the present work we proposed a method based on the object-oriented ADV/ADO framework which comprises visual approch to design (based on Petri Nets and its extentions) and parametric design (based on objects). Bottom-up and top-down approaches are nested in a way that allow the designer to control and document the design process using the same graph formalism applied to artifacts (Silva J., 1992). PEER database supports the parametric and object-oriented composition of models and also may provide a basis for object-oriented design reuse.

In the future we plan to build some realistic applications in PEER and to combine in the same design environment, PEER database, a PFS/MFG object-oriented simulator and a software agent to control and generate queries to PEER according to the needs of the design process.

7 REFERENCES

Afsarmanesh H. and McLeod D. (1989) The 3DIS: An Extensible Object-Oriented Information Management Environment, ACM Transaction on Information Systems, 7:339–377

Afsarmanesh, H., Tuijnman, F., Wiedijk, M. and Hertzberger, L.O. (1993) Distributed Schema Management in a Cooperation Network of Autonomous Agents. In Proceedings of the 4th IEEE International Conference on "Database and Expert Systems Applications DEXA'93", Lect. Notes in Computer Science (LNCS) 720, pages 565–576, Springer Verlag.

Afsarmanesh, H., Wiedijk, M. and Hertzberger, L.O. (1994) Flexible and Dynamic Integration of Multiple Information Bases. In Proceedings of the 5th IEEE International Conference on "Database and Expert Systems Applications DEXA'94", Athens, Greece, Lect. Notes in Computer Science (LNCS) 856, pages 744–753. Springer Verlag.

Alencar, P.S., Carneiro-Coffin, L.M., Cowan, D.D., Lucena, C.J.P. (1994) The Semantics of Abstract Data Views: A Design Concept to Support Reuse-in-the-Large, In Procedings of the Coloquium on Object-Oriented in Databases and Software Engineering (to appear), Kluwer Press.

Bruno, G., Balsamo, A. (1986) Petri Net-Based Object-Oriented Modelling of Distributed Systems, OOPSLA'86 Proc.

DiCesare, F., Mu der Jeng (1993) Synthesis for Manufacturing Systems Integration, in Practice of Petri Nets in Manufacturing, DiCesare, T., Harhalakis, G., Proth, J.M., Silva, M., Vernadat, G.B., (eds) Chapman & Hall.

Cowan, D.D., Ierusalimschy, R., Lucena, C.J.P. (1993) Abstract Data Views, Structured Programming, 14 (1), 1-13.

Cowan, D.D., Lucena, C.J.P. (1993a) Abstract Data Views: A Model Interconnection Concept to Enhance Design for Reusability, Technical Report 93-52, Cjmputer Science Department and Computer System Group, University of Waterloo.

Cowan, D.D., Lucena, C.J.P. (1995) An Specification Concept to Enhance Design for Reuse, to appear in IEEE Transactions on Software Eng.

Fields, B., Harrison, M., Wright, P. (1993) From Informal Requirements to Agent-Based Specification: An Aircraft Warning Case Study, in, Procc. of the Workshop on Specification of Behavioral Semantics in Object-Oriented Information Modelling, IIM (Inst. of Inf. Modelling), Robert Morris College.

Kagohara, M., Toledo, C., Silva, J.R., Miyagi, P.E. (1994) Automatic Generation of Control

Programs for Manufacturing Cells, IFIP Transactions: Applications in Technology, B-19, pg. 335-343.

Lucena, C.J.P, Silva, J.R. et al. (1989) The Specification of a Knowledge Based Environment for the Design of Production Systems, 6th. Sym. on Information Control Problems in Manufacturing Technology, INCOM, Madrid.

Marca, D. (1988) DADT: Structured Analysis and Design Techinque, McGraw Hill.

Miyagi, P.E, Hasegawa, K., Takahashi, K. (1988) A programming Language for Discrete Event Production Systems Based on Production flow Schema and Mark Flow Graphs, Trans. of the Soc. of Instrument and Control Engineers, vol 24, no. 2.

Proth, J. M. (1993) Principles of System Modeling, in Practice of Petri Nets in Manufacturing, DiCesare, T., Harhalakis, G., Proth, J.M., Silva, M., Vernadat, G.B., (eds) Chapman & Hall.

Ranta, J., Tchijov, I. (1990) Economics and Success Factors of Flexible Manufacturing Systems: The Conventional Explanation Revisited, IJFMS, 2, pg. 142-154.

Silva, J.R. (1992) A Formalization to the Design Process Based on Theory of Metaphors: Its Application to Discrete Events System Automation, Ph.D. thesis, (in Portuguese) University of São, Brazil.

Silva, J.R., Pessoa, F.J.B. (1992a) Análise Semi-Automatica de Mark Flow Graphs, Ibero-American Workshop in Autonomous Systems Robotics and CIM, Lisbon.

Silva, J.R., Cowan, D.D., Lucena, C.J.P (1994) Case-Based Approach to the Design of Flexible Manufacturing Systems, (to apear).

Silva, M. (1993) Introducing Petri Nets, in Practice of Petri Nets in Manufacturing, DiCesare, T., Harhalakis, G., Proth, J.M., Silva, M., Vernadat, G.B., (eds) Chapman & Hall.

Sommerville, I. (1992) Software Engineering, Addison-Wesley Pub. Co.

Tempelmeier, H., Kuhn, H. (1993) Flexible Manufacturing Systems: Decision Support for Design and Operation, Wiley Series in System Engineering, John Wiley & Sons.

Tomiyama, T. et. al. (1992) Systematizing Design Knowledge for Intelligent CAD Systems, Human Aspects in Computer Integrated Manufacturing, G.J. Olling and F. Kimura (eds.), Elsevier Science Publishers.

Tuijnman, F., Afsarmanesh, H. (1993) Management of Shared Data inFederated Cooperative PEER Environment, Jour. Intelligent and Cooperative Inf. Sys. (IJICIS).

10

Integration of Object Oriented Programming and Petri Nets for modelling and Supervision of FMS/FAS

J. Barata; L.M. Camarinha-Matos
Universidade Nova de Lisboa - Dep. de Engenharia Electrotécnica
Quinta da Torre, 2825 - Monte Caparica, Portugal
Tel +351-1-3500224 / 2953213   Fax +351-1-2957786
E-mail: {jab,cam}@uninova.pt

W. Colombo; R. Carelli
Instituto de Automática - Universidad Nacional de San Juan
Av. San Martin-oeste 1109, 5400 San Juan, Argentina
Tel +54-64-213303   Fax +54-64-213672
E-mail: rcarelli@inaut.edu.ar

Abstract

An hybrid approach combining frames/object oriented and Petri Nets paradigms for modelling, evaluating and supervising FMS/FAS is presented. Application examples in the context of a pilot FMS/FAS system are discussed.

1. INTRODUCTION

The aim of this paper is to provide a hybrid approach for dealing with a class of control problems occurring in manufacturing flow control. In modern production systems, the introduction of automated flexible manufacturing and assembly systems (FMS/FAS) has increased the need for efficient production planning and control techniques. FMS/FAS can produce multiple types and variants of products using various resources, such as robots, multipurpose machines, etc. While the increased flexibility of a FMS/FAS provides a greater number of choices of resources and routings, it imposes a challenging problem; i.e. the allocation of given resources to different processes required in making each product, the planning of the activities to accomplish the best efficiency and the actual link between a global control program and the real controllers of each manufacturing resource.

Traditionally, the routing of a part to complete a sequence of required processes is considered planning, and the assignement of resources according to the determined routing is considered scheduling. In a FMS/FAS, the planning and the scheduling should be collectively carried out to take advantage of the flexibility of the system. Since a machine in the FMS/FAS can be used for multiple jobs and there are choices of resources to be used, the assignement of resources

-scheduling- must be considered when the routing of a part is planned [1].

In this paper we present a hybrid platform to assist in designing, modelling and analysis of FMS/FAS, which uses a Petri Net formulation and frame/object oriented modelling. Petri Nets constitute a mathematical and graphical tool for modelling and analysis of discrete-event systems, which allow to easily represent parallelism and synchronization of activities [2]. Frames/Object oriented programming is introduced to model the structural and operational facets of the manufacturing resources and to establish a link between the model and the local controllers. The proposed platform allows to link object oriented models of FMS/FAS developed by UNINOVA [3-5]with a Petri Net based simulation and evaluation software (SIMRdP) developed by UNSJ [6, 7] The structure is designed to be interactive with the real world, to perform monitoring and control tasks on the actual FMS/FAS system.

2. GENERAL ARCHITECTURE

Figure 1 shows the general structure of the proposed system.

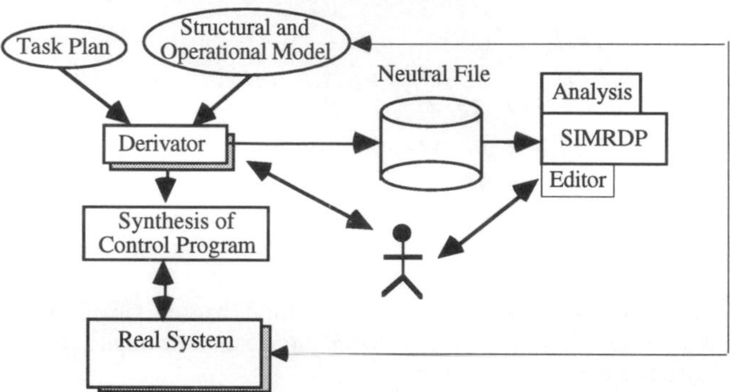

Figure 1 - System Architecture.

The system consists of the following parts:

1. A model for derivation of a Petri Net from a task plan. This plan is a graph, represented in frames, that describes the production tasks at the operational level.
2. A module for creation of structural and operational model of the FMS/FAS, based on a frame/object oriented representation and a subsystem for derivation of Petri Net models of the FMS/FAS system.
3. A module for editing, evaluation (simulation, structural and quantitative analysis) of FMS/FAS system modelled by means of Petri Nets.
4. A module to synthetize a control supervision program from the analyzed Petri Net.
5. A neutral file for linking the above mentioned modules. This data structure is important as the various modules are separetely developed by the two universities.

3. STRUCTURAL AND OPERATIONAL MODELS

In our approach the frame representation paradigm, with some facets of OOP, is adopted to model the structure and operational behaviour of the manufacturing system.

3.1 Structural aspects

The various examples to be discussed in order to introduce the main modelling concepts use a cell as the basic modelling unit. A cell is a composite entity that is capable of making some transformation, movement or storage related to some product or part. In structural terms, each cell has components to support the input of parts, an agent to perform the transforming actions and components to support the output of products/processed parts.

The input, output and agent elements will be supported by manufacturing components. Some components can only support one function but there are others components which can support more than one function. Components adapt themselves to the roles they can perform. Some components are more adaptable than others. For instance, a Conveyor belt is very flexible because it can perform an input, output or agent role, but a CNC machine only can play an agent role.

The generic cell concept can be specialised by activity. There can be cells specialised in assembly, painting, welding, storage, machining, transportation, etc. A shop floor is just a set of specialised cells.

Metaknowledge can be associated to each specialised cell to represent the specificities of its application domain. For each domain, the specific cell has the same structure as the generalised Cell concept (Input Agent, Processing Agent, Output Agent) but the domain and cardinality of the implementing components is different in each specialisation (Figure 2). For example, in a Painting or Welding Cell, a vibrator feeder is not a valid Input item, but this component is valid in an Assembly Cell. The Metaknowledge seems to be a very important element at the configuration phase, assuring the validity of cells.

```
FRAME  CELL
name:
base_coordination_system:
processable_products:
input_parts:
connected from:
processor:
connected to:

FRAME  ASSEMBLY-CELL
is-a: CELL
val-inp-ag: vibratory_feeder,
            buffer,
            gravitic_feeder,
            Index_Table, agv,
            conveyor
val-out-ag: conveyor, agv, buffer,
            index_table
val-proc-ag: robot
```

```
FRAME  ROBOT_COMPONENT
is-a: manufacturing_component
Base_coordinate_system:
Controlled by:
Applications: assembly, gluing, ..
DOF: 6
Working_area:
Load:
Repeatability:
Current_position:
Cost:
Cycle_Time:
Next_maintenance:
N_working_hours:
Weight:
Max_speed_by_axes:
```

Figure 2 - Example concepts of cell, assembly cell and component.

At this stage it is convenient to clarify the concepts of agent, input and output, and their relationships to the components/manufacturing resources.

Components are entities which participates in the productive process with a specific function and can be controlled by a computational entity. Components models are context independent descriptions of their static and dynamic characteristics. A robot component model (Figure 2), for instance, includes all the characteristics which completely characterize its structural and dynamic aspects.

A robot **agent** is a model of a robot and associated resources, like tools or auxiliary sensors, when inserted in a particular context. A robot can play different **roles** (Figure 4) in different **contexts**. The (expected) behaviour of a robot in an Assembly context is different from its behaviour in a spot welding context.

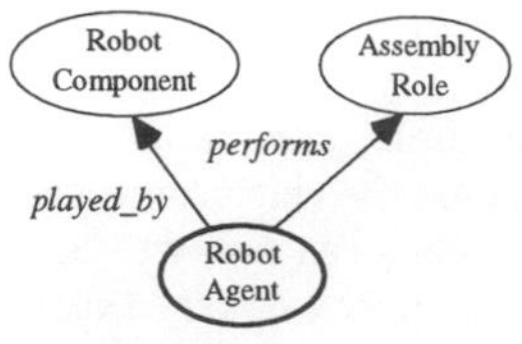

Figure 3 Structure of a robot agent.

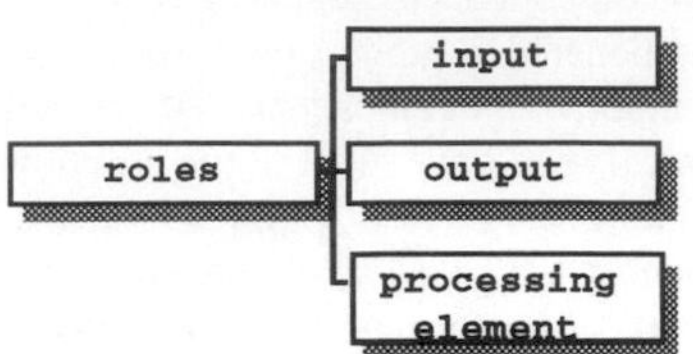

Figure 4 Role taxonomy: main level.

On the other hand, when a robot is performing a given role, it may resort to auxiliary resources, like tools, sensors, buffers, etc., that extend the robot's functionality in order to fulfil the functionality required by this role. A robot agent is, therefore, a model of the robot when playing a particular role and extended by selected attributes inherited from the auxiliary resources.

The entity that effectively participates as an assembly robot, for instance, is one which has those characteristics from the robot component model to perform the assembly role.

The agent entity ASSEMBLY_ROBOT is a structure which is supported by two relations: *performs* and *played_by* (Figure 3). The relation *performs* assures the inheritance of the role characteristics to the structure while *played_by* assures the inheritance of those agent relevant aspects, from the component. *Main_attributes* and *component_attrib* are attributes to be used by *played_by* and *performs* relations.

```
FRAME  ASSEMBLY_ROBOT
is-a: agent
performs: ASSEMBLY_ROLE
played by: ROBOT_COMPONENT
main_attributes:force_sensor,
                current_tool,
                available_tools,
               available_resources
component_attrib:Base_coord_system,
                 Controlled_by,
                 Working_area,load,
                 Current_position
```

```
FRAME  AG_ROBOT_ASSEMBLY_ROLE
is-a: role
tools_domain:(grippers,
              screwdriver)
aux_res_domain: (buffers)
force sensor:
current tool:
available_tools: gr1, gr2, sd2
aux_resource: buf1, buf2
assembly_device: fixture1
```

Figure 5 Example concept of an agent.

The slots, *tools_domain* and *aux_res_domain* represent domain-knowledge that is important during configuration time. The slot *current_tool* is a relation that associates the main player of this role (robot component) to a particular tool. By the "inclusion" restriction, only tool_operations will be inherited by the ag_robot_assembly_role. *Assembly_device* is an attribute describing where assembly operations are really done. *Fixture1* is an instance of a component specialised in holding parts.

```
RELATION  PERFORMS
type: intransitive
inherit_slot: main_attributes
inverse_relation: performed_by

RELATION  CURRENT_TOOL
type: intransitive
inherit_slot: tool_operations
inverse_relation: attached_to
```

```
RELATION  PLAYED_BY
is-a: relation
type: intransitive
inherit_slot: component_attributes
inverse_relation: play
```

Figure 6 Definition of relations *performs ,played_by* and *current_tool.*

A cell is made of entities that are playing different roles.

This modular approach to cell representation facilitates the creation of complex systems by simple "concatenation" of cells. A particular manufacturing unit is made of several subsystems (Transportation Cells, Painting Cell, Assembly Cell, ...). A manufacturing unit could be modelled by a FMS/FAS entity, which has access to all characteristics and functionalities of the subsystems involved in the Unit.

The frame representation paradigm, specially when it allows the definition of new relations, is quite adequate to model this kind of structures.

3.2 Dynamic or Operational Aspects

Dynamic aspects are related to the components internal state changes. The dynamism presented by components is achieved through controller actions. Every component with dynamics must have a controller associated to it. The way the model reflects physical changes on the component and the way physical component reflects model changes is the most important point when discussing dynamic aspects.

Dynamic aspects can also be discussed with two different views: (1) considering the components as isolated entities or (2) considering complex structures, like cells, made of components. In the first view the key point is how components are actuated, without any concerns about their interrelationships. In the second view, aspects related to synchronisation are the most important ones (this will be analysed in the PETRI NETs chapter). In this point the concern is with the first view.

```
RELATION  CONTROLLED_BY
is-a: relation
type: intransitive
inherits: inclusion (move_wc,
                     move_jc,
                     hardhome,
                     acceleration,
                     speed)
inverse_relation: controls
inverse_relation: controls
```

```
FRAME  ROBOT_CTRL_COMPONENT
is-a: controller
move_wc:method move_wc_fn(x,y,z,q)
move_jc: method
          move_jc_fn(m1,m2,m3,m4)
hardhome: method hardhome_fn
acceleration:demon if write
               accel_dem
speed: demon if write speed_dem
input: byte demon if needed
input_dem
output:byte demon if write
output_dem
```

Figure 7 Model of a component and its controller.

Every component model with dynamic behaviour should have a controller model. This model should be like an image of the real controller. Using a frame oriented paradigm, the controllers functionalities could be defined by methods or demons. In this way most of the controller's model is a list of methods, a method for each functionality.

A component is related to its controller by a *controlled_by* relation while the controller relates to its component by a **controls** relation.One of the most important points in this discussion is the way a controller model is connected to the local controller. This connection is sometimes not easy because it involves the cooperation of two different computational worlds: the computational world where the model runs and the real controller. To make things even more difficult, sometimes, real controllers have closed architectures. From our experience, a lot of effort has usually to be put in trying to open real controllers architecture, and implies the production of an interpreter that runs on it. This interpreter accepts commands from an image that runs in the modelling world.

The methods of a Controller model implement the actions that are needed to send the right commands to the real controller. The real controller image should be developed using a client-server approach. In this way, implementation methods can ask this server to perform the needed actions. These methods hide the underlying hardware structure from the application, i.e., any application using a robot component doesn't need to know anything about the real robot controller and its image or server. The applications only know what functionalities are provided by the robot component model. This approach could be very suitable to integrate existing controllers, making the integration of legacy systems an easier task.

Figure 8 illustrates this approach that was followed in the Uninova's FMS/FAS system (NovaFlex).

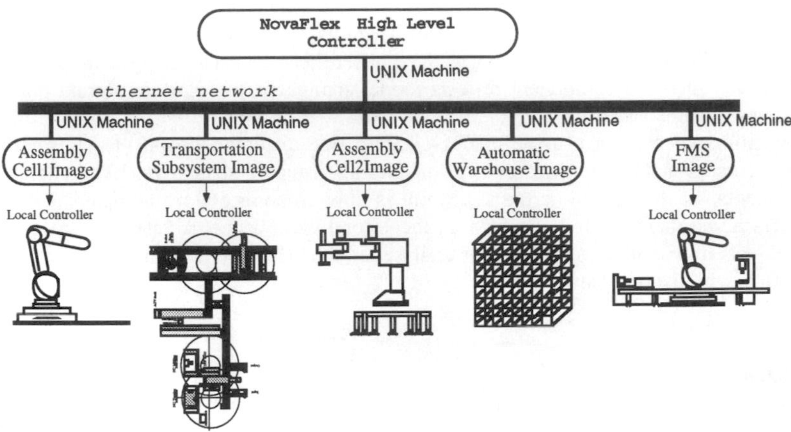

Figure 8 - NovaFlex physical infrastructure.

The Flexible Manufacturing and Assembly System -- NovaFlex -- installed at the Center for Intelligent Robotics (CRI) of UNINOVA was conceived as a demonstration unit able to handle a set of typical activities of a Computer Integrated Manufacturing (CIM) system [4].

Besides the machining and assembly subsystems, the Pilot Unit includes a storage component, an input section for raw materials, a delivery section for finished products and a transportation subsystem that links all the other components. The transportation medium is a pallet-based conveyor belt. Each pallet can be adapted to transport different kinds of parts and products.

As one of the basic design goals, the system was required not to be restricted to a particular type of product. The objective was to build a relatively generic infrastructure, that could be adaptable to a range of products with minimal setup effort. The requirement was for a flexible infrastructure, with a representative set of manufacturing resources, and not a special purpose system.

Another very important aspect is the possibility of different groups being simultaneously using different subsystems of the Unit for separate experiments. As a matter of fact, this situation is expected to be the most common practice during the system's life time.

These requirements led to an architecture in which NOVAFLEX can be operated either as an integrated FMS/FAS system or as a set of isolated subsystems (machining, assembly, transportation and storage, etc.). This last aspect has particular consequences on the design of the control architecture.

Therefore, the need to support these different research areas implied the design of a flexible architecture, from the topology to the control points of view. An easy reconfiguration of its operating mode is an important requirement to support concurrent research activities.

4. FMS/FAS EVALUATION USING PETRI NETS

In this chapter it is proposed to use Temporized Petri Nets (TPNs), as a discrete event specification tool for modelling the FMS/FAS under study. Petri Nets (PN) have been designed specifically to model FMS/FAS [8-10] They provide suitable models due to the following reasons:

- PNs capture the precedence relations and structural interactions of unpredictable, concurrent, and asynchronous events. In addition, their graphical nature helps to visualize such complex systems.
- Deadlock, conflicts, and buffer sizes can be modelled easily and efficiently.
- PNs models represent a hierarchical modelling tool with a well-developed mathematical and graphical foundation.
- PNs allow the use of different abstraction level models and the use of powerful modelling methodologies: refinement and modular composition, each of them with its corresponding validation analysis.

It is very important to take into account that the main functions of production/assembly systems, be them flexible or not, are to input raw material, to perform a certain number of transformation tasks, assemble and disassemble of initial parts and, finally, to output finished parts. Therefore, the designer has to identify the input and output sequences of material parts, at the same time all the elementary operations which have to be applied on these parts, and the main characteristics of the resources (turning, milling machines, manipulators, AGVs, etc.) involved in each of these operations.

According to the above considerations, the Petri Net based simulator -SIMRdP- has a logical model where:

- One place is associated to one state of the produced part in its operating sequence, or the state of a resource (iddle or busy by a part).

- One transition is associated to only one operation on the produced part, or to the assembly operation of several parts.
- One token is associated to one material part, or information about the resources of the FMS.
- To each place of the net, a capacity place or capacity monitor is associated to indicate the maximum number of tokens this place may have [11].This capacity is introduced to indicate that a buffer, a storage area, or another resource can only contain a limited number of parts at one time.

In addition, all the resources and operations verify the following set of constraints:

- The operations are completed in finite time, and they can be decoupled from the time point of view.
- One part is submitted to only one operation at a time, and each resource can perform only one task at a time.
- One separate part can be submitted only to transformational or informational functions. A transformational function consists in modifying the physical attributes of the part (shape, constitution, surface, etc.). They are the machining functions: turning, milling, manipulation, and conditioning functions: termical treatment, painting, washing, etc.. An informational function consists in verifying that the operations have been accomplished correctly.

4.1 Functional Analysis

After the modelling, the designer must develop a complete study and proof for the correct behaviour of the modelled system specifications. In order to derive the functional (i.e., logical) properties, the structure of the Petri Net model is considered for the analysis, which is called qualitative validation in the literature.

For developing the qualitative validation, it is possible to use some of the methods which are extensively described in the literature of Petri Nets [8, 12]. According to the structural characteristics of the Petri Nets used in this approach, it has been developed a software tool [13],which includes the following Petri Net analysis methods: the Coverability Graph method by means of simulation of the Petri Net token-game [12] and Structural Analysis for obtaining the p- and t-invariant relationships of the net [12].

After performing the simulation of the net evolution, and obtaining the invariants relationships, very important conclusions on the net behaviour, concerning the dynamics of the modelled system can be stated, such as: boundedness, liveness, and reversibility of the net; reachability of some special state -marking- of the net; mutual exclusion among marking of places; capacity of storage zones and other modelled resources; the development of managing strategies for the optimal firing of transitions -optimal sequence of modelled operations- and refining of the modelled control processes, by adding, for instance, a new resource.

Only if the Petri Net model has good behaviour properties, like the ones above described, the designer can reach the next stage of the design methodology, using a temporized Petri Net model. This new stage corresponds to the performance validation of the modelled system.

SIMRdP enables to work with the following kinds of Petri Net models: ordinary Petri Nets, generalized Petri Nets, Petri Nets with predicates associated to their transitions and temporized Petri Nets with time associated to their transitions.

Remark: For modelling with each one of the above named kinds of Petri Nets, it is necessary to take into account the use of monitor places, when the net has places with capacity [11].

4.2 Performance analysis of FMS/FAS modelled through Temporized Petri Nets

4.2.1 Basic concepts

A Flexible Manufacturing/Assembly System (FMS/FAS) -whether simple or complex- includes any raw materials, tools, operators, and control policies for the operation of the equipments. At the sector level, it also includes flow of material and information through a collection of tools. It can even extend to the procurement, production and distribution networks of suppliers, plants, and distributors. Using computer simulation for modeling and analyzing such systems helps the designers to predict and improve a system's performance, as measured by such elements as capacity, cycle time, inventory, utilization, service level, and costs.

Once a model of the system is developed and running, the simulation qualitative analysis tool come to play. Information about the dynamics and kinematics of the involved equipment of the FMS/FAS, the product routing, and the production and parts-availability schedules must be inputs to the software simulator. Using the graphical user interface of the simulator, these inputs could be described and modified easily without programming.

Information such as the production capacity of machines on the FMS/FAS, the duration of various operations performed by different machines on different products, buffer sizes, and the staffing must be processed before they entry to the simulator, in order to allow their adaptability to the signal handling of the simulator.

Key outputs of SIMRdP are the system throughput (the quantity of products produced in each time period), a work-in-process inventory (the quantity of products being worked on in the system), the cycle time (the time from release of jobs into the system to completion), and machine utilization (the proportion of time the machines work on products). These results can be displayed as bar or line charts. By examining the above results, the designers are able to readily identify the system bottlenecks, and to suggest answers to how to alleviate the problems that may arise.

Since a simulation imitates the behaviour of a system as it evolves over time, basic to the approach of this simulation is the building of a model that highlights the vital characteristics of the system. Good models are needed to capture the characteristics of the Flexible Manufacturing Systems [10], namely: concurrency or parallelism, asynchronous operations, deadlock, conflicts and event driven.

4.2.2 Performance Indexes defined for the Flexible Manufacturing Systems

Taking into account the statistical information obtained from the simulation of the temporized Petri Nets, new performance indexes can be defined. Nevertheless, this new information must be refered to the dynamical behaviour of the modelled systems and their functioning specifications.

At this stage, performance evaluation in FMS/FAS allows to answer [8]., how many machines and/or automatic guided vehicles (AGVs),which transport topology and routing strategy are best suited to obtain, among others: small makespan and/or more balanced flows, optimal work in progress, minimal bottleneck situations, minimal death time for resources,etc.

According to the above considerations, the TPN model proposed in this work and the algorithm for its simulation, the following performance indexes for Flexible Manufacturing/Assembly Systems can be introduced [7]:

1. *Production period for a part (PPfP).* The total time necessary for the FMS to produce a part, calculated by taking into account the duration of all operations performed over the raw material taken from storages, over the intermediate manufactured parts, and the final

processed part stored into the product storages. From the TPN point of view, this index can be reached by computing the *Cycle time of the net* corresponding to the evolution from an initial marking characterized by: raw material storage busy, all resources of the system free, and product storages free, and another marking corresponding to a state having the product storage busy by the first manufactured part.

2. *Production period for a batch (PPfB).* The total time necessary for the FMS to produce a batch of parts, calculated by taking into account the duration of all operations performed n the raw material picked from the storages, over the intermediate manufactured parts, and the final processed parts deposited into the product storages. From TPN point of view, this index can be reached by computing the *Cycle time of the net* corresponding to the evolution from an initial marking characterized by: raw material storage busy, all resources of the system free, and product storages free, and another marking corresponding to a state having the product storage busy by all parts composing the batch.
3. *Percentual use of a resource related to a part (PURP)*: By analyzing the Gantt diagram of the temporized evolution of the TPN, and taken into account the production cycle for a part (PPfP), and all operations performed by the resource (modelled through transitions, this index is defined as:

$$PURP = \{[\sum_i \sigma_i^r . \theta(t_i^r)] / PPfP\}.100 \qquad (1)$$

where

σ_i^r it is the number of occurrence of transition t_i^r, which models the operation op_i^r of the resource r over the part.

$\sum_i$ this summation takes into account all the transitions which model the operations of the resource under study.

4. *Percentual use of a resource related to a batch (PURB)*: By analyzing the Gantt diagram of the TPN temporized evolution, and taken into account the production cycle for a batch (PPfB), and all operations op_i^r performed by the resource r (modelled through transitions t_i^r), this index is defined as:

$$PURP = \{[\sum_i \sigma_i^r . \theta(t_i^r)] / PPfB\}.100 \qquad (2)$$

where

σ_i^r it is the number of occurrence of transition t_i^r, which models the operation op_i^r of the resource r over the batch

$\sum_i$ this summation takes into account all the transitions which model the operations of the resource under study.

5. *Manufactured Parts per Time Units(MPpTU)*: Following the definition of "Percentual use of a resource related to a batch (PURB)", and considering the number of parts to be processed in a batch (b), the index is defined as:

$$MPpTU = b / PPfB \qquad (3)$$

By considering the application of the above defined analysis methodology, the designers can now draw major conclusions about the functioning of the modelled Flexible Manufacturing Systems. The optimization of the systems will depend on the final decisions imposed by cost factors, which can be derived from the above presented performance index. For example, if additional capital expenditures are made, what is the time frame for the increased throughput to pay back the investment? Cost analysis will be divided into three activities: assigning cost attributes to modelled resources, defining new performance-cost indexes, and analyzing the results after running one or more simulations.

5. SYNTHESIS OF CONTROL PROGRAMS

In previous chapters we discussed modelling aspects in terms of the functionality of the FMS/FAS. Once a system is evaluated and a good functional model is obtained, we have a platform on top of which we can run application programs.

Such programs or plans can be generated either manualy or resorting to an automatic or interactive planner. Independently of the process used, lets now discuss some aspects related to task plan representation.

5.1 Task Representation

A task plan can be represented by an hierarchical structure, following a NOAH - like model. High level operators are expanded into lower level operator [14]. This operator expansion ends when the lowest (primitive) operators are reached. Each primitive operator is performed by a resource agent (physical component). Grasp and ungrasp operations, for instance, will be supported by current tool controller.

In order to determine precedence between operations (precedence graph) two phases are normally considered:

1. product related constraints (determined during process planning)
2. manufacturing system constraints (determined during scheduling)

Once such ordering of actions is determined, we'll have a precedence graph for each level of the hierarchical representation of the plan.

Figure 9 illustrates a partial ordering for an intermediate level.

From each level a Petri Net which represents the assembly operations to be carried out, can be directly derived from the precedence graph. Figure 10 illustrates a Petri Net which represents the same partial assembly plan as in figure 9.

Each graph node can be expanded to another graph, where nodes correspond to next lower level operators. The same kind of expansion can be applied to the Petri Net model, where each place can be expanded to a new Petri Net

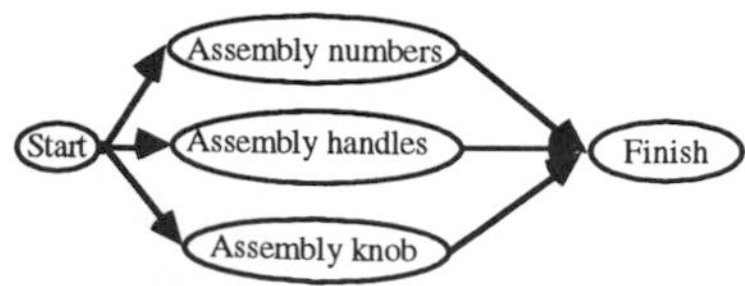

Figure 9 - Graph Plan.

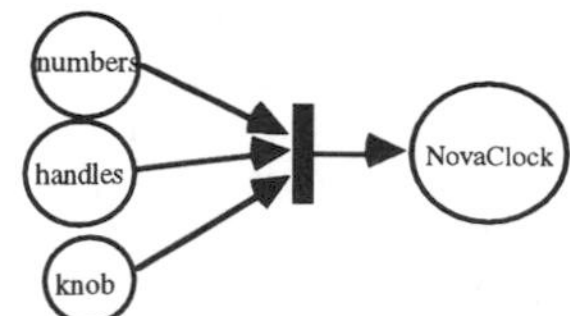

Figure 10 - Petri Net

5.2 Controller Synthesis

As mentioned before, Petri Nets are an important tool to model and evaluate the structure and behaviour of controllers and application programs in a manufacturing environment. Complex system dynamics can be described and analysed in a structured way (mathematical methods).The benefits of using a high level modelling tool, like a Temporized Petri Net (TPN) or a Predicate Transition Net (PTN), shouldn't end in their descriptive characteristics but there should be a direct connection between the description and the real controllers. This means that besides supporting the analysis and evaluation, the model could also drive directly its associated local controller. To assure this connection two different approaches could be followed: (1) using PTN to directly program the local controller, which implies a support by its manufacturer or (2) using a PTN translator which converts PTN to the language of the local controller.The first approach is unrealistic, at current stage, because the concept of PTN is not well disseminated among controllers' manufacturers, which makes the second approach a better one. The PTN description should be compiled in order to generate a program which interacts with the real system in the way described by the PTN.

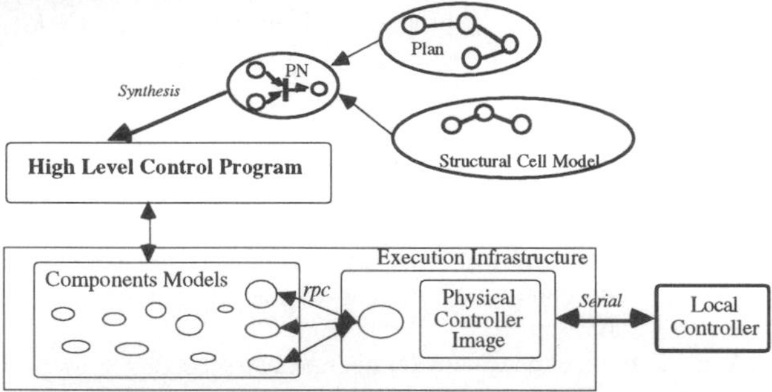

Figure 11 - Interactions between HLCP and Execution Infrastructure

The application program or High Level Control Program (HLCP) is generated from a PTN which describes components behaviour and interactions (Figure 11). The HLCP interacts with the execution infrastructure, mainly with the components models. These models are described using an Object/Frame paradigm. As mentioned before, the components operational behaviour is implemented by methods, which interact with the component's local controller through a server which supports an image of the local controller functionalities.

It should be noted that a Component Model can interact with more than one server, depending on the number of needed controllers to control the components being used. For instance there can be different servers to control de robot, the gripper, gravity feeders, etc.

In order to generate the HLCP from a PTN some considerations should be made about PTN. Components actions can only be done in places with marks. Predicates associated to transitions specify conditions. In order to match places to components actions, places' names include components' name and action name separated by an underscore (for instance, the name for the place that describes the grasp action of a robot is named **robot_grasp**).

The HLCP program generated is like a simulator of the PTN being modelled. Different PTNs may have the same kind of simulator, differing only in which order transitions will be fired and which actions will be done. Program generation module was developed using Prolog, and the generated program is also described in Prolog with a frames extension developed by UNL - Golog [15].

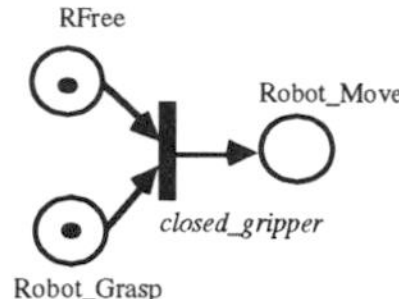

Figure 12 A Petri Net example.

The generated program obtained from the Petri Net described in Figure 12 can be seen below. Each section of the generated program is described.The first section is concerned about place definition; every place is defined by an object/frame whose main attribute is the slot mark to store the place's mark value. During this phase the program that contains the components model is consulted.

```
:- consult('models.pl').
:- new_frame(places), new_slot(places, mark).
:- new_frame('RFree'), new_slot('RFree', isa, places), new_slot('RFree', mark, 1).
:- new_frame(robot_grasp),new_slot(robot_grasp,isa,places), new_slot(robot_grasp,mark,1).
:- new_frame(robot_move),new_slot(robot_move,isa,places), new_slot(robot_move,mark,0).
```

After this, transitions are defined. Each transition is defined by checking its enabling condition. When this occurs, input places are updated and transition fires with output place updating and the corresponding method activation: *call_method(robot, move, [true])*. This method's code will send a message to the server which will react by sending the command "move" to the robot.

```
t1 :- get_value('RFree', mark, X0), X0 > 0,
     get_value(robot_grasp, mark, X1), X1 > 0,
     NVal0 is X0 - 1, new_value('RFree', mark, NVal0),
     NVal1 is X1 - 1, new_value(robot_grasp, mark, NVal1),
     call_method(robot, move, [true]),
     get_value(robot_move, mark, VOal0), NVOal0 is VOal0 + 1,
     new_value(robot_move, mark, NVOal0).
```

The main program consists of a forever cycle that continuously apply the existing transition names and randomly choose one which will be checked for its enabling condition.

```
rep_run([]).
rep_run(List) :- length(List, Tam), Pos is ip(rand(Tam)), position(Pos, List, Tr),
                 remove(Tr, List, RList), call(Tr, Success), !, fail == Success, rep_run(RList).
run :- repeat, rep_run([t1]).
```

This generated program runs with a similar behaviour as the PTN shown in figure 12.

6. CONCLUSIONS

In this paper we have presented a joint research work between UNL/UNINOVA and UNSJ, aimed at developing an hybrid platform for designing, evaluating and controlling Flexible Manufacturing/Assembly Systems. At present, the individual modules have been developed and separately tested and the main concepts related to their linking/integration have been designed.

A first applicative case was made for a subset of the NovaFlex pilot FMS/FAS system. Next phase will be the integration of individual parts through normalized models representation.

ACKNOWLEDGEMENTS

This work has been partly funded by the European Commission (ECLA Flexsys Project), JNICT (SARPIC Project), CYTED-D Programme and CONICET (Arg).

REFERENCES

1. Lee, D.Y. and F. DiCesare, *Scheduling Flexible Manufacturing Systems using Petri Nets and Heuristic Search,* in *IEEE Transactions on Robotics and Automation*199
2. Silva, M., *Las Redes de Petri en la Automática y la Informática.* 1985, Madrid - Espana: Editorial A.C.
3. Barata, J., L.M. Camarinha-Matos, and J.F.R. Chavarría, *Modelling, Dynamic Persistence and Active Images for Manufacturing Processes,* in *Studies in Informatics and Control*1994, p. 173-183.
4. Barata, J. and L.M. Camarinha-Matos, *Development of a FMS/FAS System - The CRI's Pilot Unit,* in *Studies in Informatics and Control*1994, p. 231-239.
5. Barata, J. and L.M. Camarinha-Matos. *Dynamic Behaviour Objects in Modelling Manufacturing Processes.* in *CAPE'95 - The Fifth International Conference on Computer Applications in Production and Engineering.* 1995. Beijing - China:
6. Tello, R. and A. Martínez, *Software for Analysis, Simulation and Validation of FMS,* E.E. Graduation Thesis, Universidad Nacional de San Juan, 1994
7. Colombo, A.W., *Modelling and Analysis of Flexible Production Systems,* MsC Thesis, Universidad Nacional de San Juan, 1994
8. Silva, M. and R. Vallete, *Petri Nets and Flexible Manufacturing,* in *Advances in Petri Nets.* 1989, p. 374-417.
9. Zhou, M., F. DiCesare, and A. Desrochers, *A Hybrid Methodology for Synthesis of Petri Net Modells for Manufacturing Systems,* in *IEEE Transactions on Robotics and Automation*1992, p. 350-361.
10. David, R. and H. Alla, *Petri Nets for Modelling of Dynamic Systems - A Survey,* in *Automatica - IFAC*1994, p. 175-202.
11. Ezpeleta, J. and J. Martínez. *Petri Nets as Specification Language for Manufacturing Systems.* in *IMACS World Congress on Computation and Applied Mathematics.* 1991. Dublin - Ireland:
12. Murata, T., *Petri Nets: Properties, Analysis and Applications,* in *Proceedings of the IEEE*1989, p. 541-580.
13. Colombo, A.W., *et al. Simulador de Sistemas Flexibles de Manufactura usando Redes de Petri Temporizadas.* in *6º Congreso Latino-Americano de Control Automático.* 1994. Rio de Janeiro - Brasil:
14. Camarinha-Matos, L.M., *et al., Interactive Assembly Task Planning and Execution Supervision,* in *Studies in Informatics and Control*1994, p. 185-193.
15. Lopes, L.S., *GOLOG - Um gestor de Objectos em Prolog* - 1993, DEE - UNLisboa,

BIOGRAPHY

Dr. Luis M. Camarinha-Matos received his Computer Engineering degree and PhD on Computer Science, topic Robotics and CIM, from the New University of Lisbon. Currently he is auxiliar professor (eq. associate professor) at the Electrical Engineering Department of the New University of Lisbon and leads the group of Robotic Systems and CIM of the Uninova's Center for Intelligent Robotics. His main research interests are: CIM systems integration, Intelligent Manufacturing Systems, Machine Learning in Robotics.
Dr. Ricardo Carelli is auxiliar professor (eq. associate professor) at the Electrical Engineering Department of the University of San Juan. His main research interests are: CIM systems , Manufacturing Systems.
Walter Colombo is a Phd student. His main research interests are: CIM systems, Manufacturing Systems.
José Barata received his Computer Engineering degree on Computer Science, from the New University of Lisbon. Currently he is junior assistant at the Electrical Engineering Department of the New University of Lisbon and belongs to the group of Robotic Systems and CIM of the Uninova's Center for Intelligent Robotics. His main research interests are: CIM systems integration, Intelligent Manufacturing Systems, Machine Learning in Robotics.

11

Design of FMS: Logical architecture[1]

Alejandro Malo & Antonio Ramírez
Dpto. de Ingeniería Eléctrica - Sección de Control Automático
CINVESTAV-IPN
Av. IPN 2508, México D.F.

Abstract

The work in this article is within the framework of the design of flexible manufacturing systems. Its aim is to represent the different ways the processes and the assignment of the resources can be done; preserving the flexibility found in this kind of systems and allowing the calculus of resources. The different manufacturing sequences found in the system are represented by state machines, while the manufacturing resources are shared. All this information is used, in this report, to define a Petri net class, that in a single net permits the specification, in a single net, jobs that can be done in different ways, that is, the redundancy that exist within a manufacturing system, not found in other representations.

Keywords

FMS Design, Petri Nets

1 INTRODUCTION

The design of flexible manufacturing systems is usually decomposed in several steps, e.g.; specification of the requirements, definition of the dynamic model, organization of the information, selection of the elements of the system, plant layout and validation. In this article we are concerned with the definition of a dynamic model specified by the following data:

- description of the different jobs as operation sequences.
- desired throughput (number of jobs per unit of time).

Within the design of FMS we found different views [Kouvelis 92, ProHil 90, SilVal 89]. In this article we use the Petri Net approach because it is a formal tool that can represent the structure and dynamics of a flexible manufacturing system [Silva 85], and with the addition of time, timed Petri nets, can be used to calculate the the resources needed to fulfill a desired throughput [ProHil 90, HilPro 89, LafProXie 92].

The article is organized as follows. Section 2 briefly presents structural aspects of Petri nets, for more information an interested reader can consult [LafProXie 94, Murata 89, Silva 85, RamHo 80]). Section 2.1.1 presents the modeling of the manufacturing sequences and the machine job assigment. Section 3 presents an aproximation to the modeling of processes and control circuits when we have a redundant system. The load balancing problem and the selection the minimum number of resources that fullfil with the desired throughput is solved. The last section presents an example.

---

[1]This paper presents work done for the ECLA/FLEXSYS project

2 PETRI NETS

In this section the basic concepts related with Petri nets are presented.

An ordinary Petri net is a directed bipartite graph represented by the fourth-tuple $N = (P, T, Pre, Post)$ where: $P = \{p_1, p_2, ..., p_n\}$ is a set of elements called places; $T = \{t_1, t_2, ..., t_m\}$ is a set of elements called transitions; $P \cap T = \emptyset$, $P \cup T \neq \emptyset$; Pre (Post) is the pre (post) incidence function that represents all the input (output) arcs to (from) a transition, $Pre, Post : P \times T \rightarrow 1, 0$. The places are drawn as circles and the transitions as bars. The pre-post incidence functions can be represented by matrices $PRE = [a_{ij}]$ and $POST = [b_{ij}]$ respectively, where $a_{ij} = Pre(p_i, t_j)$ and $b_{ij} = Post(p_i, t_j)$.

The incidence matrix $C = [c_{ij}]$, $i = 1, .., n$ $j = 1, ..., m$, is defined as $c_{ij} = b_{ij} - a_{ij}$. All the vectors X such that $C \cdot X = 0, X \geq 0$ are called T-semiflows; all the vectors Y such that $Y^T \cdot C = 0$, $Y \geq 0$ are called P-semiflows.

2.1 Time in Petri Nets

A Timed Petri Net $TPN =< N, D >$, where $D : T \rightarrow I\!R$. $D(t_i) = d_i$ is called the delay of the transition t_i and is the time needed to accomplish the firing of the transition. We prefer to assign the time to the transitions rather than to the places [LafProXie 94], since for us transitions represent the activities of the system.

2.1.1 Some Results in Strongly Connected State Machines

A state machine is a PN or a TPN where $|^\bullet t| = |t^\bullet| = 1, \forall t \in T$. This kind of PN allows the representation of conflicts, so in order to specify the optimum cycle time π (the inverse of the throughput) we must specify the visit ratio vector v_k. The following equation computes the optimal cycle time:

$$\pi \, 1^T \cdot M_0 = 1^T \cdot Pre \cdot D \cdot v_k$$

where $D = Diag[d_i]$, and v_k is the visit ratio vector.

2.1.2 Computing the visit ratio vector

The information in this section was taken from [CamChiSil 91]. In a state machine, those transitions that have the same predecessor are in conflict since the firing of one of them disables the others. Conflicts can be solved assigning each transition a routing ratio, i.e., the ratio each one is fired. Suppose t_i has the routing ratio r_i and t_k the routing ratio r_k. then for σ a periodic sequence, and $\phi(\sigma, x)$ the number of times that the transition x appears in the sequence σ, we have

$$\lim_{length(\sigma) \rightarrow \infty} \frac{\phi(\sigma, t_i)}{\phi(\sigma, t_k)} = \frac{r_i}{r_k}$$

Petri nets that we are using are bounded and live, so σ is generated by flows that are executed several times. It means that $\phi(\sigma)$ is a vector that can be decomposed as a linear combination of T-semiflows.

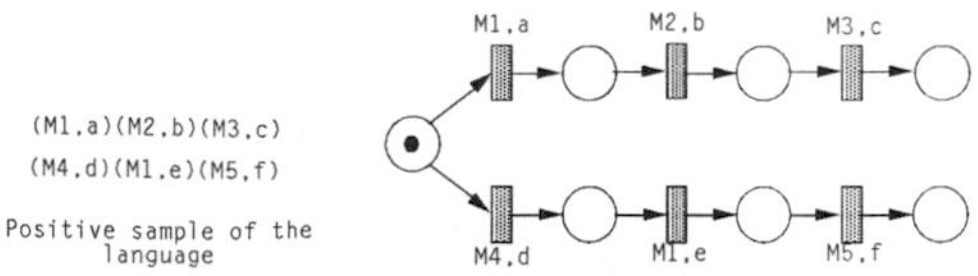

Figure 1: The job (as a positive sample and state machine).

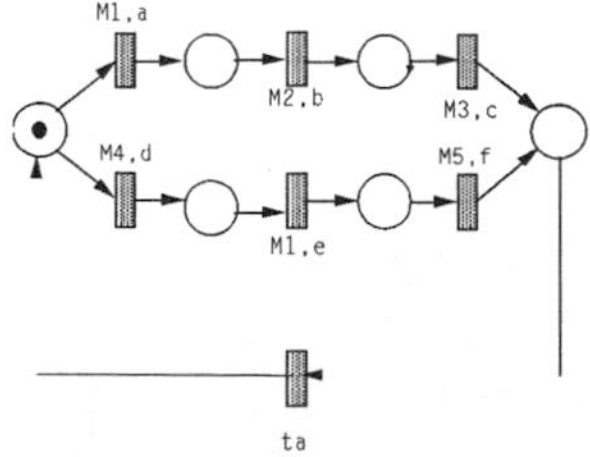

Figure 2: The job (as a strongly connected state machine).

$\phi(\sigma)$ is the Parikh vector of the σ sequence, so its elements are the elements of visit ratio vector. Then the previous equation can be rewritten as:

$$v(t_i) \cdot r_k - v(t_k) \cdot r_i = 0$$

Also, we have that the visit ratio vector is a combination of T-semiflows of the Petri net, we have

$$\begin{pmatrix} C \\ R \end{pmatrix} \cdot v = 0 \tag{1}$$

The vector is usually normalized for one transition, that is $v_k(t_j) = v_j/v_k$. After this brief introduction to Petri Nets, lets present how we use them to design FMS.

3 MODELING FLEXIBLE JOBS

Marked graphs have been used to model FMS[HilPro 89, ProHil 90, LafProXie 92], however, decisions exist in an FMS, so marked graphs are not a realistic approach. Flexible jobs have different production paths and these paths must be executed in some proportion in order to fulfill load balance constraints. We need a tool that captures this characteristic. In this paper we propose the use of a subclass of Petri nets that captures synchronization and decisions in its structure.

The job model is the result of the union of two nets: a processing net and a command net, both modeled as state machines. When we fuse both circuits, however, the resulting Petri net is no longer a state machine.

3.1 Processing Nets

A job processing net is described by a positive language sample [Angluin 87], each sample is an operation sequence (and a T-semiflow of the net), these operations are performed by machines; it is possible that one machine performs more than one operation in any sequence of the sample. A finite automata can be built from the positive language sample; this paper considers the case where the automata can be described as a state machine Petri net. In the left of figure 1 we have the operation sequence of a product represented by a positive sample. The sample says that the product can be done performing operation a on machine

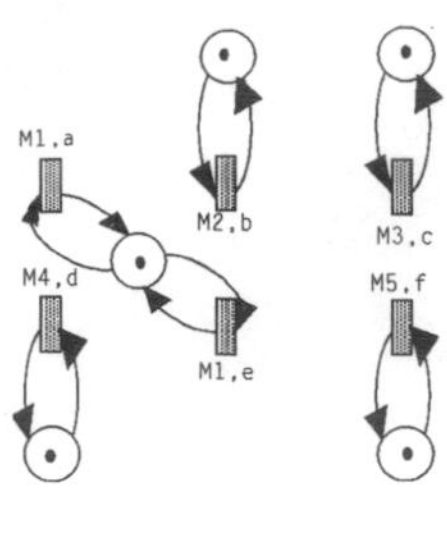

Figure 3: The command net.

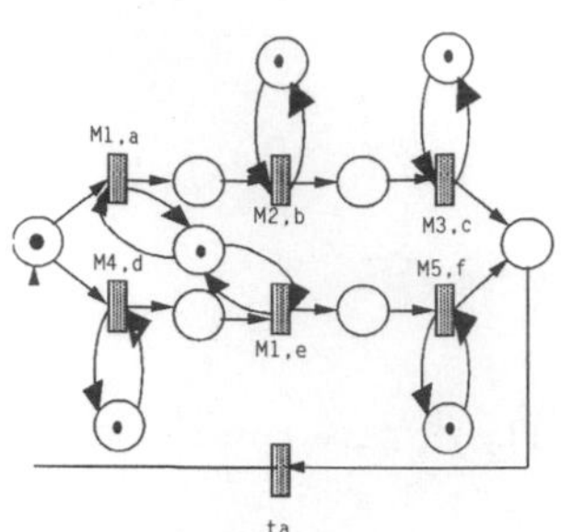

Figure 4: The global model of a job.

one, then operation b on machine 2 and finally operation c on machine 3; or by operation d on machine 4 followed by operation e on machine 1 and by operation f on machine 5. On the right side of figure 1 a Petri net model of these operation secuences is presented. The reader can observe that this Petri net model is a state machine.

From figure 1 we get, fusing all final states and adding the feedback transition t_a, figure 2 a strongly connected state machine that generates the same positive sample, so it also specifies the same job. We use the later model to solve the load balance and the throughput problems. The subclass N, of Petri nets, that models the process can be described as follows:

- $N = (P, T, I, O)$ is a strongly connected state machine
- $\exists t_a \in T \mid \|{}^\bullet t_a\| = \|t_a^\bullet\| = 1$
- $\exists p_i \in P \mid {}^\bullet p_i = \{t_a\}$
- $\exists p_f \in P \mid p_f^\bullet = \{t_a\}$
- $\forall |X_i|$ such that $C \cdot X_i = 0,\ t_a \in \|X_i\|$

Since the process net model is a strongly connected state machine, it has only one P-semiflow and the throughput and minimum initial marking can be computed in polynomial time. All P-components include p_i (initial state or raw material) and p_f (final state or product).

3.2 Command nets

The job command net is represented as a positive sample of operations. Each element of a sequence is an ordered pair (figure 3), where the first item represents the machine and the second item represents the operation. A command net will also be represented by a state machine Petri net, however, this state machine is not strongly connected (Figure 3). Since the command and processing nets perform the same activities, the set of transitions T will the same for both.

The set of places changes in the command net case. Each type of machine is represented by one place. These places are connected to transitions that represent the activities performed by the machine (represented by the place). Figure 3 shows a Petri net model of a command circuit for the positive sample (job) presented in figure 1.

3.3 Joining processing and command circuits

A global model (that includes both processing and command nets) for a job is built using both models (processing and command models). It is an easy task because both models have the same transitions. The processing circuit model is the net $N_p = (P_p, T, I_p, O_p)$ and the command circuit model is the net $N_c = (P_c, T, I_c, O_c)$, then the global model is the net $N = (P_p \cup P_c, T, I_p \cup I_c, O_p \cup O_c)$. Figure 4 presents the global model for the job presented in figure 1.

3.4 A model for all jobs

In a system we usually have several concurrent jobs. For job j_k, we have the model $N_k = (P_k, T_k, I_k, O_k)$. Each job has its own model, but they can share machines, so the places of command circuits can be shared by different jobs. Formally it means that if $N_k = (P_k, T_k, I_k, O_k)$ is a net modeling job k and $N_j = (P_j, T_j, I_j, O_j)$ is a net modeling job j, then $P_k \cap P_j \neq \emptyset$; $T_k \cap T_j = \emptyset$; $I_k \cap I_j = \emptyset$; $O_k \cap O_j = \emptyset$. Places representing machines, that are shared by different jobs, are merged into a single place, and the models can be merged into one. Figure 5 shows how a family of jobs can be modeled.

We remark that the final model is flexible. Since it allows us to model a family of jobs processed by shared machines, and, at the same time, jobs can be realized in different ways.

3.5 Computing the process net minimum initial marking

In this report we suppose the load balance problem solved, so we know, in the long term, the times each machine must be visited in relation to the remaining machines, i.e., we know the routing ratio numbers r_i, from which the visit ratio vector can be computed.

Unfortunatelly, the resulting Petri Net model of a job is not easy to analyse. However, if we suppose that the resources are infinite, then we can consider that a job model is a state machine and the result on section 2.1.1 can be used, since the resources do not introduce any constraints.

We have as input data the job specified as a State Machine; we also have the throughput and the proportion in which the different production paths must be executed.

Our problem consists in finding the minimum number of tokens in the Petri net that allows to reach the desired throughput. From [Campos 90] we know that Little's law can be expressed as:

$$Y_{job}^T \cdot M_{0_{job}} \; / \; Throughput_{job} \geq Y_{job}^T \cdot Pre \cdot Diag[d] \cdot v_a \tag{2}$$

then we can compute the minimum initial marking, i.e., the work in process, from:

$$Y_{job}^T \cdot M_{0_{job}} = \lceil Y^T Pre \cdot Diag[d] v_a \cdot Throughput_{job} \rceil$$

3.6 Computing the minimum number of resources

In the previous section we considered the resources infinite. The previous computation gives us the work in process, for the given throughput. Now it is necessary to compute the minimum number of resources.

Unfortunately, when we introduce the resource constraints, the job model, figure 4, is no longer a state machine. However, if we only see the work of a machine, the command subnet model is strongly connected state machine, so we can use the same procedure to compute the minimum marking, this can be done considering one or several jobs.

The throughput for a machine p_c is guaranteed, without scheduling problems, if the sum of the resources needed for individual job is equal to the number of resources needed for doing simultaneously all the jobs, that is,

$$\sum_{job}\lceil Y^T(p_c)\cdot Pre\cdot Diag[d]\cdot Diag[T_{job}]\; v_a\cdot Throughput_{job}\rceil$$

$$-\lceil Y^T(p_c)\cdot Pre\cdot Diag[d]\cdot \sum_{job} v_{job}\cdot Throughput_{job}\rceil = 0$$

where $v_k = \sum_{job} v_{job}$. A feasible schedule that satisfies the throughput constraints can be found if the available time is greater than the utilization time, that is, if

$$(Y^T(p_c)\cdot M_0\cdot \sum_{job} Throughput_{job})^{-1} \geq Y^T(p_c)\cdot Pre\cdot Diag[d]v_k$$

otherwise there will be. With this considerations, the initial marking can be calculated.

4 EXAMPLE

Suppose that we need to design an FMS to manufacture the products given by the following positive sample.

Job 1: $(M_1, a)(M_2, b)(M_3, c)$ and $(M_4, d)(M_1, e)(M_5, f)$

Job 2: $(M_6, g)(M_7, h)(M_8, i)$ and $(M_4, j)(M_2, k)(M_9, l)$

The throughput must be [1/4 1/2], i.e., one product of type 1 must be manufactured every 4 time units and one product of type 2 must be manufactured every 2 time units. The operation times for the first job are:

$(M_1, a) = 1$; $(M_2, b) = 2$; $(M_3, c) = 1$; $(M_4, d) = 1$; $(M_1, e) = 1$; $(M_5, f) = 2$
and for the second job:

$(M_6, g) = 1$; $(M_7, h) = 1$; $(M_8, i) = 1$; $(M_4, j) = 1$; $(M_2, k) = 1$; $(M_9, l) = 1$

In this case each job can be performed in two different ways. In both cases we will suppose that the load is equally balanced, i.e., all sequences must be performed once. The model of the first job is shown in figure 4, in this case, the visit ratio vector is given by $v_{a_1} = [1\ 1/2\ 1/2\ 1/2\ 1/2\ 1/2\ 1/2]$ (where $v_{a_1}(1) = 1$ and corresponds to t_{a_1}; $v_a(2)$ corresponds to $t_{(M_1,a)}$; $v_{a_1}(3)$ corresponds to $t_{(M_2,b)}$ and so on).

The minimum initial marking for the processing net for the first job is computed from:

$$Y^T_{job_1}\cdot M_0 = \left\lceil Y^T Pre\cdot Diag[d]v_k\cdot Throughput\right\rceil$$

so the work in process needed to satisfy the throughput is $Y^T_{job_1}\cdot M_0 = 1$.

When we consider the command nets, the minimum number of manufacturing resources can be computed. In the case of the first job, one resource of each type is enough to fulfill throughput constraints.

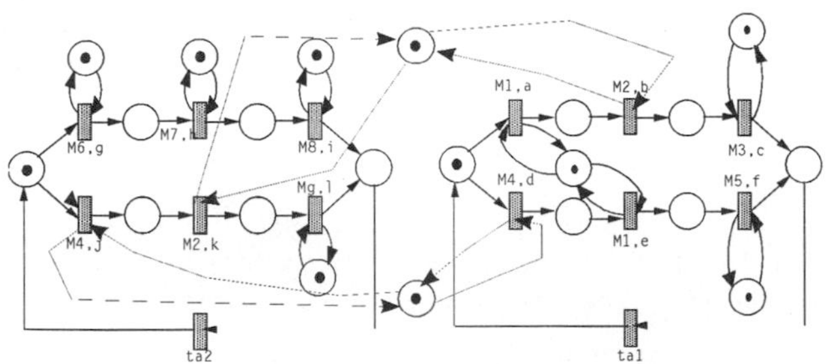

Figure 5: Model of a system performing two jobs.

The requirementes for the second A job can be calculated in a similar fashion. In this case we need two tokens to satify the throughput, that is, $Y_{job_2} \cdot M_0 = 2$. And one token for each kind of machine. So the initial marking is:

$$M_0 = [1\ 0\ 0\ 0\ 0\ 0\ 2\ 0\ 0\ 0\ 0\ 0\ 1\ 1\ 1\ 1\ 1\ 1\ 1\ 1]^T$$

Now, both jobs are merged to get the final model. In this case, places representing same machines are merged. Figure 5 represents the final model.

When we consider both jobs we obtain a similar marking. However, due to the coupling between jobs, we must be aware that this minimum marking might not be enough to satisfy the throuhgput, and it will depend of the desired scheduling.

5 CONCLUSIONS

This paper presents a methodology to model FMS. The FMS is modeled as processes to be done and as machines assigned to these processes. Since, in many cases a product can be done in several ways, we are using state machines to represent them. We also use state machines to represent the work done by each kind of manufacturing resource. The flexibility is retained by the representation. And allow us to calculate the minimum marking of the net, that is, the resources needed to fulfil the throughput.

6 REFERENCES

[Angluin 87] D. Angluin. Learning regular sets from queries and counterexamples. *Information and Computation*, (75):87–106, 1987.

[CamChiSil 91] J. Campos, G. Chiola, and M. Silva. Properties and performance bounds for closed free choice synchronized monoclass queueing networks. *IEEE Transactions on Automatic Control*, 36(12):1368–1382, December 1991.

[Campos 90] J. Campos. *Performance Bounds for Synchronized Queueing Networks.* PhD thesis, Universidad de Zaragoza., María de Luna 3 E-50015 Zaragoza, España., 1990.

[HilPro 89] H. Hillion and J. Proth. Performance evaluation of job-shop systems using timed event-graphs. *IEEE Transactions on Automatic Control*, 34(1):3–9, January 1989.

[Kouvelis 92] P. Kouvelis. Design and planning problems in flexible manufacturing systems: a critical review. *Intelligent Manufacturing*, (3):75–99, 1992.

[LafProXie 92] S. Laftit, J. Proth, and X. Xie. Optimization of invariant criteria for event graphs. *IEEE Transactions on Automatic Control*, 37(5):547–555, May 1992.

[LafProXie 94] S. Laftit, J.M. Proth, and X.L. Xie. Petri nets for modeling of dynamic systems - a survey. *Intelligent Manufacturing*, 30(2):175–202, 1994.

[Murata 89] T. Murata. Petri nets: properties, analysis and applications. In *Proceedings of the IEEE*, pages 541–580, April 1989.

[ProHil 90] J. Proth and H. Hillion. *Mathematical Tools in Production Management.* Plenum Press New York, 233 Spring Street, New York, NY 10013 USA, 1990.

[RamHo 80] C. Ramamoorthy and G. Ho. Performance evaluation of asynchronous concurrent systems using petri nets. *IEEE Transactions on Software Engineering*, 6(5):440–449, September 1980.

[Silva 85] M. Silva. *Las Redes de Petri: en la Automática y la Informática.* AC, Madrid, España, 1985.

[SilVal 89] M. Silva and R. Valette. Petri nets and flexible manufacturing. In Rozenberg; editor, *Advances in Petri Nets 1989*, pages 375–417, Springer-Verlag, Berlin, Germany, 1989.

7 BIOGRAPHY

Alejandro Malo Born at Guadalajara, México in 1953. An Aeronautical engineer from the Esime-IPN México, specialized at l'ENSAE, Toulouse, France; with a Master Degree from the Cinvestav-IPN, México. Currently pursues Ph.D. studies in Electrical Engineering at the Cinvestav-IPN. His current research interests are the Design of Flexible manufacturing, Dynamic Discrete Event Systems.

Antonio Ramírez was born in Monterrey, Nuevo León, México in 1964. He obtained his B.S. degree in Electrical Engineering from the Metropolitan University, México; his M.S. degree from the Cinvestav, México; and his Ph.D. from the University of Zaragoza, Spain. He is presently with the Electrical department of Cinvestav, México. His current research interest are: discrete event systems, Petri net theory, real-time control and control applications

PART FIVE

Anthropocentric Systems

12

Issues on the anthropocentric production systems

I. Kovács
Section of Sociology, Dep. Social Sciences, ISEG-UTL, Faculty of Economics and Management-Technical University of Lisbon
Rua Miguel Lupi, 20, P-1200 Lisboa, Portugal, tel. 3912537, fax. 3951885 e-mail: cs0015@keynes.iseg.utl.pt

A. Brandão Moniz
Industrial Sociology Group, FCT-UNL,
Faculty of Sciences and Technology, Universidade Nova Lisboa
Quinta da Torre, P-2825 Monte da Caparica, Portugal, tel. 3500225, fax. 2954461, e-mail: abm@uninova.pt

Abstract

This paper analises the problems and trends of the introduction of anthropocentric production systems (APS), specifically in small less industrialised member states of the European Union. The aim of this paper is to characterize APS and to present some special considerations related to the socioeconomic factors affecting the prospects and conditions for APS that is defined as a system based on the utilization of skilled human resources and flexible technology adapted to the needs of flexible and participative organisation. Among socioeconomic factors, some critical aspects for the development of APS will be focused, namely technological infrastructure, management strategies, perceived impact of introduction of automated systems on the division of labour and organisational structure, educational and vocational training and social actors strategies towards industrial automation. This analysis is based on a sample of industrial firms, built up for qualitative analysis, and on case studies analysis that can be reference examples for further development of APS, and not just for economic policy purposes alone.

We have also analysed the type of existing industrial relations, the union and employer strategies and some aspects of public policies towards the introduction of new technologies in the order to understand the extente to which there exists obstacles to and favorable conditions for the diffusion of anthropocentric systems. Finally some recomendations are presented to stress the trends for the implementation and development of anthropocentric production systems.

Keywords

Anthropocentric production systems, labour skills and qualifications, organisation, vocational training, automation, industrial relations.

1 SOME OF THE KEY CHARACTERISTICS OF APS

Many terms are employed to illustrate the central features of new production systems: one-of-a-kind production, skill-based systems, flexible specialisation, customised quality-competitive production, human centred system and anthropocentric production system (APS).

Although, the use of the APS designation is recent, a lot of their principles and ideas may be considered both as a development and integration of models recommended by social science specialists since the fifties and practised by innovative firms since the seventies.

APS can be defined as a production system which improves skills, participation in the decision-making processes and the quality of working life. In this system new technologies are moulded to valorise specific human capacities and to meet the needs of organisational structures designed to increase the participation of people in decision-making and the control of production processes, thus leading to a better quality of working life. However, there is no universally accepted definition. For this reason we decided to mention some of the APS definitions.

APS as a *coherent set of technological and organisational innovations* to improve productivity, quality and flexibility: "The production system that fits this condition is a computer-aided production system strongly based on skilled work and human decision-making combined with leading-edge technology. It can be called 'an anthropocentric production system." (Lehner, 1992: VII). The *essential components* of these systems are:

- flexible automation, supporting human work and decision-making
- a decentralised organisation of work, with flat hierarchies and a strong delegation of power and responsibilities, especially at shop-floor level;
- reduced division of labour;
- continuous, product-oriented up-skilling of people at work;
- product-oriented integration within the broader production processes.

Within this approach the combination of advanced technology and skilled work in a decentralised, product-oriented organisation leads to an intelligent manufacturing system able to support high quality and technological sophistication, rapid adjustment to change as well as diversification and the efficient use of resources (Lehner, 1992: IX).

According to another approach, APS represents a *step in the evolutionary process of the production systems*. In this sense, a new paradigm of production systems is emerging (Piore and Sabel) which will gradually displace the old mass production system in sectors whose activity involves advanced technology. The Fordist-Tayloristic approach is becoming increasingly inadequate in view of present economic, social and cultural conditions. The ability to adapt the products to costumer requirements by increasing variety, quality and short delivery times, are becoming the most important competitive factors. APS is seen as a competitive tool for the modernisation of European industry. For supporters of APS, Europe with its tradition for small batch production is in a comparatively more favourable position to improve APS than the USA, with its highly Tayloristic-Fordistic traditions (Brödner).

In this paper we consider APS to be synonymous with the concept of a human centred production system. APS is an *alternative response* but it is not "the one best way" to respond to the requirements of changing market conditions calling for flexibility, innovation, diversification, short delivery times and customisation. It is, however an adequate response to the new expectations and attitudes of people towards work.

At present there are many solutions - both technical and organisational - for improving firms' competitiveness. This is a new trend, insofar as, for some time there would appear to have been a strong tendency for production systems to converge in most of the economic activities of industrialised or industrialising countries. The Taylorist-Fordist principles were felt to be universally applicable. However, we disagree with this approach which considers APS to be new universal model displacing the old Taylorist-Fordist model. From our standpoint APS is an alternative strategy and a question of choice rather than "the one best way" of ensuring the best performance.

The technology-centred strategy is another choice for the improvement of highly automated production systems. There are underlying ideas, namely that economic superiority is based on technological sophistication in which competitiveness pressuposes hierarchical and centralised organisation. Technology is regarded as a mean of replacing people reduced to machine components in the automated system whose role is becoming more and more reduced through higher automation, leading to increased replacement of human skills. There are many other possible strategies arising from different combination of the principles of two basic strategies we know as human-centred and technology-centred strategies.

Table 1 Comparison of the technology-centred and human-centred approach

Technology-centred Approach	Human-centred Approach
Introduction of new technologies with a view to reducing human roles on the shop floor, and labour costs	Introduction of new technologies as a complement to specific human capacities, aimed at increasing funcional flexibility, quality of products and of working life
Replacement of skills by technology, leading to an increase in the de-skilling and de-motivation of shop floor employees.	Improvement of the quality and stability of human resources at all levels for the improved exploration of the potentials offered by new technologies
Centralised technical solutions	Decentralised technical solutions
Rigid work practices based on principles such a centralisation, vertical and horizontal separation of tasks and competence specialisation	Flexible work practices based on principles such as decentralisation, multi-valence, vertical and horizontal integration of tasks, participation and co-operation
Rigid hierarchic and professional boundaries	Supple boundaries
Passive role at operational level: execution of simple tasks	New professionalism at operational level: autonomy to perform different, complex tasks, capacity for problem solving, creativity and autonomy, at individual or group level
Integration of units of firms by way of computer assisted centralisation of information, decisions and control	Integration of parts of firms by way of training, socialisation, communication-cooperation and information accessibility, participation in decision making and self-control

The human-centred approach is directed towards the development of a flexible and decentralised production system. Here the potentials of technology complement human skills and specific human abilities are valorised.

In other words technology should not replace people, but rather improves their competence and decision making capacity. Flexibility is achieved using intuition, imagination, individual and collective know-how, existing skills and working methods enriched through new knowledge and methods.

These specific human abilities related to the management of the unexpected, are based on information which cannot be formalised and on an understanding of complex and non-structured situations. The job is designed according to *socio-technical principles*: the improvement of variety, identity, sense of fulfilment and autonomy at work. Their aims are:

- the use of skills and abilities including tacit knowledge,
- the creation of favourable conditions for development and learning,
- the improvement of collaborative work.

These principles involve the integration of conception and execution, intellectual and manual functions through work enrichment with discretion in selection of work methods (in low automated work areas) or by the integration of planning, programming, processing and maintenance tasks (in highly automated areas).

The work is structured in *work groups* with a high level of autonomy and self-control. The work group activity focuses on the main type of product, or on a small group of related products. The group tasks include planning and allocation of work: loading, setting, unloading the machines; programming, maintenance, quality and performance control. Various skills are required and job rotation among group members is used. At *factory level* the basic principles are:

- de-centralisation of the company, to form autonomous production units,
- collaborative relationship between departments,
- strong communication links between the groups, including informal and personal communication,
- co-operative relationship between specialists (engineers, technician) and operators (workers).

The *technological dimension* is developed by taking account of a desired decentralised organisational structure, requirements linked to team work, people's needs and motivations rather than ergonomic criteria alone. Technology should make the best use of human beings by developping tools to support skills and competence, should allow group work by grouping machines and software to support planning, control and scheduling activities as a group responsibility, and should support group autonomy by decentralised information, communication and transport systems.

The anthropocentric approach may be exemplified by ESPRIT-CIM projects 1199 (Human Centred CIM) and 534 (Development of a Flexible Automated Assembly Cell and Associated Human Factor Study), 6896 (CONSENS-Concurrent/Simultaneous Engineering System) and 8162 (QUALIT-Quality Assessment of Living with Information Technology) and ESPRIT 2338 (Integrated Manufacturing Planning and Control System, oriented to develop a decentralised system architecture with emphasis on shop-floor scheduling). These projects have been multi-disciplinary projects covering the technical, psychological and organisational aspects, with the co-operation between engineers and specialists from different social sciences. They recognise that "the joint optimisation of human and technical criteria is a pre-requisite to the development and successful implementation of technology" (Kidd, 1988, 297-302). Their aim is to improve economic results, as well as the quality of working life.

The development of an anthropocentric productive system, in accordance with human-centred principles, may be undertaken by implementing all principles in a complete system, using all the abovementioned elements, or by the introduction of some changes. The first case involves the shaping of new plants, whilst the second case requires modifications in accordance with APS principles. Such alterations may consist of the formation of working groups and/or "production islands", task re-organisation for their enrichment and de-centralisation methods.

Models and methods are necessary to analyse and design integrated socio-technical systems and evaluate the relationship between people-organisation-technology, based on interdisciplinary work and co-operation among technologists and social scientists. This interdisciplinary approach enables one to take into consideration the organisational structure chosen, users' needs and motivations in the development of production systems (Kidd, 1992 37).

2 ADVANTAGES OF APS

In a humanistic approach the promotion of APS is always desirable. However, we may well ask: is APS at present feasible in the context of the competitive imperatives of economic life?

The experiences reported in many studies show that the APS is not only a desirable model from a humanistic perspective, but can also lead to increased productivity, improved quality and greater effectiveness.

Market conditions have become unstable, very differentiated and extremely dynamic. Advanced technologies offer new opportunities, such as higher technical flexibility, a greater degree of quality and precision and the integration of different areas of activity. At the same time, people with a higher level of education and professional training expect jobs with enriched content besides participation possibilities in the decision-making processes. In this context the APS provides psychological, social and economic benefits. There are many case studies elaborated within the FAST Framework Programme that demonstrate the advantages of APS (LEHNER, 1992, 47). For example one German experience shows the following results:

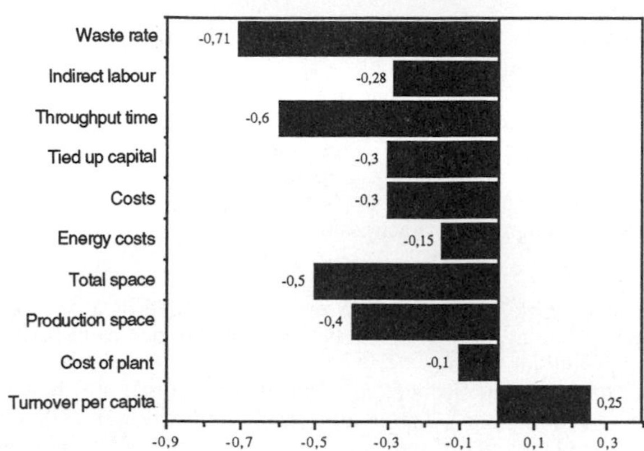

Figure 1 Performance and anthropocentric production systems (Kidd, 1992, 47).

Some examples of Flexible Manufacturing Systems show the economic advantages of the application of APS principles:

Table 2

Reduction in %	Freudenberg (D) case No. 6	Sealectro (GB) case No. 11	Felten & G. (D) case No. 19	Volvo (S) case No. 23	Lukas (GB) case No. 26
lead times	80	50	50	80	55
stock	15	45	50	50	50
production costs	no data available	75	10	25	no data available

Source: Brandt, 1991, 40.

Among the *technical and economic benefits* one may consider the following:

a) improved quality (less rejects and flaws)
b) increased responsiveness,
c) shorter throughput times,
d) lower indirect costs,
e) easier planning and control of production processes,

f) simplified material flows,
g) smaller production areas
h) swifter response to quantitative and qualitative changes in demand,
i) less breakdowns,
j) increased capacity for innovation and continuos improvement.

Among the *social-human benefits* one may consider:

a) the increasing quality of working life,
b) higher job satisfaction through meaningful rewarding tasks,
c) higher degree of motivation and involvement,
d) greater personal flexibility and adaptation,
e) improved ability, creativity and skills of shop floor personnel,
f) enriched direct interpersonal communication and social relations,
g) increased collective and co-operative spirit,
h) greater capacity for collective learning of new practices.

We can thus conclude that within a flexible economy, APS may be regarded as a strategic answer to new economic requirements. Today, competitive advantages are gained from factors related to quality, flexibility, creativity and innovation.

The excellent results obtained by Japanese firms stem from organisational and human resource related factors *. According to Jaikumar's comparative study on FMS in the US and Japan metal sector, Japanese firms make better use of information-intensive technology than the US. firms. "With few exceptions, the flexible manufacturing systems installed in the US show an astonishing lack of flexibility. In many cases, they perform worse than the conventional technology they replace. The technology itself is not to blame; it is management that makes the difference. Compared with Japanese systems, those in US plants produce an order-of-magnitude less variety of parts" (Jaikumar, 1986, 69). The US firms use FMS for the high-volume production of a few parts, whereas the Japanese firms use it for high-variety production of many parts at lower per unit costs.

Table 3 Comparison of Japanese FMS studies in US. and Japan

Results	US.	Japan
System development time (years)	2,5 to 3	1,25 to 1,75
Number of machines per system	7	6
Types of parts produced per system	10	93
Annual volume per part	1727	258
Number of parts produced per day	88	120
Number of new parts introduced per year	1	22
Utilisation rate (two shifts)	52 %	84 %
Average metal-cutting time per day (hours)	8.3	20.2

Source: Jaikumar, 1986, 70.

* Altough the productive systems developed in Japan are in some key areas similar to anthropocentric system, as we shall mention later, in view of their specific features, they are regarded as "lean production".

The performance disparity was mainly due to differences concerning the workforce's level of skill and type of work organisation. In US firms, management mastered the production system on the principles of scientific management. Here skilled blue-collar machinists were replaced by trained operators whose tasks were specified by management. The operators did not have the discretion to change procedures. In Japanese firms, highly skilled people with multi-functional responsibilities work in small teams. Operators on the shop floor can introduce programming changes and are responsible for writing new programs.

In the "lean production" model the emphasis is on the relations between the firm and suppliers and customers. In the APS model the emphasis is above all on internal competence and the achievement of functional flexibility, qualified versatile people. In the "lean production" model, technology is accepted piecemeal whereas, in the APS model, technology is specifically adapted to peoples and organisational needs. The so-called Japanese "lean production" model is in some aspects similar to that of APS, differences do though exist:

Table 4 Comparison of APS and "lean production"

Characteristics	Lean Production	APS
Aims	Increasing productivity, industrial modernisation, based on human resources and organisation	Same
Qualifications	Training	Education/training
Technology	No need for a specific technology	Technology should be specifically adapted
Organisational principles	Organisation of business , plant and shop floor	Organisation of plant and shop floor
Organisation	Work in groups, integration of groups, complex tasks, responsibility at execution level, collaboration between different departments	Same
Volume of production	Volume of production close to large batch production	Small batch production in small series close to prototype production and large series production
Industrial sectors	Automobile	Mechanical engineering and related industries
Professional relations	Leadership	Participation

Source: Wobbe, 1992, 49.

The advantages of APS systems are evident though this model cannot be universally applied. Many factors must be taken into account, such as the type of production, firm size, type of organisation, power relationships, management practices with regard to human resources, existing skills and social competence, etc.

The APS model may be particularly appropriate when the level of product variety is high, and the level of quantity is low (prototype and small batch production), and/or when high standards of development and social needs are called for.

The diffusion of APS can lead to an increasing level of quality of working life of people who have jobs. But it is necessary to have awareness about the limited effect of APS in the quality of life in society as a whole. The APS promotion does not solve the unemployment and precarious job problems. The diffusion of APS can co-exist with the increase of unemployment and with the precarious jobs.

3 CONCLUSIONS

Diffusion of APS at European level requires changes in existing research programs. There should be a greater emphasis on the human and organisational factors rather than on technical considerations. Education and training programms should be reoriented towards human-centred strategy.

The human-centred orientation for industrial modernisation can also be adopted in a less industrialised country, like Portugal. The APS is not a new specific model for advanced European countries only. This orientation can be particularly recommended in sectors and firms with no Tayloristic traditions. Whilst, in a Taylorised sectors and firms many obstacles must be overcome.

Therefore is we feel it is imperative to prepare an industrial strategy within a development plan to serve as a point of reference for economic agents' decisions and specific policies (namely for scientific & technologic, employment or vocational training policies) in orther to make these policies coherent. A strategy designed to raise the technological level of Portuguese industry must take into account the following:

a) modernisation of traditional industries, to ensure they remain competitive and become coherent on the strenght of quality requirements and flexible specialisation;
b) increase in technological level through more advanced technology transfers (and not just the more mature ones offering less risk factors), not forgetting the development of endogenous growth capability and the increase in the assimilation capability of more advanced technology, and specially of APS;
c) participation in co-operation at EU level (Information Technologies, EUREKA, Production Technologies, Telematics, etc.), in the development of new technical systems for the creation of new growth centres leading to reduced Portuguese dependency on equipment, foodstuff and energy sectors;
d) experimentations supported by Public Programmes to promote APS in mixed capital enterprises;
e) training programs for and dissemination of publications (books, booklets, videos, reports) among the social partners on APSt themes;
f) training programmes for all who are involved in the labour world, must include the human and social issues of production.

On the other hand, in view of the absence of a coherent financial policy supportive of research (which restricts the role of R&D government agencies), the more active groups in this area of flexible automated production. Should also to take part in European projects, where the Portuguese companies participation have been weak. It is therefore necessary to raise the level of R&D to match APS technological requirements through local development efforts. If such efforts are not combined with some commercial strategy, there is a risk of poor results.

State administration must show determination in supporting laws designed to establish a basic framework of participation and co-operation between entrepreneurs and the work force, promote programs to facilitate technical refrecher and training, and also to relocate workers affected by technological advances. In the same way, experimental programmes should be introduced to innovate management techniques in mixed capital enterprises. Company managers and union leader feel it is indispensable to define a development plan at national level.

Finally, there is an urgent need to carry out empirical research on the socio-economic consequences of Tayloristic-technocentric systems (hidden costs of absenteeism, lack of quality, etc.), and to discover new organisational forms; specially those derived from or which facilitate implementation of anthropocentric automated production systems. Essentially, this implementation requires the knowledge of the socio-cultural reality of the industrial environment. The lack of innovative experimentations, also explains why new forms of work organisation are poorly known besides the fact that motivation and the human factor have not received the attention they deserve.

4 REFERENCES

Adler, P. S.: "Automation et qualification. Nouvelles orientation" (1987) *Sociologie du Travail*, N° **3**;

Altmann, N. e Düll, K. (1978) New Form of Work Organisation and Job Design in the Federal Republic of Germany: A Survey of Measures, Studies and Sociological Requirements, in: EFILWC: *New Form of Work Organisation in the European Community - Germany*. Munich: European Foundation for the Improvement of Living and Working Conditions.

Brandt, D. (1992)*Advanced Experiences with APS*, 30 European Case Studies, FOP 246, Brussels, FAST/CEC.

Brödner, Peter (1990) Technocentric-anthropocentric Approaches: Towards skill-based Manufacturing, in: Warner, M.; Werner, W.; Brodner, P. (Ed.), *New Technology and Manufacturing Management*, John Wiley & Sons .

FAST: *Human Work in Avanced Technological Environment*, CEC, June 1989

Gélinier, O. (1984) *Stratégie de l'entreprise et motivation des hommes*, Paris, Hommes et Téchniques.

Hertog, J.F.; Schroder, P. (1989) *Social Research for Technological Change: Lessons from national programmes in Europe and North America*, MERIT - Maastricht Economic Research Institute on Innovation and Technology, University of Limburg, Maastricht.

Jaikumar, R. (1986) Post-industrial manufacturing, *Harvard Business Review*, **6**.

Jones, B.; Kovács, I.; Moniz, A.B. (1988), 'Understanding What Is the Human Factors in CIM Systems: Some International Evidence' in B.W. Rooks, *CIM-Europe 1988 Conference: Present Achievements, Future Goals*, Bedford, IFS Publication.

Kern, H.; Schumann, M. (1988) *El Fin de la Division del Trabajo?* Ministério de Trabajo y Seguridad Social, Madrid.

Kidd, P. (1992) *Organisation, people and technology in European manufacturing*, CEC, FAST, Brussels.

Kidd, P. (1988) Technology and engineering design: shaping a better future or repeating the mistakes of the past?, *IEE PROCEEDINGS*, Vol. 135, **5**.

Kovács, Ilona et al. (1992) *Sistemas flexíveis de produação e reorganização do trabalho*, CESO I&D, PEDIP, DGI.

Kovács, I. and Moniz, A.B. (1988) 'Aspects sociaux de l'automation industrielle au Portugal: Analyse de quelques cas' (Social Aspects of Industrial Automation in Portugal: Analysis of some Cases), in A.C. Gonçalves, A.T. Fernandes and C.L. d'Epinay (eds.), *La Sociologie et les nouveaux défis de la modernisation*, Oporto, AISLF/SSFLP.

Kovács, I.; Steiger-Garção, A. and Moniz, A.B. (1987) 'Flexible Production Systems and Work Organisation: The Portuguese Situation for the Nineties', *Proc. 8th EGOS Colloquium Technology as the Two-Edged Sword of Organisational Change*, Antwerpen, EGOS.

Kovács. I.; Moniz, A.B.; Mateus, A. (1991) *Prospects for anthropocentric production systems in Portugal*, CEC/FAST, Vol. 16, FOP 260, CEC.

Kovács. I.; Moniz, A.B. (1994) Trends for the development of anthropocentric production systems in small less industrialised countries: The case of Portugal, *Proc. European Workshop in Human Centred Systems*, Univ. Brighton-ERASMUS, Brighton.

Lehner, F. (1992) *Anthropocentric production systems: the European response to advanced manufacturing and globalization*, Brussels, CEC.

Moniz, A.B. et al. (1989): Occupational Structure, Training and Labour Relations in the Metal Industries in Portugal, Lisbon, CESO I&D/CEDEFOP.
Moniz, A.B.; Kovács. I.; Pereira, Z.L. (1994) "Quality and Work Organization with Advanced Automation in Portugal" in Kidd, P.; Karwowski, W.: *Advances in Agile Manufacturing*, Amsterdam, IOS Press, 675 - 679.
OCDE (1988) *Nouvelle technologies, une estratégie socio-economique pour les années 90.* OCDE, Paris.
Woomack, J. P. et al. (1990) *The Machine that changed the world*, Rawson Associates, New York.

BIOGRAPHIES

Ilona Kovács, is Full Professor of Sociology at the Faculty of Economics and Management of the Technical University of Lisbon, received her PhD in Economics Sociology at the University of Budapest, and is author of several books and articles on organisational innovation, vocational training for new competences and management strategies. She is also responsible of post-graduated course on Socio-organisational systems of the economical activities, and is consultant of industrial firms and social partners.

António Brandão Moniz, is Professor of Industrial Sociology at the Faculty of Sciences and Technology of University Nova of Lisbon, where received the PhD on Sociology of Industry and Work, and is author of books and articles on technological aspects, organisational design for FMS, and job design for automated systems. Is also researcher at UNINOVA Institute and consultant of industrial firms and social partners.

13

A Complementary Approach to Flexible Automation

R. Roshardt, C. Uhrhan*, T. Waefler**, S. Weik***
**Institute of Robotics, **Work and Organizational Psychology Unit*
Swiss Federal Institute of Technology (ETH) Zürich
8092 Zürich, Switzerland
Tel: ++41 1 632 5553, Fax: ++41 1 632 1078, e-mail: roshardt@ifr.mavt.ethz.ch

Abstract

In the early days of automation the engineer designed the task to be automated in a manner where technical solutions were available. The rest of the work had to be done by the operator. These technically oriented systems failed because they were not flexible and the human had no possibilities to intervene or override the programmed process. In this paper a method for the design of automated, sociotechnical work systems is presented. It is based on a concept of complementary allocation of functions between human operators and machines. The method was used to design a robot system for automated charging of a press brake within a computer integrated sheet metal forming cell. This leads to extended requirements of the technology that are briefly presented.

Keywords

balanced automation, complementary function allocation, flexible press brake automation

1 INTRODUCTION

In former times it was tried to increase the flexibility in manufacturing by full automation and computer integration of the production lines. Therefore computer aided design systems (CAD), computer aided planning systems (CAP) and numerical controlled machines (CAM) were used and coupled together by data networks. The problem of such fixed planned production models are that uncertainties, failures and disturbances which occur daily in every factory cannot be handled automatically. These tasks are left to the operator. The result of this automation concept is a so called "CIM-ruin" (CIM - Computer Integrated Manufacturing), where expensive installations do not work because of small errors. New approaches suggest that only with the simultaneous planning of human resources, technology and organization economic automation is possible. This judgement yields to concepts like the "fractal company" (Warnecke, 1993) and the "MTO - Man Technology Organization - approach" (Ulich, 1993). In this paper we describe an approach to automating a flexible bending cell according to the global concept of "MTO".

Bending with press brakes is a very flexible manufacturing technology. Multiformed parts can be produced with one press brake and some fixed installed standard tools. In fact, sheet metal bending with press brakes is predestined for low batch sizes. Combined with appropriate automation bended sheet parts become an economic alternative to welded, casted or cutted parts.

Human work at non automated press brakes consists of several tasks: the definition of the bending sequence from workshop drawings, choice and installation of tools, programming the press brake and workpiece handling at the press brake. Nowadays, automation at the planning level is already available - for example CAP-systems that generate bending sequences from a CAD model (e.g. Reissner & Ehrismann, 1990). Nevertheless, the computer assisted work at press brakes is only semi-automated because the manual feeding of sheets in the press brake is still necessary.

There are already some research projects as well as industrial applications working on automating the charging of press brakes with robots (Uhrhan, 1994). The industrial applications are normaly used for high batch sizes. With decreasing batch sizes and more complex final part shape, advanced robot features like sensor guidance and sensor supported programming are needed. Also interaction of an operator with tasks and decisions must be possible for optimization and to keep the system running.

The planning and valuation of our production system was carried out by a method which is based on a complementary design of sociotechnical systems named KOMPASS (Weik, Grote & Zölch, 1994; Grote, Weik, Wäfler & Zölch, in press). The method helps to define which functions should be automated and how they should be automated.

2 FUNCTION ALLOCATION - BASIC TOPIC IN REALIZING BALANCED AUTOMATION

Nowadays it seems to be widly accepted, "...that automated systems still are man-machine systems, for which both technical and human factors are important..." (Bainbridge, 1983, 129). That is why in the course of every system design process - especially in those which are concerned with the introduction of automated manufacturing technology - sooner or later the question will arise how functions between operator and machine should be allocated.

Despite the omnipresence of that very central, work psychological and technical question (generally) design methods are only available on a theoretical level but are difficult to apply in practice. What different authors consistently plead for is that "...the correct process for balancing human and machine actions should become an institutionalized part of system design" (Bastl et al., 1991, 19). Very often the technical system is developed first or even exists already. The social system just gets adapted to it, which means that psychological and also organizational aspects which cannot be systematized in the same way are usually not given due regard or even are not taken into consideration at all.

30 years ago Jordan (1963) pointed out already, that for really do justice to human features in a work system one should turn to a *complementary* view of the relationship between human and machine. Such a view accommodates that the capabilities of human and machine differ also in a *qualitative* manner, meaning they cannot replace but only *complement* each other.

In the following sections we intend to show, how this concept of complement - which is cited quite often but still is rather poorly operationalized - might be applied in practice.

2.1 The instrument KOMPASS (complementary analysis and design of production tasks in sociotechnical systems)

KOMPASS wants to provide industrial practitioners (e.g. planners, engineers, production managers, human factors specialists), who are concerned with the development and adaptation of automated production processes. Special attention is paid to the allocation of functions between operator and machine (Grote, 1994; Weik et al., 1994). The project has two main objectives:

(1) Definition, operationalization, and validation of an integrated set of evaluation criteria support complementary system design.

(2) Development of a heuristic which helps multidisciplinary design teams to find and evaluate different kinds of allocation scenarios in a participating manner. Initiation of a learning process which leeds to complementary system design.

Criteria for complementary system design (objective 1) have to cover the three domains: human resources, technology, and organization as demanded by the MTO-approach. Design guidelines which cover the *organizational aspect* are nowadays widely developed (cf. Ulich, 1994). They demand independent organizational units, interdependent tasks within the organizational unit and the unity of product and organization. Criteria for the analysis, evaluation, and design of *humane working conditions* have been described in detail by the psychological action regulation and activity theories (e.g. Hacker, 1986; Ulich, 1994). Requirements for humane tasks are, for instance: Completeness, planning and decision making requirements, variety of demands, cooperation requirements, learning possibilities and autonomy.

As far as the *technological aspect* is concerned, the design of human-machine systems is very often reduced to the design of optimal (i.e. user friendly) interfaces. KOMPASS concentrate more on the question of how to allocate the actual *contents* of *complete* working tasks to either human, machine or both (Ulich et al. 1991). Therefore one of the objectives is to develop criteria which enable a design team to evaluate whether an already existing or planned human-machine system complies with the demands of complementary design. The second objective of KOMPASS is the development of a design heuristic which enables design teams to integrate work psychological findings - such as complementary function allocation - into their activities.

2.2 Stepwise description of the design process

A first version of the KOMPASS design heuristic was used to structure the design process of an automated sheet bending press. The design team comprised two computer scientists, one engineer and two work psychologists. Five steps were worked through in rather an iterative than a sequential process!

Step 1: Definition of the primary function and the subfunctions of the planned work system

Step 2: Participatory development of a shared evaluation concept to differentiate between successful and unsuccessful system design

Step 3: Identification of the potential contributions of human operator, technological system and organizational conditions to a successful system performance

Step 4: Discussion and definition of design objectives for each subfunction, focusing on selected evaluation criteria

Step 5: Definition of allocation scenarios for each subfunction

In the *first step* the primary function and the subfunctions of the planned work system have to be defined. The work system 'sheet metal bending cell' consists of technical components and human operators. Therefore the fulfilment of the operative system function 'sheet metal forming' depends on the quality of the joint performance of operator and technology. The definitions of the operative function and the subfunctions of the work system are to be made on the system level, i.e. without allocating functions to either human or machine at this early stage of the design process. The functions defined for the automated sheet bending cell are presented in Table 1, column 1.

In the *second step* the design team is asked to charaterize how the planned system should perform. For that purpose the team members should jointly develop and define criteria that allow them to measure the system performance and to compare different scenarios of the planned system. Criteria that have been defined are for instance: profitability/productivity (effectiveness: e.g. quality of process result, efficiency: e.g. quality of process), flexibility/adaptability (fast and easy change of jobs, fast and easy tool setting), reliability/availability, acceptance by operators and usability (easy handling and maintenance).

In *step 3* it is discussed which specific contributions to a successful performance of the planned system are provided by the human operator, the technical system and the organizational conditions. The purpose of this step is to direct the attention of the team members towards the qualitatively different influences of human operators, technology and organization to the performance of the work system.

In *step 4* it is now to define design objectives for the functions (cf. step1). Based on assumptions about which criteria are the most critical in any design process (Grote, 1994), some of the criteria developed by the research project KOMPASS are selected for further discussion. These are in our case planning and decision-making requirements, transparency, decision authority, and completeness of tasks. The results of this discussion are summarized in Table 1, columns 2-5. It follows a brief description of the criteria:

The criterion *planning and decision-making requirements* is used to identify those subfunctions with a high potential for planning and decision making. Based on this potential, it should be discussed what consequences it might have for human efficiency if this planning and decision making is not delegated to the operator, but to the technical system or the organization.

Transparency: It has to be discussed how the technical components of the human-machine system have to be designed to enable the operator to understand (cognitive level) what happens "in" the machine he is working with. Additionally it has to be clarified whether the future operator will get the technical opportunity to develop tacit knowledge.

Automation potential and decision authority: In a first step the design team has to mark all those subfunctions whose automation is (at least partly) technically feasible or where a technical support of any kind is possible. In a second step it has to be discussed how the decision authority within these functions should be allocated between human operator and machine.

Completeness of tasks: Each subfunction should be marked, according to it's preparing, planning, executing, controlling and/or maintenance elements. In order to create meaningful and intrinsic motivating working tasks, the future operator should be confronted with a task, that contains all those elements.

In the final *step 5* a first variant of an allocation scenario has to be developed. For each subfunction it has to be defined which parts of the function are allocated to the operator, the machine or both of them. If other persons or organizational units are involved their part has also to be described. The variant developed by our design team is presented in Table 1, columns 6 - 8.

Table 1 - Partly automated metal sheet bending cell with sensor supported robot programming; the evaluation process

functions at press brake	planning&decision requirements (potential)	transparency (requirement)	decision authority (requirement)	Completeness of tasks	machine	operator	other persons
receive orders fine planning determine start of execution, batch size and production time	*high* relevant decisions regarding the optimization of set-up times and the minimisation of down-times	*high* to provide the operator with all information relevant to the order of jobs	*human* the operator should decide whether he takes over, changes or refuses the automatically generated data	*planning*	Cell control station to plan the orders. Enables input to define priority especially for shifts without operators.	The orders are downloaded to the operator's computer. He plans the priority and inserts express jobs.	Interaction with people from the production planning department concerning dates and terms.
Programming: get information drawing data, bending sequence (BS), gripping points (GP) tool data, material data	*high* relevant decisions regarding procedural strategies, i.e. bending sequences and gripping points	*high* to enable the operator to develop tacit knowledge and react flexibly to disturbances	*human* the operator should decide whether he takes over, changes or refuses the automatically generated data	*planning*	Cell control station networked with CAD-CAP system. Downloading of the drawing data, the automatically generated BS, GP and tool data. The generated BS can be simulated graphically. The system enables the operator to access all relevant data (on a display or on paper).	The operator can access all generated data. He can browse and print them. He has the possibility to see the documents at the author's office. If the BS or the GP cannot be generated automatically the operator must be able to complete this information (see coding).	Cooperation with designers and planners, which should be located near the cell. Suggestions to the designer (know-how feedback) Between operator and designer, no other organizational unit exists to discuss techn. questions.
Programming: coding write programm in the machine specific language	*low* only transformational knowledge, i.e. to translate procedural strategies into NC-code	*low* transformation of procedural strategies into code is not relevant for bending process knowledge	*human* the code is generated automatically, but the operator should have the opportunity to program the system making the bending process independent of the CAP system (integrate rush orders or complex bending forms)	*executing*	Automatic program generation with tests of plausibility (is the program executable by the robot?). Sensor supported teach-in programming of movements and GP. Visualisation of the programmed robot movements Interface for off-line editing/defining of BS, GP and tool data	The operator decides whether to use the automatically generated programs or not. Possibility to program the robot himself by teach-In or predefined macros (process-level) and to edit BS, GP (product-level).	Coordinated with the designer.
setting up the machine mount tools download programs	*low* only procedural knowledge	*low* function is not relevant for bending process knowledge	*machine* total automation	*executing*	Automatic loading of programs. Automatic change of gripper and bending tool. Both are dependent on the fine planning.	Operator checks the set-up and supervises the first run.	
sheet handling and supervising of the process grip, feed, support, unload	*middle* handling of work orders, but with the possibility to real time process optimization	*high* to enable the operator to develop tacit knowledge regarding BS and GP	*human* the operator should decide whether he or the robot should handling the workpieces	*executing, controlling*	Automatic handling according to the program, automatic supervision and security checks, log-file.	Possibility of manual handling. Supervision in case of exceptions.	
reaction to disturbances (robot & machine) stop process, diagnosis, alarm	*high* relevant decisions regarding utilisation of the machine and the design of the processes	*high* to enable the operator to develop tacit knowledge regarding the technical system	*human* as much technical support as possible (e.g. error messages)	*maintenance*	Automatic stop of the robot Entry in log file alarms (acoustic, visual)	Operator stops process, diagnoses and repairs the failure if he can; otherwise calls expert.	Call of experts
suggestions know-how feedback optimised BS, GP and handling	*high* creative potential	*high* mainly regarding possibilities to ameliorate production strategies	*human* as much technical support as possible (e.g. error messages)	*planning*	Possibility to back up, and document the changes (BS, GP, robot motions)	The operator backs up, and documents the changes (BS, GP, robot motions). Quality circles with designer, programmers and operators	designer programmer operators

3 TECHNICAL REQUIREMENTS OF THE ROBOT SYSTEM

To realize the automation solution as described in Table 1, the technical system must be designed in an appropriate manner. Therefore we designed a robot system including a programming and interaction environment, a special manipulator and a highly integrated sensor gripper.

The user-interface: interaction and programming

To fulfil the requirements of transparency and decision authority machines and particular the machine controllers must be designed in a way that data access and interaction in tasks and decisions are supported. Therefore a control system was developed which supports efficient robot programming and interaction at four different levels (Uhrhan & Roshardt, 1994).

The *system programming level* is highly specific to the robot and is to be used by the robot engineer at development time.

The *motion level* is the basis for the definition of more application oriented commands. Motions consist of a geometrical and eventually kinematic description of a path to be moved. This level is used to program on-line new robot motions or tune and edit those automatically generated by a CAP-system. The generated paths are presented in a graphical way, see Figure 1. The point and click interface with drag and drop operations is running on a pen computer with a joystick. This allows the operator to move around in the bending cell and to look at the point of interest while programming in an intuitive way.

At the *process level* the system is programmed in a macro like language with application oriented high level commands. It uses the same graphical user interface as the motion programming but the level of abstraction is higher. The modular units can be composed into complex tasks. The parameters of the macros may be given from a previous planning level, by teach-in, by the description of sensor values in the goal position or by manual editing at the product level.

At the *product level* programming is done by describing the product in an appropriate manner. In our case by describing the flat sheet size and the bending sequence for manufacturing the sheet part. Robot programming is carried out automatically with this data and the result can be simulated graphically.

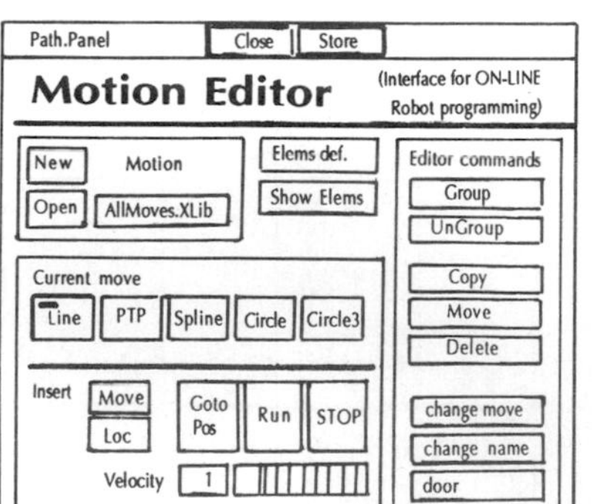

Figure 1 User interface of the motion level programming.

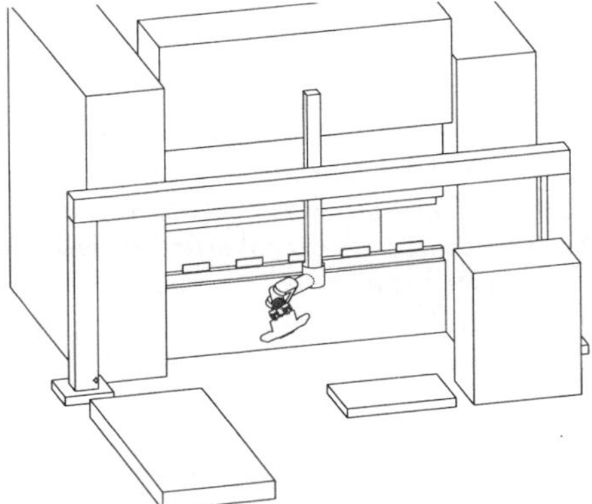

Figure 2 Simulation of the automated press brake.

The manipulator and sensor gripper

As we had established, automated workpiece handling should be aimed at with the constraint that also manual charging of the press brake should be possible. To support this we designed a robot structure consisting of a gantry robot with two prismatic and four revolute joints (Figure 2). This configuration enables manual sheet handling without any restriction and it reduces the dangerous area in front of the automated press brake.

To guarantee reliable execution of the sheet handling we developed a sensor gripper which perceives the deviations and interactions with the environment. The sensor information is also used to support manual robot programming. For teaching the robot needs only to be moved in the neighbourhood of the destination. The final location is found sensor guided.

4 CONCLUSION

KOMPASS is a heuristic for the design of automated, sociotechnical work systems. It is based on a concept of complementary allocation of functions between human operators and machines. The concept's main assumptions enclose the notion that humans and machines can not replace but complete or support each other. Thus the allocation of function must ensure that the human's working task provides him with possibilities to improve and bring in his tacit knowledge. Therefore his task has to be prospectively analysed and assessed according to it's potential planning and decision-making requirements, it's transparency, it's decision authority and it's completeness. The bending cell presented in this paper has been analysed according to the KOMPASS heuristic. Analysing the tasks in advance allows to design the technology in a way to fulfil the needs of flexible function allocation. The main spots in realizing our system for press brake automation are the design of a manipulator, the integration of sensors, the user interface and the programming environment and concept.

The project is part of the "ZIP - Zentrum für integrierte Produktionssysteme" (Centre for Integrated Production Systems), which is composed of several institutes of the ETH. The construction of the forming cell is carried out in cooperation with the Institute for Forming Technology of the ETH. The aspects belonging the operators work are analysed together with the Work and Organisational Psychology Unit.

5 REFERENCES

Bainbridge, L. (1983). Ironies of automation, in *Analysis, design and evaluation of man-machine systems* (Eds: G. Johannsen & J.E. Rijnsdorp). Proceedings of the IFAC / IFIP / IFORS Conference Baden-Baden 1982. Pergamon Press, Oxford.

Bastl, W., Jenkinson, J., Kossilov, A., Olmstead, R.A., Oudiz, A. & Sun, B. (1991). Balance between automation and human actions in NPP operation, in *Proceedings of an international symposium of balancing automation and human action in nuclear power plants.* International Atomic Energy Agency, Vienna, 11 - 32.

Grote, G. (1994). A participatory approach to the complementary design of highly automated work systems, in *Human factors in organizational design and management* (Eds. G. Bradley & H.W. Hendrick). Elsevier, Amsterdam.

Grote, G., Weik, S., Wäfler, T. & Zölch, M. (in press). Criteria for the complementary allocation of functions in automated work systems and their use in simultaneous engineering projects. *International Journal of Industrial Ergonomics.*

Hacker, W. (1986). *Arbeitspsychologie.* Schriften zur Arbeitspsychologie (ed. E. Ulich), Band 41. Huber, Bern.

Jordan, N. (1963). Allocation of functions between man and machines in automated systems. *Journal of Applied Psychology*, 47 (3), 161 - 165.

Reissner, J. & Ehrismann, R. (1990). Einsatz von regel- und algorithmenbasierten Verfahren bei der Bestimmung von Biegefolgen. *VDI Berichte Nr. 867.* VDI-Verlag, Düsseldorf.

Uhrhan, Ch. (1994) Automated Sheet Bending with Press Brakes, in *Proc. of the 25th Int. Sym. on Industrial Robots (25th ISIR).* 25.-27. April 1994, Hannover.

Uhrhan, Ch. & Roshardt, R. (1994) User Oriented Robot Programming in a Bending Cell, in *Proc. of the Int. Conference on Intelligent Robots and Systems*, Munich, Vol.2.

Ulich, E. (1993). CIM - eine integrative Gestaltungsaufgabe im Spannungsfeld Mensch, Technik, Organisation, in *CIM - Herausforderung an Mensch, Technik, Organisation*, Band 1 (Eds G. Cyranek & E. Ulich). vdf, Zürich; Teubner, Stuttgart.

Ulich, E. (1994). *Arbeitspsychologie.* Verlag der Fachvereine, Zürich; Poeschel, Stuttgart.

Ulich, E., Rauterberg, M., Moll, T., Greutmann, T. & Strohm, O. (1991). Task orientation and user-oriented dialog design. *Int. Journal of Human-Computer Interaction*, 3 (2), 117 - 144.

Warnecke, H.J. (1993). *The Fractal Company, A Revolution in Corporate Culture.* Springer, Berlin.

Weik, S., Grote, G., Zölch, M. (1994). *KOMPASS Complementary Analysis and Design of Production Tasks in Sociotechnical Systems.* IOS Press.

Weik, S., Grote, G. & Zölch, M. (1994). KOMPASS: Complementary analysis and design of production tasks in sociotechnical systems, in *Advances in agile manufacturing* (Eds. P.T. Kidd & W. Karwowski), IOS Press, Amsterdam, 250 - 253.

6 BIOGRAPHY

The authors are working as research assistants at the Institute of Robotics and the Work and Organizational Psychology Unit of the ETH Zurich. R. Roshardt is computer scientist, C. Uhrhan industrial engineer and T. Waefler and S. Weik are work psychologists. All are members of the Centre for Integrated Production Systems (ZIP), of the ETH.

PART SIX

Computer Aided Process Planning

14

AI for manufacturing: Some practical experience

J. Lažanský, V. Mařík
Czech Technical University, Faculty of Electrical Engineering, Technická 2, CZ-166 27 Prague, Czech Republic
Phone: (+42 2) 293107, Fax: (+42 2) 290159
E-mail: {marik, lazansky}@lab.felk.cvut.cz

Abstract

The paper gathers authors' experience with industrial applications of AI methods. Many tasks from the area of manufacturing motivate investigations in AI and vice versa. Two particular examples of software systems for computer support for manufacturing in small and medium sized enterprises are discussed. The first series of systems is introduced, and the problem of constrained sequencing of operations is discussed based on the classical problem solving theory. The other example involves operative planning as a discrete optimization task. An attempt to solve this problem using genetic algorithms is introduced and some experimental results are provided. The conclusions summarize authors' observations on AI methods in the design of software for manufacturing.

Keywords

CIM, artificial intelligence, computer-aided production planning

1 INTRODUCTION

Besides of the individual AI methods and techniques (like state-space search, knowledge representation, uncertainty processing) which are rarely directly applicable in manufacturing, the AI has brought a new philosophy of software development, integration and reusability. This philosophy may be even more important outcome of AI than the individual techniques.

We would like to demonstrate in this paper that the individual AI methods and techniques proved to be also useful and applicable in solving routine industrial tasks. But these methods are usually too general to be directly applied. Their application requires a detailed analysis of the task, followed by an *ad hoc* design and development of the software in which modified AI techniques and their appropriate combinations are involved in a natural way.

Two examples of larger software systems for manufacturing are briefly presented in this paper with the goal to summarize some experience concerning the industrial AI applications and to show their impacts on SMEs.

2 EXAMPLE I: CAPP SYSTEM TEPRO

The Computer Aided Production Planning (CAPP) systems bridge the gap between CAD and the workshop production needs. These systems are used to transform product definition data which have been generated by CAD into data necessary for the production process. This transformation has to consider many physical constraints given by the manufacturing environment (Bray, 1988). For example: to manufacture a certain part, a sequence of production steps (operations) on various machines has to be done.

2.1 Manual versus CA planning

In a classical manufacturing environment the process planner (a person) makes the plan 'by hand' using various tables, having in mind all manufacturing facilities in the particular factory. By tradition, the planner normally makes use of a set of predefined forms with various items about material, working site to be used, textual description of the operations, tools, etc.

A detailed analysis of requirements on CAPP systems has been done in several Czech companies. All these companies were interested in CAPP systems for supporting mainly the logistic problems of the production planning. To illustrate the problem, consider the manufacturing of simple mechanical components. Each of them is produced in a series of operations. Here the 'operation' is considered as any processing done at one working site (machine, workcell), no matter how complicated the processing is.

Various working sites in the workshop differ substantially in their versatility and processing attributes. To choose particular machine, many properties have to be taken into account. These include: capability of the machine to provide the particular type of operations, their effectivity and accuracy, capability to process the given material at the current size, etc.

Moreover, particular machines are not involved in the manufacturing process as isolated units. Since the given product usually cannot be finished on one machine it is often necessary to think about the possibilities of in-process transport which is very often accomplished by making use of in-process buffering stores.

When a machine finishes its operation the product (or a batch of products) is forwarded into the corresponding store. Thus the next operation on the same product can be carried out only on a machine that is accessible from that store. Although the in-process transport system may be organized in another way, it should always be considered as a restriction for the sequence of operations. A graph structure can be used to model the transport system. Nodes of this graph represent single workplaces and the edges play the role of transport connections. This graph is not a complete one thus imposing non-relaxable constraints on the workplace sequence.

The planner is expected to create the plan using the product drawing (or the outputs of a CAD system) and having a general idea on the ordering of necessary operations. The user has to make many decisions during the detailed process plan creation. There are also many routine calculations so that the process plan be full and correct. Many data from various tables and standards have to be filled in corresponding form items.

The analysis lead to the following conclusion agreed with the customers. The CAPP systems are expected to deliver production plans similar to those prepared manually. This suggested to design the CAPP systems as interactive knowledge-based screen editors of predefined forms. Creating a new production plan can be considered as editing of a dynamically concatenated series of empty forms. Each form is displayed and edited on the terminal screen.

The editor offers and accepts only those user's actions that are in concordance with the knowledge base of the CAPP system. This mechanism of the constraint-directed planning keeps the user within the constraints preserving his/her freedom in decision making. The editor interprets the user's decisions and transfers all of the relevant data from the corresponding data base to the production plan form.

2.2 State space approach

From the viewpoint of AI, the task of computer-aided planning may be considered as a planning with constraints in a large state space. The state space reflects possible sequences of operations performed on machines existing in the given manufacturing environment. The process plan forms a trajectory in this space. The knowledge on the manufacturing environment (describing, e.g., capabilities of machines, in-process transport restrictions, etc.) can reduce dramatically the problem solving process in a very large state space.

Let us illustrate the situation of building a linear plan by an example of choosing a suitable workplace (machine). This is one of the most complicated points of the process plan design. The machine has to be capable to carry out the desired operation on given material while current dimensions of the semi-product have to be taken into account, too. Moreover the workplace has to be accessible from that one where previous operation has taken place due to in-process transport system. When all these restrictions are put together, they reduce the set of admissible machines to several ones from original high number of workplaces.

A family of CAPP systems, called TEPRO, developed at the Czech Technical University in the last decade is a group of knowledge-based decision-support systems for complex and flexible manufacturing environments. In these systems, the human operator plays the role of a supervisor. At the beginning of the work, the operator presents the material and a rough sequence of basic (macro-)operations (like cutting, drilling, punching, turning) by use of a set of menus. Next the rough sequence is being refined by assigning the operations to suitable working sites based on predefined textual descriptions of the operations. Now the system offers only acceptable alternatives for each operation by making use of all available information.

The system considers especially machine capabilities, transport flow constraints, material restrictions, dimensional limits, precision requirements, and the machine load statistics. All these components are used to generate a sorted list of admissible working sites on each step of the planning. The user makes the final choice among those alternatives or recommends to consider several of them.

To reuse the results, the authors have decided to find a formal model of the task which is described in (Lažanský, 1995) in more detail.

The AI-based philosophy of the CAPP systems issues from the consideration that the desired sequence of technological operations can be considered as a trajectory in a state space.

The idea of the state-space search controlled by knowledge seems to be a good solution. But the problem and its formalization is not simple.

Two different, but tightly linked, dual state spaces have been designed for this purpose.

- State space F_1 of *physical manufacturing of real (semi-)product* (the states represent the current states s_n of the semi-product, the transitions o_j correspond to manufacturing operations);
- State space F_2 of *the semi-product transport* (this can be described by the topological graph of transport paths with nodes representing individual workplaces w_i and transitions t_j corresponding to transport links.)

The production planning may be described in the following way. At the very beginning, the user has a rough idea about the first operation to be performed. Thus, the choice of the first operation (o_1 in F_1) is made. All the acceptable transitions are immediately projected to F_2, with the restrictions being considered. The user chooses among these acceptable transitions. Thus both the workplace w_1 and the connection path (in-process store) are determined. Whenever the user chooses the operation o_2, all possible transitions in F_2 are immediately displayed and offered for selection. The user stepwise chooses one of them until the final state s_n is reached.

Each step in one state-space is followed by one step in the other space.

The problem-dependent knowledge substantially restricts the acceptable transitions in F_2: it deals with knowledge on transport paths, manufacturability of the material and/or semi-product, restrictions given by dimensions etc. Thus, the state space search in F_2 is dramatically reduced to one or small numbers of alternatives at each of the steps.

It is quite understandable that the formalization of the state space approach by the two dual spaces is rather complicated. The state space F_2 describes the topological graph, but the fact that this graph is reduced in each step by imposing knowledge on it justifies the state-space formalization. Moreover, the restrictions put on the planning process cannot be expressed efficiently in the natural state-space of real manufacturing.

Alternatives in plans and replanning

A very important possible extension of the methodology described above includes alternatives in the plan. In the case the alternative machines are considered, a set $W_i = \{w_i\}$ of mutually interchangable workplaces is handled instead of w_i in both the state spaces. The possibility to derive alternative plans is very important in the phase of creating the schedules. When a cost is assigned to each of the alternatives, it may be used during the execution planning (scheduling) for the estimation of schedule goodness.

New plans may be derived from already existing and verified plans. This is a very important way of fast and efficient creating of new plans. There are three types of actions which could be performed by the user. It is possible

- to change some parameters of an already existing operation.
- to delete an operation,
- to insert a new operation,

After each change in the current plan, it is necessary to check the global correctness of the new plan, i.e. to test

- whether all preconditions of the inserted or changed operation are satisfied, as well as,
- whether the operations subsequent to the deleted, inserted or changed operation are executable.

Such a checking is a specific replanning task - it can be considered as an analogy to the truth-maintenance (Sundermeyer, 1991) in a wider sense. It is solved by the simplest and the most efficient way: After each change of a plan, the global checking procedure may be applied to the new, derived plan. This procedure tries to pass successfully through both the state-spaces F_1, F_2 starting from the initial states quite automatically (without any interaction with the user). If some preconditions or restrictions are not satisfied, the procedure stops its activity and informs the user immediately. The plans which were not successfully proved by the checking procedure are prevented from being handled as correct plans (i.e. stored in the list of correct plans, etc.).

2.3 Implementation of TEPRO systems

A family of the knowledge-based CAPP systems TEPRO is based on the philosophy described above. The systems have been used in different factories in the Czech Republic.

The TEPRO system is supported by an information base which is split into two parts: data and knowledge. All the necessary facts and the available knowledge is used for decision making support. The data part of the information base consists of the manufacturing data relevant to the scope of the production type. The knowledge expresses the restrictions resulting from particular workshop or factory organization and facilities and corresponds to knowledge of a highly skilled process planner. Moreover the knowledge also describes how to manipulate the data.

The TEPRO-systems have been developed for various computer architectures and for diverse manufacturing environments. As a matter of fact, they are bound together mainly by the common AI planning philosophy.

The first implementation of the TEPRO system was developed for a flexible manufacturing system with about 50 machines for manufacturing mechanical parts of flat material in TESLA Kolín Company.

The system TEPRO-ČKD represents the largest TEPRO implementation covering all the demands on production planning on the whole factory level. It has been developed for the ČKD-TATRA Comp. which is the world's largest manufacturer of trams (street cars) with an important subsidiary in Brazil. Besides manufacturing, it was necessary to capture also the problem of creation of assembling plans which are - in principle - non-linear and very often treelike. Such plans are organized as piecewise linear ones, the ‘connection points’ being located in the assembling cells.

The TEPRO-ČKD has been directly linked to the huge, already existing data-base of the ČKD Comp. and the information base has been integrated in it.

From the AI point of view, an interesting approach called **prototype plans** has been introduced. For instance, all the plans for manufacturing doors of different trams are very similar. Prototype plans stored in the database represent ‘typical paths’ through the state space. When creating a new plan (for manufacturing a little bit different door), it is possible to use the typical plan and to carry out some slight changes in it. The main problem which remains, concerns the proof of admissibility (manufacturability) of the new path in the state space.

The TEPRO-ČKD has been run daily. About 35,000 production plans consisting of nearly 200,000 operations have been created using the system.

3 EXAMPLE II: SYSTEM PACEB

3.1 Formulation of the task

As another example, let's present a CIM case-study system called PACEB (an acronym of a Czech company). The PACEB system solves a complete CIM task in a small-sized factory that manufactures building prefabricates. The system includes data management on product assortment, inventory management, reception of orders with the chance of customer specified changes of the standard products, invoicing with payment checking, and many other daily administrative routines.

From the AI point of view, the most interesting part of the system was the demand to solve the tasks of operative planning. The task is highly constrained.

The standard assortment of prefabricates (panels) consists of about 550 types differing in dimensions, shape, construction, material, and edge profile. The factory receives orders containing desired numbers of customer chosen products. The delivery dates are estimated and the orders are to be processed to satisfy the customer requirements. Many technological and managerial constraints have to be taken into account:

(1) Each prefabricate has to harden 28 days before it can be dispatched.
(2) The planning has to consider limited capacity of the factory.
(3) Due dates satisfaction on whole orders without superfluous stocking is highly desirable.
(4) The production operates in 24 hours cycles.
(5) Panels are manufactured on special mobile bases on which boarding is fixed to create the prefabricates shape. The factory possesses bases of 4 different dimensions.
(6) A set of prefabricates is planned for each base to achieve maximal utilization of the base area. Only certain combinations of prefabricates can be scheduled for each base. The admissibility of prefabricates arrangement is stated in terms of their dimensions, edge profile and material composition.
(7) Care must be taken to minimize the necessity of boarding rearrangements on each base compared to 'yesterday' layout. This is a soft constraint.

By a deeper analysis, the conditions can be split into two groups: conditions involving temporal dependencies (1, 2, 3, 4) and constraints of geometric nature (5, 6, 7). The first group of conditions can be used to form a priority queue of panels waiting for manufacture.

The interesting part of the problem is the task of planning the prefabricates from a queue onto the production bases. This problem is a typical task of **discrete optimization**. The objective function can be formulated as a combination of two antagonistic requirements: maximization of daily profit and minimization of the boarding rearrangement. The constraints for this task can be hardly expressed analytically.

3.2 Discrete optimization using genetic algorithms

We have decided to test the applicability of genetic algorithms (GA) in this optimization task. GA represent a very promising AI technique based on simulation of evolution in nature.

Our first experiments with GA applications for classical optimization problems were very encouraging. Thus, we decided to use this approach for a simplified version of the task described in the previous section.

The crucial problem of the GA application for a constrained optimization task is the problem of representation. The question is how to represent the points of the search space so that reasonably simple and efficient genetic operators can be used. Simply represented discrete problems generate a huge majority of candidate solutions that are infeasible. For constrained cases many sophisticated *ad hoc* representations and corresponding genetic operators have been proposed. Their aim is to reduce the ratio between the number potentially generated candidate solutions and that of feasible ones. Some of those methods can be found in (Michalewicz, 1992).

The GA have a very important property: They can run infinitely in a good hope of finding a yet better solution. The best-so-far solution cannot get lost.

3.3 The GA-based approach

The first testing task was formulated for one-dimensional case as a problem of cutting a rod of length L. It is necessary to cut the rod into short pieces of various lengths l_i and given number n_i of each of them. The objective is to minimize the waste. The task can be considered as a one-dimensional knapsack problem. The experimental results have been surprisingly good and their detailed description can be found in (Kouba, 1994).

The results encouraged further extension to two dimensions with constraints very close to those in the original PACEB system. To simplify the GA task, some preprocessing of the queued panels has been done. The panels have been grouped so that prefabricates that can fit in a row on the mobile base are in one group (cf. condition (6) in the previous paragraph). A separate GA run was taken for each group, forming some possible rows arrangements. In the next step, the previously obtained rows were combined using GA to achieve maximum coverage of the mobile bases area.

For an experimental evaluation, a typical set of panels was chosen and the coverage of mobile bases was computed by a very time-consuming branch-and-bound method. Than, the briefly described two steps GA was run 10 times. In 9 of those runs, the optimum was achieved in less than 20% of the time needed for the branch-and-bound computation.

4 CONCLUSIONS

Two examples of the AI-based solutions aimed to support mainly the SME manufacturers in the Czech Republic have been presented. A lot of experience has been gathered during both the development and exploration phases of these solutions.

1. Every application of AI ideas requires a very deep analysis of the industrial problem to be solved. This analysis typically brings new information to the manufacturer and helps him to recognize, define and understand the information flows in the production environment. As a consequence, some changes in the organization of production and supporting activities can be immediately accomplished by the company.
2. Nearly all the solutions based on the requirements analysis lead to knowledge-based structures and control algorithms designed *ad hoc*. This is partially true also for implementation. All of the TEPRO implementations (even though based on the same philosophy) always required a new environment analysis which resulted in substantial changes in architecture and knowledge representation formalisms, which finally lead to programming from scratch. The requirement of software re-usability in the CIM area leads

to multi-agent solutions based on the object-oriented programming. Real solutions based on the ideas and technology of an *agent factory* (O'Hare, 1993) are still far from being widely applicable in industry.

3. The problem-oriented, environment specific solutions are typically based on some AI principles which should be modified, combined and extended in a creative way. The creativity is used mainly for transforming and re-formulating the problem into a form enabling the use of some AI (even modified or combined) algorithms. Standardized prototype solutions making the transformation problem easier are not available.
4. As the software solutions are very expensive, the manufacturers are introducing their CIM software step by step: The projects usually cover only a part of the global CIM visions. This fact imposes new constraints on the solutions: (**1**) They should be compatible with the 'rest' of the company activities and (**2**) they should respect the already existing traditions, habits and style of work in the company.
5. The management on all factory levels prefer transparent and understandable software solutions which offer acceptable solutions leaving the decision itself at a human operator. By our experience: A fully automated decision making is rarely appreciated and hardly acceptable by the end-user.
6. The software developers are welcome to use the latest AI technology. The use of up-to-date (and from the AI viewpoint attractive) terminology is not recommended even if the corresponding technology is being used. The people in industry do not trust these methods which they consider as too difficult to understand. This fact could bring an artificial barrier between the AI system designers and end-users (having, hopefully, natural intelligence).

5 REFERENCES

Bray, O. H. (1988) *CIM: The data management strategy*. Digital Press, New York

Kouba, Z., Lažanský, J., Mařík, V., Vlček, T., Ženíšek, P. (1994) Experiments with Genetic Algorithms in a CIM Task, in: *Cybernetics and Systems Research* (ed. R. Trappl), World Scientific Publ., Singapore, Vol. 2, pp. 1833-1841

Lažanský, J., Kouba, Z., Mařík, V., Vlček, T. (1995) Production Planning Systems with AI Philosophy. Int. Journal on *Expert Systems with Applications*, Elsevier Science, Vol. **8**, No. **2**, pp. 255-262

Mařík, V., Lažanský, J. (1994) Applying AI for the Development of CIM Software, in: *Proceedings of 13th World Computer IFIP Congress 94* (ed. K. Duncan and K. Kruger), Elsevier Science, Vol. 3, 1994, pp. 416-423

Michalewicz Z. (1992) *Genetic Algorithms + Data Structures = Evolution Programs*, Springer-Verlag, Berlin

O'Hare, G.M.P. and Wooldridge, M.J. (1993) A Software Engineering Perspective on Multi-Agent System Design, in *Distributed AI: Theory and Praxis* (ed. M.N. Avouris, L. Gasser), Kluwer Academic Publ., Dordrecht, pp. 109-128

Sundermeyer, K. (1991) *Knowledge-based Systems*. BI Wissenschaftsverlag, Mannheim, Germany

15

A manufacturing support system for industrial part process planning

Adilson U. Butzke, Joao C. E. Ferreira
Federal University of Santa Catarina,
Mechanical Engineering Department, GRUCON, CaixaPostal 476, 88040-900, Florianopolis, SC, Brazil, Tel:(+55)(48) 2319387, Fax: (048) 2341519, E-mail: jcf@grucon.ufsc.br

Abstract

Much research work has been carried out (and still is) on the development of Computer-Aided Process Planning (CAPP) systems as a means of bridging the gap between design and manufacture, aiming at an increase in production, higher production quality and lower costs. However, most of these works have been developed at research centers outside the factory. This commonly leads to the resulting CAPP system not being of efficient use on the shop-floor. Described within this paper is the architecture of a Manufacturing Support System for Process Planning, referred to as MSSPP, which is being developed as a cooperation between the Federal University of Santa Catarina (UFSC) and Schneider Logemann Cia. (SLC), a Brazilian company which produces machines for agricultural harvesting and planting. The system will support the process planner in the preparation of consistent and feasible process plans, taking into account the resources available at the shop-floor. It will also provide the simulation of the tool paths for machining.

Keywords

Process Planning, Design and Manufacture, Features, Databases

1 INTRODUCTION

Some years ago the industrial sector used to manufacture their goods and launch them into the market independent of the customer's desires. Since there was almost no competition among producers, the customer had no alternative but to purchase the product. With the technological developments on communications, machine tools, cutting tools and computers, there was an increase on competition, and a greater variety of products was available for the customer, who obviously would choose the one which was cheaper, and of good quality and reliability. The companies used to be "manufacturer-oriented", and became "customer-oriented" [1].

These companies purchased state-of-the-art equipment, and reorganized themselves in order to cope with this new type of production. However, some bottlenecks are usually present after the modernization, which prevent the company from attaining even higher levels of production, together with lower scrap and higher operator satisfaction. One of these bottlenecks is *process planning*.

Process planning is an activity which takes place in most factories, and is responsible for decisions such as: *machine tool selection, tool selection, workholder selection, machining parameter selection* and *determination of the operations and their sequence*. Process planning is therefore a crucial activity in product manufacture. It is traditionally carried out manually, which means that company production depends on the experience of process planners.

The process planning activity, performed in that way, can lead to a longer time for part manufacture. Some of the problems related to that are as follows:

- The process planner must have many years of experience. If him/her retires, a significant amount of time will be needed for the substitute planner to be as productive as his antecedent.
- The amount of information to be considered by the process planner is very high. Even with a vast experience, a long time is spent by him/her on process plan preparation.
- Process planning is traditionally a human activity, and as a result different planners normally prepare different process plans for the same part. Production organization may be difficult due to this inconsistency. Since on customer-oriented production the variety of parts is relatively high, it is difficult to manage these process plans.
- With the available equipment, it may be difficult to manufacture the part. In this case, the process planner would suggest a design change to the designer, so that the part can be manufactured on the shop-floor. This leads to a longer time for part production. Moreover, the different points of view of these professionals may result in clashes between them.
- It is a common practice to file the part drawing together with the process plan associated with each part. In the case of demand for a certain part, the process plan associated with it is followed through. However, this process plan may not be feasible with the resources available at the time, due to reasons such as machine maintenance, machine replacement, tools not available, etc. In this case, the process plan would have to be altered in order to adapt to the available resources. This process takes time, and thus delays part production.

Many companies, despite presenting a certain amount of production organization and automation, and also being susceptible to market change, have the deficiencies presented above, with regard to the important activity of process planning.

Research on Computer-Aided Process Planning (CAPP) systems has started in the 70's, and since then the efforts made in that direction have been academic [2,3], although some researchers report the use of their systems by industry [4,5]. In spite of the benefits attained by the CAPP systems developed to date, very few companies implement these systems. According to ElMaraguy [6], approximately 85% of all process plans produced in industry today are created manually, and optimized process plans are rarely produced.

Knowing that process planning is a bottleneck in many companies, it is important that a CAPP system is developed in such a way that the culture and organization of the company are taken into account. As a result, a cooperative work has started between the Federal University of Santa Catarina (UFSC) and Schneider Logemann Cia. (SLC), a Brazilian company which produces machines for agriculture harvesting and planting, aiming at the development of a Manufacturing Support System for Process Planning, referred to as MSSPP, whose architecture is described within this paper. This system is intended to assist the process planner in choosing the most appropriate process plan for part manufacture. With this system it is expected that the problems pointed out above are reduced or eliminated.

Although the automation of process planning has been aimed at by many authors (e.g. [7-9]), these systems lack an efficient interface between CAD and CAPP. One of the main objectives of MSSPP is to link CAD and CAPP through "feature" technology, i.e., the part is modelled based on features in the CAD module. Since features carry geometrical and technological attributes which aggregate process planning information, such as tools and fixtures, they enable the efficient process planning for the part's manufacture.

2 METHODOLOGY FOR THE DEVELOPMENT OF THE MSSPP

The activity of process planning is usually very complex, and that is the case at SLC. In order for the development and implementation of the MSSPP, the process planning problem has been subdivided into smaller problems. This subdivision consisted of: 1-reduction of the process planning domain; 2-decomposition of the system into modules. An explanation of these two strategies is given below.

Reduction of the Process Planning Domain

SLC produces machined and sheet-metal parts, and for these types of parts there are related departments. Most of the machining department is organized into manufacturing cells, which present advantages such as: shorter material transportation and handling times, easier part finding, etc. Since each cell manufactures a significant amount of parts, they can be considered as "small factories" (see Figure 1). Rarely a part from one cell is introduced into another cell. So, for the present work, one cell has been chosen for the development and implementation of the CAPP system, which is a reduction in the process planning domain. It is expected that the same approach can be applied successfully to other cells.

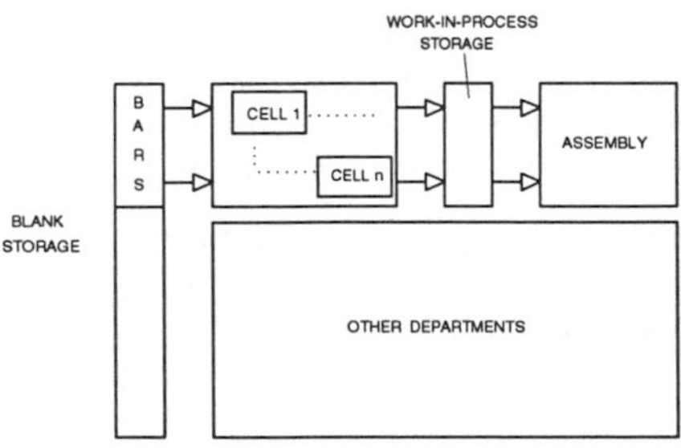

Figure 1 Flow of parts within SLC (Storage → Cell → Assembly)

It has been chosen a cell where approximately 480 parts are manufactured, and the machine tools in this cell are: four CNC lathes with bar feeder, one conventional Universal lathe, and two sawing machines. The tools used by each machine are placed on shelves beside the machines, as well as the workholders. This reduces transport time of tools and workholders, and facilitates their management. The transport of parts, tools and workholders within the cell is done manually.

System Modularization

In software development, the technique of separating specific tasks into modules usually gives good results, mainly because different groups of people may develop each module, independent of the other modules, which decreases the time for development. Also, if modifications to a specific task are necessary, the related module is altered, without need to change the other modules.

Process planning is an activity where various tasks are performed. According to Wang and Li [1], an effective CAPP system should: 1-communicate with a CAD module, 2-select the blank, 3-select the operations, 4-select machine tools and cutting tools, 5-determine intermediate surfaces, 6-select machining parameters, 7-select workholders, 8-estimate time and cost, and 9-generate the cutter paths

Therefore, it is recommended that a CAPP system should be modular for ease of development and maintenance.

The proposed architecture of this system is shown in Figure 2. This architecture has been defined taking into consideration the attributes pointed out above.

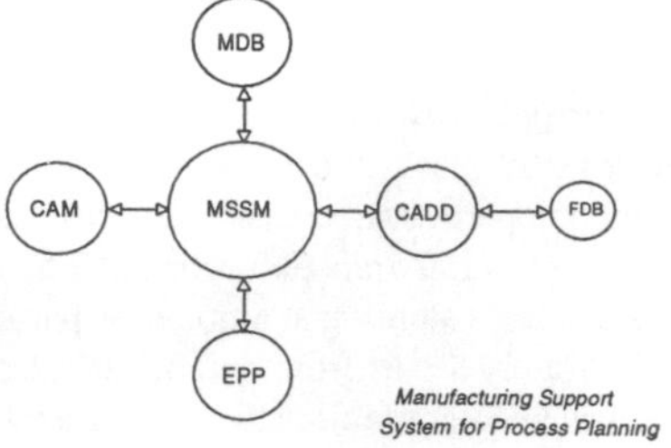

where:
CADD: design and drafting module
FDB: feature database
MDB: manufacturing database
EPP: expert process planning module
CAM: tool path generation and simulation module
MSSM: manufacturing support system manager

Figure 2 Proposed architecture of the Manufacturing Support System for Process Planning

A description of each of the modules, and how they interact with each other is given below.

3 THE CADD MODULE

This module has as objective to assist the designer in the construction of the part in the computer. The part representation must be complete and unambiguous, so that the constructed part corresponds to the desired one.

In this module, "features" are used as the building blocks for the parts. Many different feature definitions have been found in the literature, but there has not been a consensus yet. In this work, "features" are defined as "regions or volumes in machined parts, being important for design, process planning and other activities" [10]. Some examples of features in this work are shafts, holes, chamfers, grooves, etc. Figure 3 illustrates how process planning information (e.g., machine, tool, workholder) can be selected, having a feature as the origin.

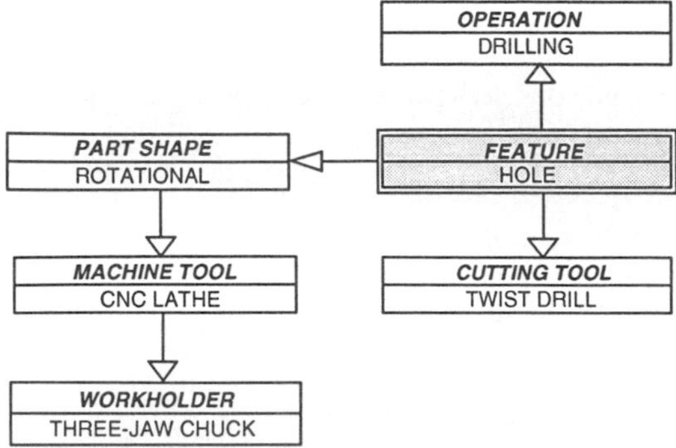

Figure 3 Manufacturing decisions based on features

Two advantages of feature-based design are:

- Features are elements which carry geometrical, functional and technological information, and therefore the designer is more acquainted with them.
- Features facilitate reasoning about the manufacturability of the part, and consequently its process plan.

Much research work has been developed on feature-based design (e.g. [11,12]), and features have been used in many CAPP systems, such as TURBO-MODEL [7], FIRST-CUT [13] and Juri's system [14].

In the TURBO-MODEL system [7], the methodology applied consists of representing various types of information about the part (i.e., geometrical, technological and global information). It models rotational parts, and part's design is carried out in a pre-determined way. That system includes a sub-program to verify the feasibility of the instantiated features, with respect to geometrical, technological and functional attributes. It also has an interface with AutoCAD. A limitation of that system is that it does not provide much flexibility to the designer during feature-instantiation.

FIRST-CUT [13] is a feature-based design system that aims at integrating CAD and CAPP. In this system, the process analysis for product manufacture is incorporated into the design phase, in order to attain an efficient process plan.

Juri [14] investigated the feature-based approach for modelling rotational parts in a product modelling system. A product is modelled through form features, which can be related to a manufacturing process, and a structure based on frames is used to represent the features. A technique based on production rules is applied in order to perform operation selection and sequencing.

Since the design methodology at SLC is based on features, the CADD module is feature-based, and thus a feature database (FDB) has been constructed. In order to construct the feature database, information has been gathered on the features present in the drawings of the parts machined at the cell. Based on this information, features have been classified, and a portion of this classification is illustrated in Figure 4. This classification is based on the geometry of the features, as it has been done by Jasthi [7].

Detailed information on two features present in parts at SLC is given in Figure 5. Notice that in order to decrease time for search and disk space usage, the features are parameterized. It should also be noticed that none of the geometric constraints shown in the figure can be violated, otherwise the part would be inconsistent. This is a means of making the software robust and reliable.

In order to construct a part, the designer must go through the steps described in the pseudo-code below:

```
Feature_number = 0;
Repeat
    Feature_number = Feature_number + 1;
    Search the feature database;
    Extract the desired feature;
    If (Feature_number = 1) then
        Place the feature at any position;
    Else If (Feature_number ≥ 2) then
        If (Axial Feature) then
            Place the feature with its reference point on the centerline of the previous
              feature;
        Else If (Radial Feature) then
            Place the feature with its reference point on the outer surface of the previous
              feature;
        End If
    End If
Until (Part Construction is finished);
```

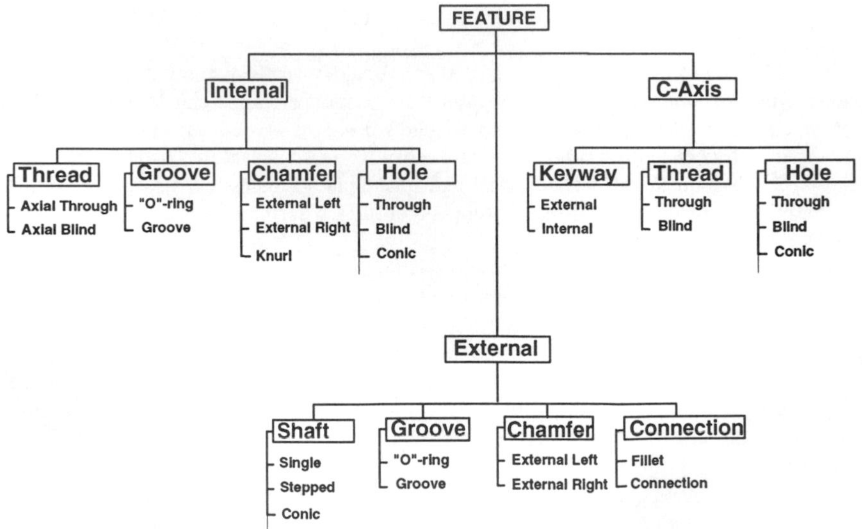

Figure 4 Portion of the feature classification

Feature type: Symmetrical internal
Feature name: Axial eccentric through hole
Feature material: SAE 1020 Steel

Geometric Parameters:
LF = Hole depth
LDF = Distance between hole center and shaft center
DF = Hole diameter
A = Angle between two adjacent holes
Z = Distance between two adjacent holes
DS= Shaft diameter
pt = Reference point
Geometric Constraints:
$0 < DF < (DS/2)$
$(DF/2) < LDF < (DS/2 - DF/2)$
LF = shaft length
$Z = LDF\sqrt{2(1+\cos A)}$; $Z > DF$
pt should coincide with shaft's reference point
Surface Finish:assigned to internal surface of the hole
Dimensional Tolerances:assigned to DF, LF,LDF
Geometric tolerances: Straightness, Circularity, Cylindricity, Concentricity, Runout

Feature type: Symmetrical external
Feature name: Shaft with thread
Feature material: SAE 1020 Steel

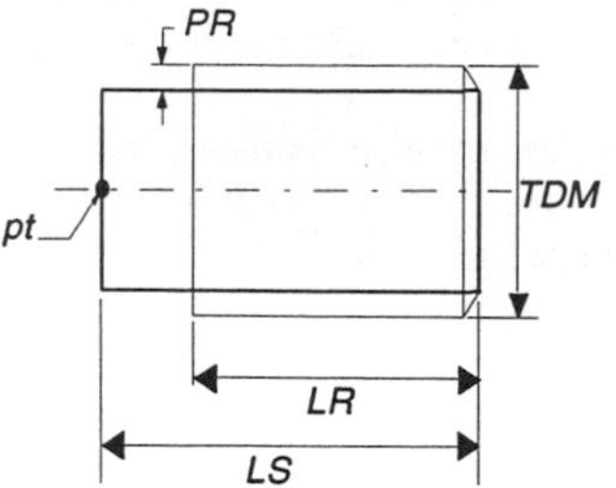

Geometric Parameters:
LS = Shaft length
LR = Thread length
PR = Thread depth
TDM = Thread diameter
pt = Reference point
Geometric Constraints:
Standard DIN 13
pt should coincide with shaft's reference point
Surface Finish: assigned to external surface of the thread
Dimensional Tolerances: assigned to PR, LS, LR and TDM
Geometric Tolerances: Straightness, Circularity, Cylindricity, Concentricity, Runout

Figure 5 Detailed structure of two features present in parts manufactured at the cell

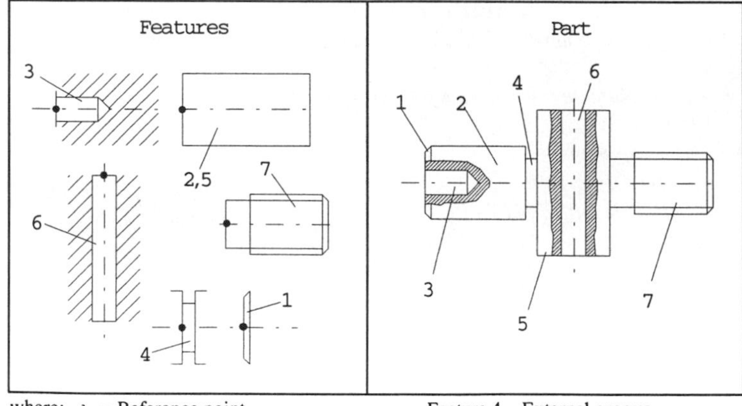

where: ı = Reference point
Feature 1 = External chamfer
Feature 2 = Shaft
Feature 3 = Axial concentric blind hole
Feature 4 = External groove
Feature 5 = Shaft
Feature 6 = Radial through hole
Feature 7 = Shaft with thread

Figure 6 Example of part modeling with features

In Figure 6 an example is given of a part and the features used in its construction.

It should be noticed that the axial features 2,3,4,5 and 7 have their reference point lying on the centerline of feature 1. On the other hand, feature 6, which is a radial feature, has its reference point lying on the outer surface of feature 5.

Two important characteristics of the CADD module concerning the construction of the part are:

- If the positioning of a feature related to the part violates at least one of the geometric constraints, the CADD module notifies the designer about the inconsistency.
- The designer can choose whether to input the tolerances and surface finish during or after part construction.

The philosophy of the CADD module is similar to the TURBO-MODEL [7], which was also developed taking into account an industrial problem, but CADD gives more flexibility to the designer to construct the part.

FIRST-CUT [13] and the CADD module are different, since in the latter process planning is carried out after the design of the part, which is the approach to design and process planning applied at SLC.

The CADD module of MSSPP is currently being implemented at SLC. This system is being tested by the designers at the company, and it is expected that eventual problems identified by them will be corrected, and suggestions for improvement of the software will be incorporated into the software.

4 THE EPP MODULE

The aim of this module is not to automate completely the process planning activity, but rather to provide the process planner with a means for him/her to elaborate consistent and feasible process plans. This module will be responsible for providing the user with the following information: 1-selected blank, 2-selected machine tool, 3-selected cutting tool and workholder, 4-selected machining parameters, 5-time and cost estimates, and 6-process plan.

In order to build the knowledge base in the EPP module, it was necessary to carry out an interview with the process planners of SLC. An example of these rules is given below:

```
Axial Hole (D=diameter; F=surface finish)
If (D < 12mm) then
  If (F > 10 µ m) then
          Center Drill;
          Drill Hole with Twist Drill;
  Else (0 < F < 10 µ m) then
          Center Drill;
          Drill Hole with Twist Drill;
          Ream;
  End If
End If
```

There have been some attempts to standardize process planning information [15,16]. In spite of these efforts, in the development of the EPP module, priority will be given to the information obtained from the SLC process planners, which represent the culture of the company. However, recommendations from outside the company will also be represented in the knowledge base, but they will be applied only when the production rules obtained from SLC process planners fail.

The gathering of information necessary for process planning (e.g. machines, tools and fixtures) has been completed, and also the link between this information with the features through their geometrical and technological attributes. However, the EPP module is still to be developed.

5 THE MANUFACTURING DATABASE

The information necessary for process plan preparation at SLC is included in the Manufacturing Database (MDB), which is composed of blanks, workholders, machine tools and cutting tools. The structure of the information contained in each of these elements is shown in Figure 7.

The machine utilization parameter is concerned with the percentage of time that the machine tool is being utilized. This information is updated after a certain period of time (e.g. one week, one month, three months, etc.), depending on the company. With this information it is possible to know whether the machine is available at a certain moment or not. This is important for the MSSPP, because the generated process plan should be feasible.

Notice that at the end of its development the manufacturing database will contain the items in the cell considered. In the case of other cells (e.g. cells for prismatic part manufacture), the database will have to be adapted to the blanks, machines, tools and workholders present at the specific cell.

6 THE CAM MODULE

This module is concerned with the generation of the tool paths to machine the part. Given the volume to be machined, the machine tool and the cutting tool, this module will generate automatically the speeds, feeds and depths of cut to machine the volume, both for roughing and finishing operations. Also, the tool paths will be simulated on the computer screen. The method to be applied consists of minimizing the cost (or maximize production), having the following as constraints [17,18]: 1-tool life, 2-cutting forces, 3-machine power, 4-cutter deflection, 5-part deflection, 6-surface finish, and 7-workpiece holding stability.

It should be noticed that at SLC, as well as in many other companies, the machining

BLANK	*MACHINE TOOL*	*WORKHOLDER*	*CUTTING TOOL*
Code:	Code:	Code:	Code:
Name:	Name:	Name:	Name:
Cutter diameter:	Power:	Type:	Cutting Speed:
Length:	Table feed:	Max. Clamping Dia.:	Feed-rate:
Chemical composition:	RPM:	Workholder diameter:	Depth of cut:
Manufacturing method:	No. turret positions:		Nose radius:
Shape:	Bed ways dimensions:		Diameter:
Quantity in storage:	Machine utilization:		Hardness:
			Operations:
			Recommended material:

Figure 7 Information on blanks, machines, workholders and tools

parameters are determined typically from the following sources: tool manufacturer catalogs, machining data handbooks or process planner's experience. However, these parameters are usually conservative for the company's specific equipment. This is why an optimization program will be developed in the CAM module, which will take into account the equipment and the environment at SLC. This will result in more appropriate parameters for the operations, and it is expected that higher production rates and lower costs will be attained.

7 THE MANUFACTURING SUPPORT SYSTEM MANAGER

The main attribute of the Manufacturing Support System Manager is to enable a proper communication between each of the modules. This is necessary because the information in each module is represented in different ways, which are usually not compatible. For example, the CADD Module will produce drawing files in a specific CAD format, which may not be standard (e.g. AutoCAD *dwg* format), and so the CADD module would then have to write the drawing information in a standard format (e.g. *IGES*). The same applies to the Manufacturing Database, whose information can be created and managed by any commercial database management software. Since the EPP module needs information both from the CADD module and from the MDB module, the MSSM would read the standard part file, which would contain feature information, and also the MDB file. These files would then be translated into an "understandable" format by the MSSM and fed to the EPP module.

If a process plan is successfully generated by the EPP module, the MSSM can then proceed to the generation of the tool path. In the case the EPP module fails in generating a process plan for the given part, the system would notify the user that it failed, and why it happened.

8 CONCLUSIONS

The cooperation between UFSC and SLC has been quite encouraging on both sides. On the company's side, the resulting process planning system will assist their process planners to prepare consistent, feasible and reliable process plans in a shorter period of time. Moreover, there will be no need for the process planner to check frequently for the items in storage, machines, tools and workholders available, resulting in a higher job satisfaction and productivity.

On the university side, students become more motivated to work on a project which will result in a product that will be implemented and used in a day to day basis in industry. The students have a practical problem in their hands, and search for realistic solutions.

The feature-based design in the CADD module is one of the most important tasks in the development of this system, because it is the system's "front door". Since this system is to be used in a company, in order for the designers and process planners to became quickly acquainted to it, it should be as friendly as possible. It is thought that with the approach described in this paper this target will be met.

When comparing the systems pointed out above with the CADD module of MSSPP, the latter presents the following advantages:

- It allows a complete definition of the parts present in the application domain, both geometrically and technologically;
- It does not allow a redundant model, as well as a model with geometrical errors;
- Feature-instantiation is easy, which gives flexibility to the designer;
- Since it is connected to a manufacturing database, it enables the designer to look into the raw materials and the manufacturing resources during the design phase;

• The CADD module can be used in any system that utilizes AutoCAD for part design.

One could argue that the reduction of domain to one cell in the whole factory may hinder the portability of the MSSPP to other cells at SLC. That may be true when considering for instance a cell for manufacturing prismatic parts, where probably in the CADD module the parts would be represented by solid modelling. However, it is thought that, with the knowledge of the company's culture and organization, and after the experience of applying the methodology above to this system, other systems could be developed for other cells, in a shorter time. The system's modularity facilitates the adaptation of each module to the new situation.

In case another company wishes to use the software, it is important that its production is of the batch type, which is the case of most industries today. It is likely that the software would need to be altered in order to incorporate the culture of the new company (e.g. feature information, machine tool and cutting tool information, etc.).

Since the MSSPP is under development, no results have been obtained so far, and therefore effective results of the system, after its implementation at the company, will be reported when that occurs.

9 EQUIPMENT AND SOFTWARE USED

The system as a whole is being developed in a 386 microcomputer, under DOS operating system. AutoCAD is being used in the CADD module. IGES is the standard format for interfacing MSSM and CADD. The EPP module will be developed in an expert system shell, based on the "C" language, which provides the tools necessary for attaining the required efficiency of process planning. The other modules are being developed in the "C" language.

10 ACKNOWLEDGEMENTS

We would like to thank the staff at the Schneider Logemann Cia. (SLC) for their support, without which this work would not have happened. The first author would like to thank "CNPq" (Conselho Nacional de Desenvolvimento Cientifico e Tecnologico) of Brazil, for the scholarship provided.

11 REFERENCES

1. Wang, H.P. and Li, J.K. (1991) *Computer-Aided Process Planning*. Elsevier Science Publishers.
2. Wysk, R.A. (1977) *An Automated Process Planning and Selection program: APPAS*. Ph.D. Thesis, Purdue University, West Lafayette, Indiana, U.S.A.
3. Van't Erve, A.H. (1988) *Generative Computer Aided Process Planning for Part Manufacturing - An Expert System Approach.* Ph.D. Thesis, University of Twente, Enschede, The Netherlands, January.
4. Link, C.H. (1976) CAPP - CAM-I Automated Process Planning System, in *Proceedings of the 13th Numerical Control Society Annual Meeting and Technical Conference*, Cincinatti, Ohio, U.S.A., March.
5. Eversheim, W. , Fuchs, H. and Zons K.H. (1980) Automatic Process Planning with Regard to Production by Application of the System AUTAP for Control Problems, in *Computer*

Graphics in Manufacturing Systems, 12th CIRP International Seminar on Manufacturing Systems, Belgrade, Yugoslavia.

6. ElMaraguy, H.A. (1993) Evolution and Future Perspectives of CAPP. *Annals of the CIRP*, **42**, 2, 739-751.
7. Jasthi, S.R.K. et al. (1994) A Feature-Based Part Description System for Computer-Aided Process Planning. *Journal of Design and Manufacturing*, **4**, 67-80.
8. Wang, H.P. and Wysk, R.A. (1988) A Knowledge-Based Approach for Automated Process Planning. *International Journal for Production Research*, **26**, 6, 999-1014.
9. Hinduja, S. and Barrow, G. (1986) TECHTURN: A Technologically-Oriented System for NC Lathes, in *1st Conference on Computer-Aided Production Engineering*, Edinburgh, UK, 295-305.
10. Korde, U.P. (1992) Computer-Aided Process Planning For Turned Parts Using Fundamental and Heuristic Principles. *Journal of Engineering for Industry*, **114**, February, 31-40.
11. Shah, J.J. and Rogers, M.T. (1988) Functional Requirements and Conceptual Design of the Feature-Based Modelling System. *Computer-Aided Engineering Journal*, February, 9-15.
12. Gu, P. (1994) A Feature Representation Scheme for Supporting Integrated Manufacturing. *Computers in Industrial Engineering*, **26**, 1, 55-71.
13. Cutkosky, M.R. and Tenembaum, J.M. CAD/CAM Integration Through Concurrent Process and Product Design.
14. Juri, A.H. et al. (1990) Reasoning About Machining Operations Using Feature-Based Models. *International Journal for Production Research*, **28**, 1, 153-171.
15. Butterfield, W.R.; Green, M.K.; Scott, D.C. and Stoker, W.J. (1986) Part Features for Process Planning. *CAM-I Technical Report R-86-PPP-01*, Arlington Texas, U.S.A.
16. ISO TC184/WG3 N324 - T7 (1994) ISO 10303 - Part 224 - Mechanical Product Definition for Process Planning using Form Features, South Carolina, U.S.A., 27/June.
17. Huang, H. (1988) *A Generative Process Planning System for Turned Components.* Ph.D. Thesis, UMIST, Manchester, UK.
18. Lau, T.L. (1987) *Optimization of Milling Conditions*. Ph.D. Thesis, UMIST, Manchester, UK.

16

Fuzzy logic for similarity analysis

J. A. B. Montevechi and G. L. Torres
Escola Federal de Engenharia de Itajubá
CP 50 - Itajubá - MG - 37500-000, Brazil, tel.: +5535 629 1212,
fax: +5535 629 1148, e-mail: arnaldo%efei.uucp@dcc.ufmg.br

P. E. Miyagi and M. R. P. Barretto
Universidade de São Paulo - Escola Politécnica
CP: 61548 - São Paulo - SP - 05508-900, Brazil

Abstract

In this article a procedure for obtaining similarities using fuzzy logic is discribed. It permits taking into account uncertainties and ambiguities usually present in manufacturing. Aspects of the relational data base are presented which can aggregate design and manufacture information. How to assign membership to the features that will be analysed, the similarity analysis to the resemblance relation and part processing information are also presented. This procedure makes it possible the elaboration of one software which will be a interesting tool to the manufacture of small lots, it integrates the informations of design and manufacture and makes possible a rationalization of resources.

Keywords

Group Technology, part family, fuzzy logic, membership attribution

1 INTRODUCTION

Most papers about part family formation assume that information about cost and processing time, demand of part, etc.. are accurate. It is usually supposed that a part only belongs to one family. Nevertheless, in many cases it does not occur. The analysis of grouping, making use of fuzzy logic, can provide a solution to this problem. However, few articles have been published dealing with the problem of uncertainty, the formation of manufacturing cells and the part families. Likewise, those articles consider such questions isolately neglecting the development of methods to be shared by all company users (Montevechi, 1994).

The part similarities, which is the basic aspect for family formation, consist of a close classification in geometry, function, material and /or process. It may be not sufficient to describe part features using yes or no labels, when accurate classification is required (Montevechi, 1994). In order to obtain an efficient and flexible classification which considers uncertainties, thus eliminating the shortcoming of the currently employed methods, this article describes a procedure that makes use of fuzzy logic for the part family formation. Fuzzy membership function permits taking into account the inherent uncertainties into part features description. Thus producing more realistic results. The use of this technique, will make the part family formation more sensible. The membership value, which lies between 0 and 1 can express what extension of the feature the part has. The closer the value is to 1, more quantity of feature the part has.

First are described details of a relational data. The grouping principle employed is also described which consists of choosing one threshold value to the similarity. Once this threshold value is chosen, two elements will be in the same group if, for example, in case of similarity function, the similarity between them is greater than the threshold value. Since the similarity relationship is not necessarily transitive, it is necessary to employ the fuzzy matrix theory to form the closest structure which permits the separating data in exclusive and separated groups which are, in essence, an equivalent class over a certain threshold value.

For process similarity a procedure is shown to search information of similarities that should guide the formation of manufacturing cells.

Another important aspect described is the possibility to use qualitative data, such as, complex, easy, hard, high surface roughness, etc.... How to translate this information into numerical values, which is essential to similarity analysis, is also shown.

The object of this methodology, is to develop an alternative procedure to traditional methods for obtaining similarity. To this purpose, it is necessary to integrate apropriate approachs that can incorporate the uncertainty which nowadays serve isolated aspects of similarity analysis.

2 DATA BASE

An extremely important component of the proposed methodology to obtain the part families, is the data base providing the features that will be classified according to their similarities. In the data base all important information about several component features of the company have to be available, including design and manufacturing features. However, it should be borne in mind that no coding procedure is proposed, which is common in the Classification and Codification System (CCS). The part input information is made according to discription of its features. This fact is very important because it permits data handling by all company user, fact that does not happen with most CCS. Features that may be in the data base are shown in Figure 1.

Development a relational data base permits adding more data and features. It is not possible in CCS, because after the coding is obtained the insertion of new features will be difficult.

DESIGN ATTRIBUTES		MANUFACTURE ATTRIBUTES	
External shape	Main dimensions	Main process	Machine tool
Internal shape	Material	Largest dimension	Lead time
Tolerances	Superficial Finishing	Superficial Finishing	Lot size

Figure 1 Example of data base design and manufacturing features.

3 MEMBERSHIP ATTRIBUTION

The parts for the classification by their similarities is represented in a **n** x **m** matrix form (part x feature). This matrix will be formed with the data base features important for the grouping analysis. Basically in the data base it is possible to have two types of features, namely quantitative and qualitative.

Since the features can be quantitative or qualitative, it is important to develop a procedure for the grouping which can deal with these two kinds of features in a unified way. It is necessary to transfer the data to these features in the same unit. Otherwise, there will be a scale problem. In this methodology the data of different features are expressed by memberships. To this purpose, each part feature is given a membership between 0 and 1, which will put an end to the scale problem.

Membership values of quantitative features can be expressed directly in function of values obtained from the data base. For example, the length feature whose values for 7 parts are given by the vector [10.00; 8.5; 5.5; 3.75; 6.25; 8.00; 7.50], the membership values for this feature, can be calculated by dividing each length value by the largest value of the vector (i.e. 10.00). This procedure results in another vector, that is the membership vector, given by [1.00; 0.850; 0.55; 0.375; 0.625; 0.800; 0.750]. These values also can be given by any of the expressions suggested by Xu and Wang (1989).

Likewise, it is possible to use graphs with a more suitable membership function, which automatically gives the values of membership to the selected features.

With qualitative features, now, the attribution of membership is not so easy. How to transform a qualitative information, such as part complexity or with high roughness in a number?

To solve this problem, such as proposed by Arieh and Triantaphyllou (1992), it is possible to utilize the AHP (Analytic Hierarchy Process) method (Saaty, 1977), which by means of a comparative analysis permits the calculation of membership for qualitative attributes. These comparative analysis are made in pairs between the attributes of the feature in question. It is necessary to make a matrix of comparison for the attributes for each feature. Features are, for example, roughness, shape complexity, length, etc, whereas the attributes are the feature designation as small, large, complex, little complex, high roughness, etc. Saaty proposed that to provide attribute comparison should be used values from the finite set: {1/9,1/8,...,1,2...,8,9}. These matrices are calculated by the evaluation of the importance of an attribute over the other, using an appropriate scale (Saaty, 1977).

Each entry of the matrix is a pair of judgment. After the matrix of comparison is defined, the eigenvalue (λ_{max}) and its respective eigenvector will be calculated. The eigenvector will represent the memberships that can be used for the attributes in question and the eigenvalue is the measure or rate of consistency of the result.

To illustrate this method, one of the features which may be important for the obtaining similarity is the complexity of shape evaluated by a analist. The attributes of this feature originating from the data base range from very complex, to complex, mean complexity, low complexidade and very low complexity. All the parts, of the data base, have one of these qualitative values for its feature of complexity. By means of the scale of priority shown, a specialist will provide the matrix **A** of Figure 2.

	very complex	complex	mean complexity	low complexity	very low complexity
much complex	1	3	5	7	9
complex	1/3	1	3	5	7
mean complexity	1/5	1/3	1	3	5
low complexity	1/7	1/5	1/3	1	3
very low complexity	1/9	1/7	1/5	1/3	1

Figure 2 Matrix **A** of comparison by pairs for the feature shape complexity.

In matrix **A** of comparison, entry a_{ij} indicates the number that estimates the relative membership of attribute $\mathbf{A}_i$ when it is compared with the attribute $\mathbf{A}_j$. Obviously, $a_{ij}=1/a_{ji}$. With the eigenvector, the memberships that can be used for the similarity classification are available, obviously after testing the consistence of the result to conclude that the answer is good (Saaty, 1977). In this case, the memberships (one of eigenvectors normalized for the greatest weight to be equal 1), after the calculation, for the attributes values will be to very complex: 1, complex:0.517, mean complexity: 0.254, low complexity: 0.125 and very low complexity: 0.064. These values of memberships now will give a weight for each of the parts from the data base. These memberships represent the importance of complexity of shape for each part.

After the memberships for the several features (qualitative and quantitative) are calculated, now, it is necessary to have a procedure to obtain the clusters of similar parts.

4 ANALYSIS OF SIMILARITY FOR THE FEATURES

A principle that can be used for the grouping is to choose a threshold value for the similarity. Once this threshold value is chosen, two elements will be in the same grouping if the similarity between them is larger than the value of comparison.

To estimate the resemblance between pairs of data, it is possible to use the convention of arranging the data in form of matrix. Each entry in this matrix will represent the proximity between two parts. This relationship, called **S** which is a matrix **n** x **n**, represents the similarity between different parts. To obtain this matrix it is possible to utilize several formulaes for the calculation of similarity, as examplified by the expression (1).

The symmetric matrix can be used directly in the analysis of fuzzy grouping. The similarity of parts consist of a very close classification in geometry, function, material and /or process. The similarity as seen, will be obtained throught the membership manipulation.

$$S(x_i, x_j) = \frac{\sum_{k=1}^{p} min(\mu_k(x_{ik}), \mu_k(x_{jk}))}{\frac{1}{2}\sum_{k=1}^{p}(\mu_k(x_{ik}) + \mu_k(x_{jk}))} \qquad (1)$$

But the matrix does not have the propriety of being transitive. Here what occurs is that there is a relationship of resemblance. Then, the following conclusion, that if **A** is similar to **B**, and **B** is similar to **C**, then **A** is similar to **C** will not be possible. To deal with this problem, it is necessary to transform the matrix of similarity into a transitive matrix. It is possible to use the fuzzy theory to transform this matrix (Montevechi, 1994). The transitive matrix is the matrix fuzzy equivalent.

Finally, given one ∝ level, the groupings of similar parts are obtained for the level chosen. With different ∝ values, different classifications will appear. The greater is the ∝ value, less parts will be classified in each family, thus more families will be formed.

An example of decomposition, for obtaining the families can be better understood through Figures 3 and 4. Figure 3 shows the attribution of some ∝ values and Figure 4 shows, for each one of the (∝) levels, the groupings formed. For example, for ∝ = 0,9, there are three groupings, the first one consisting of parts **A**, **D** and **E**, the second one, of part **B** and the third one, of part **C**.

$$R = \begin{bmatrix} 1 & 0{,}8 & 0{,}7 & 1 & 0{,}9 \\ 0{,}8 & 1 & 0{,}7 & 0{,}8 & 0{,}8 \\ 0{,}7 & 0{,}7 & 1 & 0{,}7 & 0{,}7 \\ 1 & 0{,}8 & 0{,}7 & 1 & 0{,}9 \\ 0{,}9 & 0{,}8 & 0{,}7 & 0{,}9 & 1 \end{bmatrix} = \vee \left(0{,}7. \begin{bmatrix} 1 & 1 & 1 & 1 & 1 \\ 1 & 1 & 1 & 1 & 1 \\ 1 & 1 & 1 & 1 & 1 \\ 1 & 1 & 1 & 1 & 1 \\ 1 & 1 & 1 & 1 & 1 \end{bmatrix}, 0{,}8. \begin{bmatrix} 1 & 1 & 0 & 1 & 1 \\ 1 & 1 & 0 & 1 & 1 \\ 0 & 0 & 1 & 0 & 0 \\ 1 & 1 & 0 & 1 & 1 \\ 1 & 1 & 0 & 1 & 1 \end{bmatrix}, 0{,}9. \begin{bmatrix} 1 & 0 & 0 & 1 & 1 \\ 0 & 1 & 0 & 0 & 0 \\ 0 & 0 & 1 & 0 & 0 \\ 1 & 0 & 0 & 1 & 1 \\ 1 & 0 & 0 & 1 & 1 \end{bmatrix}, 1. \begin{bmatrix} 1 & 0 & 0 & 1 & 0 \\ 0 & 1 & 0 & 0 & 0 \\ 0 & 0 & 1 & 0 & 0 \\ 1 & 0 & 0 & 1 & 0 \\ 0 & 0 & 0 & 0 & 1 \end{bmatrix} \right)$$

Figure 3 Decomposition of a similarity relationship.

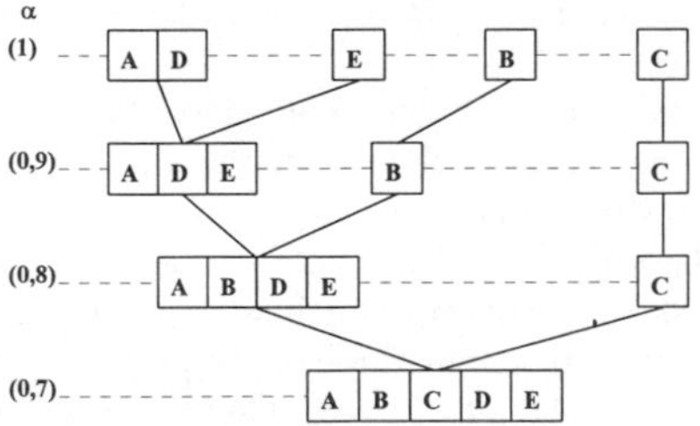

Figure 4 Tree shaped decomposition.

5 ANALYSIS OF PROCESS SIMILARITY

A problem that may happen, in the formulation for obtaining the similarity seen in the last item, is that the information about cell formation may not be enough. To prevent this from happening, it is possible to think in such a way to obtain the matrix of membership between part **x** machine, where the vagueness will also be considered. In this way an algorithm for checking the grouping of machines may be used.

If **n** parts and **m** machines are being considered to obtain manufacturing cells, usually the representation of the machines that processes each part is given by a binary matrix, where de entries are 1 or 0. Due to the inflexibility of this matrix, another matrix should be developed, which should be called of nonbinary, represented by (2).

$$\text{Nonbinary Matrix} \Rightarrow \begin{array}{c} \\ Y_1 \\ Y_2 \\ Y_3 \\ \vdots \\ Y_m \end{array} \begin{array}{c} \begin{array}{ccccc} X_1 & X_2 & X_3 & \cdots & X_n \end{array} \\ \begin{bmatrix} u_{11} & u_{12} & u_{13} & \cdots & u_{1n} \\ u_{21} & u_{22} & u_{23} & \cdots & u_{2n} \\ u_{31} & u_{32} & u_{33} & \cdots & u_{3n} \\ \vdots & \vdots & \vdots & \cdots & \vdots \\ u_{m1} & u_{m2} & u_{m3} & \cdots & u_{mn} \end{bmatrix} \end{array} \quad (2)$$

in (2), X_j is a part and j = 1, 2, 3, ..., n; Y_i is a machine and i = 1, 2, 3, ..., m and u_{ij} represents the relationship between part **j** and machine **i**. In (2) it is possible to observe the property (3).

$$0 \leq u_{ij} \leq 1 \text{ for } i = 1, 2, 3, ..., m; j = 1, 2, 3, ..., n; \quad (3)$$

The property defined by (3) indicates the intensity with which a machine is designated to process a determined part, a number near of 1 meaning a great potentiality to process the part, while with a number near 0, the machine would definitely not be appropriate.

The elements of matrix (2) are calculated from mixed functions between machines and components, which is an interesting proposition from Zhang and Wang (1992).

To illustrate this procedure, for a small example, if the pertinence functions are those of Figure 5, for a hypothetic case of machine 1 of **m** available, for each one of the 7 parts that are to be grouped, with the **x** value (from a data base) for the feature in question, it is possible to do a chart of Figure 6.

In the same way that the values of Figure 6 are given, the process should be repeated for all the machines that will be analyzed, thus **m** matrices analogous to this is obtained. The minimum values of each row of Figure 6 result the vector {machine 1 [1 0 1 0.8 0 0.1 0.9]}, which expresses the membership for the 7 parts of machine 1.If the process for the **m** machine is repeated, **m** vectors will be calculated, which will consist of the nonbinary matrix that should be studied for obtaining the similarities of the process. If for the 7 parts example a universe of 7 machines is available, after the execution of the procedure, a nonbinary matrix as (4) may be obtained. The matrix (4) is the one that should be analyzed to have a solution of similarity of process.

Obtained the non-binary matrix, it is now necessary to run a proper grouping algorithm to get the possible manufacturing cells. Zhang and Wang (1992) shows the use of the Rank Order Clustering (ROC) (King, 1980) to analyze matrix (4). The result for this matrix is that cell 1 is composed of machines Y_1, Y_6 and Y_5 and family 1 of part X_3, X_1, X_7 and X_4, cell 2 is composed of machines Y_7, Y_2, Y_3 and Y_4 and family 2 of part X_2, X_5 and X_6.

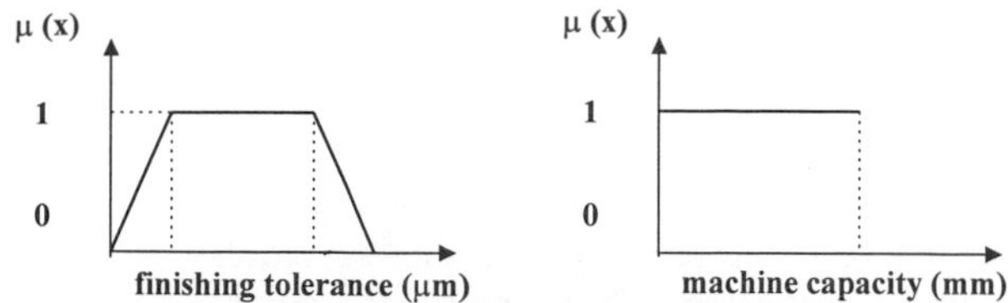

Figure 5 Example of membership function relative to machine 1.

Parts	finishing tolerance	machine capacity
1	1	1
2	1	0
3	1	1
4	0.8	1
5	1	0
6	0.1	1
7	0.9	1

Figure 6 Memberships given to each pair part x feature for the machine 1.

$$\begin{array}{c} \\ Y_1 \\ Y_2 \\ Y_3 \\ Y_4 \\ Y_5 \\ Y_6 \\ Y_7 \end{array} \begin{array}{c} \begin{array}{ccccccc} X_1 & X_2 & X_3 & X_4 & X_5 & X_6 & X_7 \end{array} \\ \begin{bmatrix} 1 & 0 & 1 & 0.8 & 0 & 0.1 & 0.9 \\ 0 & 1 & 0.3 & 0 & 0.7 & 0.8 & 0 \\ 0.3 & 0.7 & 0 & 0 & 0.8 & 0.8 & 0 \\ 0 & 0.7 & 0 & 0.3 & 0.7 & 0 & 0.3 \\ 0.6 & 0.1 & 0.5 & 0 & 0 & 0 & 0.5 \\ 0.7 & 0.2 & 0.8 & 0.8 & 0.3 & 0 & 0.7 \\ 0 & 1 & 0.5 & 0 & 0.8 & 0.9 & 0 \end{bmatrix} \end{array} \tag{4}$$

The importance of utilizing the nonbinary matrix is that, after the grouping of machines is obtained, there is now the possibility of analyzing the machines that are more appropriate to process the part families. Furthermore, it makes sense to eliminate the machines that process similar operations. This characteristic is not possible if the binary matrix is used.

Other algorithms for the grouping can be adapted so that the nonbinary matrix will be used for the cell formation.

6 CONCLUSIONS

This paper is the synthesis of what is possible to do with fuzzy logic in order to deal with the problem of obtaining similarities, an important aspect for part families formation, considering the uncertainty present in the manufacturing environment. This procedure groups techniques that can cope with the problem of similarities isolately, and more comprehensively, making use of the same data base, a fact that usually is necessary, although it is not possible in the most currents tools.

With the procedure described here it is possible to provide a new contribution for the Group Technology.

The development of this model can also supply a solution for the problem of setup time decrease, once it is possible to retrieve information of process and geometry similarities together. In this way, is possible to profit by these similarities so that the preparation time of machines will be shorter. It is possible to observe then that with the rationalization, which is possible with the identification of similarities, work is being done to save the company resources. Finally, this methodology may be transformed into a software, which will be an interesting tool to the manufacture small lots. This consists of a different proposition as an alternative to current methods available.

7 REFERENCES

Arieh, D.B. and Triantaphyllou, E. (1992) Quantifying data for group technology with weighted fuzzy features. *Int. J. Prod. Res,* **30**, 1285 - 1299.

King, J.R. (1980) Machine-component grouping in production flow analysis: an approach using a rank order clustering algorithm. *Int. J. Prod. Res.* , **18**, 213 - 232.

Montevechi, J.A.B. (1994) Formação de famílias de peças prismáticas utilizando lógica Fuzzy. *Doctorate Qualifying Examination,* São Paulo, EP-USP.

Saaty, T. L. (1977) A Scaling Method for Priorities in Hierarchical Structures. *Journal of Mathematical Psychology* , **15**, 234 - 281.

Xu, H. e Wang, H.P (1989) Part family formation for GT applications based on fuzzy mathematics. *Int. J. of Prod. Res.*, **27**, 1637 - 1651.

Zhang, C. and Wang, H.P (1992) Concurrent Formation of Part Families and Machine Cells Based on the Fuzzy Set Theory. *Journal of Manufacturing Systems*, **11**, 61 - 67.

8 BIOGRAPHY

J.A.B. Montevechi is Assistent Professor, received the master degree from Florianópolis University, Brazil, in 1989. Research areas include: Group Technology, expert systems and fuzzy logic.

G.L. Torres is Full Professor, received the degree of Ph.D. from École Polytechnique, Canada, in 1990. Research areas include: expert systems, fuzzy logic and neural networks to power systems problems.

P.E. Miyagi is Associate Professor, received the degree of Dr.Eng. from Tokyo Institute of Technology, Japan, in 1988. In 1994 received the degree of L.Doc. Research areas include: discrete event, dynamic systems, sequential control and systems programming.

M.R.P. Barretto is Assistent Professor, received the degree of Dr.Eng. from University of São Paulo, Brazil, in 1993. Research areas include: production planning and control, production order programming.

PART SEVEN

Scheduling Systems

HOLOS : a methodology for deriving scheduling systems

Ricardo J. Rabelo [a] ; *L.M. Camarinha-Matos* [b]

[a] *scholar of CNPq - Brazilian Council for Research - at New University of Lisbon - Portugal (e-mail : kadu@uninova.pt)*

[b] *New University of Lisbon and UNINOVA - 2825 Monte da Caparica - Portugal (e-mail : cam@uninova.pt)*

Abstract

This article presents a methodology - HOLOS - for deriving particular multiagent dynamic scheduling systems from a generic architecture. An object-oriented approach is used to support information modeling and knowledge representation. The integration aspects are addressed since agents are heterogeneous and have to access external information sources. A brief explanation about the HOLOS System Generator, an interactive system deriver based on the HOLOS methodology, is also given, as well as a general description of the prototype being developed at the UNINOVA FAS/FMS pilot system. Finally, some comments are made on the current status and next steps of the work.

Keywords

Dynamic Scheduling, Multi-Agent System, Architecture Derivation, Negotiation, CIM-OSA, Integration, Virtual Manufacturing.

1 MOTIVATION AND PROBLEM DEFINITION

The current industry scenario is changing very fast due to the need of industries being competitive. After an initial period in which a simple replacement of the human resources by machines was pursued, human-centered approaches are arising as a balanced alternative in face of some negative results of total automation (Nakazawa,1994).

The dynamic scheduling activity (dynamic task assignment to production resources along the time) is directly related to resources management, which includes not only equipments but humans too. Various industrial technological waves have emerged as a result of the competitiveness requirements, and the architectures of the scheduling systems cannot stay immune from that. From an emphasis on scheduling optimality, the industry has passed to scheduling flexibility, and turning fast to scheduling agility, i.e., a scheduling system which can support an agile manufacturing towards the extended enterprise paradigm. However, even within this current dynamic scenario, industries keep demanding for custom tailored dynamic scheduling systems, which besides being open, should support a flexible (re)configuration of themselves so that they can be adapted once new production methods, scheduling control procedures, layout of the production resources, etc., are changed.

This paper introduces HOLOS : a methodology for deriving open, agile and (re)configurable dynamic scheduling system for a particular enterprise based on a Generic Architecture. The HOLOS System Generator - a computer aided derivation system - is briefly explained as well.

2 A MULTIAGENT APPROACH

The development of a dynamic scheduling system which copes with the all mentioned aspects above is still a challenge. The Multi-Agent Systems (MAS) paradigm (Huhns,1987) has arisen as a powerful approach to develop a supporting framework, specially when the following points are taken into account :

- scheduling domain is intrinsically distributed;
- scheduling domain requires a joining of different expertises;
- MAS supports a dynamic domain;
- MAS potentially supports both agents autonomy and decentralized schedules, two key factors for reaching the desired scheduling agility;
- MAS can support the integration of different problem-solving (represented in different paradigms) in the same framework.

The Generic Architecture (GA) created within the HOLOS approach - and which will be the base for a derivation - is a logical collection of distributed agents which can perform a schedule and be supervised during its execution. Applying a holistic approach in the proposed distributed scheduling architecture signifies getting a 'state of harmony' within the enterprise via a complementary and cooperative relationship between the agents involved in the scheduling (Rabelo,1994a). Such state of harmony in an agile context is a tough task to be reached. The manufacturing environment is dynamic, normally over-constrained and unpredictable to some extent. Besides that, current constraints are commonly conflicting to each other so that a trade-off or requirements relaxations have to be *negotiated.* A contract net protocol / Negotiation paradigm (Davis,1983) appears to be a suitable mechanism for supporting the desired flexibility in conflicts resolution on that holistic relationship during a schedule generation and execution. Figure 1 illustrates such a Negotiation process in our approach for scheduling. It consists of a process that leads agents to exchange information with other agents about a given business processes' requirements (Figure 1a) until a production resource agent is selected to execute it (Figure 1b).

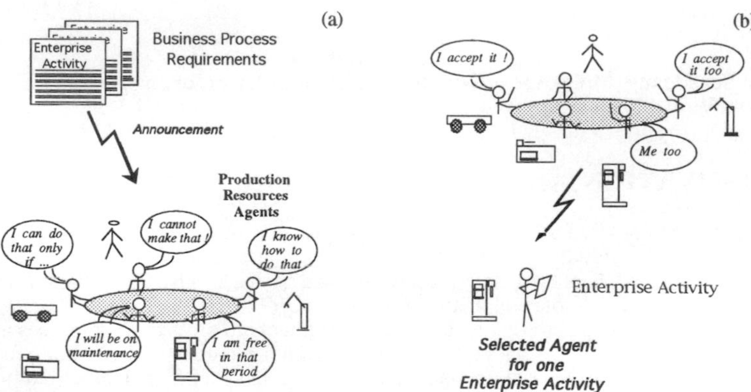

Figure 1 Negotiation in scheduling.

The general intended scenario is illustrated in the Figure 2, which can be viewed as an example of a derivation from the GA. It represents a dynamic view on the cooperative and complementary agents' behavior so that all enterprise's production structure can flexibly adapt itself to attend the arrival of a business process. In fact, this illustration joins the three vectors under which the HOLOS GA is based :

- Virtualization of the enterprise's production structure.
- Integration and information modeling.
- Multiagent distributed control.

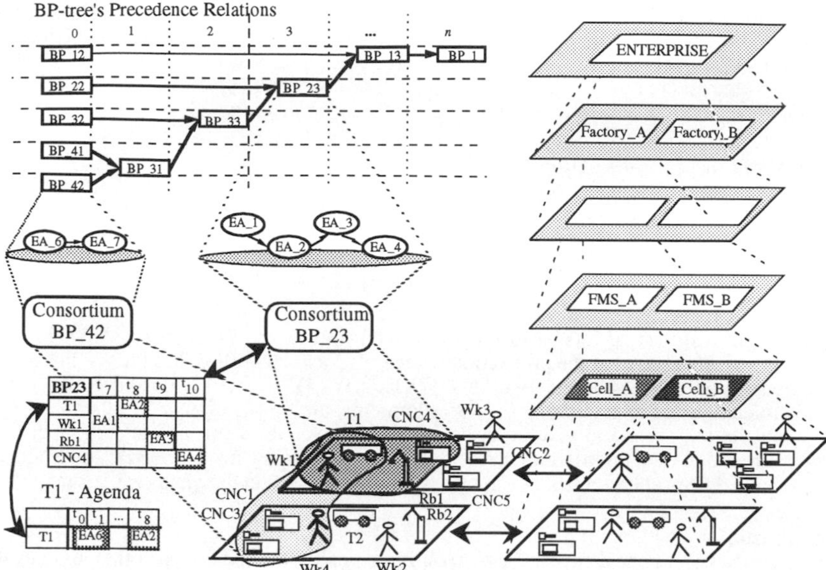

Figure 2 A Virtual Manufacturing scenario.

Scheduling cannot be seen as an isolated CIM activity. A dynamic scheduling system needs reliable and timeliness information from several sources. A global CIM Information System (CIM-IS) (Osorio,1993) containing common information models is the main vehicle through which all heterogeneous subsystems can communicate to each other, and hence the source of all information needed for the scheduling (Figure 3).

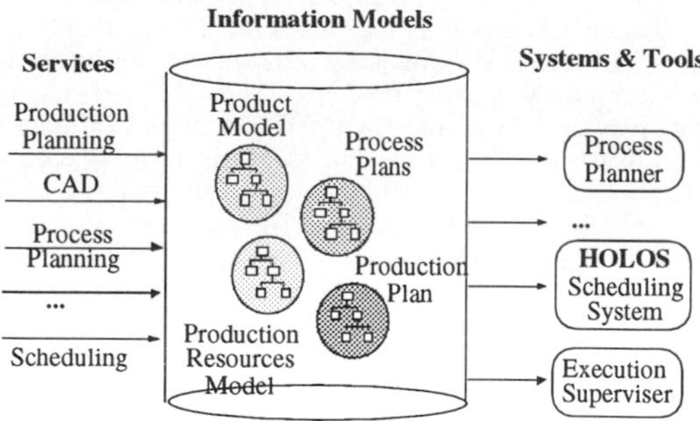

Figure 3 A CIM Information System.

Four types of agents are used in the GA. They are hierarchically related, heterogeneous, have different levels of autonomy, and have an explicit control link between each other. Briefly :

Scheduling Supervisor agent (SS)
It is a semi-hierarchical supervisor agent with the following basic functions : loading both sets of business processes (BPs) to be executed (BP-trees) and involved information models from the CIM IS; definition of business processes' requirements and their sending to the Local Spreading Centers; creation of Consortia; high level changes on planned BPs; high level actions for conflicts resolution; and visualization and scheduling evaluation.

Local Spreading Center agent (LSC)
It is a decentralized control structure for spreading the BPs announcements (requirements) through a network of production resources agents. Its main function refers to negotiate with these agents about BPs' requirements in order to select (based on some criteria) the more adequate ones for the execution of each BP.

Enterprise Activity Agent (EAA)
It is the agent responsible for executing a task itself (an enterprise activity - EA). In fact, it represents a virtualization of the production resource's local controller. In Figure 2 this agent is illustrated as human resources (workers Wk1,...,Wk4), robots (Rb1 and Rb2), etc. Its essential functions are receiving BPs/EAs requirements, their evaluation and further answer to a LSC about its temporal and technical capabilities to execute them, and a 'self-supervision' activity (in order to guarantee an EAA will only participate in a negotiation when it is 'operational'). The problem related to legacy systems is tackled in the chapter 5.1.

Consortium
It is a temporary and logical clustering of EAA dynamically selected (via negotiation) to execute a whole BP. Each Consortium has its own and local schedule, which means the global scheduling is decentralized. In the example shown in Figure 2, there is a set of business processes (BP_i) with a precedence relationship between them. BP23 for instance is composed by four enterprise activities, EA1 : EA4. Each one requires a specific type of production resource for its execution. Thus, Consortium BP23 represents the team of production resources selected to execute BP23 (T1 ,Wk1, Rb1 and CNC4). However, the Consortium BP42 needs T1 and Wk1, which in turn are also assigned to Consortium BP23. Therefore, due to the precedence relation between those two BPs, T1 has to execute EA6 before EA2. An EAA can belong to several Consortia along the time, which generates EAA contention and hence temporal constraints. Since an EAA finishes the execution of contracted EA(s) for some Consortium, this EAA becomes free both to execute other EA(s) already contracted for another Consortium and to look for more EAs, which are still waiting for execution proposals. At the end of an entire BP execution the Consortium agent kills itself.

The Consortium improves the traditional Group Technology Cell concept since it supports several types of flexibility (classified in (Chryssolouris,1992)), such as internal routing, product, volume and production. In other words, it provides the base to support a virtual manufacturing (Hitchcock,1994). Other concepts, like the 'logical cell' (AMICE,1993) and 'virtual production area' (Hamacher,1994) seem to be equivalent to the notion of Consortium. However, the basic difference is on the control flexibility, and on how it is managed in rescheduling situations. Due to the close link with the EAA agent, the Consortium is able to find a substitute EAA (via negotiation) when someone else fails.

3 INFORMATION MODELING

Information modeling and knowledge representation represents an extremely important aspect to support the mentioned MAS architecture. Presented work resorts to object oriented technology. It means that all agents and information structures are modeled as objects; i.e., by means of attributes (*slots*) and functionalities (*methods*).

Scheduling needs to have access to several information sources for / during its execution, as well as to other ones directed related to the architecture's approach. Two of essential sources for scheduling are the process plans and production resources. Their models composition have been inspired on some international projects (like IMPPACT (Gielingh,1993) and CIMPLATO (Bernhard,1992)) and on the results of the STEP community to some extent (Schenck,1994). Figure 4 shows an example of a process plan, whereas Figure 6 illustrates (in the EAA agent model) part of a production resource model.

CIM-OSA (AMICE,1993) concepts have been used for modeling the dynamic processes of an enterprise. In this sense, a production plan is modeled as a set of business processes and enterprise activities (EAs), and Procedural Rules Sets (PRSs) as the link between them. These entities form a *BP-tree* when seen as a whole (Figure 5). The negotiation process between agents is mainly based on the BP-tree's entities. However, these entities' models have been extended in order to improve the efficiency and negotiation flexibility (Rabelo,1994a).

Open solutions also include people talking a common language (terms and their precise (semantic) meaning) in order to avoid misunderstanding and, to some extent, to take the local culture into account. Some efforts in creating 'standard glossaries' have been made (Camarinha,1991). Thus, the glossary is applied on all interactive interfaces and reports.

The four types of agents used in the generic architecture are modeled as *classes* of agents, from which instances of agents are created and filled in during a derivation process. Examples of the agents classes are shown in Figure 6.

4 HOLOS - THE METHODOLOGY

The process of deriving (an instance-of) a dynamic scheduling system from a generic architecture is not 'anarchical' but based on a method. The HOLOS, a methodology to support such derivation, has been in development at UNINOVA. It corresponds to a sequence of interdependent general procedures and considerations which are to be followed by a 'human deriver' towards the implantation of a dynamic scheduling system for a particular enterprise.

CIM-OSA appears as the most prominent and wide methodology related to derivation of CIM architectures from an abstract and general reference model. HOLOS methodology is more restricted in scope than CIM-OSA, since it specifically addresses scheduling systems development as well as it is tightly biased by with the multiagent approach.

A scheduling system derivation process is normally too complex. It comprises lots of parameters and information about production, engineering and scheduling control, which in turn may be combined to each other. An open solution (the derivation) for that scenario requires an exhaustive discussion between all people engaged in, as well as the evaluation of its impacts on existing technology and human resources management. Such discussion can involve not only technical points (like production system, integration, heterogeneity of existing systems, production resources layout, solution costs, etc.), but also the enterprise's organizational culture and its work organizational methods (like team work, decentralization levels and autonomy for decision making, human resources qualifications, their functions (re)definition, training policies, etc.) as well as national singularities (like work shifts related to special holidays or local tradition, efficiency criteria specifications, etc.). The consideration of all these questions may determine the success of the system and of its implantation (Jones,1992), i.e., that the expected particular system can be achieved.

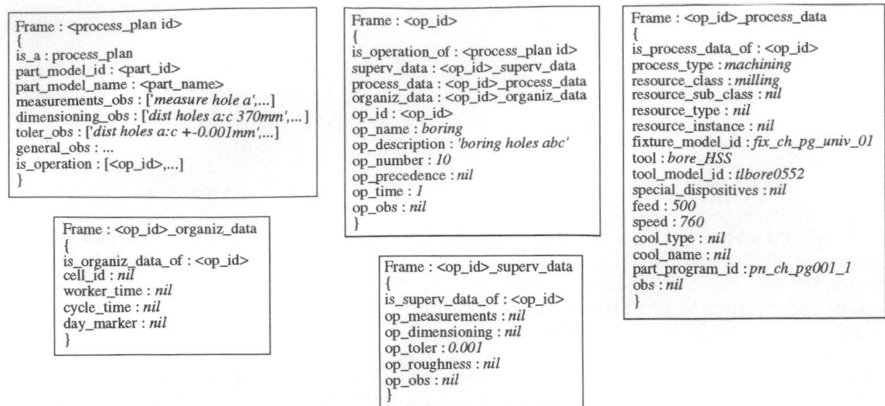

Frame : <process_plan id>
{
is_a : process_plan
part_model_id : <part_id>
part_model_name : <part_name>
measurements_obs : ['measure hole a',...]
dimensioning_obs : ['dist holes a:c 370mm',...]
toler_obs : ['dist holes a:c +-0.001mm',...]
general_obs : ...
is_operation : [<op_id>,...]
}
Frame : <op_id>_organiz_data
{
is_organiz_data_of : <op_id>
cell_id : nil
worker_time : nil
cycle_time : nil
day_marker : nil
}
Frame : <op_id>
{
is_operation_of : <process_plan id>
superv_data : <op_id>_superv_data
process_data : <op_id>_process_data
organiz_data : <op_id>_organiz_data
op_id : <op_id>
op_name : boring
op_description : 'boring holes abc'
op_number : 10
op_precedence : nil
op_time : 1
op_obs : nil
}
Frame : <op_id>_superv_data
{
is_superv_data_of : <op_id>
op_measurements : nil
op_dimensioning : nil
op_toler : 0.001
op_roughness : nil
op_obs : nil
}
Frame : <op_id>_process_data
{
is_process_data_of : <op_id>
process_type : machining
resource_class : milling
resource_sub_class : nil
resource_type : nil
resource_instance : nil
fixture_model_id : fix_ch_pg_univ_01
tool : bore_HSS
tool_model_id : tlbore0552
special_dispositives : nil
feed : 500
speed : 760
cool_type : nil
cool_name : nil
part_program_id : pn_ch_pg001_1
obs : nil
}

Figure 4 Example of a process plan model.

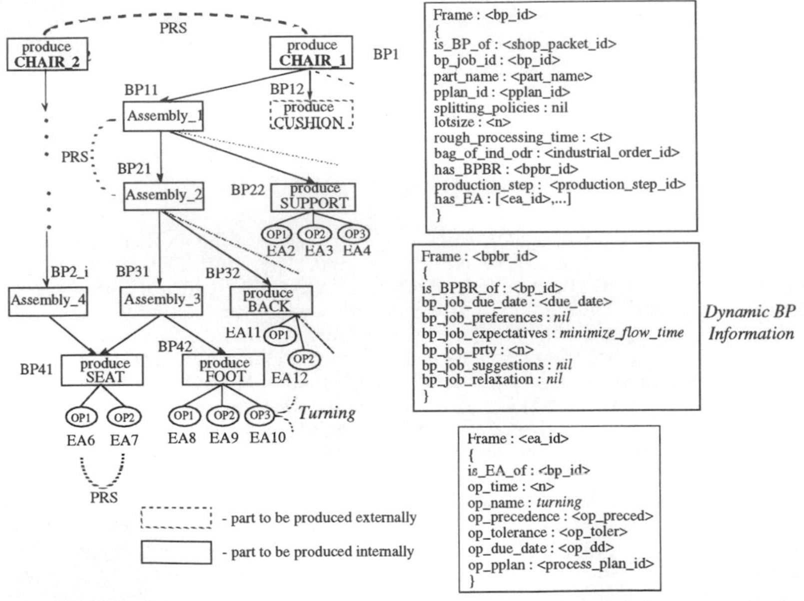

Figure 5 Example of a BP-tree model.

Although all these concerns are pointed as crucial, they are out of scope of the HOLOS methodology itself. Hence, the HOLOS methodology assumes the whole phase mentioned above is done **before** starting a derivation process. In other words, it presupposes that a preliminary analysis and a rough system evaluation is made and their results are incorporated in the form of :

- information modeling of all entities directly related to the scheduling activity (production resources and their layout and topology, process plans and production plans);
- information modeling and knowledge representation of all entities directly related to the scheduling architecture (agents' classes and the control flow between them, glossary, communication protocol, negotiation entities and events to be treated);
- identification of the information flow between dynamic scheduling with supervision and planning actions;
- set of procedures to be used in the particular system according to the parameters and information about engineering, production and scheduling control (EPS).

```
Frame : <manager_id>
{
manager_of : <EAA_id>
class : manager
has_mailbox : <mailbox_id>
has_agenda : <agenda_EAA_id>
has_server_info_model :
        <server_id_info_model>
has_team : [<server_id>,...]

managEAA_add_EA_Agenda :
    mt_managEAA_add_EA_Agenda
managEAA_add_msg_Mailbox :
    mt_managEAA_add_msg_Mailbox
managEAA_answer_EA_LSC :
    mt_managEAA_answer_EA_LSC
managEAA_decompose_packet :
    mt_managEAA_decompose_packet
managEAA_evaluate_EABR :
    mt_managEAA_evaluate_EABR
managEAA_evaluate_EAR :
    mt_managEAA_evaluate_EAR
managEAA_get_EAR_from_IS :
    mt_managEAA_get_EAR_from_IS
managEAA_receive_EABR :
    mt_managEAA_receive_EABR
managEAA_receive_answer_EA :
    mt_managEAA_receive_answer_EA
managEAA_remove_EA_Agenda :
    mt_managEAA_remove_EA_Agenda
managEAA_remove_msg_Mailbox :
    mt_managEAA_remove_msg_Mailbox
}
```

```
Frame : <server_id_info_model>
{
is_model_of : <EAA_id>
wc_id : milling_nc_1
type : pgm_milling_nc
sub_class : pgm_milling
class : milling
super_class : machining
behavior : active_resource
planning_info : <server_id_planning>
techno_info : <server_id_technological>
capab_info : <server_id_capability>
topol_info : <server_id_topology>
}
```

```
Frame : <server_id_technological>
{
is_component_of : <server_id_info_model>
control_name : siemens_7M
cool : 4
ntools : 20
precision : 10
speed_max : 2500
type : cnc
wp_weight_max : 20
feed_max : [800,1000,1500]
rapid : [1000,1000,1500]
wp_dim_max : [80,15,1]
wp_dim_min : [15,0,15]
zero_position : [-100,100,212]
}
```

```
Frame : scheduling_supervisor
{
registered_LSC : [[<LSC_id>, [<LSC_responsibility>],
        <LSC_org_id>],...]
registered_EAA : [[<EAA_id>, <manager_EAA_id>],...,]
registered_Consortium : [[<consortium_id>,<BP_id>,
        <status_BP>],...,]
param_Consortium_eval : [completion_time, tardiness,
        lateness, lead_time, slack_time, idle_time, ...]

load_EAA : mt_create_EAA
attach_server : mt_attach_server
load_LSC : mt_create_LSC
load_EAA_to_LSC : mt_load_EAA_to_LSC
reg_EAA_in_LSC : mt_reg_EAA_in_LSC
load_BP_tree : mt_load_BP_tree
send_EA_to_LSC : mt_send_EA_to_LSC
create_Consortium : mt_create_Consortium
evaluate_Consortium : mt_evaluate_Consortium
create_temp_BP_control : mt_create_temp_BP_control
show_resources : mt_show_resources
show_agenda : mt_show_agenda
show_all_consortium : mt_show_all_consortium
show_consortium : mt_show_consortium
get_EA_status : mt_get_EA_status
get_BP_status : mt_get_BP_status
scheduling_execution : mt_scheduling_execution
}
```

```
Frame : <lsc_id>
{
lsc_id : <lsc_id>
lsc_functionality : [machining,milling]
lsc_org_id : nil
registered_EAA : [[<EAA_id>,<manager_EAA_id>,
        <availability>,<maintenance>],...]
received_EA : [[<EA_id>,<BP_job_id>], ... ]
selection_criteria : [[due_date,1],[minimum_path,2],[less_numb_EAA,3],
        [greater_EAA_cost,4], ... ]
time_out : 25

check_inexistence_candidates_BP : mt_check_inexistence_candidates_BP
check_inexistence_candidates_EA : mt_check_inexistence_candidates_EA
choose_EAAs_to_send_EABR : mt_choose_EAA_to_send_EABR
receive_answer_EAA : mt_receive_answer_EAA
select_EAA : mt_select_EAAs
send_EAA_to_Consortium : mt_send_EAA_to_Consortium
send_EABR_to_EAA : mt_send_EABR_to_EAA
send_answer_to_EAA : mt_send_answer_to_EAA
send_answer_to_EAA : mt_verify_relaxation
send_answer_to_EAA : mt_evaluate_candidates
}
```

```
Frame : <consortium_id>
{
controlled_by : <consortium_i>
bp_id : <BP_id>
type : make
composed_by :
[[<manager_id><server_id><EA_id><start_time><end_time>],...]

consortium_receive_EAA : mt_consortium_receive_EAA
scheduling_execution : mt_consortium_scheduling_execution
modify_BP_information : mt_modify_BP_information
find_substitute : mt_find_substitute
get_EAA_status : mt_get_EAA_status
give_EA_status_to_SS : mt_give_EA_status_to_SS
give_BP_status_to_SS : mt_give_BP_status_to_SS
}
```

EAA agent class / LSC agent class **SS agent class / Consortium (C) agent class**

Figure 6 Example of the agents' classes.

In the HOLOS approach, a derived dynamic scheduling system is represented as a particular configuration of agents organized for a concrete scenario, and that can be supervised during the execution of a scheduled production plan. A derivation basically consists in creating instances of the agents' classes and then filling their *skeleton* along some derivation phases (discussed later). In fact, these phases represent a stepwise way through which the HOLOS methodology is utilized. Thus, hidden under those phases, the methodology's procedures are applied. Briefly, they are :

a) Agents Specification

This procedure is related to the knowledge to be incorporated into each type of HOLOS agent (SS, LSC, Consortium (C) and EAA) when instances of them are created. This knowledge is represented by attributes and functionalities, which in turn can be :

- Generic : those which each agent should have, independently of the particular site *
- Customized : the generic attributes and functionalities which need to be customized (within a set of options) for a particular scheduling system, but still independently of the particular physical site.
- Particular : the attributes and functionalities which should exist and/or have to be customized taking into account the particular physical site.

a1) Selection of EPS Criteria

The EPS aspects are directly or indirectly indicated (selected) via customized attributes. In general terms there is a *method* associated to each indicated EPS aspect. Nevertheless, the specification of an attribute may be a result of a combination of EPS aspects.

a2) Consistency Verification

The indication of the EPS aspects can be a difficult task, specially when they have to be combined to each other. A wrong specification and/or combination can provoke a situation of domain inconsistency. Two consistency verification levels exist : a simple check to guarantee that all terms indicated are defined in a glossary; and a more sophisticated analyses to guarantee a valid combination between those aspects (based on a 'derivation map', which could model all possible combinations for each aspect). Due to the complexity, this last level can suggest a decision support module may be used to help the user.

b) Agents Implantation

Implanting the agents means to make them exist in the 'world'; i.e., they can be recognized in the system, can communicate to the other agents and can execute actions. In this sense, once the logical agents instances are completed created, they do a self-announcement making use of their respective communication channels previously assigned. The 'compilation' of all agents' functionalities and other programs, the creation of libraries and adjustment of graphical interfaces are other steps to be pursued.

c) Agents Integration

In the HOLOS approach, agents have to communicate to external and heterogeneous entities in order to execute a scheduling. Such entities are the CIM-IS, other sub-systems (specially those related to the planning and supervision activities) and the production resources' local controllers. There are three possible integration layers to be made (see Figure 11) :

1- PLC (or other local controller) : Server - it aims at transforming the local production resource's controller (its PLC) in a server; i.e., creating a higher level client (in DOS or Unix for instance) which can communicate with the server. The communication process can make use of RPC, for instance.

---

* As already mentioned, the HOLOS methodology assumes that all agents' classes and information models are already composed before starting a derivation. However, a class concept can be changed (by the user), and this may be done due to some requirements of the particular site.

2- Server : EAA Manager - it aims at allowing the server to be integrated and representable into the community of agents. The paradigm client (EAA Manager / Unix) : server (Server / Dos or Unix) is applied. The communication process can make use of RPC, sockets, etc., in UDP or TCP, depending on the server's communication services.

3- EAA Manager : other agents - it aims at allowing the EAA Manager to make a conversation (in an abstract 'multiagent scheduling language') to the other agents of the architecture, to other subsystems and to the CIM-IS. A high level protocol can be utilized for that, which can be supported by RPC, sockets, etc., in UDP or TCP.

d) Agents Reconfiguration

An open solution implies giving the user the possibility to refine the Particular Architecture Infrastructure, after its generation, for the specific target system. This can be done through modifications (or extensions) and/or insertions of attributes and/or functionalities not generated during the derivation.

e) Architecture Reconfiguration

Production domain and scheduling policies and their control structures can change along the time, either due to its obsoleteness or due to some adaptation for a specific derivation (as mentioned in the Agents Specification procedure). Thus, an open architecture has to contemplate the possibility to the user for changing the agents' classes and the domain knowledge.

4.1 The Derivation Phases

Five derivation phases are utilized in the HOLOS methodology (Figure 7). Briefly they are :

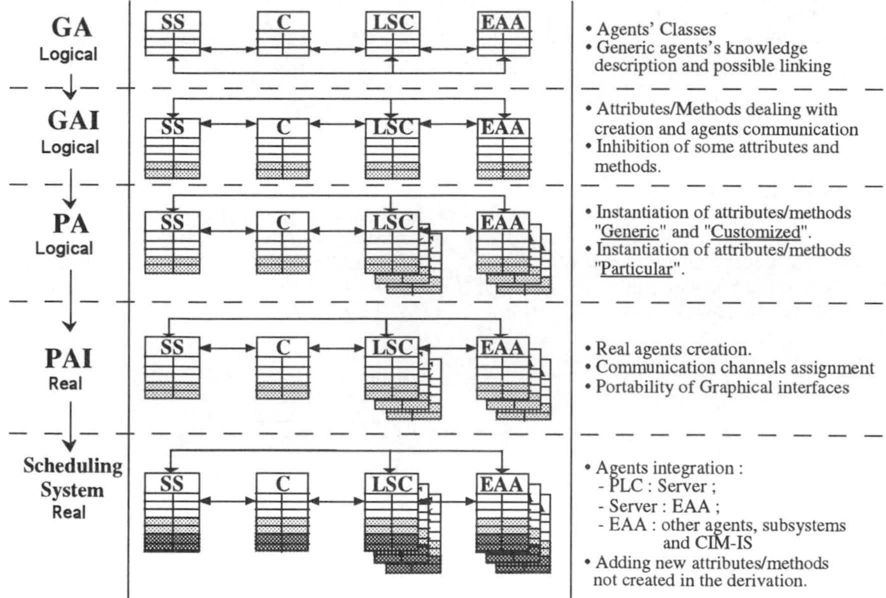

Figure 7 HOLOS derivation phases.

Generic Architecture (GA)

In this initial phase all classes of agents are abstract objects. Examples of these classes were shown in the Figure 6. They represent the 'genesis' of the scheduling system.

Generic Architecture Infrastructure (GAI)

It is the first stage of the derivation process. The main goal in this phase is to give the agents the first 'seeds of life', i.e., a set of primitives related to their creation and communication (including with the CIM-IS). The RPC protocol has been used to support the agents communication. Due to the difficulties for calling RPC services from a program (a method) written in Prolog, C programs are used as an 'intermediate binding' for that (Figure 8). Generic attributes and/or methods can be inhibited for the particular system.

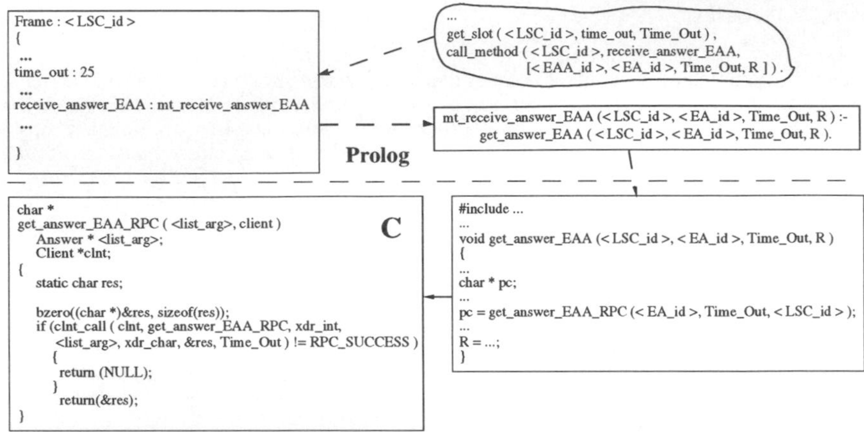

Figure 8 Example of an agent LSC in the GAI phase.

Particular Architecture (PA)

In this phase the logical instances of the agents' classes are originated. The Generic functionalities and attributes are added (by inheritance) to the respective instances, the Customized functionalities and attributes can be chosen and/or indicated from a set of options (library), and the Particular ones are specified (Figure 9). This is the first step in direction to the particular system instantiation.

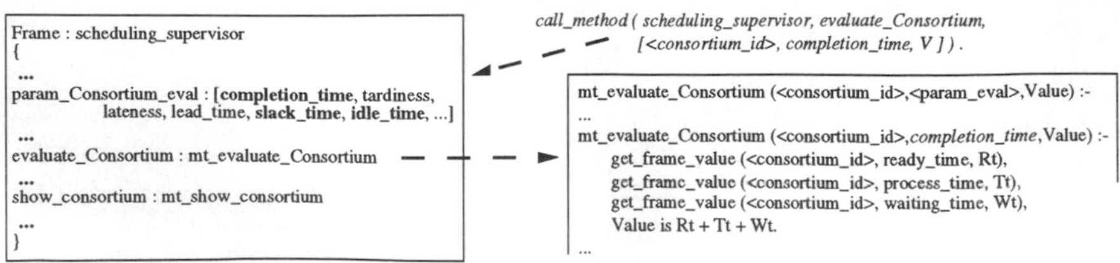

Figure 9 Example of an agent SS in the PA phase.

Particular Architecture Infrastructure (PAI)

This phase is responsible for the real creation of all logical agents instances composed in the previous phase. The communication channels have to be assigned to the agents according to their topology and enterprise model in order to connect them to each other. The adequacy of user interfaces is another aspect to take into account. In this phase, the agents become 'real' entities.

Dynamic Scheduling System

The user takes the PAI and makes the necessary adaptation for the particular system as well as handles the system (agents) implantation. An old derivation is replaced by the new one. This last phase represents a PAI completely instantiated and implanted.

5 HOLOS - THE SYSTEM GENERATOR

Likewise the other areas, the systems development technology has also suffered the 'Pendulum Law' effects. From the one extreme situation in which the systems were designed all custom-tailored and hence with high costs in development, it has passed to the other extreme in which the systems became generic, less expensive but 'black boxes'. More recently, due to the increasing of the development complexity of industrial systems and, at the same time, their need being open, modular, reusable and integratable, the 'derivation approach' has stood in significance and seems to be a balanced trend (Dietrich,1994). Its basic idea corresponds to create generic systems architectures and then to create particular 'instances-of' from that. The HOLOS System Generator (HOLOS-SG) (Rabelo,1994b) can be seen as an example in that mentioned direction. It represents an automatic way to guide a derivation. In fact, the HOLOS-SG appears to be a 'computer aided derivation' tool. By means of a strong interaction with an *user deriver* the HOLOS derivation phases are passed so that at the end of the process a PAI is generated.

It is not the objective here to describe it in details, but just to give a rough idea about it and its philosophy. Figure 10 shows its generic architecture. Some aspects deserve a brief explanation. The first one refers to the user intervention. Beyond his/her position as a deriver (and as a decision maker to some extent), he/she can alter / have access to the system concepts and libraries. Further, once the PAI is generated and in order to generate the particular scheduling system, it is necessary to implant, to integrate and, possibly, to reconfigure agents. The second one is just related to the CIM-IS role. Basically, it is the source of all information models needed for a derivation as well as the repository of a derivation representation (old ones, current one, or even one in progress). The last aspect is concerned with the rules to guide a derivation, which is supported by other structures (a Help, derivation maps and a decision support system) for consistency verification, specially in the PA phase.

5.1 Prototype under development

A test case for the NOVAFlex (Barata,1993), the UNINOVA's FMS/FAS pilot system, is in development. The objective is to derive a dynamic scheduling system based on the HOLOS methodology. NOVAFlex is composed by three robots (one *scara* and two 6 dof), two numerical control machines (a lathe and a milling), an automatic warehouse and a pallet based transport system with sensors. The current prototype has been in development and being implemented in Prolog for Aix language with an object-oriented extension (Seabra Lopes,1994), in an IBM Risc 6000 workstation.

The integration aspect is vital in manufacturing. Apart this prototype, other works on integration have been in development at UNINOVA. In short, we have faced with the legacy system problem. The production resources' controllers are quite heterogeneous, and they need to be recovered in such a way they can be integrated into the architecture infrastructure, i.e., they can be represented within the community of intelligent agents. The UNINOVA's approach is the development of encapsulating layers (as mentioned in Agents Integration in chapter 4).

The first integration layer (PLCs-Servers) is already finished for all NOVAFlex's servers, in PCs, with implementations in C and C++ languages, Linux and DOS operational systems, and X11 and TCL Tool-Kit for graphical interfaces. It means that all servers can 'offer their services' to the other agents / applications. Other works were made on how to integrate agents with other subsystems and with the CIM-IS (third integration layer). We are now concentrated in the second layer, i.e., the integration of these servers with their 'managers'. Thus, an EAA is modeled as a logical clustering of two basic interacting processes, a Manager and a Server - a tandem architecture (Figure 11). The Server, representing the resource's local controller, is a slave process which gives its Manager an allowance for offering services which it is capable to execute. The Manager 'represents' this Server within the manufacturing environment. Its basic function is 'selling' (via negotiation) the Server's services. In fact, a Manager can also represent more than one Server, depending on the production resources' topological model.

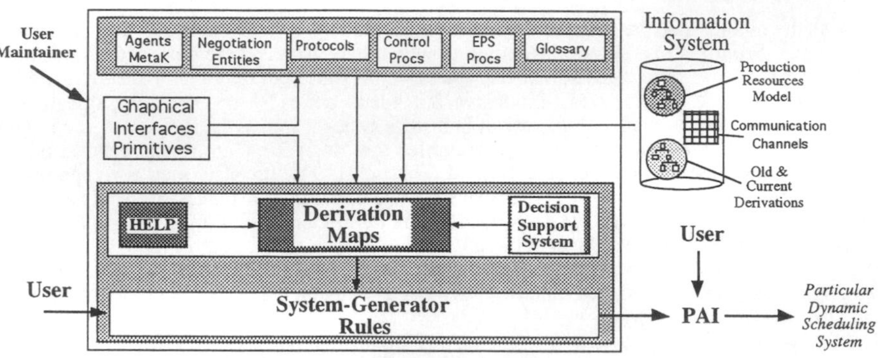

Figure 10 HOLOS System Generator architecture.

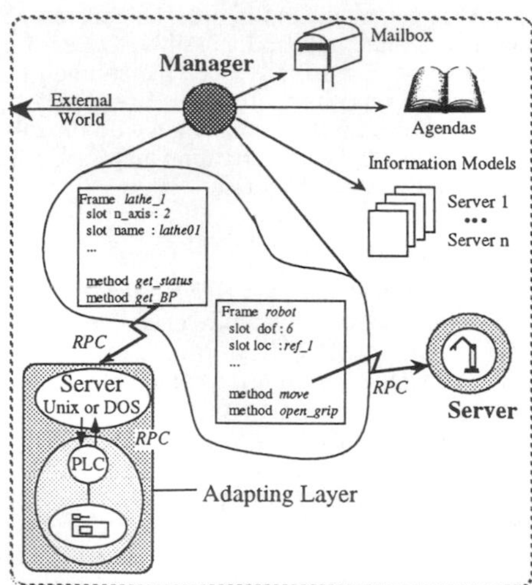

Figure 11 An example of the EAA architecture.

6 CONCLUSIONS

The HOLOS, a methodology for derivation particular dynamic scheduling systems from a generic architecture, was presented. It makes a matching with most of the characteristics considered as trends and emerging concepts in scheduling (Szelke,1994). It utilizes some anthropocentric concepts in applying a decentralized control and in exploiting the autonomy of the production resources. The negotiation between intelligent agents is used as the support technique for that.

The Engineering-Tool-Kit (Hirsh,1994) and OPIS (Smith,1994) represent two other related works. In spite of they are older, apply different approaches and are already more established than HOLOS, their general objective is equivalent : generating particular architectures and systems based on generic concepts. However, due to the HOLOS potentiality in terms of flexible modeling and control, system modularity and expandability, integration and MAS approach, a HOLOS instance system appears to fit in a more suitable way with the requirements of the industries which have envisaged for a virtual manufacturing towards the extended enterprise paradigm.

It does not intend to cover all kind of industries. The first prototype is directed to discrete and jobshop manufacturing. Further, this prototype assumes that the generic architecture is 'good enough' and that the human deriver has all knowledge on how the particular dynamic scheduling (the instance-of) has to be. However, the architecture can be modified along the time, and the instance can be adapted by the user after a derivation.

The HOLOS System Generator, a 'computer aided derivation' tool, was briefly presented, allowing the user to be assisted during a derivation.

A prototype has been in development in order to generate an instance-of for NOVAFlex. After the first implantation, the evaluation and methodology validation correspond to the main next steps to be pursued. Further, in being this work a cooperation between UNL and Federal University of Santa Catarina / Brazil, the intention is also to derive a particular system for its manufacturing cell.

7 ACKNOWLEDGMENTS

We would like to thank the UNINOVA Institute for the general infrastructure, the support provided by the JNICT CIM-CASE and ECLA Cimis.net projects, and Francisco Bernardes and Roberto Espenica for their support in the implementation of HOLOS-SG. The first author also would like to thank CNPq - Brazilian Council for Research - for the scholarship, Mafalda Leitão for her comments about HOLOS from the sociological point of view, and Gentil Lucena for his *holistic* way of being.

8 REFERENCES

AMICE (1993) CIM-OSA : Open Systems Architecture for CIM. 2nd revised and extended version, Springer-Verlag, Berlin.

Barata, J. and Camarinha-Matos, L.M. (1993) Development of a FMS/FAS System - The CRIs Pilot Unit. *Proceedings of ECLA-CIM93*, Lisbon, Portugal.

Bernhard, R., editor (1992) CIM Systems Planning Toolbox - Project Survey and Demonstration. *Proceedings of CIMPLATO Workshop on CIM Planning Tools*, University of Karlsruhe, Germany.

Camarinha-Matos, L.M., Pinheiro-Pita, H. and Moura-Pires, J. (1991) CIM Glossary - 4th Revision. UNL-Report.

Chryssolouris, G. (1992) *Manufacturing Systems : Theory and Practice*. Springer-Verlag, New York.

Davis, R. and Smith, R. (1983) Negotiation as a Metaphor for Distributed Problem Solving. *Artificial Intelligence*, **20**, 63-109.

Dietrich, B. (1994) Automation in Manufacturing, Control versus Chaos, in *Advances in Agile Manufacturing* (ed. P.T. Kidd and W. Karwowski), IOS Press.

Gielingh, W. and Suhm, A., editors (1993) IMPPACT Reference Model : An Approach to Integrated Product and Process Modelling for Discrete Parts Manufacturing. Springer-Verlag, Berlin.

Hamacher, B., Klen, A. and Hirsh, B. (1994) Production Management Elements for the Learning Enterprise. *Proceedings of IFIP WG5.7 Conference on Evaluation of Production Management Methods*, Gramado, Brazil.

Hirsh, B., Kuhlmann,T. and Marciniak, Z. (1994) Engineering Tool-Kit for Implementation of Shop Floor Control Systems. *Proceedings IFIP WG5.7 Conference on Evaluation of Production Management Methods*, Gramado, Brazil.

Hitchcock, M. (1994) Virtual Manufacturing - A Methodology for Manufacturing in a Computer. *Proceedings of Workshop on The Automated Factory of the Future : Where do we go from here ? / IEEE 1994 International Conference on Robotics and Automation*, San Diego.

Huhns, M., editor (1987) *Distributed Artificial Inteligence.* Pitman Publishing / Morgan Kaufmann Publishers, San Mateo, USA.

Jones, B. (1992) Essential Cultural Aspects, Strategies and Techniques - A Comparative View of Work Technology and Flexible Production, in *Flexible Manufacturing Systems and Work Reorganization* [in portuguese] (ed. Ilona Kovács at all), Lisbon, Portugal.

Nakazawa, H. (1994) Human Oriented Manufacturing System, in *Advances in Agile Manufacturing* (ed. P.T. Kidd and W. Karwowski), IOS Press.

Osorio , A. and Camarinha-Matos, L.M. (1993) Information based control architecture for CIM. *Proceedings IFIP Conference Towards World Class Manufacturing*, Phoenix, USA.

Rabelo, R. and Camarinha-Matos, L.M. (1994a) A Holistic Control Architecture Infrastructure for Dynamic Scheduling. *IFIP KBRS94 - Knowledge-Based Reactive Scheduling Workshop,* Budapest, Hungary. In press by Chapman & Hall.

Rabelo, R. and Camarinha-Matos, L.M. (1994b) Generation of Multi-Agent Infrastructures for Dynamic Scheduling and Control Architectures. *Proceedings 27th ISATA / Conference on Lean/Agile Manufacturing in the Automotive Industries*, Aachen, Germany.

Schenck, D. and Wilson, P. (1994) Information Modelling : The EXPRESS Way. Oxford University Press.

Seabra Lopes,L. (1994) GOLOG 2.0- A Frame Engine in Prolog.Technical Report UNL12-94.

Smith, S. (1994) Configurable Systems for Reactive Production Management, *in IFIP Transactions - Knowledge-Based Reactive Scheduling* (eds. E. Szelke and R. Kerr), North-Holland.

Szelke, E. and Kerr, R. (1994) Knowledge-Based Reactive Scheduling, *in IFIP Transactions - Knowledge-Based Reactive Scheduling* (eds. E. Szelke and R. Kerr), North-Holland.

9 BIOGRAPHY

Mr. Ricardo J. Rabelo received his degree on Computer Science in 1984, worked as consultant for several Brazilian companies as a collaborator of GRUCON / Federal University of Santa Catarina, and he is actually taking his Ph.D. at New University of Lisbon / UNINOVA on Robotics and CIM. His main interest are : dynamic scheduling and virtual manufacturing.

Dr. Luis M. Camarinha-Matos received his Computer Engineering degree and Ph.D. on Computer Science, topic Robotics and CIM, from the New University of Lisbon. Currently he is auxiliary professor (eq. associate professor) at the Electrical Engineering Department of the New University of Lisbon and leads the group of Robotics Systems and CIM of the UNINOVA's Center for Intelligent Robotics. His main research areas are : CIM systems integration, Intelligent Manufacturing Systems, and Machine Learning in Robotics.

18

Evaluation of on-line schedules by distributed simulation*

S. R. Jernigan, S. Ramaswamy, K. S. Barber
The Laboratory for Intelligent Processes and Systems, The Department of Electrical and Computer Engineering, The University of Texas at Austin, Austin TX 78712-1084, USA

Abstract
A new algorithm for the distributed simulation and evaluation of on-line schedules is presented. Generally, on-line scheduling has often been restricted to scheduling activities on a single machine or workcell. The exploratory research reported in this paper expands on-line scheduling to encompass several machines or workcells. Branch and bound search techniques are used in the simulation to reduce the number of simulations simultaneously in execution. The algorithm is applied for the distributed simulation of on-line schedules for a manufacturing example.

1 INTRODUCTION

Several obstacles prevent current assembly and manufacturing lines from capitalizing on the promises of complete automation. Among these are the difficulties in scheduling[1] resources and ordering processes such that optimal use is made of the current configuration. Fox and Kempf [FoKe85] distinguish the differences between planning and scheduling and suggest that a distinction is both essential and beneficial. While planning is often done off-line, scheduling can be best performed at run-time [FoKe85, XiBe88, ChAl86, MeSa91, Lyon90]. Planning is an activity that produces a minimally constrained plan[2] and provides estimates for raw materials requirements and product yields. Scheduling is an activity that uses the plan developed by the off-line planner and develops a schedule at run-time, incorporating the current state of the factory floor. Such a schedule will be maximally constrained and it will specify the necessary details for executing a schedule. Furthermore, the scheduler can reactivly adapt the schedule in response to unexpected outcomes or states. Previously, on-line scheduling has been restricted to a single machine or workcell. The exploratory research reported in this paper expands on-line scheduling to encompass several machines or workcells. Various representation schemes have been used to represent operational orderings of a plan. These include: (i) AND/OR graphs [FoKe85], [MeSa91] , (ii) Hierarchical hypergraphs [XiBe88], (iii) process algebra [Lyon90],

---

* This research was supported in part by the Texas Higher Education Coordinating Board under grant ATP-115.

[1] Scheduling can be classified as either on-line or off-line scheduling. In this paper, scheduling is used to refer to on-line scheduling, unless otherwise explicitly stated. For more discussions on scheduling, the reader is referred to [FoKe85] and references therein.

[2] A minimally constrained plan is a plan that will not require replanning in the future. Constraints known apriori are used to generate a minimally constrained plan off-line. A maximally constrained plan is a plan that is generated taking into consideration all available constraints. A maximally constrained plan is generated at run-time and can be directly executed.

(iv) directed acyclic graphs (DAGs), (v) Petri Nets [KrSr887], and, (vi) frames [Haye93]. For the example in this paper, a listing of partial orders, a termination set of processes, and a set of previously executed processes will form the part representation. A distributed simulation scheme is used to evaluate possible schedules. The reader is refered to [Jern95] for a detailed survey of the literature on search techniques and simulation.

Finding efficient heuristics to reduce the size of the search space to make the simulation tractable is a primary objective of this work. Although the search is bounded, the number of simulations is large for even moderately complex environments. The significant characteristics of the simulation described in this paper include: (i) Each resource on the factory floor is assumed to have some computational element that can make process decisions with respect to its current status. Typically, this computational element is combined with other such elements in a central process. However, this tight coupling between the central process and the individual machines makes expansion, changes, and upgrades difficult. In the distributed simulation environment described in this paper, the computational element for a machine, the machine's controller, and the physical resource are modeled as a single agent. Given this model, each machine has the added capability to reason by itself. Once the system is decoupled in this manner, machines may now reason very differently. For instance, one machine may use intensive geometric reasoning while another may use simple semantic reasoning. The part and schedule representations are the only constraints on the method of reasoning for each agent, (ii) Upgrades can be incorporated into the scheduling unit of a single machine without affecting the rest of the system. Thus, each module must have a minimum set of constraints to be integrated with the system, while allowing the maximum amount of freedom to perform its job in the manner it chooses. On the other hand, the modularity introduces a serious architectural problem. Since no machine can assume anything about the operation of the other machines, each module cannot directly assess the extended results of its actions. In essence, each machine is blind with respect to other machines or its role in the system. The absence of a global view makes the application of heuristics considerably more difficult, (iii) The maintanence of a central time keeping process helps to provide a global view of the simulation, (iv) The simualtion considers the generation of on-line schedules for a group of machines or workcells, and, (v) The method proposed has the flexibility to exploit process dependent and process independant heuristics [Jern95].

This paper is organized as follows. Section 2 presents the new distributed environment. Section 3 analyzes a complete trace of a simulation. Section 4 concludes the paper with issues for future development and research.

2 THE NEW SIMULATION METHOD

In this research, simulation is employed to both generate and evaluate schedules. Simulated schedules are generated and evaluated based on the processes performed, in comparison to the input plan, and the current state of the system. Thc number of potentially feasible schedules to simulate is a dynamic function of the changing system state. The state of the system cannot be predicted off-line due to unexpected and random events. An on-line simulation of schedules can react to the current state when evaluating plans. The primary modification made to basic sequential simulators [Misr86] for this research is the ability to handle points of non-determinism. In each new simulation created, one of the valid responses to the event that triggered the duplication will be simulated. Each of these simulations can continue in parallel since they are completely duplicated.

2.1 The Central Algorithm

Each simulation can be in one of four states: ready (to process an event), waiting (for the previous event to finish processing), terminated, successfully completed. While the simulation is active, it cycles between the ready and the waiting states. When a simulation reaches a point of non-determinism, all the possible choices must be considered. At this point, a message is sent to the central process that requests the simulation to be split into a number of duplicate simulations. The central processor duplicates the event list for the simulation and signals all agents to duplicate any information held for that simulation. The requesting agent can then reflect each possible path in one of the duplicate simulations. The algorithm for the central process is given in Figure 1.

```
T = T0, T1, ... T(x-1)
where x is the number of simulations and Tw is the global clock of simulation w. The global clock can be
defined as: Tw = MIN (t)
where t = t0 t1, .. t(y-1), y is the number of factory floor processes, and ts is the next expected time in which
factory floor process s will generate an event.

Central
        Create simulation 0, T= 0, invalidate all expected event times
        Send A Start Simulation Message
        Bound = ∞
        Repeat
                From subset of simulation that have all agents registered, T',
                find the simulation with Tv = MIN(T')
                        Advance Tv to MIN(t)
                        If Tv > Bound then
                                Locally delete simulation v
                                Broadcast delete simulation v message
                        Notify agent with registered time Tv of the current time and mark that agent as
                        invalid
                If any validation messages have been received for agent z in simulation q
                        For simulation q
                                register new time for Tz
                                invalidate any other agents included in the message
                                mark agent z as valid
                If any delete messages have been received for simulation b
                        Locally delete simulation b
                        Broadcast delete simulation b message
                If any split message received
                        Create the simulation to the correct number of new simulations
                        Broadcast split message with old and new simulations
                Search any simulations that have all registered times as ∞ and no work in progress
                        If Th is < Bound then Bound = Th and save simulation h
                                Until no simulations are still running.
```

Figure 1 The Central Algorithm.

2.2 Floor Agents

Until this point in discussion, it was assumed that all factory floor processes are identical. In actuality, there exist four distinct groups of factory floor process. Each represents a group of physical entities on a real factory floor. These are: (i) queues, (ii) inspection stations, (iii) materials handling processes, and, (iv) machines.

- Queues: Queues are the simplest form of floor process and represent deterministic material handling methods. For instance, a conveyor belt can be modeled deterministically because it cannot change the state or ordering of the parts it contains. Queues must be specified with upstream and downstream neighbors as well as capacity and transfer rates.

- Inspection stations: The simulator assumes that no errors will be generated by the inspection station. The inspection stations may introduce delays and exhibit some buffering. Inspection stations can be a form of subgoaling to reduce the search space.
- Materials handling process: Material handling processes are non-deterministic. For instance, a robotic manipulator may have a choice of one of many machines to which it can deliver the part. The material handling process must know its downstream and upstream neighbors. Furthermore, its capacities and capabilities must be thoroughly defined.
- Machines: Machines are floor agents that can cause a change in the state of the product and are the most complicated floor process. Machines must match their own capabilities with the needs of the current part. This involves solving the partial ordering problem. The specific algorithm used depends on the part representation. Each machine must also know its capabilities as well as its upstream neighbor, downstream neighbor, and central process.

3 AN EXAMPLE

3.1 System Description

This section provides a complete trace of a single simulation for a sample computer assembly factory. In the example, three separate workstations are being assembled. The workstations produced by the factory can have several different configurations. New worsktations start at the beginning of the factory as an empty case. Each workstation must have a motherboard (MB) and a power supply (PS). The motherboard must be installed first. Cards that are available for installation are: a network interface card (NI), a monochrome monitor card (BW), a color monitor card (CR), and a drive controller (DC). Any subset of these cards may be installed in a workstation. Three drives are available for installation: a hard drive (HD), a floppy drive (FD), and a tape drive (TD). Drives have to be installed after the installation of the drive controller.

Workstation A is a diskless workstation with a motherboard, power supply, monochrome monitor card, and a network interface card. Workstation B is a standalone workstation with a motherboard, power supply, drive controller, monochrome monitor card, hard drive, and floppy drive. Workstation C is a high-end workstation with a motherboard, power supply, drive controller, color monitor card, network interface card, hard drive, and tape drive. Table 1 summarizes the configuration of each workstation. At the given moment, t = 0, workstation A has just arrived and the other two workstation's assembly are already in progress. Workstation B is having a motherboard installed and has 2.4 time units remaining on the operation. Workstation C already has a motherboard and power supply and has 3 time units remaining on the installation of its color monitor board.

Table 1 Workstation configurations.

Workstation	MB	PS	NI	BW	CR	DC	HD	FD	TD
A	√	√	√	√					
B	√	√		√		√	√	√	
C	√	√	√		√	√	√		√

The example factory is shown in Figure 2. It contains all of the four fundamental types of factory floor agents; queues (Q0, Q3, Q5, Q8, and Q10), material handlers (MH1 and MH7), machines (M2, M4, and M9), and an inspection agent (I6). Notice the MH7 has the option of routing parts along a reentrant pathway. The capabilities for each machine are described below.

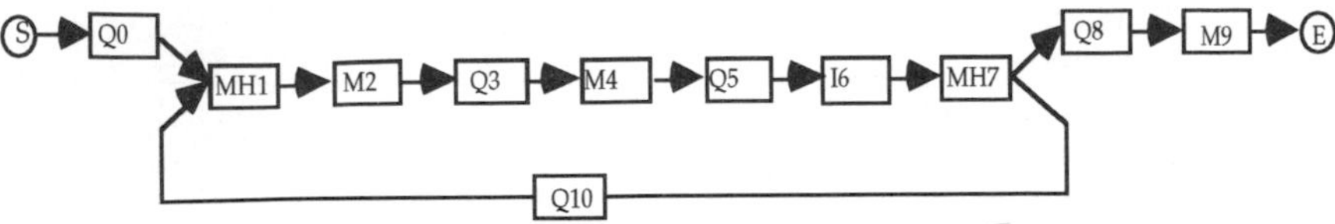

Figure 2 Block diagram of the example factory.

Although varying in length, all the queue agents in this example have similar characteristics. For instance, each queue takes .2 time units to advance a part one position in the queue and 1.3 time units to transfer a part out of the queue. Table 2 shows the specifications for the queues in the example.

Table 2 Queue descriptions.

Position to position transfer rate	.2
Exit transition time	1.3
Length of Q0	3
Length of Q3	3
Length of Q5	3
Length of Q8	3
Length of Q10	10

Two material handlers are present in the system. MH1 can transition a part from one of two sources to a single destination. Conversely, MH7 can transition a part from a single source to one of two destinations. Material handlers will always introduce a point of non-determinism becuase of the multiple sources or destinations. Both machines take a finite amount of time to grasp a part and perform a transport operation. If the destination is currently full, the material handler will attempt to accomplish the grasp, but will wait for the destination to empty before performing transportation operation. Table 3 shows the specific characteristics of the material handlers in this example.

Table 3 Material Handler descriptions.

Grasp time for MH1	1
Transport time for MH1	1.3
Grasp time for MH7	.3
Transport time for MH7	.7

In the physical system, inspection stations serve to synchronize the knowledge bases with the real world. If the state predicted by the knowledge base and the state of the real world are sufficiently different, a process error may have occurred. Once the knowledge base is updated, the next scheduling phase will begin with a new initial state and reactively adapt to the changes this introduces. Since contingency schedules are not created, the inspection system will not find errors. The inspection station requires 2 time units to inspect a part in the example.

Three machines are present in the system. Each of these machines have different capabilities. M2 can install cards, but it is primarily used for installing the motherboards and power supplies. M4 is a card installation machine only. M9 installs drives only. Each machine can determine if any of its capabilities match those needed by the current part. Machines may perform several sequential operations on the same part without the expense of fixturing and transferring the part between operations. When a part leaves a machine in this example, there is a one time step delay for grasping and transfering the part (not listed in the Table 4. Table 4 gives the capabilities for each machine and process times for the installation of each option.

Table 4 Machine capabilities.

Machine	MB	PS	NI	BW	CR	DC	HD	FD	TD
M2	3.2	3	2	2	2	2			
M4			3.7	3.7	3.7	3.7			
M9							4.2	4.2	4.2

If a part reached M9 in a simulation and did not have all its cards installed, the part would be passed on to the termination point and the simulation would fail. Given the current configuration, there was no way for the cards to be installed before the part exited the assembly line. However, reentrant pathways allow parts to move against the flow of the assembly line. As in the example below, the pathways can also be used to alter the ordering of the parts in the assembly line. Parts with little remaining work can pass jobs with a substantial amount of remaining work to reduce the number of parts in the system.

3.2 Simulation

The trace shown in Table 5 is a complete simulation generated in the system described above. For clarity, this simulation is always marked U0. Time T=2.4 is a point of non-determinism. M2 requests to split the simulation because it has to perform one of the following possible operations: install the power supply, install the drive controller, install the monochrome monitor card, or transfer the part out with no further operations. The simulation was split into four duplicate simulations. In each of the duplicates, M2 simulated one of the possible operations. As noted in the comments for that line, the power supply was installed in simulation U0. At simulation time T=1.9, another type of non-determinism occured. MH1 had the choice of drawing the next part from Q0 or Q10. For simulation U0, the part was drawn from Q0. In the alternate simulation, MH1 will wait for a part to become available from Q10.
At time step 20.6, two event are shown occurring at the same instant. While this is rare, it does not pose a simulation problem because of the decoupled nature of the event processing. Furturemore, notice that workstation C has chosen to follow the reentrant pathway at T=7.9. Because of the volume of work remaining to be performed on C, it would have retarded the progress of workstations A and B. In effect, the smaller jobs passed the larger job.

Table 5 Simulation trace.

Time	Q0	MH1	M2	Q3	M4	Q5	I6	MH7	Q8	M9	Q10	MIN.	Comment
0	1.9	∞	2.4	∞		∞	∞	∞	∞	∞	∞	1.9	
1.9	∞	2.9										2.4	Split U0= 0»1 Ux= 10»1
2.4			5.4									2.9	Split U0= PS Ux=DC,BW,xfer
2.9		∞†										3.0	
3.0					4.0							4.0	Split U0= xfer Ux= NI,DC
4.0					∞	5.9						5.4	
5.4			6.4									5.9	Split U0= xfer Ux= BW,DC
5.9						∞	7.9					6.4	
6.4		7.7	∞	8.3								7.7	
7.7		∞	10.9									7.9	Split U0= MB Ux= xfer
7.9							∞	8.9				8.3	Split U0= 7»10 Ux= 7»8
8.3				∞	12.0							8.9	Split U0= BW Ux= DC,xfer
8.9								∞			12.2	10.9	

Time	Q0	MH1	M2	Q3	M4	Q5	I6	MH7	Q8	M9	Q10	MIN.	Comment
10.9			13.9									12.0	Split U0= PS Ux= BW,NI,xfer
12.0					15.7							12.2	Split U0= DC Ux= xfer
12.2		13.2									∞	13.2	Split U0= 10»1 Ux= 0»1
13.2		∞†										13.9	
13.9			14.9									14.9	Split U0= xfer Ux= BW,NI
14.9		16.2	∞	16.8								15.7	
15.7					16.7							16.2	
16.2		∞	18.2									16.7	Split U0= NI Ux= DC,xfer
16.7					∞	18.6						16.8	
16.8				∞	20.6							18.2	Split U0= BW Ux= NI,xfer
18.2			20.2									18.6	Split U0= DC Ux= xfer
18.6						∞	20.6					20.2	
20.2			21.2									20.6	
20.6					24.3							20.6	Split U0= NI Ux= xfer
20.6							∞	21.6				21.2	Split U0= 7»8 Ux= 7»10
21.2			∞	23.1								21.6	
21.6								∞	23.5			23.1	
23.1				∞†								23.5	
23.5									∞	27.7		24.3	Split U0= HD Ux= FD,xfer
24.3					25.3							25.3	
25.3				26.6	∞	27.2						26.6	
26.6				∞	27.6							27.2	
27.2						∞	29.2					27.6	
27.6					∞	29.5						27.7	
27.7										31.9		29.2	Split U0= FD Ux= xfer
29.2							∞	30.2				29.5	Split U0= 7»8 Ux= 7»10
29.5						∞	31.5					30.2	
30.2								∞	32.1			31.5	
31.5							∞	32.5				31.9	Split U0= 7»8 Ux= 7»10
31.9										32.9		32.1	
32.1									∞†			32.5	
32.5								∞	∞††			32.9	
32.9									34.2	∞		34.2	B OUT
34.2									∞†	35.2			
35.2									36.5	∞		36.5	A OUT
36.5									∞	40.7		40.7	Split U0= HD Ux= TD,xfer
40.7										44.9		44.9	Split U0= TD Ux= xfer
44.9										45.9		45.9	
45.9										∞		∞	C OUT

† part waiting to exit

4 FUTURE DIRECTIONS AND CONCLUSIONS

In this work, a new algorithm for distributed simulation and on-line evaluation of schedules has been presented. The algorithm can be applied for multiple machines and is flexible enough to handle most manufacturing scheduling applications. The algorithm is to evaluate heuristics to reduce the complexity of the search procedure. This work has proposed a distributed simulation approach that offers the following advantages: (i) Simulation helps in reducing the number of schedules evaluated to provide a satisfactory response time, (ii) The design of the simulation environment is flexible and can be tailored to the constraints of a factory environment, (iii) The implementation of the distributed simulation environment reflects the requirements foreseen in future schedulers (e.g. scheduling with machines that reason in a variety of ways), and, (iv) Without additional problem-specific heuristics, the system is designed to take a plan and perform a bounded search on generated schedules.

Future research will be directed to additional heuristics, such as: (i) Not allowing parts to cycle through the reentrant pathways unless there is a state change, (ii) Methods for identifying and associating subgoals with parts. These subgoals will determine if a part can transition beyond a specific machine. For instance, in all cards should be installed before a part can enter Q8. (iii) Methods for establishing the initial bound on the simulations. (iv) Alteration of the branching policy to favor attractive solutions.

5 REFERENCES

[ChAl86] H. Chochon and R. Alami (1986) "NNS, A Knowledge-Based On-line System For An Assembly Workcell", *Proc.of IEEE International Conference on Robotics and Automation*, pages 603-9.

[FoKe85] B.R. Fox and K.G. Kempf (1985) "Opportunistic Scheduling for Robotic Assembly", *Proc. of IEEE International Conference on Robotics and Automation*, pages 880-9.

[Haye93] B. Hayes-Roth (1993) "Opportunistic Control of Action in Intelligent Agents", *Proc. of IEEE Transactions on Systems, Man, and Cybernetics*, Vol. 23, No. 6, pages 1575-87, November/December.

[Jern95] S.R. Jernigan (1995) "A Reactive, Distributed Simulation Method for Scheduling flow through a Factory Floor", submitted as master's thesis, The University of Texas at Austin, Summer.

[Lyon90] D.M. Lyons (1990) "A Process-Based Approach to Task Plan Representation", *Proc. of IEEE International Conference on Robotics and Automation*, pages 2142-7.

[MeSa91] L. S. Homem de Mello and A. C. Sanderson (1991) "A Correct and Complete Algorithm for the Generation of Mechanical Assembly Sequences", *Proc. of IEEE International Conference on Robotics and Automation*, pages 228-40.

[Misr86] J. Misra (1988) "Distributed Discrete-Event Simulation", *Computing Surveys*, Vol. 18, No. 1, March.

[XiBe88] X. Xia and G.A. Bekey (1988) "SROMA: An Adaptive Scheduler for Robotic Assembly Systems", *Proc. of IEEE International Conference on Robotics and Automation*, pages 1282-7.

19

Integration of Process Planning and Scheduling using Resource Elements

J.D.A. Carvalho and N.N.Z. Gindy
Department of Manufacturing Engineering and Operations Management, University of Nottingham, Nottingham NG7 2RD, United Kingdom, Tel +44-115-9514031, Fax +44-115-9514000, Email epxjdc@unicron.nott.ac.uk

Abstract

This paper outlines a new approach utilising generic capability units termed 'resource elements' (REs). REs describe the capabilities of machine tools and machining facilities, and represent the process planning information. A prototype integrated environment for process planning and scheduling in which REs are used for both the generation of nonlinear process plans and production scheduling is described. It is shown to offer significant improvements in manufacturing system performance and REs are seen to provide a powerful and practical approach to integration.

Keywords

Process planning, scheduling, integration, resource elements, nonlinear process plans

1 INTRODUCTION

Manufacturing companies are divided into specialist departments performing computer aided design (CAD), computer aided process planning (CAPP) and production planning and control (PPC). These activities are normally performed sequentially by departments, with most performance-improving effort directed towards refining and perfecting individual functions rather than performance across functions.

Today, CAPP and scheduling remain essentially separate activities. Process planning is considered a time-independent activity that generates an optimised set of sequenced operations and resources for transforming part design into a finished product, while scheduling is treated as a time-dependent function aimed at the utilisation of resources to satisfy the process plans for a number of products (Chryssolouris and Chang 1985, Kuhnle 1990).

Difficulties in integrating CAPP and PCC relate either to the functionality of these systems or to their use of data. While process planning systems focus on the operations to be performed on single parts, scheduling systems deal with both multiple parts and multiple products to be manufactured within the same system. Conventional scheduling systems also adopt a global view and thus aim to optimise machine utilisation and maximise throughput to meet production targets. Schedulers can normally only handle linear sequential process plans and resources are often combined into capacity groups for the purpose of load balancing (ElMaraghy 1993).

Production disruptions and bottlenecks occur on the shop floor due to resource shortages and routes have to be changed to suit demand; however, this is done without full consideration of its implications for the overall operation of the shop floor. The lack of communication between process planning and scheduling therefore leads to higher costs and is a serious obstacle to achieving effective integrated manufacturing systems.

Many shortcomings in current CAPP and scheduling industrial practices are the result of several assumptions: (i) Process planning is a static activity, influenced only by component design and its technological constraints, and thus focuses on technical manufacturing analysis of products to satisfy their design intent. (ii) Scheduling is strictly sequential, to be performed only after process planning, and therefore focuses solely on capacity constraints and efficient resource time-allocation. (iii) There is unconstrained resource availability during component manufacture. Moreover, process planning is performed such that some 'desirable' routing normally utilising the most capable machine tools is repeatedly selected for each component.

In modern, highly dynamic, responsive industrial environments, none of the above assumptions is realistic or applicable. Process planning and scheduling may have conflicting goals, but a significant potential for improvement in the CAD/CAM information integration can be achieved if the gap between CAPP and scheduling can be bridged.

There are three basic approaches towards integration of CAPP and PPC (ElMaraghy 1993). The global integration approached is based upon high level integration of functions and CIM modules. Each CIM module (CAD, CAPP, MRPII) maintains its own database and a global update scheme is devised to achieve information integration. This method is data-intensive and does not address the need for nonlinear plan representation (ElMaraghy and ElMaraghy 1993).

The second approach attempts unification, merging CAPP and PPC into one system with a common structure. Petri-nets are commonly used to model logical and temporal relationships between system entities. The issues of realtime events and related feedback from PPC to CAPP in response to shop floor 'disturbances' are not fully considered, e.g. FLEXPLAN-ESPRIT Project 2457 (Toenshoff et al. 1989, Toenshoff and Detand 1990). Extensions to the approach are made in the new EPRIT sponsored project COMPLAN (Kruth et al 1994) to develop an integrated system for concurrent process planning, scheduling and shop floor control based on nonlinear process plans.

The third approach to integration can be considered an intermediate solution between the first two approaches (ElMaraghy 1993, ElMaraghy and ElMaraghy 1993). This approach is essentially modular where CAPP and PCC are not treated as one system. However, the CAPP and PPC systems need to have the ability to interact with the shop floor disturbances, nonlinear process plans, and dynamic resources and constraints. In this approach, integration schemes utilising a separate module called the 'integrator' were developed to bridge the functional and data gaps between CAPP and PPC (Jack and ElMaraghy 1992).

Real improvement in production control can only be achieved by integration of a CAPP system and a scheduling system that fully utilises the benefits of manufacturing alternatives provided by nonlinear process plans. The actual utilisation of machines is significantly determined early on in process planning and to meet the flexibility requirements of workshop control, the process plan itself has to be more flexible. It is recognised that CAPP and PPC systems need to be able to generate and utilise nonlinear process plans, interact with the shop floor disturbances, and dynamically take into account resource availability constraints on PPC (Jack and ElMaraghy 1992).

It is the contention of the authors that one of the main directions for improving manufacturing performance lies in improved interaction and integration between process planning and scheduling. This will lead to a dynamic manufacturing environment capable of reacting to real factory conditions, and will have a significant impact on reducing cost, lead time and inventory as well as improving manufacturing responsiveness.

The reported work is based upon utilising generic (machine-independent) capability units termed 'Resource Elements' both to represent component process planning information and to act as the basis for the generating the production schedule. An integrated environment for process planning and scheduling is currently under development, and is described here.

2 SYSTEM DESCRIPTION

The input to the integrated system (Figure 1) is a 'manufacturing job' list of the components to be produced within a scheduling period. It includes: component identification, due date and a priority value associated with each component. The list can be sorted according to component priority or earliest due date depending on the desired manufacturing strategy. The way the components are sorted represents the sequence that they are processed through the proposed system. (1) Each component goes through a generic planning module where machine-independent (RE-based) component process plans are generated. (2) Processing and setup times are generated for each process plan. (3) A simulating model is created based upon the process plans, and (4) this model is simulated for different dispatching rules and different shop floor performance measures. The dispatching rule resulting in a better performance indicator is selected. The final component process plans and schedule are those obtained using the selected dispatching rule.

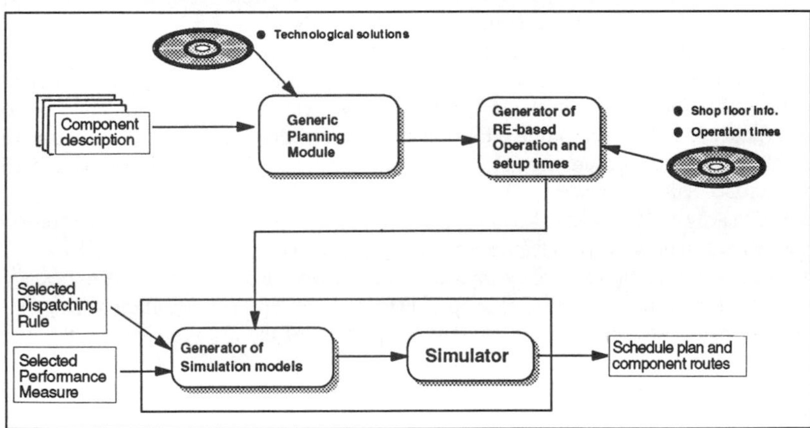

Figure 1 System Overview.

The results obtained from the simulation module shown in Figure 1 as the output (schedule plan and component routes) are both machine-based although the simulation model is created using RE-based information.

The number of components in the list depends on the company's scheduling period. Each manufacturing system has its own minimum limit for the scheduling period representing the time the system needs to check the feasibility of making each component. It depends on the time necessary to prepare and test fixturing devices, tools, NC part programs, to verify resources availability, etc., as well as the frequency that the shop floor's status database is updated. Manufacturing systems that can perform all these tasks rapidly would have very short scheduling periods. Vice versa, long scheduling periods normally imply that the real allocation of jobs to machines will not follow the scheduling plan in order to cope with the unexpected (machine breakdown, resource shortages, etc.), thereby wasting all the effort to 'optimise' the shop floor performance at the planning stage. It is important to have short scheduling periods in order to gain the advantages of the integration.

2.1 Generic Planning Module (GPM)

An overview of the GPM is shown in Figure 2. A feature-based component data model is used to represent component information (Gindy and Ratchev 1991). Each component is treated as a set of connected features, with each feature (e.g. hole, slot, step, pocket) relating to a component region which has significance in the context of machining operations. Based on their geometric attributes, features are classified into categories, classes and sub-classes, which may be followed by secondary forms to describe component geometry (Gindy 1989). Technological requirements (dimensions, accuracy and surface finish etc.), which influence the selection of processing methods, are attached to each feature. Features relate to local component regions and alone are unable to describe the structural aspects of component geometry, i.e. the relationships between component features which can influence the selection of its methods of manufacture. Feature relationships are represented using feature connectivity graphs which relate each feature to adjacent features and to the potential processing directions of the component (Gindy et al. 1993).

During generic planning, each component is treated as a set of features that require machining using the available processing system resources. Feature connectivity provide the constraints that may exist on the formation feature clusters that can be machined from a common component direction (potential component setups). Each feature has a multiple set of technological solutions capable of producing the feature geometry and satisfying its technological requirements in terms of accuracy, surface finish, tolerance.

For a machine-independent but machining-facility-specific description of the processing requirements of components, each feasible solution is represented by the necessary REs.

REs can be considered as elementary capability units collectively representing the full capabilities of the machines contained in a machine shop (Gindy et al. 1995). They are machine-shop-specific and describe the distribution (commonality and uniqueness) of the resources among the machine tools contained in a manufacturing cell or machine shop (Figure 3). REs can belong to several machine tools in a manufacturing facility and thus provide a basis for a generic definition of the processing requirements of components. Each component can be represented by a set of unique REs obtained from alternative machine tools without having to specify the actual tools to be used.

During feature level optimisation, the set of REs needed to produce each component feature is determined. An algorithm based upon minimising the variety of resources needed for each component is used for optimising the resource set required for individual components. The most appropriate technological solution is then attached to each feature. During resource optimisation the system is capable of selectively avoiding the use of REs which are in heavy demand.

For each component in the manufacturing job list, the output of the GPM is a machine-independent process plan expressed in terms of the REs needed for its execution.

2.2 Generation of production schedule

Although the physical entities upon which the production schedule is ultimately generated are machines, component routes are defined generically in the simulation model based upon their RE requirements. The components wait in RE queues for a machine with the required REs to become available. REs act as parallel processing resources thereby increasing the flexibility of utilising system resources. Dispatching rules are used as the basis for simultaneously generating the production schedule and the final process plan for the components in the manufacturing job list.

The discrete simulation package SIMAN (Pegden et al. 1990) generates component routes and schedules and calculates system performance based upon the actual resources used for processing each component in the job list. The system is based on simulation models for capturing the characteristics of the manufacturing system in a mathematical form using a simulation language.

At the start of the simulation, generic plans are created sequentially for the components in the manufacturing job list and the simulation model is executed. The simulation model assumes the use of RE-oriented dispatching rules, to drive the forwards allocation of jobs to machines on a time basis. The system output is a production schedule and the process plans for the components in the job list as well as the system performance indicators.

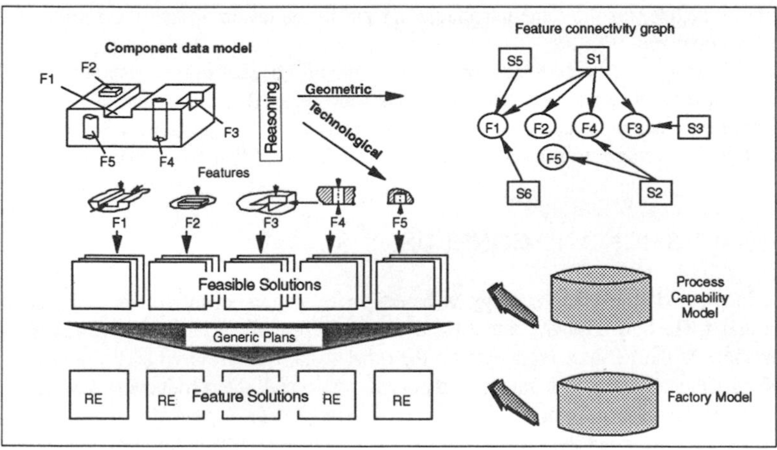

Figure 2 Overview of Generic Planning Module (GPM).

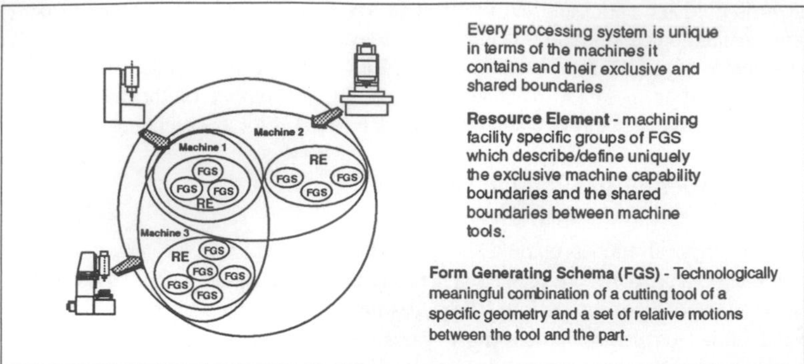

Figure 3 Definition of resource element.

3 RESULTS

The results presented here relate to a machining facility containing 22 machine tools operating as a one-of-a-kind production environment. The manufacturing job list within the scheduling period contained 25 components, with 4-12 features per component.

To evaluate the RE approach against the current practice of separately generating component process plans then scheduling whole machine tools assuming infinite resource availability, a production schedule, for the same manufacturing job list, is produced and the results of the two approaches compared.

Three dispatching rules were tested: first in first out (FIFO), shortest processing time (SPT) and earliest due date (EDD). The best values obtained were selected and are presented below. System performance was measured using five performance indicators: average flowtime (AFT), maximum flowtime (MFT), average tardiness (AT), maximum tardiness (MT) and machine utilisation (MU).

4 DISCUSSION AND CONCLUSIONS

The machine-based scheduling strategy is based on fixed component routes determined at the planning stage. The only flexibility here is the number of repeated machine tool types available in the machine shop. In the RE-based strategy, on the other hand, REs are used all the way through the process of simultaneously generating component plans and the production schedule for the components in the manufacturing job list.

RE-based integration achieves significant improvements in system performance (Figure 4) since REs offer a sub-machine capability units for use during schedule generation. It therefore increase the choices for optimising the allocation of jobs to available machine tools. Machine-based scheduling, however, has its number of choices limited by the number of the repeated machines and the rigidity of the routes decided at the planning stage.

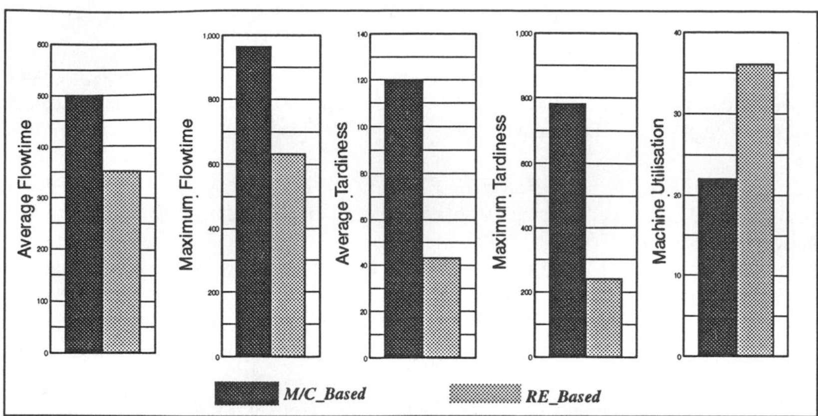

Figure 4 M/C-based strategy versus RE-based strategy.

The case study results show that integrating process planning and scheduling significantly improves manufacturing system performance. The RE concept is shown to be a powerful and practical approach towards this integration and can significantly improve shop floor performance.

The proposed system is a straightforward tool that can easily be applied to manufacturing systems of any size or any level of automation. This system uses the new RE-based way of describing a shop floor which proves to be very efficient. Any existing CAPP and scheduling system can be transformed into a RE-based integrated system, by (i) adding a module between the CAPP system and the shop floor database that transforms the machine-based information into RE-based information; (ii) adapting the scheduling system to schedule machines using RE information; (iii) feeding the scheduling system with the generic plans created by the CAPP system.

5 AKNOWLEDGMENTS

J D A Carvalho would like to aknowledge the grant from "Programa Ciencia, BD/2103/92-IB, JNICT, Portugal", under which his part of the work was carried out.

6 REFERENCES

Chryssolouris, G. and Chang S. (1985) An integrated approach to integrate process planning and scheduling. Annals of the CIRP, 34(1), 413-417.

ElMaraghy, A. (1993) Evolution and future perspectives of CAPP. Annals of CIRP, 42(2), 739-751.

ElMaraghy, H.A. and ElMaraghy, W.H. (1993) Bridging the gap between process planning and production planning and control. Proc. 24th CIRP International Seminar on Manufacturing Systems, June 11-12, 1992. Copenhagen, Denmark, pp. 1-10; also published in Manuf. Sys., 22(1), 5-11.

Gindy, N.N.Z. (1989) A hierarchical sutructure for form features. Int. J. Prod. Res., 27(12), 2089-2103.

Gindy, N.N.Z. and Ratchev, T.M. (1991) Product and machine tools data models for computer aided process planning systems, in Computer Applications in Production and Engineering: Integration Aspects (ed. Doumeingts, Browne and Tomljanovich), Elsevier Science Publishers B.V., IFIP.

Gindy, N.N.Z., Huang, X. and Ratchev, T.M. (1993) Feature-based component model for computer-aided process planning systems. Internation Journal of Computer Integrated Manufacturing, 6(1&2), 20-26.

Gindy, N.N.Z., Ratchev, T.M. and Case, K. (1995) Component grouping for cell formation using resource elements. Int. J. Prod. Res., submitted.

Jack, H. and ElMaraghy, W.H. (1992) A Manual for Interprocess Communication with the MPS (Message Passing System). DAMRL Report No. 92-08011, the University of Western Ontario, London, Ontario.

Kruth, J.P., Detand. J., VanZeir. G., Kempenaers, J. and Pinte, J. (1994) Opportunistic process planning: A knowledge based technique for CAPP applications. Computers in Engineering, Proceedings of the International Conference and Exhibit 94, 1, 227-233

Kuhnle, H. (1990) Prerequisites for CAPP. 22nd CIRP Seminar on Manufacturing Systems, University of Twente, The Netherlands.

Pegden, C.D., Shannon, R.E., Sadowski, R.P. (1990) Introduction to Simulation Using SIMAN. McGraw-Hill, ISBN 0-07-049217-4.

Toenshoff, H.K., Beckendroff, U. and Anders, N. (1989) FLEXPLAN - A Concept for Intelligent Process Planning and Scheduling. CIRP Intr. Workshop on CAPP, Hannover University, Sept. 21-22, pp. 87-106.

Toenshoff, H. K. and Detand, J. (1990) A Process Description Concept for Process Planning, Scheduling, and Job Shop Control. 22nd CIRP Seminar on Manufacturing Systems, University of Twente, The Netherlands.

7 BIOGRAPHIES

José Carvalho, Lic., Msc, is currently a PhD student at the Department of Manufacturing Engineering, University of Nottingham, UK. He worked as a lecturer at Universidade do Minho Portugal until 1992. Major interests include CAPP, PPC and CAPP/PPC integration.

Nabil Gindy, Bsc, Msc, PhD, is professor of Advanced Manufacturing Technology and leads the responsive manufacturing research group at Nottingham University. Research interests include machining and tooling technology, feature-based CAD/CAM systems, generative process planning, integrated planning and scheduling systems, CNC part programming and the design and confguration of manufacturing systems.

PART EIGHT

Decision Support Systems in Manufacturing

20

A Decision Making Tool in FMS Design

G. Perrone, G. Lo Nigro, S. Noto La Diega
Dipartimento di Tecnologia e Produzione Meccanica
Università degli Studi di Palermo
Viale delle Scienze 90128 -Palermo, Italy
tel(+39) 91 422142 fax (+39) 91 599828
e-mail: noto@power1. dtpm.unipa.it

Abstract

In FMS design one of the initial fundamental decision concerns the flexibility level of the workstations to be implemented. New machining centers offer the possibility to carry out several operations on the workpieces to be processed, thus reducing the workload both of the material handling system and of the resources used to control the part flow in the FMS. In spite of advantages related to the use of the machining centers with high flexibility, the better typology of each workstation must be evaluated taking into account its influence on the tools fleet and on the workload of the handling and tool room subsystems.

This paper proposes a fuzzy decision support model to select the flexibility level of the workstations facing the uncertainty related to the dimension of the subsystems to be implemented in order to obtain a reliable tool flow.

Keywords

FMS design, fuzzy optimization

INTRODUCTION

In the Flexible Manufacturing Systems (FMS) design two main issues have to be considered: the flexibility and the integration level among the resources in the system.

The flexibility can be measured according to the amount of time and cost that the manufacturing system will spend to react to external changes, such as market changes, and internal changes, such as resources breakdown; the integration level can be evaluated by the influence of the management policies on the system performances (Alberti, N. et al. 1988).

The above characteristics make the FMS design more complex than a traditional system both for the larger number of design alternatives which have to be considered and for the uncertain impact of the above alternatives on the integration among the system resources and therefore on their performances.

Moreover the FMS design is characterized by an higher entrepreneur risk caused by the economic importance of each design alternative and by the wider integration due to higher use of shared resources when flexibility and production capacity are increased.

The FMS design parameters able to influence the system flexibility are the amount and the typology of the technological operations which will be processed by the manufacturing system; these parameters are usually evaluated taking into consideration the actual production plan in the short period and the opportunity production plan in the medium and long period (Perrone, G. et al., 1994a) (Perrone, G. et al., 1994b). The flexibility level gained will depend on the technological resources effectively installed in the system able to perform the above operations. This configuration impacts the operative execution of the operations in the production plan and then times and costs required to react to the changes.

One of the main design alternatives concerns the machine flexibility of each resource to install, that is:

- dedicated technological resources which are able to perform a small number of technological operations;
- general purpose technological resources able to perform a large number of technological operations in the same machining center.

Even if the emerging trend seems to encourage flexible manufacturing systems consisting of general purpose work-centers (Semeraro, Q. et al., 1993) in order to reduce the workload of the material handling system, to improve the utilization rate of the machining centers and finally to make easier the modularity of the production system, it is necessary to consider both the alternatives because they require different investment and running costs due to the number of workstations to be installed, to their utilization rates and to the tool fleet design.

A system characterized by dedicated technological resources asking for an almost static tool management allows to reduce the tool fleet dimension and the tool handling system, but on the other side it penalizes the utilization rates of the machines and it requires an higher investment cost of the material handling system; on the contrary, a system characterized by general purpose machine allows to improve the resources utilization rates, but it asks for a wider tool fleet and a more complex tool handling system (La Commare, U. et al., 1993).

The paper proposes a model able to support the FMS designer in getting a satisfied cost compromise solution between the two alternatives above discussed allowing the designer to put into the model some vague information that characterize the design phase.

MODEL ASSUMPTIONS

The tool fleet and management tool system investment costs

Tools, tool fixtures and tool handling system are resources whose cost can be compared with workstations in a FMS environment; late technical and economical estimations have been figured out that for each CNC machine center installed tools accounts for 29% and fixture for 28% of the total invested capital; 16% of scheduled production cannot be met because of tooling is not available; 30-60% of tooling inventory is somewhere on the shop (Gaalman, G. J. C. et al., 1994). Therefore, in order to avoid high production costs caused by idle times on machine centers due to tools unavailability, it is necessary to dimension the tools fleet ant its management system in order to guarantee an high reliability level of the tools system.

In order to evaluate all of the costs involved with the tool system in the early phase of the FMS design the proposed model considers the following assumptions.

A different tool code for each technological operation

This statement allows to assign both the tools and the machining operations to each workstation and therefore to figure out the number of tools with the same code duplicated in each workstation tools storage. If the workpiece part program in a given workstation requires many times the use of the same tool code, all of those technological operations are considered as a unique operation with a tool time equal to the sum of the partial tool times.

In this way the minimum number of tools, with the same code in the tool fleet, depends on the typology of the installed workstations being necessary to foresee that code in each workstation where that operation is performed.

The designer should be able to approximately evaluate the amount of time required to regenerate each tool and to handle it from/to each workstation

The tool fleet have to be designed in order to assure the availability of the tool on the machine storage when it is required by the part program. This condition it is not of easy achievement because of the great number of variables involved in it, such as the production policies and the detailed working conditions of the tool system.

Furthermore, while it is well known that the demand rate of spare tools on each workstation depends on technological parameters, on tool service times and on temporal distribution of the part type demand, and that the regeneration tool time, which affects the tool lead time between tool room and workstation, depends on the amount of resources available on the tool room and on the utilized tool handling system, the estimation of an analytical function able to express such dependencies is almost impossible.

A rough estimation of the number of spared tools n_g, coded with g, can be figured out using the Little law, i.e.:

$$n_g = d_g \cdot T_g \qquad (1)$$

where d_g is the mean demand rate, whose value can be evaluated referring to the total tool service time and to its mean useful life for specific machining conditions, and T_g is the mean throughput tool time between tool-room and work center.

The model assumes the hypothesis that the designer has enough experience in the early design phase for an approximate estimation of the variability range of T_g and so of the spare tools n_g to foresee in the tools fleet.

The uncertainty in the estimation of n_g is approached through a fuzzy number (Zadeh, L. A., 1965) which is an easy way to express in mathematical way vague concept such as "*approximately*".

Because of both d_g and T_g increase with the number of workstations which require the same tool code, the annual investment cost of the tool fleet and management tool system can be expressed as:

$$C_t = \sum_k y_k \cdot \tilde{n}_k \cdot \sum_i \beta_{i,k} \cdot z_{i,k} \qquad (2)$$

where:

- $\beta_{i,k}$ is the equivalent annual investment cost, comprehensive of the direct cost of the tool fixture and of the quota of investment cost concerning the tool management system; this latter quota is assigned to the tool code required for the operation i performed on the workstation k;

- y_k is the number of workstations of type k to be installed in the production system;
- $z_{i,k}$ is a binary coefficient equal to 1 if the operation i (i = $1,...,I$) is performed on the workstation k, 0 otherwise;
- $\tilde{n}_k$ triangular fuzzy number (a,b,c) whose mean value b is equal to the most possible value and the extreme values a and c depend on the variability range of the dimension of fleet tools estimated by the designer.

Opportunity costs

Opportunity costs are those investment costs which are not necessary for production activities. They are caused by the unbalancement among the workstations in the FMS that determines an incomplete utilization of the technological resources.

This under utilization of the technological resources is due to the necessity to process simultaneously some part types with different process plans; therefore, these costs can be configured like fixed costs that the company decide to burden with in order to deal with the competitive advantages coming from a flexible manufacturing system.

These costs reduce the operative profit of the company and the investment net present value and they increase the investment payback period of the investment, so that they have to be minimized choosing the system configuration that better allows to utilize the resources assuring an high flexibility value.

The global opportunity cost to be considered in the choice of the technological resources is:

$$C_O = \sum_{k=1}^{K} y_k \cdot C_k \cdot \left(1-u_k\right) \tag{3}$$

where:

- C_k is the annual equivalent investment cost of the workstation k (k=$1,.....,K$), comprehensive of the direct costs and of the quota of investment cost concerning the logistic and control resources;
- u_k is the mean utilization rate of the workstation k.

The model

The model assumes a scenario in which the designer can choose the typology k (k= $1,...K$) of each workstation inside a range K of possible alternatives characterized by different machine flexibility or abilities to perform various machining operations (i =$1,.....,I$) involved in the mix production plan.

The criterion driving the model is the minimization of the investment cost of the workstations with the related material handling system, of the tool fleet and its management system cost and the opportunity cost. Since these fixed costs have different life cycle they have to be capitalized in a proper way. The model assumes for each one the annual equivalent cost. Let us indicate with:

- Wl_i the total annual service time required by the machining operation i for realizing the required production plan;
- $x_{i,k}$ the percentage of WL_i to be assigned to the workstation k.

The workload for each workstation can be computed as:

$$T_k = \sum_{i=1}^{I} x_{i,k} \cdot WL_i \qquad (4)$$

so that using the static allocation procedure the utilization rate for each workstation can be expressed as:

$$u_k = \frac{T_k}{y_k \cdot T} \qquad (5)$$

where T is the annual availability time of each technological resources k and y_k is the number of resources k to be installed in the designing FMS.

With the above assumptions the analytical decisional model can be formulated as:

Model A (Decision variables: $y_k; x_{i,k}$)

$$\text{Min}(C_T) = \text{Min}(C_w + C_t + C_o) \qquad (6)$$

subject to:

$$\sum_{k=1}^{K} x_{i,k} = 1 \ \forall i;$$

$$\sum_{i=1}^{I} x_{i,k} \cdot WL_i \le y_k \cdot T \ \ \forall k \qquad (7)$$

$$z_{i,k} \ge x_{i,k} \ \forall i,k$$

with:

- $C_w = \sum_k y_k \cdot C_k$ annual equivalent investment cost of the workstations and the related logistic and part control flow system.
- $C_t = \sum_k y_k \cdot \tilde{n}_k \cdot \sum_i \beta_{i,k} \cdot z_{i,k}$ annual equivalent investment cost of the tool fleet and the related management system.
- $C_o = \sum_k y_k \cdot C_k \cdot (1 - \frac{\sum_i x_{i,k} \cdot WL_i}{y_k \cdot T})$ opportunity costs.

In order to get a solution of the model A it is necessary to apply a method able to compare triangular fuzzy numbers each other; in order to obtain a crisp linear model that can be resolved using an integer linear programming resolutor the Integral Value Method (Liou, T. S. et al., 1992) is suggested. Moreover, the uncertainty about the tool fleet dimension

$\left(\tilde{n}_k = \left(a_{n_k}, b_{n_k}, c_{n_k}\right)\right)$ can be observed in the expression of the cost C_T that becomes a triangular fuzzy number too: $\tilde{C}_T = \left(a_{C_T}, b_{C_T}, c_{C_T}\right)$, where a_{C_T} and c_{C_T} are the values assumed by C_T with the workstations typology corresponding to the solution of the crisp model when the minimum a_{n_k} and the maximum c_{n_k} values of $\tilde{n}_k$ are considered.

In order to control the investment risk due to the uncertainty the following complementary linear objective function can be introduced:

$$\text{Min}(U) = \text{Min}(c_{C_T} - a_{C_T}) \qquad (8)$$

Adding the above objective function to the model A, it becomes a multiple objective model. A fuzzy multiple objective programming method able to find a compromise solution with Pareto optimum will be used (Lee E.S. et al.,1993). The new linear model can be formulated as:

Model B (Decision variables: y_k; $x_{i,k}$)

$$\text{Max}\left[\gamma \cdot \lambda_1 + (1-\gamma) \cdot \lambda_2\right] \qquad (9)$$

subject to:

$\lambda_1 \leq \dfrac{C_{T\,max} - C_T}{C_{T\,max} - C_{T\,min}}$ λ_1 pushes the cost C_T to the minimum value

$\lambda_2 \leq \dfrac{U_{max} - U}{U_{max} - U_{min}}$ λ_2 pushes the uncertainty U to the minimum value

Other constraints Equations (7) of the model A

where:

- U_{max}, U_{min} are the maximum uncertainty considered obtained with the model A and the minimum uncertainty obtained minimizing eq. (8) with constraints (7) respectively;
- $\gamma \in [0,1]$ is the weight measuring the relative importance of the two objectives.

NUMERICAL EXAMPLE

The proposed model has been tested by a proper numerical example in which it has been supposed to design an FMS able to manufacturing a mix of parts requiring 15 different machining operations. The global machining times (Wl_i) for the operations, referred to the annual availability (T) of the FMS, are reported in the table 1.

To perform the operations the designer can choose 7 different typology of workstations ($k=1....7$). An annual equivalent investment cost decreasing with the flexibility has been assigned to each typology; in fact the increment of the investment cost of a machining center with higher flexibility is absorbed by the lower quota of the logistic and part flow control resources costs for that station.

The annual equivalent investment cost ($\beta_{i,k}$) assigned to each tool code increases with the flexibility of the machining center as consequence of the greater investment cost of the tool handling system due to the dynamic allocation of the tools in the workstation tool storage. All input costs data reported in table 1 are referred to the annual equivalent investment cost C_7.

Table1 Design data

		$C_1,C_2,C_3=1{,}4\cdot C_7;\ \tilde{n}=2,2,2.5$				$C_4,C_5,C_6=1{,}3\cdot C_7;\ \tilde{n}=2,2,3$				$C_7;\ \tilde{n}=2,2,4$	
Op.	*WI_i/T*	$\beta_{i,k}$	*k = 1*	*k=2*	*k=3*	$\beta_{i,k}$	*k=4*	*k=5*	*k =6*	$\beta_{i,k}$	*k=7*
1	1,042	0,010	x			0,011	x	x		0,012	x
2	0,146	0,020	x			0,022	x	x		0,024	x
3	0,404	0,040	x			0,044	x	x		0,048	x
4	0,537	0,020	x			0,022	x	x		0,024	x
5	0,958	0,120		x		0,132	x		x	0,144	x
6	0,175	0,100		x		0,110	x		x	0,120	x
7	0,962	0,080		x		0,088	x		x	0,096	x
8	0,717	0,150		x		0,165	x		x	0,180	x
9	0,383	0,190		x		0,209	x		x	0,228	x
10	1,058	0,210		x		0,231	x		x	0,252	x
11	0,116	0,080			x	0,088		x	x	0,096	x
12	0,445	0,090			x	0,099		x	x	0,108	x
13	0,446	0,110			x	0,121		x	x	0,132	x
14	1,300	0,080			x	0,088		x	x	0,096	x
15	0,104	0,100			x	0,110		x	x	0,120	x

Table 2 Model results

	K=1	*K=7*	*K=1*	*K=2*	*K=3*	*K=2*	*K=3*	*K=4*	*K=5*
Op.	$X_{i,1}$	$X_{i,7}$	$X_{i,1}$	$X_{i,2}$	$X_{i,3}$	$X_{i,2}$	$X_{i,3}$	$X_{i,4}$	$X_{i,5}$
1	100%		100%					6%	94%
2	11%	89%	100%					100%	
3	100%		100%						100%
4	100%		100%					100%	
5		100%		100%		100%			
6		100%		100%		100%			
7		100%		100%		100%			
8		100%		100%		65%		35%	
9		100%		100%		100%			
10		100%		100%		100%			
11		100%			100%		100%		
12		100%			100%				100%
13		100%			100%		90%		10%
14		100%			100%		100%		
15		100%			100%				100%
	Results ($\gamma=1$)		Results ($\gamma=0$)			Results ($\gamma=0{,}5$)			
y_k	2	7	3	5	3	4	2	1	2
	$\left(\frac{C_T}{C_7}\right)^*=12{,}78$		$\left(\frac{C_T}{C_7}\right)^*=19{,}6$			$\left(\frac{C_T}{C_7}\right)^*=13{,}72$			
	U = 2,36		U = 0,3			U = 0,44			

Observing the results shown in table 2 the relation between the typology of the workstations and the objective employed by the decision maker can be pointed out; considering only the economic objective in the model B ($\gamma = 1$) machining centers with higher flexibility are suggested, while reducing the risk due to the uncertainty ($\gamma = 0$) more and more dedicated resources are preferred.

CONCLUSIONS

The proposed model can be considered a tool to support the selection of the workstations typology to be considered in the early phase of a FMS design when a lot of variables are affected by uncertainty. The model assumes that workstation flexibility scenario is correlated to the social environment in which the company is operating; this kind of social constraints can bound the possible technological alternatives reducing the capability of choice in selecting appropriate manufacturing technologies with related economical implications.

The fuzzy approach presented gives the opportunity to evaluate in an interactive way the economical consequences of each alternative offering to the designer the possibility to choose between the minimum investment cost and the minimum uncertainty investment cost, or a trade-off solution.

The research has been supported by MURST 40%

REFERENCES

Alberti, N., Noto La Diega, S., Passannanti, A. and La Commare, U. (1988) Cost Analysis of FMS Throughput. CIRP Annals, Vol.37/1, 413-416.

Gaalman, G. J. C. and Nawijin, W. M. (1994) Tool Sharing in Parallel Part Production. 8th International Working Seminar on Production Economics, Innsbruck, Austria, 95-122.

La Commare, U., Noto La Diega, S. and Perrone, G. (1992) FMS Management Considering Tool and Part Flow Coordination. Manufacturing Systems, Vol. 22, 339-343.

Lee E.S. and Li, R.J. (1993) Fuzzy Multiple Objective Programming and Compromise Solution with Pareto Optimum. Fuzzy Set and Systems, Vol. 53, 275-288.

Liou, T. S. and Wang, M. J. (1992) Rankings Fuzzy Numbers with Integral Value. Fuzzy Set and Systems, Vol. 50, 247-255.

Perrone, G. and Noto La Diega, S. (1994a) Strategic FMS Design Under Uncertainty: A fuzzy Set Theory Based Approach. Proceedings of the 8th Int. Working Seminar on Production Economics, Innsbruck, Austria, 385-404.

Perrone, G. and Noto La Diega, S. (1994b) Approccio Fuzzy Multi Obiettivo alla Progettazione di Massima di Sistemi Flessibili di Produzione. Proceedings of Production Systems Design & Management - Workshop and Summer School, Varenna, Italy, 141-154.

Semeraro, Q. and Tollio, T. (1993) Tool Management and Tool Requirement Planning for FMSs. I ° Convegno AITEM, Ancona.

Zadeh, L. A. (1965) Fuzzy Sets. Information vol, Vol. 8, 338-353.

AUTHORS BIOGRAPHY

Sergio Noto La Diega is full professor of Mechanical Technology at the Department of Tecnologia e Produzione Meccanica of the University of Palermo where he held the chair for 6 years since the 1986.
At the beginning he focussed his research interest to production engineering particularly aiming at: Metal Forming, Metal Cutting and Machine Tool Design. At present his main interest field is Design and Operation of Advanced Manufacturing Systems with emphasis on Economic issues. In the above areas he has published over seventy papers.
He is corresponding member of C.I.R.P., member of IFAC Group for Manufacturing Management and Control.

Giovanni Perrone achieved a Ph.D. in Production Engineering in 1994 and now is assistant professor of Production Management at the Department of Tecnologia e Produzione Meccanica of the University of Palermo. His main research interest activities is on the application of Soft Computing Techniques to the Design and Operation of Advanced Manufacturing Systems and in the above areas has published almost 20 papers. He is member of TIMS and of an international research team on Fuzzy Constraint Networks.

Giovanna Lo Nigro is a Ph.D student on Production Management at the Department of Tecnologia e Produzione Meccanica of the University of Palermo. Her main research activities concerns the application of Neural Networks Technology to the Design and Operation of Advanced Manufacturing Systems and she is member of the Society of Italian Engineers (SIE).

21

Multipurpose Layout Planner

[1]A. Gomes de Alvarenga, F.J. Negreiros-Gomes, Hannu Ahonen+
[2]H. J. Pinheiro-Pita , L. M. Camarinha-Matos
[1]Universidade Federal do Espírito Santo
Av. Fernando Ferrari S/N - CEP 29060-900, Vitória- ES, Brasil
E-mail: [gomes,negreiro,hannu]@ufes.br
[2]Universidade Nova de Lisboa
Quinta da Torre, 2825 Monte Caparica, Portugal
E-mail:[hp,cam]@uninova.pt

Abstract

There are several kinds of layout problems, most of them concerning the allocation of a set of planar objects over a delimited planar area. There are several forms of representation of such problems, depending on their application. In order to treat these layout problems in a unified way, we propose an object-oriented approach that decomposes the problems in generic entities needed by a problem processing engine, the Layout Planner, to form a problem solver instance. This paper concerns the description of a tool for development of decision support systems for layout planning, where the entities are mapped into objects that own a dynamic behaviour. The user describes his/her problem instance in a graphical way using visualisation forms of the objects. A fertile application field of such DSS is in CIM activities such as machine layout and cutting. First, we identify models and approaches for layout problems and present an architecture for the DSS generator and then discuss a case study of the leather cutting problem.

Keywords
Layout Optimization, Interactive Planning, CIM systems, Knowledge-Based Systems, Problem Visualisation.

1 INTRODUCTION

Industrial cutting and layout problems are N P-complete and are characterised by several specific constraints. As such, exact solution approaches for them are not usually suitable. On the other hand, heuristical approaches can incorporate knowledge in the form of preferences and constraints guiding the search process.
The Multipurpose Layout Planner proposed in this study is based on the observation that in spite of their different contexts, these problems have several common features. Thus the proposed system is not restricted to factory floor layout problems as the expert system presented in [4]. Another feature of the proposed system is the generalised concept of algorithm: instead of an algorithm simply based on the optimization approach , techniques belonging to problem solving and search approaches are also considered. In fact, proposed system includes a catalog of several algorithms to be selected according to the specificities of

---

+ On leave from VTT Information Technology, Espoo, Finland

each application scenario. In terms of implementation, the definition of the system follows the object-oriented paradigm, which makes it possible to implement a flexible and extendible problem solving system.
Concerning the level of automation of such system, it can be viewed as a computer aid tool to help an expert planner to design the desired layout, monitoring her/his actions with the objective to provide real time consistency (reactive behaviour) of the design and to improve her/his performance. For example, in the case of leather cutting, there are several handly operations, e.g. pieces' allocation, that obey unstructured rules that can be built in the system. These rules are the representation of the layout problem knowledge that, together with the problem visualisation form, the user supplied parameters and a subset of the algorithm catalogue, identify a problem solver instance.
It shall be noted that the intended user of such system is not necessarily an expert in optimization and problem solving areas. However, he is supposed to be an expert in the application domain, i.e., able to specify constraints of the problem domain, optimization criteria and to evaluate solutions. Therefore, a decision support system is devised to help this user in "configuring" the layout planner for his particular needs.
The paper is organised as follows. In the section 2, we analyse some classes of layout problems concerning the representation, and the approaches for their solution. Section 3 deals with the architecture of the layout planner presenting a description of its functionality. In section 4, we describe, as application case study, the leather cutting problem.

2 CLASSES OF LAYOUT PROBLEMS

The purpose of this work is to develop an integrated tool for decision support in the layout context. In this section, we identify some problem instances that in some way can be characterised as problems of this context.

2.1 Nesting for Two Dimensional Shapes

A problem of theoretical importance as well as of industrial interest is that of optimization of two dimensional layout. In this problem, we are concerned with the allocation of a specified number of two dimensional regular or irregular (non-rectangular) pieces on a finite dimensional stock plate, the objective being to minimise the amount of produced waste.
This problem appears in the context of several production processes like steel, clothing manufacturing, wood, and shipbuilding. An interesting application is found in the shoe manufacturing industry, where moulds are cut from leather plates. The problem here belongs to the more generic class of cutting and packing problems in that the leather plates tend to be highly irregular in shape and to contain defective non usable areas.
We shall use a state graph to represent the problem although other approaches can be used. We will make the following assumptions about the problem

- Geometric shapes of the pieces (or leather plate) are approximated by polygons.

- Overlap of pieces is not accepted.

- Rotations of the pieces are allowed only within a bounded range.

We are thus using a representation by a state graph [1, 5] where each piece j as well as the stock plates are stored into the global database represented as polygonal regions with vertices $(x_{i,j}, y_{i,j})$
A generic state of the system is characterised by storing into the database all the information relative to a specific layout pattern as illustrated in Figure 1.

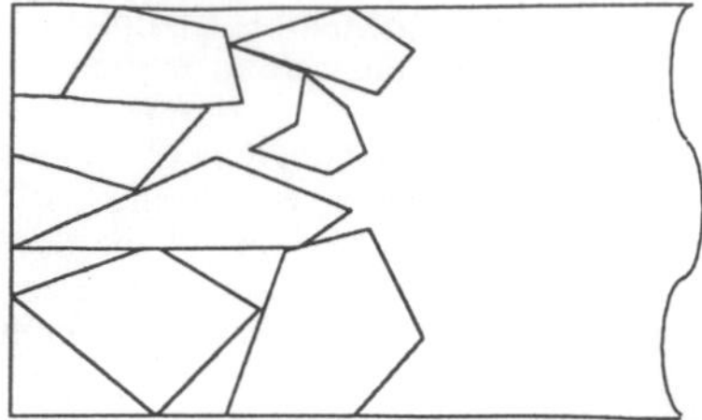

Figure 1 A Generic State of the System.

In this state graph, the production rules are defined by considering the strategies inside the universe of feasible solutions that are applicable to a current state E_k allowing the generation of a new state E_k+1 (a child state of E_k), by the nesting of a new piece to the layout corresponding to the state E_k

An algorithm to solve the nesting problem based on this approach must take into consideration the solution paths defined from the root of the graph to a generic state E_k. In this way we can associate to a generic state E_k a labelled element E(k, c, p) where k is the terminal state on the path from the root, c is the production cost of the layout associated with E_k, and p is a pointer to another labelled element E(r, v, q).

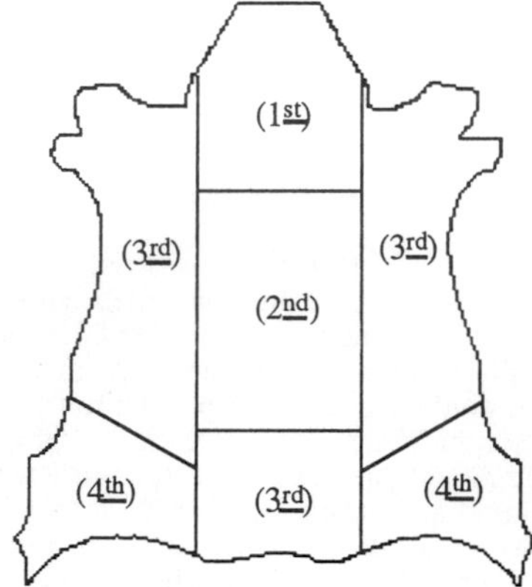

Figure 2 Classification of a leather plate.

Now, related to the nesting of leather plates, it is possible to generate a partition (e.g., the regions 1, 2, 3 and 4 in Figure 2) on the leather plates based on quality classification of the raw material.

We can apply the above approach to each sub-region of the partition in order to obtain a representation of the problem in the form of a state graph.

2.2 Machine layout problem on a factory floor

One problem encountered in the design of automated manufacturing system is the layout of machines on a factory floor. This problem can be seen as an application of the facility layout problem in automated manufacturing system [6]. The facility layout problem involves the arrangement of a given number of facilities so that the total cost required by the material handling systems to transport material among the facilities is minimised. One of the most frequently used formulations to model the facility layout problem is the quadratic assignment problem, an optimization technique which is concerned with the allocation of discrete entities to discrete locations. This problem is JVP-complete [8]. Considering the computational complexity of the facility layout problem, many heuristic methods have been developed. Kusiak and Heragu [7] classified these as

- construction methods;
- improvement methods;
- hybrid methods;
- graph theoretic methods.

We observe that the basic types of machine layout problems can be represented by a graph search like in the nesting problem. In this graph the root node represents an empty floor. The rules identify how to assign a new machine to a partial layout. So, effective search process can be used to solve the problem.

3 THE GENERAL ARCHITECTURE AND FUNCTIONAL DECOMPOSITION

3.1 General Layout Planner Overview

A layout problem can be defined as follows: Given a set of planar objects $O=\{o_1,o_2,\ldots,o_k\}$ where each object o_i has a geometry defined by $G_{o_i}=\{(x_1,y_1),\ldots,(x_n,y_n)\}$, a planar area A with geometry represented by $G_A=\{x_1^a,y_1^a),\ldots,(x_m^a,y_m^a)\}$ where $(x_\bullet,y_\bullet)$ is xy-axis coordinates, and a set of constraints R: $O\rightarrow A$ representing the objects' behaviour, the problem is to place the object $o_i, i=1,\ldots,k$ on A, subject to R, with the objective of optimising some functional $f(O,A)$.
Depending on the problem's domain as well as the objective function, there are many algorithms to solve that problem. Our approach is to investigate a kind of interactive system that, instead of doing an automatic planning, cooperates with a human expert to achieve an optimised solution for a layout problem.

3.2 The general architecture of the Layout Planner

The general architecture of the Layout Planner system is depicted in Figure 3. The system consists of five functional modules: the Taxonomy of Problem's models, the Algorithms Catalog, the Archive of Old Solutions, the User Interface Module, and the Layout Planner Module.
The Taxonomy of Problem's models organises the different classes of problems known by the system. Each problem's model is defined by a Specific Domain Knowledge and a Specific Planning Knowledge. The Domain Knowledge specifies the characteristics of the problem (e.g. application domain, shapes of parts to be allocated, etc.), while the Planning Knowledge, composed by sets of rule bases, specifies the constraints and some rules that should be observed for that problem.
The Algorithm Catalog is a library of algorithm prototypes each of which contains an executable code and an indication about the applicability of the algorithm (meta knowledge) to a problem (model). This module is designed to support maintenance, graphical browsing and explanation facilities about the algorithms.

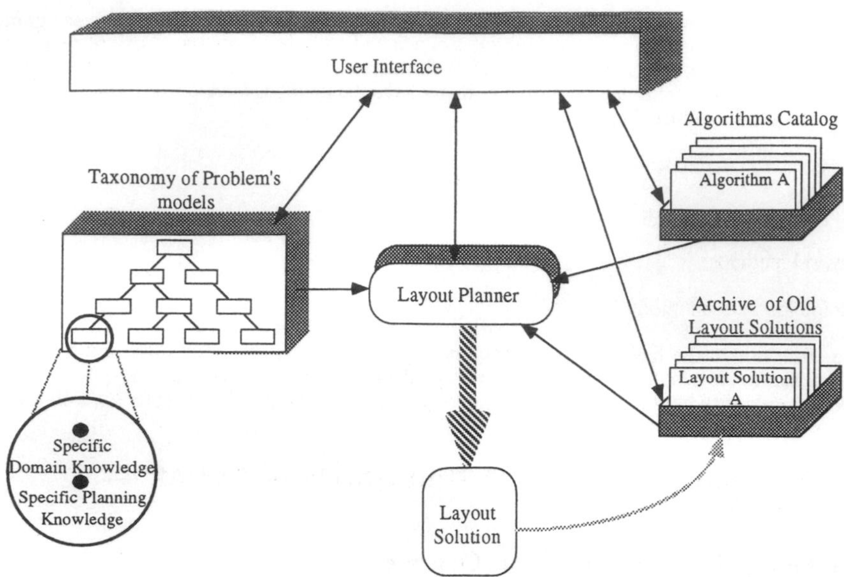

Figure 3: General architecture of the Layout Planner

The Archive of Old Solutions is composed by two archive subsystems, Archive of Problem Instances and Archive of Layout Solutions that allow the user to restart working with problems and solutions constructed before.

The User Interface Module is responsible for the management of the interaction mechanisms. As a platform to support interaction, a set of appropriate interaction forms was developed. An interaction form is a general script that characterises a graphical presentation and an interaction behaviour. In this module, each application object has a different behaviour and appearance depending on the user that interacts with it [3]. The concept of objects with dynamic behavior developed in the CIM-CASE system [9] is applied to this context.

The Layout Planner Module, described in more detailed below, is the central component of the system.

3.3 The Layout Planning Methodology

The proposed layout planning methodology is decomposed into four phases: Specification; Selection of Optimization Algorithms, Generation of Layout Solutions, and Selection of a Final Solution(figure 4).

The first phase is started with the selection of a problem's model. As an answer to this action, the Specific Domain and Planning Knowledge are instanciated and, a subset of old solutions, belonging to this model, are selected from the Archive of Old Solutions. The user can choose one of these problem's instances as the starting point or he can start a new specification. In the second case he defines the area where the shapes will be allocated and builds a draft layout, allocating the shapes to that area. Depending on the complexity of the problem, this procedure may need a hierarchical support (e. g. using the shop floor layout problem, three levels can be foreseen: shop level, cell level and component level).

As the construction of the draft layout is an activity requiring various iterations that couldn't be guided by the system, a great flexibility at the interaction mechanisms is needed and the system should also provide a set of critics that, based on the Specific Planning Knowledge, validate the user options and make suggestions. For some user actions these critics may be activated automatically (e.g. using the leather cutting problem, if a mould is allocated in the appropriate part of a leather plate) - embedded criticism. For other actions on-demand criticism is more appropriate, i.e., the criticism can only be activated when the user thinks he has achieved a solution (e.g. using the shop floor layout problem and supposing that the user has allocated an assembly robot, if the assembly position was defined).

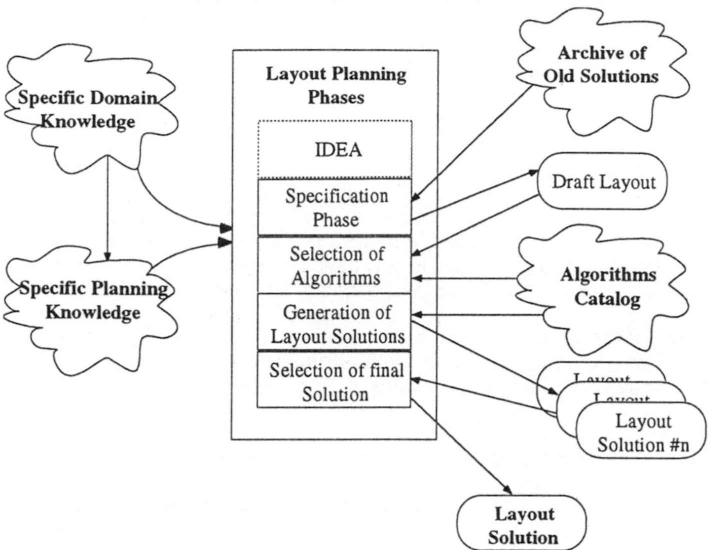

Figure 4: General model of the Layout Planner

In the phase of Selection of Optimization Algorithms, the user specifies a set of optimization criteria. On the basis of this set and taking into account the type o problem to be solved, a set of available algorithms is shown, regarding the appropriate Catalog. The user selects one, or more, of those algorithms and the system generates the formulation of the problem within the selected approaches (e.g. optimization methods or heuristical search methods). This formulation may consist of a mathematical model of the problem with an objective function and constraint equations (optimization methods) or it is given with the help of the definition of search states (heuristical search methods).

In the Generation of the Layout Solutions phase, for each selected algorithm a concrete instance of the processor able to solve the given problem using that algorithm is built. The parameters required to accomplish the computation task are initialised and, sometimes, additional features, like the estimation of path costs for the A^* algorithm, are added. After that, the processors are activated under the control of a Processor Supervisor. This module transmits, to the User Interface, information on the computation status and enables the user to decide if a given processor should be stopped or not (e.g. if it is going to take too much time). On the other hand, the Processor Supervisor saves a trace of the computation of each processor activity to be exploited later.

Finally, in the Selection of a Final Solution phase, the quality of the achieved solutions is evaluated, taking into account the Specific Planning Knowledge, and the best solution is selected. This phase sends to the user interface an explanation about the chosen solution, i.e., how the solution was achieved and which alternatives solutions could be available (if any).

At the end, it should be noted that the user is the responsible to accept or refuse the final solution. In the second case he can return to the Selection of Optimization Algorithms and selected a different set of criteria and/or algorithms and repeat the process. In any case, the draft layout is saved in the Archive of Old Solutions (Problem Instances Archive), while the achieved solution will be saved in the Layout Solutions Archive, if and only if it was accepted by the user.

4. CASE STUDY: LEATHER CUTTING PROBLEM

Lets now describe the classes of objects of the layout design for the leather cutting problem [3]. The leather plate is represented by the area A (see section 3.1), which is a two dimensional bounded region. This area has two derived classes: Graded Leather Plate (GLP) and Defective Leather Plate (DLP). An object of the GLP subclass contains additional attributes such as texture, thickness and boundary type. It also incorporates a partition that can be applied to generate, based on a raw material quality criterion, new objects of these subclasses.
DLP object contains a sub region of the leather plate that encloses a defect such as a stain, hole or stretch. A defect operator may be defined mapping the leather plate class into objects of this derived class.
The objects to be placed on the area A (see section 3.1) represent the moulds, which are abstractions of two dimensional regions with attributes that identify the pieces that will be produced by the layout planner, the quality of the required raw material and directional properties.
The classes of leather plate and mould present a set of constraints that represents the interaction of the objects of those classes. Some constraints are as follows:

Overlap: its purpose is to avoid an overlap among the moulds in the layout plan generated by the system over a leather plate.
Direction: Its purpose is to define the set of feasible directions for realising the layout of the leather plate.
Identification: These constraints represent technical specifications of the moulds concerning the leather plate. For example, there are constraints that specify whether a mould may be manufactured or not from a graded leather plate.

5 CONCLUSIONS

An architecture for a Multipurpose Layout Planner (MLP) was proposed. This system includes a catalog of optimization algorithms that can be configured to operate in different contexts. A decision support system helps the user in configuring the system for his particular application domain. Examples of applications on the leather cutting and factory layout planning have been used to test the validity of the approach. Due to the dynamic nature of leather cutting, the MLP seems to be a well-fit tool to improve the cutting process performance.
With the purpose to spread the application range of the MLP, we are currently studying the problem of the multilayer shopfloor layout.

ACKNOWLEDGEMENTS

The authors thank the European Commission in reference to the CIMIS.net project.
The Brazilian authors also thank CNPq (Brazilian Council of Research and Development) which partially supports this research, and the Portuguese authors would like to thank JNICT for supporting the projects where these ideas were developed.

REFERENCES

[1] Alvarenga, A.G.; Gomes, F.J.N, Provedel, A.; Sastron, F.; Arnalte, S. (1994) Integration of an Irregular Cutting System into CIM. Part I - Information Flows - Studies in Informatics and Control, **Vol. 3. Nos. 2-3**, pp. 157-163.
[2] Bell, P.C. (1991) Visual Interactive Modelling: the past, the present and the prospects - European Journal of Operational Research, **54**, pp. 274-286.
[3] Gomes, F.J.N; Alvarenga, A.G.; Lorenzoni, L.L.; Pinheiro-Pita, H.J.; CamarinhaMatos, L.M.(1994) Objects Dynamic Behaviour for Graphical Interfaces in VIM = An Application Case Study =, Studies in Informatics and Control, **Vol. 3. Nos. 2-3**, pp. 165-171.
[4] Heragu, S.S.; Kusiak, A. (1990) Machine Layout: an Optimization and Knowledge-based approach - Int. Journal Production Research **28**, pp. 615-635 .
[5] Nilsson, N. J. - *Principles of Artificial Intelligence* - Tíoga Publishing Company, Palo Alto, Califórnia.
[6] Kusiak, A. - *Intelligent Manufacturing Systems* - Prentice-Hall , New York.
[7] Kusiak, A.; Heragu, S.S. (1987) The Facility Layout Problem. European Journal of Operational Research **29**, pp. 229-251.
[8] Sahni, S.; Gonzalez, T. (1976) P-Complete Approximaton Problem - J. Assoc. Comput. **23**,pp.555-565 .
[9]Pinheiro-Pita, H., Camarinha-Matos, L. (1993). Comportamento de Objectos Activos na Interface Gráfica do Sistema CIM-CASE. In 4ªs. Jornadas Nacionais de Projecto, Planeamento e Produção assistidos por Computador, 1 (pp. 189-198). Lisboa: Ordem dos Engenheiros.
[10]Camarinha-Matos, L., 1993 IEEE International Conference on Robotics and Automation, 3 (pp. 63-70). Atlanta, Georgia, EUA: IEEE Computer Society Press.

BIOGRAPHY

Arlindo Gomes de Alvarenga and Francisco José Negreiros Gomes are Associate Professors at Federal University of Espirito Santo. They received D.Sc. degrees in Systems Engineering and Computer Science from Federal University of Rio de Janeiro in 1982 and 1988, respectively. Their research interests are in Combinatorial Optimization, Automated Systems for Manufacturing and Decision Support Systems.
Hannu Ahonen is Visiting Professor at Federal University of Espirito Santo, being on leave from VTT-Information Systems, Finland. He received a D.Tech. degree in Mathematics from Helsinki University of Technology in 1982.His research interest is in Combinatorial Optimization and Search Methods of Artificial Intelligence.
Mr Helder J. Pinheiro-Pita is adjunct professor in Instituto Superior de Engenharia de Lisboa. He received his B.Sc. degree by ISEL and his Lic. Eng. degree (five years Portuguese degree) by New University of Lisbon. Now he is finishing his PhD thesis on Interactive Planning applied to the generation of Control Architectures to CIM.
Dr. Luis M. Camarinha-Matos received his Computer Engineering degree and PhD on Computer Science, topic Robotics and CIM, from the New University of Lisbon. Currently he is auxiliar professor (eq. associate professor) at the Electrical Engineering Department of the New University of Lisbon and leads the group of Robotic Systems and CIM of the Uninova's Center for Intelligent Robotics. His main research interests are: CIM systems integration, Intelligent Manufacturing Systems, Machine Learning in Robotics.

22

Towards more humanized real time decision support systems

F.G. Filip
Research Institute for Informatics
8-10, Averescu Avenue 71316 Bucharest, Romania
Phone: +40-1-2223778 Fax: +40-1-3128539
E-mail: FILIPF@ROEARN.ICI.RO

Abstract

This paper aims at reviewing several results concerning the development of practical **Decision Support Systems** (DSS) for real time production control in manufacturing. Several characteristic features of DSS that are highly demanded and/or accepted by human factors **and** mimic the human problem solving are identified. The perspective includes technology , business and human factors. A practical DSS illustrates the approach.

Keywords

Decision support systems, intelligent control, human factors, levels of automation, process industry

1 INTRODUCTION

In order to make industrial enterprises survive under sharp economic competition, engineers have for many years been striving at building large integrated computer based management and control systems. Various terms and concepts such as '**plant wide control** '- **PWC** (in continuous process industries), '**computer integrated manufacturing** '- **CIM** (in discrete part manufacturing) illustrate a holistic approach based on computing and communication facilities thereby to automate/support, and integrate various automatable or less automatable engineering, production and management activities (Williams, 1990). Recently, besides communications, extensive use of AI techniques (Kusiak, 1992) appears to be an important technology driving force. This latest development is mainly tributary to a ' knowledge based production' trend that might sometimes be too difficult to reflect in case of SME which cannot afford hiring highly qualified and highly paid personnel.

Other recent concepts such as **Computer Integrated Enterprise - CIE** or **Computer Integrated Business - CIB** (Raulefs, 1994), are examples of a less technology-driven, but apparently more business-oriented perspective, and illustrate an even broader approach of automation and computer based integration concept.

Moreover recent evolutions such as globalisation of markets, better informed clients, knowledge based products, integration of SMEs into global production, new enterprise paradigms (Norman, 1994), and **business process reengineering - BPR** (Hammer and Champy, 1993), to integrate people, technologies, production and business, are supported by existing information technologies. In turn, these developments exert significant impact on **information systems (IS)** development (Granado, 1993).

In addition to technology and business efficiency factors, a third element namely the **human factor**, should be taken into account in modern IS. A human centred approach would also include human centred goals such as job satisfaction, transparency, knowledge enrichment, understandability, etc. (Johannsen, 1994).

A preliminary analysis of various interacting internal influence factors (information technologies and control methodologies) and of external factors , was reported (Filip, 1994).

A series of articles (Filip et al, 1985; Filip,1988;Filip,1993) has been produced on concepts and methods devised or adopted in evolving a family of practical **decision support systems** for **manufacturing** (DSSfM) addressing production control in continuous process industries (refineries, chemical plants, pulp and paper mills) as well as in discrete part manufacturing (DPM) workshops.

This paper focuses on **human factors** (**HF**), with particular emphasis on **DSSfM**. The paper is organised as follows. A general discussion of HF is made in Section 2. Section 3 reviews several relevant aspects of **real time** (RT) DSSfM. A practical RT DSSfM evolution is shown in Section 4.

2 HUMAN FACTORS

Johannsen (1994) remarks that 'in addition to CIM, it seems to be necessary to also consider **human integrated manufacturing (HIM)** during system development. In a human centered design approach the question how to integrate the humans on several or all levels of the plant has to be dealt with during early systems development phases'.

In general, one can identify three main groups of questions about the interaction and integration of the human and the machine (the IS): 1. how does the computer **serve** the human (process/plant manager) perform his tasks better?; 2. what is the **impact** of the man - machine system on the operational performance of the managed process/plant? ; 3. how the **human status** and working conditions are affected by the presence of computer?

Most of the early IS have not been used as expected because of being **unreliable**, **intolerant** (requiring an absolutely correct stream of directives to carry out their functions), **impersonal** (the dialogue was not consistent with user's previous experience) and **not self sufficient** (the help of the computer professional has been frequently needed) (James, 1980). While most of those early critical issues have been solved so far by technology developments and intensive ('continuous' and 'accelerated') training of users, rendering the information tool less impersonal is still an open problem, at least in process supervision and production control.

To answer the second question, a separate analysis of the performances of the information tool, as made by Hill et al (1993), starting from the management system model, though interesting , might prove irrelevant if not completed with a clear definition of user classes (**roles**) and of their tasks. A recent study in the world of SME (Halsall et al, 1994) showed that 'many smaller companies lacked a clear understanding of many of the elements of sound production management practice'. Hence a 'normative use' of IS in an attempt at changing working styles, is desirable. In addition, in the industrial process automation context, the process efficiency is less important than safety considerations. As Johannsen (1994) remarked, the tendency to continue automating in a technology-driven manner led to 'deskilling the human operators, and thus to boredom under normal operational conditions as well as to human errors in emergency situations'. Even though with Johannsen physical process control counts, it is human errors that may bear on physical safety and also economically at the higher decision levels such as production control and plant management.

This latter observation set the stage for (and partially answered) the third question. Many years ago, Briefs (1981) stated rather dramatically that the computerisation of human work seemed to imply a 'major threat against the human creativity and conscious development, since there is a tendency to polarise humans into two categories. The first group includes computer professionals who manifest and develop their skills and creativity in designing ever more sophisticated tools. The second group would include communities of computer users who fast and easily perform their jobs without getting deep insights into their comfortable production means'.

The previously presented conclusion (Filip, 1989) is that 'it is necessary to develop IS that are not only precise, cheap and easy to use, but also stimulating for users as to acquire new skills, adopt new working styles and deploy their talents and creativity', is still valid. It is difficult to propose a methodology for attaining these rather general goals. An imperfect and incomplete refinement of the above goals made from a pure human centred perspective, could now cover several subgoals/attributes of IS such as wide (not 'Procustian') range of services offered ('physiology' dimension), evolutive development and possible learning facilities ('ontogeny' dimension), and transparency on system structure and explanation features (' anatomy' dimension). A particular subclass (component) of IS, namely DSS, is discussed in the sequel, trying to outline its specificities.

3 REAL TIME DSS FOR MANUFACTURING

Most of the traditional developments in the DSS domain have addressed systems not involving any real time control (Charturverdi et al, 1993). In the sequel, real time decisions for control applications in manufacturing will be considered. Bosman (1987) stated that control problems could be looked upon as a 'natural extension' and as a 'distinct element' of planning **decision making processes (DMP)**.

Real time (RT) DMP for control applications in manufacturing are characterized by several particular aspects: 1. they involve continuous monitoring of the dynamic environment; 2. they are short time horizon oriented and are carried out on a repetitive basis; 3. they normally occur under time pressure; 4. long-term effects are difficult to predict (Charturverdi et al, 1993). It is quite unlikely that an 'econological' (Bahl and Hunt, 1984) approach, involving optimisation, be technically possible for 'pure' RT DMP. Satisfactory approaches, that reduce the search space at the expense of the decision quality, or fully automated DM systems (corresponding to the 10th degree of automation in Sheridan's (1992) classification), if taken separately, cannot be accepted either, but for some exceptions.

At the same time, one can notice that 'pure' RT DMP can be come across in 'crisis' situations only. For example, if a process unit must be shut down, due to an unexpected event, the production schedule of the entire plant might turn obsolete. The right decision will be to take the most appropriate compensation measures to 'manage the crisis' over the time period needed to recompute a new schedule or update the current one. In this case, a satisfycing decision may be appropriate. If the crisis situation has been previously met with and successfully surpassed, an almost automated solution based on past decisions stored in the IS can be accepted and validated by the human operator. On the other hand, the minimisation of the probability of occurrences of crisis situations should be considered as one of the inputs (expressed as a set of constraints or/and objectives) in the scheduling problem. For example in a pulp and paper mill, a **unit plant** (UP) stop may cause drain the downstream tank (T) and overflow the upstream tank and so, shut/slow down the unit plants that are fed or feed those tanks respectively. Subsequent UP startings up normally imply dynamic regimes, that determine variations of product quality. To prevent such situations, the schedule (the sequence of UP production rates) should be set so that stock levels in Ts compensate to as large extent as possible for UP stops or significant slowing down.

To sum up those ideas, one can add other specific desirable features to the particular subclass of IS used in manufacturing control. An effective **real time DSS for manufacturing (RT DSSfM)** should support decisions on the preparation of 'good' and 'cautious' schedules as well as 'ad hoc', pure RT decisions to solve crisis situations. An example comes to illustrate this approach.

4 DISPATCHER - A DSS SERIES

DISPATCHER is a series of DSSs, developed over a fifteen-year time period, to solve various DMPs in the milieu of continuous 'pure material' process industries. The system initially addressed the short-term production scheduling problem. Then it evolved in both function set supported and new technologies used in order to satisfy users' various requirements (see Figure 1). New supported functions such as tank sizing, maintenance planning and even order acceptance and planning of raw materials or/and utility purchasing allow a certain **degree of integration of functions** within the enterprise.

4.1 Application Area

A typical plant in process industry consists of several (tens) of **unit plants** interconnected via **tanks** or directly through pipelines. The planning decision problem lies in choosing (over a number of time intervals) the sequence of operation regimes (**configuration problem**-CP) and production rates of UP (**flow scheduling problem**-FSP) so that the stock levels should be in the vicinity of some safe and economical values and product deliveries should be made in due time, given the initial stocks and the forecasts on material inputs. In order entry applications, or in crisis situations, orders are not treated as external 'fixed'/given **consumers (C)** but as free decision variables. In a similar way, raw material and industrial utility flows are viewed as external **resources (R)** in ordinary scheduling problems, but as constrained decision variables in crisis situations and purchasing decisions. More details about the problem (including mathematical model and algorithm) together with numerical examples can be found in (Filip, 1990, 1993).

4.2 Evolution

Initially, DISPATCHER was a pure interactive optimisation program (running on PDP-DEC 11 machines) to quickly solve large scale FSP represented as linear, quadratic , discrete time, constrained, optimal tracking models. The model took stock levels for **state variables** ($\boldsymbol{x}(k)$) and production rates for control variables ($\boldsymbol{u}$ (k)). The sequences of desired safe states, recommended

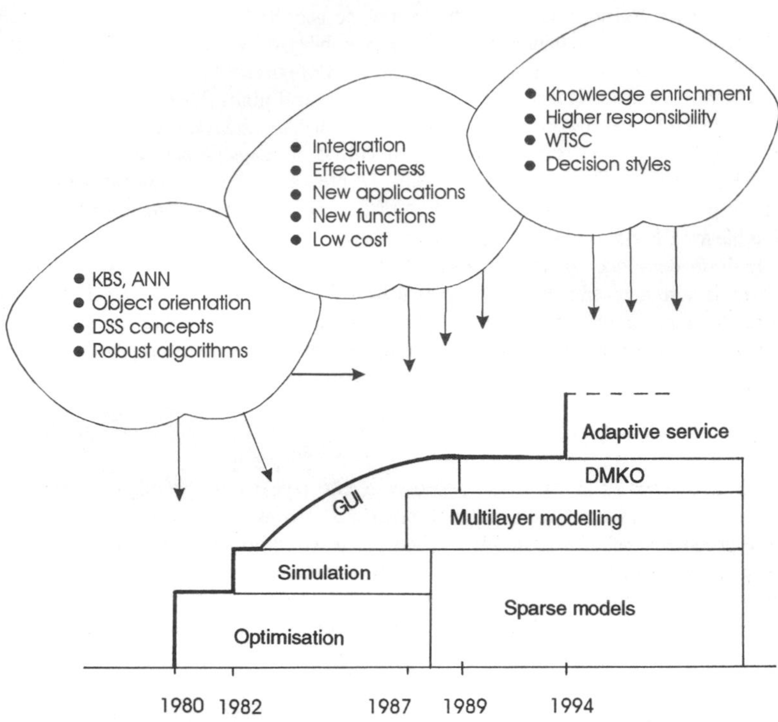

Figure 1 Technology, business and human influence factors and DISPATCHER development layers

controls, and forecast disturbances (material inputs and product deliveries) ($\boldsymbol{x}\ \boldsymbol{d}(k+1)$, $\boldsymbol{u}\ \boldsymbol{d}(\mathrm{t})\ \boldsymbol{w}(k)$ were set by the decision maker over the planning horizon with KF time intervals together with the values of interconnection matrices ($\boldsymbol{A}\ (k)$, $\boldsymbol{B}(k)$), permitted variation limits ($\boldsymbol{xl}$, $\boldsymbol{xu}$ and $\boldsymbol{ul}$, $\boldsymbol{uu}$) and penalties on deviations from the expected values as diagonal matrices ($\boldsymbol{Q}\ (k)$, $\boldsymbol{R}\ (k)$).

$$\boldsymbol{x}\,(k+1)=\boldsymbol{A}(k)\,\boldsymbol{x}(k)+\boldsymbol{B}(k)\,\mathbf{u}\,(k)+\boldsymbol{w}\,(k));\ k=1,\mathrm{KF} \tag{1}$$

$$\boldsymbol{x}\,(k+1)\in[\boldsymbol{x}\,\boldsymbol{l}\,(k+1),\boldsymbol{xu}(k+1)];\ \boldsymbol{u}(k)\in[\boldsymbol{ul}(k),\boldsymbol{uu}(k)];\ \ k=1,\mathrm{KF} \tag{2}$$

$$\min_{\boldsymbol{x},\boldsymbol{u}}\left[\sum_{k}\|\boldsymbol{x}(k+1)-\boldsymbol{xd}(k+1)\|^{2}_{\boldsymbol{Q}(k)}+\|\boldsymbol{u}(k)-\boldsymbol{ud}(k)\|^{2}_{\boldsymbol{R}(k)}\right] \tag{3}$$

This has been solved by a fast, robust, hierarchical optimisation algorithm (Filip et al, 1985) . Large scale problems with up to 80 constrained states, 60 constrained controls and 12 time intervals could be solved for practical applications within a few minutes. Simple simulation facilities to support 'What if...?' analysis (mainly concerning the alterations of computer optimal solutions or of entirely human made schedules) and negotiation processes were offered to users adopting ' implicit favourite' or ' bounded rationality' decision styles. User-friendly graphic user interface (GUI) including a specialised editor, was developed.

Many practical applications show the problem parameters (limits, desired values) which, although variable in time, keep nevertheless constant over some time segments. This and the users' demand for a reduction of the amount of data to input into the system, as well as the necessity of easily deducing one problem specification from the previous one (to evaluate several alternatives), contributed to devising a new storage method and the corresponding set of solving algorithms for **sparse models with relatively constant parameters** (Filip, 1988). At the same time, other **roles** such as application system builders (facilitators) and developers of alternative algorithms (the 'toolsmiths') had to be considered along with production planners. Consequently, a **multilevel modelling scheme** of the problem, with three layers (external/representational, mathematical/conceptual and internal/performance) and translation mechanisms were brought in, being inspired by parallel developments in database technology, and anticipating some recent issues related with knowledge engineering (Filip, 1993). The standard version of the system is presented in detail by (Filip,1990).

4.3 Towards intelligent DSS

Numerous other practical implementations of the standard version helped draw interesting conclusions. First, the system has been considered by most users as being flexible enough to support a wide range of applications and, in some cases, its utilisation migrated from the originally intended one. It has been used in crisis situations (mainly due to significant deviation from the schedule , to equipment failures or other emergencies) as well as in normal operation or in training applications. However, though the system is somehow transparent, and the users have sound **domain** ('what type?') **knowledge** (DK), they have behaved in a 'wise' or even 'lazy'(Rasmunsen, 1983) manner, mainly trying to keep their **mental load** under an average 'willing to spend capacity' (WTSC). This can be explained by the initial lack of **tool** ('how'-type) **knowledge** (TK) as well as by insufficient work motivation.

To fight the lack of TK and to stimulate users' creativity and quest for new skills, a declarative model of an 'ambitious' and **knowledgeable** operator (DMKO) was proposed (Filip, 1993). DMKO supports 1. model building for various situations (to solve CP); 2. problem feasibility testing to propose corrective measures (for example limit relaxations or transformation of fixed/known perturbations into free variables, etc.); 3. automatically building the internal model from the external description, choosing the appropriate solving algorithm ; 4. experimenting the problem model, for example by producing a series of alternatives through modifying various parameters (mainly weights) in answer to qualitative assessment of simulated solutions by the user, followed by due explanations. To handle the complexity and diversity of the technologies used, object orientation has been adopted.

Efforts are being made to introduce new intelligence into the system, especially for evaluating users' behaviour so that DMKO (originally meant for supporting a certain 'role') could dynamically adapt to specific needs of particular 'actors', in an attempt at rendering the system less ' impersonal'.

Of course, there are other reported results combining traditional numeric methods with KBS to build 'hybrid' (Nof and Grant,1991) or 'tandem' DSSfM (Kusiak, 1992) even in process industries (Dourado-Correia, 1992). Apparently such systems are primarily meant for making numerical computation easier, including heuristics so that the space search for optimisation/simulation algorithms is adapted/reduced. It should be noted that the author's approach is mainly human factor centred and aims at increasing system acceptance rather than improving its computational performances. However it is felt that this solution, though technically interesting, together with a more intensive training, cannot always overpass the motivation problem. This problem can apparently be handled by organisational measures within an BPR effort.

5 CONCLUSIONS

Problem solving (PS) process can be considered from different perspectives. Human PS is a combination of calculation, reasoning and adaptive learning. Automatic PS is a combination of such elements as numeric, symbolic, and neural computing (Kerckhoffs, 1994).

Intelligent DSS include learning facilities, perform inferences and select appropriate heuristics, rules and models (Sen, 1993). These technologies mimic the activity of human brain as a whole, and represent an important step forward in the movement towards humanizing the modern information systems, in order to make them more compatible with human decision maker. In addition to those remarkable technology developments, a human centred perspective was intended and pleaded for in this paper.

6 REFERENCES

Bahl, H.C. and Hunt, R.G. (1984) Decision making theory and DSS design. *Data Base*, **15(4)**, . 10-14.

Briefs, U. (1981) Re-thinking industrial work: computer effects on technical white collar workers. *Computers in Industry*, **2(1)**, 76-89.

Bosman, A. (1987) Relations between specific DSS, *Decision Support Systems*, **3**, 213-224.

Chaturverdi, A.R., Hutchinson, G.K. and Nazareth, D.L. (1993) Supporting complex real-time decision making through machine learning. *Decision Support Systems*, **10**, 213-233.

Dourado-Correia, A. (1992) Optimal Scheduling and Energy Management in Industrial Complexes: Some New Results and Proposals, in *Proceedings on Computer Integrated Manufacturing in Process and Manufacturing Industries*, IFAC Conference ,Espoo.

Filip.F.G. (1988) Operative Decision Making in the Process Industry, in *Preprints*, IMACS'88 12th World Congress, Paris.

Filip, F.G. (1989) Creativity and decision support systems. *Studies and Researches in Computers and Informatics*, **1(1)**, 41-49.

Filip, F.G. (1990) Decision support systems in process coordination, part 2: technology issues. *Studies and Researches in Computers* (new series), **1(4)**, 27-45.

Filip, F.G. (1993) An Object Oriented Multilayer Model for Scheduling, in *Proceedings*, European Simulation Symposium-ESS'93 (eds. A. Verbraeck and E.J.H. Kerckhoffs), Delft.

Filip, F.G. (1994) Evolutions in systems analysis, modelling, and simulation in computer based industrial automation. *Systems Analysis,Modelling and Simulation,* **15**, 135-149.

Filip, F.G., Donciulescu, D.A., Gaspar, R., Orasanu, L. and Muratcea, M. (1985) Multilevel optimisation algorithms in computer aided production control in the process industries. *Computers in Industry,* **6(1)**, 47-57.

Granado, J. (1994) Business Technology and Information Technology, in *Annex VI to Documentation* the Working Group on TBP, V 2.0, Brussels.

Halsall, D.N., Muhlemann, A.P. and Price, D.H.R. (1994) A review of production planning and scheduling in smaller manufacturing companies in the UK. *Production Planning and Control,* **3(5),** 485-493.

Hammer, M. and Champy, J. (1993) *Reengineering the firm*. Harper Collins Books, New York.

Hill, D.T., Koelling, C.P. and Kurstedt, H.A. (1993) Developing a set of indicators for measuring information-oriented performance. *Computers in Industrial Engineering,* **24(3)**, 379-390.

James, K.M. (1980) The user interface. *The Computer Journal,* **23(1)**, 25-28.

Johannsen, G. (1994) Integrated Systems Engineering: The Challenging Cross-discipline, in *Preprints on Integrated Systems Engineering,* IFAC Conference, Pergamon Press.

Kerckhoffs, E.J.H. (1992) Parallel Processing in Model-based Problem Solving, in *Computational Systems Analysis* (ed. A. Sydow) Elsevier, Amsterdam.

Kusiak, A. (ed.) (1992) *Intelligent Design and Manufacturing*. John Wiley & Sons, New York.

Nof, S.Y. and Grant, F.H. (1991) Adaptive/predictive scheduling: a review and general framework. *Production Planning and Control,* **2(4),** 298-312.

Norman, D. (1994) Emerging Models of the Enterprise, in *Technology for the Business Process,* CEC European IT Conference, 91-92.

Rasmunsen, J. (1983) Skills, roles, and knowledge; signal signs and symbols and other distinctions in human performance models. *IEEE Transactions on Systems, Man and Cybernetics SMC* **13,** 257-266.

Raulefs, P. (1994) The Virtual Factory, in *IFIP Transactions* (eds. K. Brunnstein and E. Raubold) A52, 13th World Computer Congress'94, , Elsevier Science, Amsterdam.

Sen, M. (1993) Machine learning methods for intelligent decision support. *Decision Support Systems,* **10**, 79-83.

Sheridan, T. (1992) *Telerobotics, automation and human supervisory control.* MIT Press.

Williams, T.J. (1989) *A reference model for Computer integrated manufacturing: a description from the viewpoint of industrial automation.* ISA, Triangle Park.

F.G.Filip was born in Bucharest (1947). He took the MSc. degree and Ph.D degree from the 'Politehnica' University of Bucharest in 1970 and 1982, respectively. In 1991 he was elected as a corresponding member of the Romanian Academy. He is director of the Research Institute for Informatics in Bucharest. His main scientific interests are in large scale systems, optimisation and control, DSS, CIME, man-machine systems, technologies for the business processes. He published over 120 technical papers.

PART NINE

Shop Floor Control

23

Interoperability testing in an Implementation of ISO/OSI Protocols

P. Castori, P. Pleinevaux, K. Vijayananda, F. Vamparys
Swiss Federal Institute of Technology, Lausanne
Department of Computer Engineering
EPFL-DI-LIT, CH-1015 Lausanne, Switzerland
Email:{castori, ppvx, vijay, vamparys}@di.epfl.ch

Abstract

Practice has largely shown that interoperability problems often occur between implementations of different vendors. This paper presents the results of an experiment made at EPFL-LIT with an implementation of the CNMA profile that has been tested for interoperability with a number of commercial implementations. We present the results of the interoperability tests, propose a taxonomy of errors, compare with other interoperability tests and then discuss the lessons we have learned from this interoperability test campaign.

Keywords

Interoperability testing, CNMA, OSI

1 INTRODUCTION

Integration of enterprise applications requires the provision of a communication infrastructure that will allow applications running on different machines to exchange data. Essentially two approaches can be followed: adopt proprietary networks that must then be interconnected by expensive gateways or adopt a vendor independent communication architecture like CNMA (Communications Network for Manufacturing Applications) [7], MAP (Manufacturing Automation Protocol) [9] or TOP (Technical Office Protocol) [23]. The replacement of proprietary networks by standard networks does not come for free at the beginning. As practice has largely shown, interoperability problems appear between implementations of different vendors. These problems manifest themselves as obstacles to establish communication between applications. Very often, the complexity of the ISO/OSI protocols adopted in MAP, TOP and CNMA are mentioned as the reason why interoperability problems show up.

This paper presents the results of an experiment made at the Swiss Federal Institute of Technology in Lausanne with an implementation of the CNMA profile. This profile based on ISO/OSI protocols is tailored to industrial applications and used in industrial factories at Mercedes-Benz, Aerospatiale, etc. This experiment consists in testing this new implementation for interoperability with commercial implementations of the CNMA and MAP profiles recording the problems that appeared during the tests.

The MAP and CNMA profiles are compatible in the sense that implementations of one profile are able to communicate with implementations of the other profile. An Ada implementation of the CNMA profile made at the Swiss Federal Institute of Technology, consists of approximately 150 000 lines of Ada code, this figure giving an idea of the complexity of the software tested in this experiment.

The topic of testing for communication network is covered by a number of publications concerning conformance tests [15, 20, 21, 17, 16]. These articles concentrate on the methodology that ISO has been standardizing and on tools [17] for conformance testing. Interoperability tests have been discussed for the lower layers [13], for X.400 [14], for the HP MAP 3.0 implementation [18].

The paper is organized as follows. The implementations under test are described in section 2. Section 3 describes the methodology for the interoperability testing. Section 4 presents a taxonomy of errors. In section 5 we discuss the results using the above taxonomy and present the lessons learnt from this test campaign. In section 6 we compare our results with other interoperability tests and in particular with the results of the interoperability testing for HP MAP 3.0. The conclusion is then presented.

2 CONFORMANCE AND INTEROPERABILITY TESTING

2.1 Conformance testing

The objective of conformance testing is to establish whether the implementation being tested conforms to the specification of a given protocol. Conformance tests involve two implementations: a reference implementation which is assumed to perfectly reflect the standard being tested and the Implementation Under Test (IUT). The reference implementation interacts with the IUT according to well defined scenarios and the responses or behavior of the IUT are observed and recorded. For each interaction started by the reference implementation, a reaction is expected from the IUT.

In the ESPRIT CNMA project for example, conformance tests have been performed between 1988 and 1992 with tools developed first in the project then by ESPRIT TT-CNMA. Conformance tests have been very useful with the first versions of the CNMA implementations but as these matured the need for conformance testing has strongly decreased. Some of the drawbacks of conformance testing are their cost, incomplete coverage and they do not guarantee interoperability.

2.2 Interoperability testing

Several classifications of interoperability testing have been defined in the literature [8, 2]. These classifications are based on different levels of interoperability and criteria for success/failure of the interoperability testing. We now present an informal definition of an interoperability test. This definition will be based on function (Typically in ISO terminology, function can be considered as a service) provided by any layer of the stack. We try to keep our definition as general as possible and try to cover all aspects of the functions offered by all layers.

Let us first define a *function* offered by any layer N. This definition will form the basis for determining the success and failure of an interoperability test. Suppose layer N provides a_N number of functions. A $function_i^N$ ($i \in 1..a_N$) provided by layer N consists of a *request PDU* (Req_i^N) and *response PDU* ($Resp_i^N$). $function_i^N$ is provided by the *responder* and the *initiator* requests for the function. An initiator accesses $function_i^N$ by sending (Req_i^N) to the responder. The initiator receives a $Resp_i^N$ in response to Req_i^N. The $Resp_i^N$ may be remote (sent by the responder) or local (generated by the local layer N-1).

Sometimes, $Resp_i^N$ has temporal requirements. This means that the initiator expects $Resp_i^N$ within a certain time T_i^N after it has sent the Req_i^N. This requirement is not a strict requirement for all $function_i^N$ ($i = 1..a_N$), Req_i^N and $Resp_i^N$ may be composed of multiple PDUs. For example, in the `data transfer` service offered by TP4, the request PDU may contain multiple DT PDUs (data transfer PDU). The `CMIP M-GET` function [12] response PDU also contains more than one PDU.

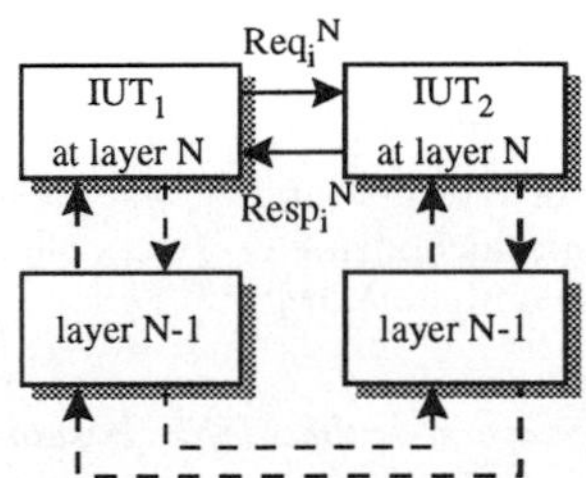

Figure 1 Interoperability test at a Layer N

The response PDU $Resp_i^N$ can be interpreted in many ways. Based on the correctness of response PDU it can be classified into the following categories.

Correct response: This includes all the responses that conform to the function definition. In some layers, the absence of a function response is also considered as a correct response.

Delayed Response: When $Resp_i^N$ includes multiple PDUs, some of the PDUs may be delayed. In other cases, the $Resp_i^N$ may not be available immediately. In this case, $Resp_i^N$ is sent by the *responder* when it is ready and available. The transport layer (TP4) retransmission function falls under this category.

No Response: When the initiator does not receive $Resp_i^N$ within a certain time T_i^N it is considered as no response. In this case, the absence of $Resp_i^N$ is a *failure* of the interoperability test. The absence of response can be due to non-availability of the responder or failure of the lower layer services.

Partial Response: This includes the cases where a $Resp_i^N$ includes multiple messages and only a few of them have been received. This type of response is considered as incorrect and results in *failure* of the interoperability test.

Incorrect Response: When a $Resp_i^N$ does not conform to its definition, it belongs to this category. The non-conformity can refer to the scope of the response (semantics) or its format (syntax).

Error Response: This type of response indicates that the responder cannot provide the requested function because an error or problem has been found in Req_i^N . The response indicates the reason for rejection or inability to provide the requested function. It results in the *failure* of the interoperability test.

The above criteria for classifying $Resp_i^N$ cover the syntax and semantics of the $Resp_i^N$ and also its temporal requirements. They also form a sound basis for determining the correctness of an interoperability test. Later on, this is also used for classifying the errors encountered during the interoperability testing.

Now we are ready to define an interoperability test for a $function_i^N$ provided by layer N. Informally an interoperability test between two implementations IUT_1 and IUT_2 for a $function_i^N$ provided by layer N can be defined as follows:

Initiator IUT_1 sends a Req_i^N to Responder IUT_2.

If IUT_2 responds with a correct and timely $Resp_i^N$ and it is received by IUT_1 then IUT_1 and IUT_2 are said to interoperate with respect to $function_i^N$

This can be represented as

$$IUT_1 \stackrel{function_i^N}{\longrightarrow} IUT_2$$

Figure 1 shows the interoperability testing of a $function_i^N$ provided by layer N.

This informal definition of an interoperability test can be applied at any layer. It is applicable to both confirmed and unconfirmed services offered by all layers.Based on this definition we now define *layer interoperability.*

If $\stackrel{function_i^N}{\longrightarrow}$ holds good for $i = 1..a_N$, then IUT_1 is said to be layer interoperable with IUT_2 with respect to layer N. This can be represented as $IUT_1 \stackrel{layer^N}{\longrightarrow} IUT_2$

This definition of an interoperability test determines the success/failure of a test based on the response PDU $Resp_i^N$. In [5], Castro has presented the success/failure of an interoperability test based on the results of conformance testing. A predictive metric based on the results of conformance tests is used to determine the success/failure of a test. The drawbacks of this method is the inability to determine the success of those test cases that do not occur during a conformance test which is not the case with our definition.

An interesting point to be noted is that $\stackrel{function_i^N}{\longrightarrow}$ and $\stackrel{layer^N}{\longrightarrow}$ are not symmetric. That is

$$(IUT_1 \stackrel{function_i^N}{\longrightarrow} IUT_2) \not\Rightarrow (IUT_2 \stackrel{function_i^N}{\longrightarrow} IUT_1)$$
$$(IUT_1 \stackrel{layer^N}{\longrightarrow} IUT_2) \not\Rightarrow (IUT_2 \stackrel{layer^N}{\longrightarrow} IUT_1)$$

It is essential to distinguish between initiator and responder associated with a $function_i^N$. These two are well separated at the user application level and are provided as separate implementation. At the user application level, the initiator is also referred to as *client* and the responder is referred to as *server.* This is not the case with layer N. Normally an implementation at layer N contains both the initiator and responder. A successful interoperability test $IUT_1 \stackrel{function_i^N}{\longrightarrow} IUT_2$ tests only the initiator of IUT_1 and the responder of IUT_2. This explains why $\stackrel{function_i^N}{\longrightarrow}$ may not be symmetric.

Most of the interoperability tests are performed at the interworking level (between two user applications). At the layer level, the tests are conducted in an indirect manner. The user application layer make use of the lower layer services. As a consequence, interoperability testing of the user application services results in the testing of those lower services that are utilized by the user application. It is interesting to note that all the lower layer services may not be tested using this strategy. It is necessary to design test strategies for interoperability testing at each layer. This will increase the confidence of the interoperability tests performed at the interworking level [14].

Interoperability tests are not perfect either, compared to conformance tests they do not cover as wide a spectrum of behaviors as conformance tests [18]. Very often interoperability tests involve two implementations from two different vendors using scenarios that are either identical or very close to those encountered in user applications. Performing a single battery of interoperability tests between any two applications is a relatively simple and straightforward process, but when many products are involved, the level of difficulty soars. For example, to test the interoperability of 12 different products, no fewer than 132 series of tests are needed ($n \times (n-1)$ because tests are not symmetric).

Interoperability tests are informal in the sense that no certificate will be issued by an accredited organization at the end of the tests. The Interoperability testing is not standardized and is still under study. Nevertheless for end users, interoperability tests are the most important tests that must be passed.

3 IMPLEMENTATIONS UNDER TEST

LITMAP is an implementation of the MAP/CNMA stack at our laboratory LIT (Laboratoire d'Informatique Technique) [22, 24]. It had been implemented on Sun Sparc workstations. Ada is the implementation language and has been chosen for its run time support and cross development facilities. LITMAP runs on embedded systems (Force Sys 68040 and MVME372A: MAP Interface Module from Motorola Inc.) [19] connected by token bus or Ethernet and on Sunsparc workstations connected by Ethernet.

Figure 2 shows the profile of LITMAP. The MAC layer supports IEEE 802.3 (Ethernet) and IEEE 802.4 (Token Bus) protocols. The network layer supports connectionless network layer protocol (CLNP). Transport Class 4 (TP4) is supported at the transport layer. The session and presentation layers support kernel services. The application layer services include ACSE and MMS (Manufacturing Message Specification) [11].

Interoperability testing was conducted between LITMAP and some commercial implementations. The following is a brief description of each implementation and its profile.

1. a Concord/SISCO implementation on PC, under DOS, attached to an 802.4 token passing bus on carrierband,

2. a Siemens S5 115 Programmable Logic Controller, attached to an 802.3 segment,

3. an SNI UNIX PC, attached to an 802.3 segment,

4. a Siemens PC, under Windows, attached to an 802.3 segment, and

5. a DEC ULTRIX DECStation, attached to an 802.3 segment. A router developed at EPFL-LIT was used to connect the 802.3 Ethernet network with the 802.4 carrierband network.

4 METHODOLOGY

In this section we present the testing methodology. The interoperability testing is done at the interworking level. The tests used for the interoperability testing are described briefly. The tools and utilities that helped during these tests are also discussed.

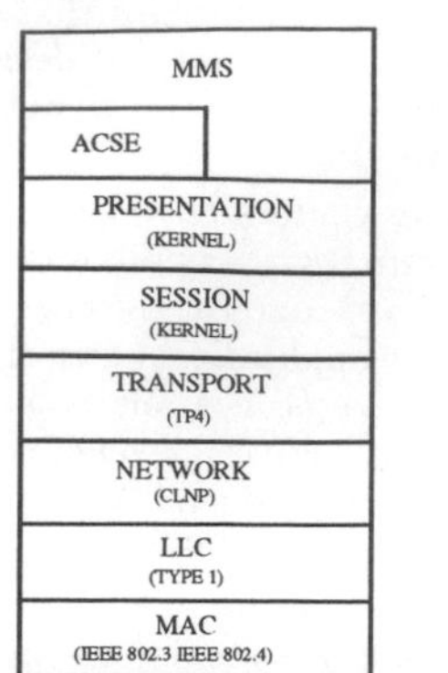

Figure 2 Profile of LITMAP

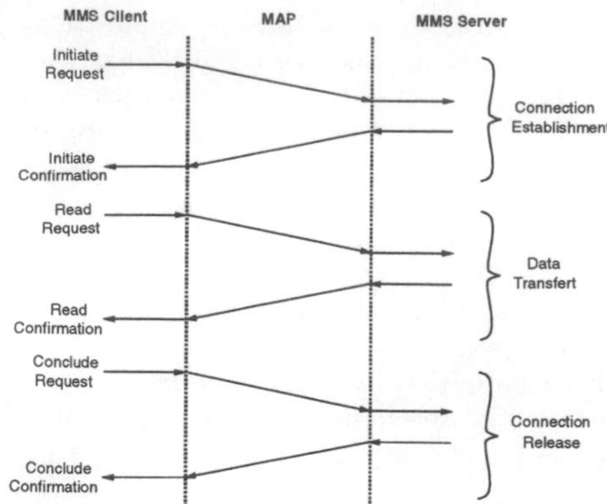

Figure 3 Testing methodology

4.1 Testing methodology

For all our interoperability tests we adopted the same methodology. It consisted of:

1. Opening an association between the MMS client and MMS server,
2. Transferring data on the opened association, and
3. Closing the association.

These various steps are illustrated on figure 3. Association establishment is the hardest part of interoperability tests. This is where we encountered the majority of problems. As a matter of fact, opening a connection means setting up all the information necessary for achieving reliable communication between two different entities. Communicating entities exchange information regarding their capabilities and requirements. A connection is then used as a safeway for data transfer. This phase is long and complex with respect to data transfer. It involves operations such as memory reservation, parameter negotiation, etc. At the user level, opening a association is done with the MMS `Initiate` service.

Data transfer is the second step of interoperability tests. Most known implementations support only a small subset of services. Only this subset could be used to perform tests at this stage. We systematically tested the well-known and classical services provided by all vendor implementations: `Identify`, `Status`, `GetNameList`, `GetCapabilityList`, `Read`, `Write` and `GetVariableAccessAttributes`. Read and and write services were restricted to simple variables. With one implementation we were also able to test MMS event management services such as `DefineEventCondition`, `DefineEventEnrollment` and `TriggerEvent`.

Finally, closing the association involves invoking the MMS `Conclude` service. This performs a gentle release of the connection environment and used resources are made available again for successive connections.

Not all vendor implementations supported both situations: acting as a server and acting as a client. But when it was possible, our implementation was tested both as client and as server. This is important because successful tests as a client (server) does not guarantee that tests will be successful as the server (client) as well.

Problems class	*Error*	*Implementation Problem*	*Interpretation Problem*	*Total*
Encoding / Decoding	1	0	0	1
Optional parameters	4	0	0	4
Incorrect Behavior	7	0	2	9
Addressing	1	3	0	4
Consistency checks	0	1	0	1
Error Code	1	0	3	4
Total	14	4	5	23

Table 1 Number of problems uncovered in interoperability tests

4.2 Limitations of our interoperability tests

The interoperability tests discussed in this paper were performed for the entire communication stack including layers 1 to 7 and even MMS servers. Only correct sequences of services and correct parameters were provided at the user interface.

Segmentation/Reassembly at the transport level was not tested because the size of the MMS PDUs sent was always smaller than the maximum PDU size negotiated at the transport level.

Error detection and recovery at the transport level was not exercised because tests were made in conditions that did not produce PDU losses such as electromagnetic noise or router overloads.

The above list shows that conformance tests are necessary and complementary to interoperability tests because they can test mechanisms that are not stressed in interoperability tests and rarely used in practice.

4.3 Utilities

During our interoperability tests, many utilities proved to be useful if not indispensable. First, we used a protocol analyzer from Siemens (K1102 for PCs). It includes decoders for the OSI protocols and MMS. A protocol analyzer is a fundamental tool for interoperability tests. It helps to discover and diagnose errors and is independent of the tested implementations. A trace facility was also used. This gives us the possibility to trace the PDUs between the different layers in LITMAP. A memory dump facility helped us to see on-line how memory was used in our stack. The dump facility allowed us to see whether memory was correctly allocated (or released) when the connection was established (or terminated).

5 ERROR CLASSIFICATION

During the interoperability tests, we encountered both problems and errors with the various implementations. We call *error* an incorrect implementation of a mechanism or parameter that is well defined in the standard or profile. We call *implementation problem* a problem related to implementation choices made by the vendor which are not incompatible with the standard. Different choices made by two different vendors may lead to impossibility to interoperate. Finally we call *interpretation problem* a problem that appears when the standard is not clear on the presence or details of parameters or mechanisms.

This first classification of problems and errors is orthogonal to a second classification we introduce and use here which makes a distinction among problems or errors according to the nature (coding, addressing, behavior, etc...) of these. In the following subsections we define these types of problems/errors.

This classification helps to identify the areas where most of the errors are found. It acts as a guideline for future tests and also helps to minimize the time taken to isolate and rectify the errors encountered during interoperability testing.

Encoding/Decoding errors: This type of error is encountered when data is encoded using different formats. When there are many choices for the encoding mechanisms, and a particular choice is not supported by an implementation, it results in a decoding error. This error is mostly encountered in the presentation layer which uses ASN.1 BER for encoding data. Another potential source of this class of error is encountered when decoding the headers of PDUs in the lower layers (session, transport and network). When certain features like *concatenation of PDUs* is not supported, it can result in a decoding error (when the decoder does not recognize the rest of the concatenated PDU). Another source for decoding errors is incorrect composition of PDUs. When octets are out of place in a PDU it can result in a decoding error.

Behavior problems: This type of error is encountered when a certain functional aspect of the protocol does not exhibit the expected behaviour. It is very difficult to isolate and diagnose this type of error. A behavior error occurs when the vendor does not implement a mechanism correctly. A behavior implementation problem occurs when a vendor imposes restrictions on a mechanism, for example forbids to have a transport credit larger than one. Finally a behavior interpretation problem occurs when the standard does not clearly define the operation of a protocol mechanism.

Addressing problems: Addressing errors occur due to mismatch in addresses (SAPs), limitation of the configuration of addresses (NSAP) or incorrect address formats (number of bytes used for SAPs). Since address configuration is an implementation detail, it can result in problems that are specific to particular implementations.

Consistency Check: When parameters are negotiated, the value of the negotiated parameters must be checked by the initiator of the negotiation. When an implementation does not perform this consistency check, it may lead to other kinds of errors (ex. cascade effect). For example, if the value of the *presentation context identifier* is not checked by the initiator of an association, it can result in the presentation data PDU being rejected by the responder.

Optional fields: Optional fields are another source of errors during interoperability testing. Most of the errors are the result of an implementation requiring a field that is optional or not handling a field that could optionally be present in the PDU. When optional fields are present in a PDU, the implementation can ignore it, but cannot reject a PDU due to its presence.

6 RESULTS AND OBSERVATIONS

6.1 Results

The results of the interoperability testing with different vendors are discussed in this section. Table 1 gives the results of the test according to the classification discussed in section 5. Table 3 presents the results for each layer. This tables gives an idea about the

Response	Layer	Application
No	1	6
Partial	0	0
Incorrect	12	5
Error	10	12

Table 2 Classification of errors based on the type of response received

number of errors detected in each layer and it also reflects on the complexity of the layer (refer to 6.2 for more details). Table 2 classifies errors based on the response received by the initiator. This classification is based on discussion in 2.2.

6.2 Observations

After this interoperability test experiment, we can make the following observations:

1. It is interesting to note that ratio of number of errors encountered to the number of implementations with which tests have been performed is very low. The total number of errors detected is also very low for the complexity of the protocols involved in the interoperability test. It is striking to see that the number of problems that have been found is as low as 23 in total and less that 4 by product tested in average.

2. Each time a new implementation is used, additional errors or problems are found. This fact reflects the large number of possible implementations of an OSI stack and the large number of ways in which these can be configured.

3. Testing as a client and as a server are two different situations. Having succeeded in client (or server) tests does not imply that tests in the reverse direction will succeed. This fact is taken into account by our definition of interoperability in 2.2.

4. Comparison of tests performed with two different implementations may point to unseen problems by one of these implementations. For example, an attempt to establish an association with one commercial implementation failed while it succeeded with another. Careful examination showed that the latter did not check a parameter, while the former did.

5. About 50% of the errors fall under the *Error* category. In particular, we can say these errors are a result of incorrect implementation and testing strategies.

6. Now that ASN.1 compilers are quite stable and ASN.1 is well understood by people, we observe a new distribution of errors different from that referenced in [5, 18] and which corresponds to the complexity of layers.

7. Addressing problems could be significantly reduced if the different vendors adopted a standard way of defining the addresses of the communicating entities. Having to deal with as many addressing definition notations as there are commercial implementations was a serious problem that delayed the interoperability testing with these implementations.

8. There is a lot of improvement to be done in standards bodies to define meaningful error codes and to precisely indicate in which circumstances these error codes must be used (Table 1). Simply defining a `Protocol Error` code is not enough and makes diagnosis of errors more difficult. The same remark is valid for MMS where the `Other` error is used by vendors in many contexts. Note that for application services like MMS, this means that an application cannot be 100% generic since it depends on the error codes returned by different vendors.

9. A considerable number of problems is due to differences of interpretation because of imprecisions, misleading definitions and the big number of options in standards and profiles. For example, when a connection is established in MMS, supported services and capabilities (S&C) are negotiated. It is not clear whether the returned S&C in the `Initiate` response are the intersection of supported S&C of both end-users or if it is the common S&C subset of the server application and the MMS-provider(The part of the application that conceptually provides the MMS service through the exchange of MMS PDUs). One of the tested implementation simply chose the former for supported services and the latter for supported capabilities!

7 COMPARISON WITH OTHER TESTS

In this section, we compare our results with other interoperability tests and in particular with the results of the interoperability testing for HP MAP 3.0[18]. These series of tests were conducted in 1990 with 8 different products from companies such as Allen-Bradley, Computrol, Motorola, etc. These tests were performed for FTAM and MMS. We will only compare the results of the MMS tests in this article.

In table 3, when we look at the number of problems by layer, we can see that the distribution of errors is very different between the two series of tests. Let's try to explain this. First, we can see that defect rates from MAC layer and session layer are low in the two columns.We can clarify this by the fact that the various implementations of these layers are quite stable and these layers are less complex compared to the others. Three of four defects found in the network layer are addressing limitations. The Swiss academic network SWITCH utilizes the *39 format* for NSAPs. The non-use of this address format accounts for the errors detected at the network layer. This explains the differing resilts observed for the network layer. In the transport layer, the larger number of errors comes from the fact that the transport class 4 protocol is complex and that for two of the implementations, it was their first interoperability test. In the case of the three upper layers, the comparison is altered by the large number of encoding errors in the HP tests. This can be explained by the multiple ways of encoding PDUs with the Abstract Syntax Notation One (ASN.1)[1]. In particular, ASN.1 permits the length of PDUs to be encoded in two different ways and several vendors had some errors in decoding both the encoding methods. At this point, it is worth mentioning that before 1990, complete and good ASN.1 compilers were scarce.

When we look at the HP tests, we see that the lower layers (up to session) are quite stable and the upper layers have several problems with ASN.1. Errors in our tests are more homogeneous in the sense that we can notice a relation between the number of errors and the complexity of the layer. The reason is certainly that some products we have tested are less mature and have been submitted to no or very few interoperability tests.

When we compare the results of the two tests, we note several identical lessons. First, interoperability testing is necessary because defects were uncovered in the tests with all vendors. Second, defect rates for a given implementation normally decrease steadily as interoperability testing is done with more vendors. We confirm that a considerable number of problems come from optional fields in network, transport and MMS layers and from negotiation of contexts in application and presentation layer. In our results, we also have several problems with the format of NSAP addresses and with the protocol error codes in network, transport, presentation and MMS layer. A small number of errors at the application level is the result of the availability of stable ASN.1 compilers[3].

Layer	*HP tests*	*LIT tests*
MMS	14	7
ACSE	14	1
Presentation	14	4
Session	2	1
Transport	1	6
Network	0	4
Data link	0	0
Total	45	23
Total/Vendor	5.6	3.8

Table 3 Number of problems uncovered in interoperability tests

8 CONCLUSION

Integration of enterprise applications increasingly relies on the use of standard communication networks. Practice has largely shown that interoperability problems occur between implementations of different vendors. In this paper we have presented the results of an experiment made at EPFL-LIT with an implementation of an OSI profile that has been tested for interoperability with a number of commercial implementations. The problems that were found during these tests have been classified and compared with other tests. Finally, we have drawn lessons on the way to proceed for interoperability testing.

Improvements can be made by both protocol specifiers and vendors. In particular, a significant effort should be made for the definition of precise error codes for all protocols. Profiles that in the context of OSI define subsets of services, parameters and mechanisms are not precise, complete and restrictive enough to ensure proper implementation by vendors. A clear documentation of a vendor implementation restrictions or particularities would certainly make the job of testers easier. Indeed, implementations may be 100 % correct and yet not be able to interoperate because of incompatible choices or implementation restrictions.

The distribution of problems and errors found in this study is different from that presented by other researchers. This difference is explained by the difference in maturity of the tested implementations and by the fact that ASN.1 compilers for example now produce error-free code for presentation and application protocols.

Work on interoperability tests will be continued in two directions: additional tests with other implementations such as those from DEC, HP and Bull; systematic exploration of interoperability problems at server level, above the application layer.

9 ACKNOWLEDGMENTS

We would like to thank our laboratory's colleagues of the CNMA group: F. Restrepo , G. Berthet and our colleagues of the ESPRIT CCE-CNMA for providing us with their implementations and their help in understanding some of the uncovered problems: A. Lederhofer and W. Blumenstock (Siemens), A. Horstmann, P. Wimmer (SNI), C. Bezencon (DEC).

REFERENCES

[1] "ISO/IEC 8824, CCITT X.208, Specification of Abstract Syntax Notation One (ASN.1", 1988.

[2] Y. Benkhellat, M. Siebert, J-P Thomesse, "Interoperability of sensors and distributed systems", *Sensors and Actuators*, Vol. 37, 1993, pp. 247–254.

[3] G. Berthet, "ASNADAC User Manual", EPFL-LIT Internal Report, 1993.

[4] G. Bonnes, "Verification d'inter-operabilite OSI, X.400 et LU6.2", *De Nouvelles Architectures pour les Communications Eyrolles*, 1988, pp. 133–140.

[5] S. Castro, "The relationship between conformance testing of an interoperability between OSI systems", *Computer Standard Interfaces*, Num. 12, 1991, pp. 3–11.

[6] "ESPRIT Project 7096 CNMA Implementation Guide Revision 6.0", November 1993.

[7] "ESPRIT CCE-CNMA, CCE: An Integration Platform for Distributed Manufacturing Application", Springer Verlag, 1995.

[8] J. Garde, C. Rohner, C. Summers and S. Symington, "A COS study of OSI interoperability", *Computer Standard Interfaces*, Num. 9, 1990, pp. 217–237.

[9] "General Motors, Manufacturing Automation Protocol, Version 3.0", August 1988.

[10] ISO 7498-1, "Information Processing Systems – OSI – Basic Reference Model".

[11] IS0/IEC 9506-1, "Manufacturing Message Specification: Service Definition", 1990.

[12] "ISO/IEC 9595-1, CCITT X.710 Common Management Information Service Definition", 1991

[13] L. Lenzini, "The OSIRIDE-Intertest Initiative: Status and Trends", *Computer Networks and ISDN Systems*, Vol. 16, 1989, pp. 243–255.

[14] L. Lenzini, F. Zoccolini, "Interoperability tests on OSI Products in the framework of the OSIRIDE-Intertest initiative", *Comp. Net. and ISDN Systems*, Vol. 24, 1992, pp. 65–79.

[15] R.J. Linn Jr, "Testing to Assure Interworking of Implementations of ISO/OSI protocols", *Computer Networks and ISDN Systems*, Vol 11, 1986, p. 277-286.

[16] R.J. Linn, "Conformance testing for OSI protocols", *Computer Networks and ISDN Systems*, Vol. 18, No. 3, pp. 203–220.

[17] R.S. Matthews, K.H. Muralidhar, S. Sparks, "MAP 2.1 Conformance Testing Tools", *IEEE Trans. on Software Engineering*, Vol. 14, No. 3, March 1988, pp. 363–374.

[18] J.D.Meyer, "Interoperability testing for HP MAP 3.0", *Hewlett-Packard Journal*, August 1990, pp. 50–53.

[19] "MVME372: Map Interface Module User's Manual", Motorola Inc., 1989.

[20] D. Rayner, "Progress on Standardizing OSI Conformance Testing", *Computer Standards and Interfaces*, Vol. 5, 1986, pp. 317–334.

[21] D. Rayner, "OSI Conformance Testing", *Computer Networks and ISDN Systems*, Vol. 14, 1987, pp. 79-98.

[22] D. Sidou, K. Vijayananda, G. Berthet, "An Architecture for the Implementation of OSI protocols: Support Packages, Tools and Performance Issue", *Proc. of SICON/ICIE'93*, Singapore, Sept. 93.

[23] "Boeing, Technical Office Protocol, Version 3.0", August 1988.

[24] F. Vamparys, P. Pleinevaux, "Architecture of the LITCommEngine", *EPFL-LIT Internal Report*, Feb. 1994.

24

Supporting the Information Management of Control Activities in a CIM Shop Floor Environment *

G. Loria, I. Mazón, F. Rojas
Instituto de Investigaciones en Ingenieria, Universidad de Costa Rica, Costa Rica
email: gmoloria@cariari.ucr.ac.cr

H. Afsarmanesh, M. Wiedijk, L.O. Hertzberger
Computer Systems Department, University of Amsterdam, The Netherlands
email: hamideh@fwi.uva.nl

Abstract

A CIM shopfloor environment can be supported by a collection of different activities, such as design, manufacturing, marketing, etc. Every activity may consist of a number of subactivities where a subactivity can be performed and represented by several agents. The "agents" are described here as components that perform activities through executing their specific "tasks". So far, for DCS (Distributed Control Systems), manufacturers have resolved the "information management", "information sharing" and "control" problems, through purchasing all instruments and control equipments from the same supplier. In real applications however, both within the CAC activity (among its sub-activities) and in relations among the CAC activity and other CIM activities, the industrial processes define differently formatted data, for instance the discrete and continuous variables that need to be shared among several multivendor control equipments. Therefore, the information sharing and management and the control problems are still the difficult technological problems without an integral solution. This paper focuses on the Computer Aided Control (CAC) activity and describes some agents involved in process control and the tasks that they perform. Furthermore, an architecture is described to support the information management requirements for agents' cooperation in a CIM shopfloor environment. This information management architecture is then applied to the definition of cooperating CAC agents.

Keywords

Integrated CIM activities, Computer Aided Control activities, Federated Information Management system

1 INTRODUCTION

With the new trend of applying the concurrent engineering methodologies to product manufacturing (Barker, 1993; Afsarmanesh et al., 1994b), regular evaluation of activities involved in the "product life-cycle" increases the productivity, improves the quality of products, and reduces

*This research is partially supported by the CIMIS.net project ECLA 004:76102.

the costs. Each activity in the product life-cycle, ranging from design and manufacturing to marketing and sales, involves a group of engineers and experts. Any improvements require that a part of the information is shared and exchanged within one activity (i.e. tightly linked) and with other activities (i.e. loosely linked) (Barker, 1993). The CAD, CAM, CAC, etc. activities support distinct phases of the product life cycle (Didic, 1993; Sepehri, 1987). Every CAx activity consists of several subactivities and executes specific tasks. A subactivity can either be executed by a single agent or in a cooperation of several agents which are, depending upon their tasks, either loosely or tightly linked. Typically, a task is supported by an agent. For instance, the CAC activity is represented by one or more agents that execute several control tasks, such as process control and quality control. Other activities associated with CIM, for instance: design, research and development, marketing, management, process and product planning, etc. can all be represented by their agents. Thus, an agent is a hardware-software basic cell that partially or totally performs one or more tasks within a (sub)activity. Quite often, these agents run on several independent and heterogeneous systems. The relationships among the CAD, CAM and CAC

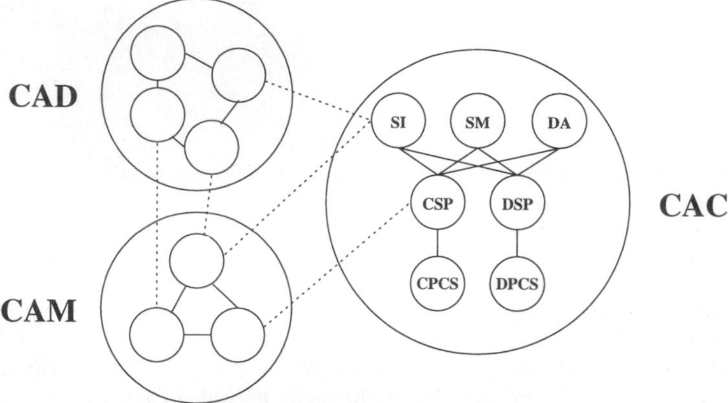

Figure 1 The tightly coupled CAC agents (represented by solid lines) and their loose coupling with CAD and CAM (represented by dotted lines).

activities, as illustrated in Figure 1, create an environment that supports the integration of the activities in the shop-floor. The CAD activity supports several subactivities, such as the geometrical design tasks and strategical control design tasks for CACSD (Computer-Aided Control System Design). CAD subactivities in turn perform certain tasks, for instance the geometrical modeling (design of parts, assemblies, etc.), simulation of control strategies (discrete and continuous), tuning (continuous process), and the timing and counting (discrete process). The CAM activity supports the supervision over geometrical product data and control of the numerical machines, FMS and the capacity evaluation of CAD parts. The main role of the CAC activity is to check and regulate the process variables on the shop floor, so that the processes have the adequate reliability and security. CAC performs some automatic tasks but mostly follows the operator supervisory recommendations in order to perform an optimal control over this important phase of the product life-cycle. The CAC activity mainly involves agents that perform control tasks such as continuous process control, discrete process control, the acquisition and monitoring of process variables, the display and management of alarms, communication services, optimization

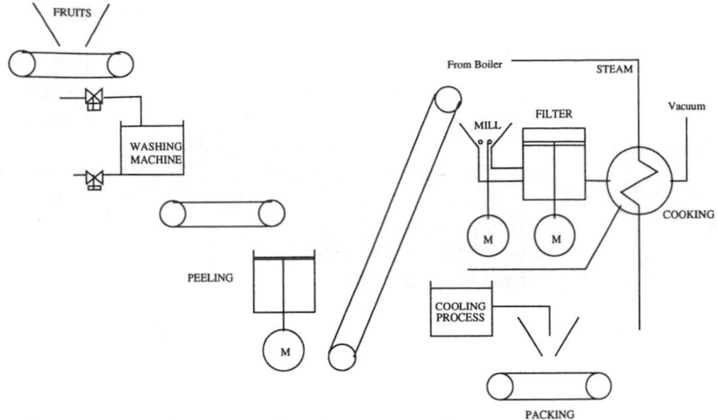

Figure 2 The marmalade production process

routines, supervision strategies, and contingency decision using expert systems. These tasks are described in more details in Section 2.

This paper discusses the requirements to achieve an integration of Shop-Floor CAC activities involved in CIM industry. Although, the focus of this paper on defining an information management architecture to support the CAC activity and the integration of its agents, this architecture is general enough to be applied to any other CAx shop floor activity. The remaining of this paper is organized as follows. Section 2 describes a marmalade application environment and the agents and tasks they perform in the continuous and discrete process control and supervision subactivities. Section 3 discusses the information management requirements for these activities, followed by a brief description of PEER's federated integration architecture in section 4. Section 5 defines some example PEER agents for the CAC activity using the PEER integration architecture, followed by an example of the information sharing among these agents.

2 A MARMALADE PRODUCTION APPLICATION

In this section we first describe an application environment; a marmalade production industry. Figure 2 shows the marmalade production processes, that consist of interrelated discrete (batch), batch/continuous and continuous processes, and their interrelationships, similar to the description of continuous and discrete events in (Eigi M., 1993) and discrete events in (Walter, 1992; Eigi, 1990). The production begins with the cleaning and peeling of the fruits. The raw material is transported to a mill system. After filtering, the glucose content is measured and a special formula will be added. This product is then transferred to a tank for dehydration and cooking. This tank works at vacuum pressure with temperatures around 60°C, the temperature is obtained by steam that comes from the boilers. The cooking process takes about 72 hours and is controlled by a specialist that knows the special aspects of the product. Once the cooking process has finished, the product is transferred to another tank. After the marmalade obtains ambient temperature the resulting product is packaged. The rest of this section describes the subactivities and agents related with the CAC activity in this marmalade industry and their interrelationships.

Subactivity	Agent	Task
Continuous Process Control	Continuous Process Control Strategies (CPCS) Continuous Step Processor (CSP)	PID Field Measurement Communication
Discrete Process Control	Discrete Process Control Strategies (DPCS) Discrete Step Processor (DSP)	 Field Measurement Communication
Supervision	Display agent (DA) Signal Management (SM) Signal Integrator (SI)	

Figure 3 Subactivities, agents and tasks in the CAC activity

Three CAC subactivities can be distinguished for the marmalade industry, namely **continuous process control**, **discrete process control**, and **supervision**. Figure 3 shows the main agents and tasks involved in each subactivity.

2.1 Continuous process control subactivity

The continuous process control subactivity involves the control of the specific hardware in the shop floor whose characteristic signals (input/output) are continuous. The main agents performing this subactivity are described below.

Continuous Process Control Strategies agent (CPCS)

The CPCS agent is responsible for the regulation of the variables of continuous processes using either local, remote, or other control setting parameters. In case of a local setting, the operation conditions are set by an operator directly on the controller hardware. However, this kind of setting is undesirable for this type of control environment. Usually, the control strategy can be changed at any time by a remote setting, through an agent in the CAC activity. CPCS usually runs on a dedicated computer and its most important task is the PID control task.

- The **PID Control Task** executes a special algorithm to control a continuous process in a feedback manner; using continuous variables from sensors (temperatures, pressure, flow, etc.), it calculates errors (regulation and tracking errors) and returns the command signals to process actuators (valves, motors, etc.). Usually, there are many continuous control loops associated to the control of this kind of shop floor environments.

Continuous Step Processor agent (CSP)

The CSP agent provides a bi-directional interface among the internal CAC subactivities, in order to communicate information of several shop floor equipments, and to transfer this information to other CIM activities. In the CIM environment, the CSP agent converts CAC activity information (such as field information) into a representation that can be used by other CAx activities; in the same way the information of the other CAx activities are translated into CAC representations. This agent translates the continuous variables to neutral files by means of STEP (Sastrón, 1993). Two tasks of this agent are described below:

- The **Field Measurement Task** reads the signal values from different kinds of equipments such as sensors, transmitters, final elements and intelligence systems. These equipments require many proprietary interfaces and protocols to link them together. An implementation can be obtained using the field bus standard ISA SB50, that enables a real multivendor instrument integration at the shop-floor level.
- The **Communication Task** enables the communication between the process control with both the CAC subactivities and with other CIM activities, while hiding the specific and distinct hardware and protocol aspects of the process control equipment, avoiding the problems related to the combination of different hardware (Sepehri, 1987).

2.2 Discrete process control subactivity

This subactivity involves the control of the specific hardware in the shop floor whose input/output signals are discrete. The main agents are described below:

Discrete Process Control Strategies agent (DPCS)

The DPCS agent realizes the control over sequential processes. The local or remote operation is similar to the CPCS agent. This agent controls numerical machines using programmable controllers or similar hardware. The main operations of this agent are: timers, counters, step sequences, combinational functions. The DPCS agent is involved in step by step operations that depend on discrete conditions. Historically the CAx tools support these kind of operations.

Discrete Step Processor agent (DSP)

The DSP agent has the same function as in the continuous process control subactivity. In the discrete process control subactivity the DSP agent operates with bidirectional discrete variables from/to the shop floor equipment and other CIM activities. This agent performs tasks similar to the CSP agent as described below:

- The **Field Measurement Task** operates in similar way as in the continuous step processor agent. However, it works with bidirectional discrete variables from/to the shop floor equipment and other CIM activities.
- The **Communication Task** communicates the information flow (on/off signals, counters and timer values) of the process control to other hardware of the CAC and CIM activities, while hiding the specific and distinct protocol aspects of the process control equipment.

2.3 Supervision subactivity

The supervision subactivity is at the highest level of the CAC activity hierarchy. It supports the coordination between the different lower level CAC agents and the relationship with other CIM activities. This subactivity has an intelligent strategy analyzer that coordinates and evaluates the process requirements and the actual state of its different controllers and generates reports and displays information at mimics and other hardware. The supervisory agent handles alarm exceptions generated by the CAx activity that requires responses from the lower level CAC activities. Based on its internal model and strategies, this subactivity minimizes the negative effects on the entire production process. Supervision subactivity is defined by the following three agents:

Display Agent (DA)

This agent presents a global overview of the process environment through a user friendly interactive graphical interface using active image techniques. It generates reports for different kinds of hardware.

Signal Management agent (SM)

The alarm agent is a traditional mimic panel for the acknowledgment of contingency conditions. This panel is built with visual and sonic indicators. The mimic panel can be emulated in the monitor using the alarm agent information.

Signal Integrator agent (SI)

The continuous and discrete control processor agents transmit information using different formats. Therefore, it is necessary to implement the integration and conversion of different information formats into a standard transportable information format.

3 INFORMATION MANAGEMENT REQUIREMENTS

There is a need to support the information sharing and exchange among different tasks in order to support the requirements for the cooperation and integration of different CIM activities. Although the focus of this paper is on the CAC activity, the cooperation and integration issues are similar for all CAx activities that will also be addressed here briefly.

The agents in the CAC activity frequently need to access the information about the state of the shopfloor. This information is managed by agents, so if an agent needs some information, it has to obtain it from other agents. Within the CAC activity, it is usually crucial for an agent to have fast (near real time) access to that information, even if it is managed by other agents, so the integration architecture must support it.

When applying a concurrent engineering approach to manufacturing, all CIM activities benefit from the information sharing and exchange both among different agents within one activity and among multiple activities. For instance, the CAD tools can then evaluate the production process and the control strategies. If a designer decides that the design is completed and ready to release, then the new models and/or parameters can be directly transmitted to the control units. The open architecture for Computer-Aided Control Engineering CACE (Barker, 1993) that includes a framework reference model, will improve the CAD capabilities and the information exchange with other CIM activities. The CAM activity can be extended to include support for the information exchange among engineers in the CAD, CAM and CAC activities. Using the time conditions and results set by the CAM and CAC activities, the design and evaluation of the control strategies in a CACE environment can be improved. A specific agent in the CAM activity can establish the link between the CACE design tasks and the actual results of the CAC activity. This agent analyzes the CAD specifications and assigns the CAC-CAM references, such as strategies of process controllers and tuning for the CAC activities, and the numerical control machine instructions for CAM activities. CAM alarm conditions could be shared with the CAC activity that will benefit from the resolution of contingencies, using the expert and optimization systems in that activity. To support the information management for the CAC activity in a CIM shop floor environment the following requirements can be distinguished:

- Integration and transformation of low level information generated and transmitted by low level control hardware.
- Sharing and exchange of information among agents within the CAC activity. The integration among these agents is usually tightly coupled.
- Sharing and exchange of information among the CAC activity and the other CIM activities such as CAD and CAM. The integration of these activities is usually loosely coupled.

In Section 4, the PEER federated object oriented database system will be described. PEER provides an integration architecture that supports the information management requirements described in this section. In Section 5.1, some example agents are defined by their PEER schemas. Finally, Section 5.2 presents how the PEER federated system supports the sharing and exchange of information among several agents within the CAC activity.

4 PEER FEDERATED INTEGRATION ARCHITECTURE

The PEER federated information management system has been designed to support the information management and information sharing and exchange in a network of both loosely and tightly coupled agents. The PEER information model is binary-based and object-oriented, in which every piece of information, both data and meta-data, are represented as objects. A PEER layer is defined for every agent that needs to share and exchange information with any other agent in the CIM environment. For example, two PEER agents in the CAC activity can be tightly coupled, while an agent in the CAD activity is loosely coupled to an agent in the CAC activity. Section 5 below describes the application of PEER to support the information management requirements of the marmalade production application in more detail.

Each PEER agent in the federation network can autonomously decide about the information that it manages locally, how it structures and represents its information, and which part of its local information it wishes to export to other agents. Each agent can import information that is exported by other agents and transform, derive and integrate (part of) the imported information to fit its interest and interpretation. PEER is a pure federated system, there is no single global schema defined on the information to be shared by all agents in the PEER network, and there is no global control. In the PEER layer of an agent, the information is structured and defined by several kinds of schemas. There is a local schema LOC (representing the information managed locally), several import schemas IMPi (representing the information shared by other agents, where integer i identifies a specific import schema), several export schemas EXPj (representing the information that this agent wishes to share with other agents, where integer j identifies a specific export schema), and an integrated schema INT (representing the integration of the local and the imported information for this agent). The powerful integration facility of PEER (Tuijnman and Afsarmanesh, 1993; Afsarmanesh et al., 1993; Afsarmanesh et al., 1994a), including the distributed schema management and the distributed query processing, makes the distribution of information and the heterogeneous information representations among different agents transparent to the user. The definition, integration and derivation of information in PEER are supported by a declarative specification using the PEER Schema Definition and Derivation Language SDDL (Afsarmanesh et al., 1993). The SDDL language supports the integration, derivation, interrelation, and transformation of types, attributes and relationships.

A prototype implementation of the PEER system is developed in the C language and includes two user interface tools (Afsarmanesh et al., 1994a), a Schema Manipulation Tool (SMT) and a Database Browsing Tool (DBT). The PEER system has been successfully used in the integration of cooperating expert systems for the control of power distribution networks (in ESPRIT ARCHON project) and the development of a planning toolbox (in ESPRIT PLATO project). At present, PEER is being applied to the integration of biomedical information systems (in DIGIS project).

5 INFORMATION MANAGEMENT OF THE CAC ACTIVITY

In this section, we focus on the definition of PEER agents representing some CAC activities described earlier in Section 2. In Section 5.1 two example PEER agents are defined that perform certain CAC activities. In Section 5.2 an example of the information sharing among these agents is defined using the PEER integration architecture described in Section 4.

5.1 Designing the PEER agents

For each CAC agent a PEER layer must be defined to support its integration and information sharing with other agents. In the PEER layer, first the local information structures (the LOC schema) must be defined. The LOC schema is comprehensive enough to support the definition of the information that must be shared with other agents (through the EXP schemas). The EXP schemas are derived from the LOC schema and support the transfer of information among different agents in CAC subactivities and other CIM activities. For instance, to support the fact that the continuous process control subactivity receives information from the low level hardware instruments in different formats. The EXP schemas defined in the PEER network can be imported by any agent that needs to access and use their remote information.

The Step Processor Agents described in Section 2 are indispensable components of the CAC activity. These agents interface between different CAC subactivities. The main task of these agents is to transform different low-level data formats into a unique and generalized format using EXPRESS from STEP (ISO/DIS 10303) definitions. EXPRESS is an excellent language that supports the translation of production information to neutral files. Similar to PEER, the data model describing EXPRESS is also binary-based, which makes PEER a suitable candidate to represent EXPRESS datastructures and entities. Using the PEER federated architecture, different CAC agents can link to other CIM activities through the Step Processor agents. Since the CAC activity is a very important component of hybrid systems, the connection of this and other activities by a PEER federated network is a step towards the entire CIM shop floor environment integration and the support for Concurrent Engineering.

The remaining of this section presents the schemas for the CSP and DSP agents defined in Section 2. The schemas are defined using the SDDL language of PEER (Afsarmanesh et al., 1993), as described in section 4. These schemas are necessary to support the sharing and exchange of information among CAC agents.

CSP agent information

Figure 4 shows the entities defined and exported by the Continuous Step Processor (CSP) agent. **Engineering_units** is a base entity, that allows the identification of instruments. The **analog_element** entity defines an analog variable and the **direct_analog_read** entity defines its specific measurement done by a direct measurement device. Export schema EXP5 in Figure 5

```
derive_schema EXP2 from_schema LOC
  type Engineering_units
    units                        : STRINGS
    zero                         : REALS
    span                         : REALS
    minimum_value_admitted       : REALS
    maximum_value_admitted       : REALS
    descriptor                   : STRINGS
    physical_address             : STRINGS

  type analog_element subtype_of Engineering_units
    input_value                  : REALS
    units                        : STRINGS
    read_quality                 : BOOLEANS

  type direct_analog_read subtype_of analog_element
    current_or_voltage           : BOOLEANS
    resolution                   : INTEGERS
    read_condition               : INTEGERS
end_schema EXP2
```

Figure 4 Export schema EXP2 of the Continuous Step Processor (CSP) agent

represents the global elements of the marmalade processing which are defined and exported by the CSP agent. The **continuous_controller** entity represents the variables and concepts of

```
derive_schema EXP5 from_schema LOC
  type _continuous_controller
    strategy                   : _strategy
    kind_of_control            : T_control
    // Other elements necessary in the definition of  a continuous controller

  type _process
    fruit                      : T_fruit
    controller                 : _analog_controller
    input_volume               : REALS

  type _washing subtype_of _process
    water_amount               : REALS
    detergent                  : REALS
    kind_detergent             : T_detergent
    duration                   : REALS

  type fruit_kind subtype_of quality_element, category_element
    name                       : STRINGS
end_schema EXP5
```

Figure 5 Export schema EXP5 of the Continuous Step Processor (CSP) agent

each controller type in the continuous step processor agent. The **_process** entity represents the process variables in the marmalade production industry. The **_washing** entity is a specialization of the **_process** entity and it represents the information specifically related to the fruit washing task. The **fruit_kind** entity defines a part of the essential physics conditions for the components involved in the marmalade industry. The quality_element and category_element are concepts associated with the particular conditions of the raw (fruit) material in the production process.

DSP agent information

Figure 6 shows the entities defined and exported by the Discrete Step Processor (DSP) agent. As the CSP agent, the DSP agent defines the **Engineering_units** entity, that allows the identification of instruments, as its base entity. The **analog_element** entity characterizes a digital variable, the

```
derive_schema EXP7 from schema LOC
  type Engineering_units
    units                         : STRINGS
    zero                          : REALS
    span                          : REALS
    minimum_value_admitted        : REALS
    maximum_value_admitted        : REALS
    descriptor                    : STRINGS
    physical_address              : STRINGS

  type digital_element subtype_of Engineering_units
     alarm_variable               : BOOLEANS
     reading_quality              : BOOLEANS

  type direct_digital_read subtype_of digital_element
    descriptor                    : STRINGS
    physics_direction             : STRINGS
    reading_conditions            : INTEGERS
end_schema EXP7
```

Figure 6 Export schema EXP7 of the Discrete Step Processor (DSP) agent

direct_analog_read entity defines the element value as a function of the physics meaning and the reading conditions.

5.2 The integration approach

An agent in the CAC activity frequently needs to access to information that is managed by other agents in the CAC activity. In this section we briefly describe an example of how the PEER architecture supports the integration of the field measurement information of the CSP and DSP agents by the Signal Integrator (SI) agent. This information includes the continuous process variables from sensors (temperature, pressure, flow etc.) transmitters, final elements, and intelligence systems, and the discrete process variables such as on/off signals, counters and timer values.

As a first step in the integration, the SI agent imports the export schema EXP2 of the CSP agent as its own import schema IMP4. The SI agent also imports the export schema EXP7 of the DSP agent as its own import schema IMP6. Then the second step for the SI agent is to create its own integrated schema (INT). The integration of the information of the CSP and DSP agents by the SI agent is defined in Figure 7 as the integrated schema of the SI agent. The first part in the INT schema represents the entities defined in the SI agent, then the second part represents the derivation specification of the SI entities in terms of the base entities as they are imported from the CSP and DSP agents. The SI agent defines its own view on the imported information and uses different names for some imported information. Namely, it interprets the Engineering_units entity as a process_variable, an analog_read as a continuous_variable, and a direct_read as a discrete_variable. The SI agent is not interested in the physical_address of an Engineering unit so that attribute is not a part of the process_variable entity. In the derivation specification, the union primitive defines the instances of the Engineering_units entities of both agents as the instances of the process_variable entity. The other entities are renamed as described above. The base entity names in the derivation specification are always extended with their import schema names to have a unique name to avoid confusion in case the entity names in different import schemas are synonyms. The integrated attributes of the process_variable entity are defined as the union of the values of the set of base attributes (enclosed by and). For a complete description of PEER's derivation and integration primitives see (Afsarmanesh et al., 1993).

```
derive_schema INT from_schema LOC, IMP4, IMP6
  type process_variable
    units                       : STRINGS
    zero                        : REALS
    span                        : REALS
    minimum_value_admitted      : REALS
    maximum_value_admitted      : REALS
    descriptor                  : STRINGS

  type continuous_variable subtype_of process_variable
    input_value                 : REALS
    units                       : STRINGS
    read_quality                : BOOLEANS

  type continuous_variable_read subtype_of continuous_variable
    current_or_voltage          : BOOLEANS
    resolution                  : INTEGERS
    read_condition              : INTEGERS

  type discrete_variable subtype_of process_variable
    alarm_variable              : BOOLEANS
    reading_quality             : BOOLEANS

  type discrete_variable_read subtype_of discrete_variable
    descriptor                  : STRINGS
    physics_direction           : STRINGS
    reading_conditions          : INTEGERS

derivation_specification

  process_variable = union(Engineering_units@IMP4,Engineering_units@IMP6)
    units = {units@IMP4, units@IMP6}
    zero = {zero@IMP4, zero@IMP6}
    span = {span@IMP4, span@IMP6}
    minimum_value_admitted = {minimum_value_admitted@IMP4, minimum_value_admitted@IMP6}
    maximum_value_admitted = {maximum_value_admitted@IMP4, maximum_value_admitted@IMP6}
    descriptor = {descriptor@IMP4, descriptor@IMP6}

  continuous_variable = analog_element@IMP4
    input_value = input_value@IMP4
    units = units@IMP4
    read_quality = read_quality@IMP4

  continuous_variable_read = direct_analog_read@IMP4
    current_or_voltage = current_or_voltage@IMP4
    resolution = resolution@IMP4
    read_condition = read_condition@IMP4

  discrete_variable = digital_element@IMP6
    alarm_variable = alarm_variable@IMP6
    reading_quality = reading_quality@IMP6

  discrete_variable_read = direct_digital_read@IMP6
    descriptor = descriptor@IMP6
    physics_direction = physics_direction@IMP6
    reading_conditions = reading_conditions@IMP6
end_schema INT
```

Figure 7 Integration of field measurement information by the Signal Integrator (SI) agent

The other agents in the supervision subactivity can be integrated in a similar way. The signal management (SM) agent needs access to the alarm information from the SI agent to acknowledge contingency conditions. The display agent (DA) needs the current state on the shop floor, for which it has to retrieve information from the step processor agents.

Similar to the methodology applied in Section 5.1, for every two agents that need to cooperate and exchange their information, we develop export schemas (EXPs) in the exporting agent. An export schema can be imported by the other interested agents as (one of) their import schemas (IMPs). The import schemas are then integrated with the local schema to generate the integrated schema (INT). Now the agent can use its integrated schema and transparently access any piece of information that is either managed locally or shared by other agents.

6 CONCLUSIONS

This paper first describes the CAC activities involved in a marmalade production industry. Each activity is represented by its agents that perform certain tasks. Then, an integration architecture, the PEER federated information management environment is described that supports the sharing and exchange of information among these agents. As an example, PEER is applied to the integration of the discrete and the continuous control activities on this shopfloor. Although the presented integration architecture is applied here mostly to Computer Aided Control (CAC) activities, the same architecture can support the information management of other multi-agent Computer Integrated Manufacturing environments.

7 REFERENCES

Afsarmanesh, H., Tuijnman, F., Wiedijk, M., and Hertzberger, L. (1993). Distributed Schema Management in a Cooperation Network of Autonomous Agents. In *Proceedings of the 4th IEEE International Conference on "Database and Expert Systems Applications DEXA'93"*, Lecture Notes in Computer Science (LNCS) 720, pages 565–576. Springer Verlag.

Afsarmanesh, H., Wiedijk, M., and Hertzberger, L. (1994a). Flexible and Dynamic Integration of Multiple Information Bases. In *Proceedings of the 5th IEEE International Conference on "Database and Expert Systems Applications DEXA'94", Athens, Greece*, Lecture Notes in Computer Science (LNCS) 856, pages 744–753. Springer Verlag.

Afsarmanesh, H., Wiedijk, M., Moreira, N., and Ferreira, A. (1994b). Design of a Distributed Database for a Concurrent Engineering Environment. In *Proceedings of the ECLA.CIM workshop, Florianopolis, Brazil*, pages 35–43. (to appear in the "Journal of the Brazilian Society of Mechanical Sciences", ISSN 0100-7386).

Barker, A. (1993). Open Architecture for Computer-Aided Control Engineering. *IEEE Control Systems*, 13(2):17–27.

Didic, M. (1993). Voice: Synergy in Design and Integration. *IEEE Software*.

Eigi, M. (1990). Modelagem e Controle de Sistemas Produtivos: Aplicaçao da Teoria de Redes de Petri. Monografia no 55/90. Brasil.

Eigi M., P. (1993). Control de Sistemas a Eventos Discretos. Technical report, Sexto Escuela de Rob tica. CINVESTAV México.

Sastrón, F. (1993). STEP, una herramienta para el CIM. *Automática e instrumentaciùn*, (233).

Sepehri, M. (1987). Integrated Data Base for Computer-Integrated Manufacturing. *IEEE Circuits and Devices Magazine*.

Tuijnman, F. and Afsarmanesh, H. (1993). Management of shared data in federated cooperative PEER environment. *International Journal of Intelligent and Cooperative Information Systems (IJICIS)*, 2(4):451–473.

Walter, A. (1992). Uso de las redes de Petri generalizadas para modelar sistemas dinámicos a eventos discretos. *Revista Telegráfica Telefónica*, (938).

25

Object-Oriented Development Methodology for PLC software

O. Durán. and A. Batocchio
orlando@fem.unicamp.br batocchi@fem.unicamp.br
Depto. Eng. de Fabricação, FEM UNICAMP FAX (0192) 393722
CP 6122 CEP 13083-970 Campinas (SP) Brazil

Abstract

This paper reports the application of an Object-Oriented Development Methodology for specifying Programmable Controllers Software. This methodology, called OOST, allows the user to specify the control logic in a natural manner, using a collection of objects that represent devices and other machines in an actual manufacturing system. The object-oriented specification technique may be used in the whole control software life cycle. It could be considered as a requirement definition language or an implementation one, since the OOST may be used as an input to an automatic PLC program generation system. A brief literature review is presented and the implementation details are discussed. In the final part of this paper some future improvements and remarks are given.

Keywords

PLCs software, Object-Oriented Development, Automatic Program Generation

1 INTRODUCTION

Programmable Logic Controllers are widely used for sequential operation control, specifically in the discrete manufacturing industry. This fact is due to the advantages and improvements obtained when compared with fixed logic systems and relays-based systems. The improvements are refereed to flexibility, safes and low maintenance costs, and the reduction in start up and operation times (Cox, 1986).

The growing up success of this equipment brought a high diversity of PLCs into the shop floor. Each day more and more manufacturers enter to the market, offering new solutions, sweeping a wide range of possibilities and features. The principal features that the manufacturers outline for winning the high competitions' levels are: throughput, Input/Outputs points, programming facilities and languages, etc.

However, there is a trouble that begins to manifest when a PLC user purchases different manufacturers' equipment. Each one of these manufacturers incorporates a proprietary programming language to their products. This fact leads to serious difficulties to the user, such as: no chance for reusability, no program sharing among solutions and so on. A solution for this problem is to have a programming specialist for each one of the programming languages used into the shop floor (Halang, 1992)

Another trouble related to PLCs software development is the extended life cycle. It is very difficult pass from one phase to another in a smoothly and natural manner, mainly because of the traditional approaches do not have been realized as a manufacturing software, but as data intense applications development tools.

Manufacturing Engineers realize the manufacturing domain as composed by different entities, such as machine tools, fixtures, pieces, material handling devices, sensors, actuators, etc. Each one of these entities has associated set of attributes and capabilities constituting abstract constructions called objects. Through this abstraction process a system definition task is simplified and made in very natural manner.

This paper reports an approach to developing PLCs software using the Object Oriented Paradigm. This approach, that is being widely used in Software Engineering, is used here for automated program generation for a PLC. The automatic generation process is performed from a textual and semi-structured description of the control logic for an automated manufacturing cell (Durán, 1993).

2 BACKGROUND

There are some researches in the literature aiming at the development of manufacturing systems modeling and control software. Menga and Morisio (1989) defined a specification and prototyping language for manufacturing systems. The language is based on high-level Petri-nets and is object-oriented. The approach uses graphical and textual tools for defining the components of the system and their behaviors. The language integrates two formalism with the object-oriented development paradigm. The first formalism is the hierarchical box and arrow graphical formalism. The second one is the Petri-net graphical formalism for the detailed specification of the control flow in objects. According the authors the use of this language allows the simulation of manufacturing systems, hierarquically structuring them and enriching the library with the new objects.

Another initiative (Boucher, 1992) addresses the definition of an interface between manufacturing system design and controller design. The high-level design methodology enhances communications between manufacturing system designers and controller designers. It also allows the automatic control code generation from that design. The high-level design methodology is based in the IDEF0 methodology, created under the development of USAF, and generates a Petri Net representation of the control logic through the use of a rule-based interpreter.

A similar approach is given by Roberts and Beaumariage in (1993) that present a methodology to design and validation of supervisory control software specification. This methodology is based on a network representation schema. The networks are composed by nodes and arcs. Messages are passed between the nodes, resembling the object-message paradigm. A version of this control specification system have been coded on a Texas Instruments Explorer and other platforms.

A textual specification technique for PLC software generation was reported by Bhatnagar and Linn (1990). The generator uses a user-defined control process specification and initially converts it into a standard control logic specification and finally generates an executable task code program for the PLC specified. The bidireccionality of the code is the main virtue of this system, providing total transportability of programs from a PLC to another.

Halang and Krämer (1992) presented an interactive system with a graphical interface for constructing and validating PLC software. It combines the Function Block Diagram with a graphical language. The approach emphasizes the description of composite PLC software from a library of reusable components and may be considered as an object-oriented approach.

Finally, Joshi, Mettala and Wysk (1992) presented a paper where a systematic approach to automate the development of control software for flexible manufacturing cells called CIMGEN. The specification is based on Context Free Grammars (CFGs) providing a formal basis for control strategies descriptions. CIMGEN generates automatically control software for workstation and cell control levels. This automated code generation is not totally satisfactory, since the generated code requires hand manipulation for completion.

There are two other papers that report Object-Oriented approaches for modeling manufacturing systems. Mize, Bhuskute, Pratt and Kamath (1992) relates the results obtained in exploring alternative approaches to the modeling and simulation of complex manufacturing systems. These results argue that is necessary a paradigm shift in developing models for manufacturing systems. Through this paradigm shift the system planner can now define models through the use of building blocks, called objects. The authors assert that this approach is an strategic opportunity for the fields of industrial engineering, operations research/management science and manufacturing systems engineering. Joannis and Krieger (1992) reported an Object-Oriented methodology for specifying manufacturing systems. The specifications are made by building successive models, each containing more details than the previous one. The desired behavior of the system is described using a set of concurrent cooperating objects and the behavior of each object is defined through the use of Communicating Finite State Machines (CFSM). According to the authors these CFSM allow the execution of the specifications. The technique has a text format.

3. MANUFACTURING CELLS CONTROL

Manufacturing systems are sets of different subsystems all working together and coordinately to get the expected results. The design of these manufacturing systems and the development of manufacturing systems controllers has become more closely linked as the manufacturing environment has become more automated (Boucher, 1992). Controller design addresses issues of communication, controller logic, sequencing, error handling and programming of programmable devices.

Usually the control of a manufacturing systems is made up in different levels of abstractions. (Menga, 1989) identified four levels to perform the manufacturing control, these levels are: plant leveel, shop level, cell level and macjine level. These levels of abstractions have different time scales, event types and decision kinds. The scope of this paper is the Manufacturing Cell control and the definition of the cell controller. The cell controller is responsible for the sequencing of activities within a cell. It is necessary to define the functions related to the major operations performed for each one of the elements that make up the cell. The control at cell level is normally implemented using PLC or combinations of PLC and a cell host computer or a factory floor computer. Programming a PLC is an error prone task, where the developed programs are very difficult to read, understand and maintain. Hence there are many efforts to define a high level programming standard to simplify the program definition and the total PLC software life cycle. There are different approaches for programming a PLC. Each one of the manufacturers incorporate one or more types of these approaches to their equipment. The most common methods for programming a PLC are:

- Ladder Diagrams
- Instructions Lists
- Structured Texts
- Sequential Function Charts
- Function Block Diagrams

4 THE APPROACH

The system that is being developed allows the programmer specify a new application control program through the use of a object oriented specification technique (OOST). This specification technique may be used from the requirement description phase to the implementation phase (Booch, 1986), since it may be used as an specification input language for an automatic PLC software generator. The development of PLC software using the object oriented specification technique can shorten the life cycle phases, even some phases are no more needed, because they are contained within the previous phases (Hodge, 1992).

The methodology when applied in the requirement analysis phase allows the natural contact between the programmer/analyst and the final user of the system. Hence, the manufacturing engineer that is in charge to project the automated system can communicate his needs and requirements with the programmer/analyst using a common language, based on abstract constructions (objects), avoiding misunderstandings and excessive documentation.

To make up the specification technique, an exploratory survey among various PLC programming languages was made (Allen Bradley, 1993; Weg, 1991; Hitachi, 1994; Modicon, 1991). This survey aimed at defining a sufficient collection of basic operations supported by any PLC performing sequence control tasks, table 1 shows the set of basic operations considered to define the OOST.

Besides the instructions showed by table 1, there is a set of instructions for data manipulation and arithmetic operations. These instructions were also considered by our research and OOST supports them. Unlike of these operations OOST does not support complex instructions, such as PID functions and Fuzzy Logic-based operations, because they are out of the scope of our research. The basic operations were grouped into five categories, considering types of devices state changes. These five categories are:

- Momentarely state changes caused by the beginning or ending of an event.
- Permanent state changes caused by the beginning or ending of an event.
- State changes caused by the existence of an event during a given time.
- State changes caused by the existence of an event a given number of times.
- State changes caused and mantained during the existence of an event.

Table 1 Basic PLCs' operations that make up the OOST

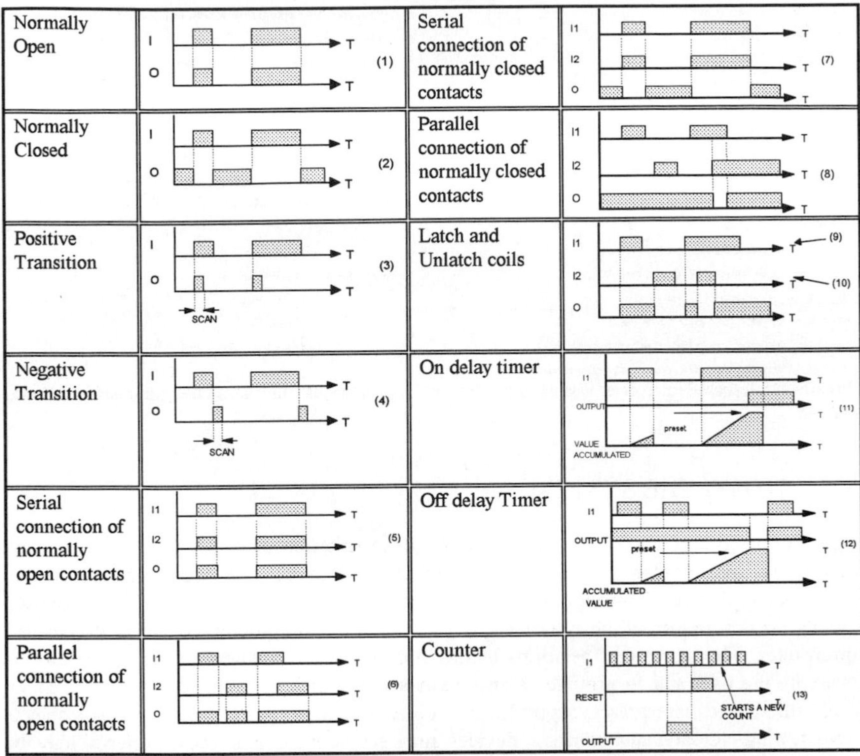

Operation	Diagram	Operation	Diagram
Normally Open	(1)	Serial connection of normally closed contacts	(7)
Normally Closed	(2)	Parallel connection of normally closed contacts	(8)
Positive Transition	(3)	Latch and Unlatch coils	(9), (10)
Negative Transition	(4)	On delay timer	(11)
Serial connection of normally open contacts	(5)	Off delay Timer	(12)
Parallel connection of normally open contacts	(6)	Counter	(13)

Next, the structures needed for describing each one of these categories was analyzed. Thus a set of textual sentences making up a description language was defined. This set of sentences was translated into a Definite Clause Grammar. Next this grammar was written using Prolog code. The following shows part of this grammar:

```
sentenças    ::= sentença, [' ; '], sentenças.
sentenças    ::= [].
sentença     ::= conjuntoA1, [','], oração_subor.
conjuntoA1   ::= conjuntoA1i.
             ::= conjuntoA1ii.
             ::= conjuntoA1iii.
conjuntoA1i  ::= conjunçãoA1i, agente, verboA1i.
conjuntoA1ii::= conjunçãoA1ii, agente, verboA1ii.
conjuntoA1iii::= conjunçãoA1iii, agente, verboA1iii.
verboA1i     ::= verbo_aux_A1ia, verbo_inf.
             ::= verbo_aux_A1ib, artigo, verbo_sustantivado.
             ::= verbo_aux_A1ic, prepos_art, verbo_sustantivado.
verboA1ii    ::= verbo_aux_A1iia,
verbo_inf.   ::= verbo_aux_A1iib, artigo, verbo_sustantivado.
```

```
               ::= verbo_aux_A1iic, prepos_art, verbo_sustantivado.
verboA1iii     ::= verbo_aux_A1iiia, verbo_inf.
               ::= verbo_aux_A1iiib, artigo, verbo_sustantivado.
               ::= verbo_aux_A1iiic, prepos_art, verbo_sustantivado.
conjunçãoA1i::= ['assim que'].
               ::= ['quando'].
               ::= ['se'].
               ::= ['sempre que'].
               ::=  ['depois que'].
               ::=  ['logo que'].
               ::= ['uma vez que'].
conjunçãoA1ii::= ['caso'].
               ::= ['desde que'].
               ::= ['contanto que'].
conjunçãoA1iii::= ['dado que'].
               ::= ['visto que'].
               ::= ['já que'].
               ::= ['uma vez que'].
sentença       ::= oração_subor, conjuntoB1.
conjuntoB1     ::= conjuntoB1i.
               ::= conjuntoB1ii.
conjuntoB1i    ::= conjunçãoB1i,  agente, verboB1i.
conjuntoB1ii::= conjunçãoB1ii, verboB1ii.
verboB1i       ::= verbo_aux_B1ia, verbo_inf.
               ::= verbo_aux_B1ib, artigo, verbo_sustantivado.
               ::= verbo_aux_B1ic, prepos_art, verbo_sustantivado.
verboB1ii      ::= verbo_aux_B1iia, verbo_inf.
               ::= verbo_aux_B1iib, artigo, verbo_sustantivado.
               ::= verbo_aux_B1iic, prepos_art, verbo_sustantivado.
conjunçãoB1i::= ['a não ser que'].
               ::= ['a menos que].
               ::= ['até que'].
conjunçãoB1ii::= ['salvo se'].
sentença       ::= conjuntoC1, [','], oração_subor.
conjuntoC1     ::= conjuntoC1i.
               ::= conjuntoC1ii.
conjuntoC1i    ::= conjunçãoC1i,agente, negação verboC1i.
conjuntoC1ii::= conjunçãoC1ii, agente, negação, verboC1ii.
verboC1i       ::= verbo_aux_C1ia, verbo_inf.
               ::= verbo_aux_C1ib, artigo, verbo_sustantivado.
               ::= verbo_aux_C1ic, prepos_art, verbo_sustantivado.
verboC1ii      ::= verbo_aux_C1iia, verbo_inf.
               ::= verbo_aux_C1iib, artigo, verbo_sustantivado.
               ::= verbo_aux_C1iic, prepos_art, verbo_sustantivado.
conjunçãoC1i::= ['contanto que'].
               ::= ['caso'].
               ::= ['desde que'].
conjunçãoC1ii::= ['dado que'].
               ::= ['visto que'].
               ::= ['já que'].
               ::= ['uma vez quue'].
```

Finally, a rule-based interpreter was written. This task was made just translating the grammar into Prolog clauses.

5 USING THE SYSTEM

The user/programmer describes any manufacturing situation using OOST and a hierachy of objects that represent a specific manufacturing domain. The manufacturing system behaviour is described through a set of textual sentences ruled by the OOST grammar. Finally the description of the system is analysed by the second module of the system called Interpreter. The interpreter performs lexical and syntactical analysis of the specifications made by the user and to provide as an output an intermediate code of the control logic. This rule-based interpreter supports the object-oriented paradigm. That is, objects representing actual manufacturing devices may be defined, and these objects may be structured in a hierarchy using the concept of superclasses and classes. Through that, a new object class or object instance can inherit attributes and capabilities of its object superclass. It is the main element of the paradigm, that allows models reusability. The intermediate code, generated by the interpreter, is used as an input for an automatic PLC program generation module. The general structure of this system is shown in figure 1.

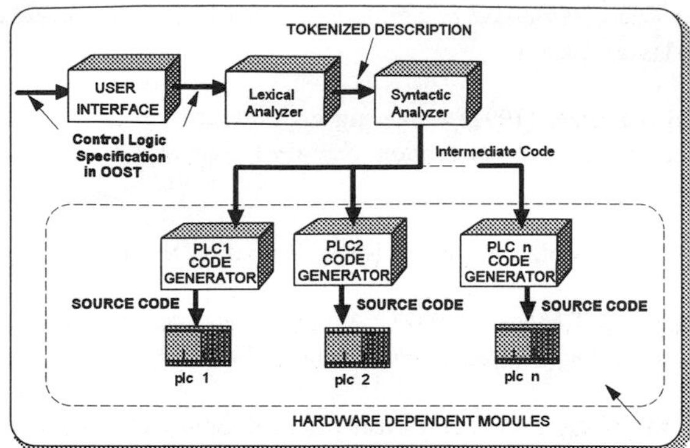

Figure 1 System Structure

The windows-based user-interface was written in Visual Basic. The other modules was written in Arity Prolog, using the concept of DCG.

6 CONCLUSIONS

This specification technique may be considered as an efficient means for performing an informal requirement description, and from it, to obtain in an automated manner the source code for an specific PLC, ready for downloading within PLC memory.

The approach allows to create object library for accelerating new applications developments and fomenting software reutilization (Shaw, 1984). This fact leads to important improvements in productivity and reliability of PLC programming tasks.

7 REFERENCES

Allen Bradley (1993) PLC-5 Programming and Operation Guide.

Bhatnagar, S. and R.L.Linn (1990) Automatic Programmable Logic Controller Program Generator with Standard Interface. *Manufacturing Review*, Vol. 3 No. 2,98-105.

Booch, G. (1986) Object Oriented Development. *IEEE Transactions on Software Engineering*, Vol. SE-12, No. 2, February, 211-221.

Boucher T.O. M.A.Jarafi (1992) Design of a Factory Floor Sequence Controller from a High-Level System Specification. *Journal of Manufacturing Systems*, Vol.11, No. 6, 401-417.

Durán, O. and A. Batocchio (1993) A high-level object-oriented programmable controller program interface. *ISIE'94, Proceedings of the IEEE International Symposium on Industrial Electronics, Santiago, Chile, May, 1993.*

Cox, B,J. (1986) *Object Oriented programming: an evolutionary approach.* Addison-Wesley Reading, Massachusetts.

Halang, W.A. and B.Krämer (1992) Achieving high integrity of process control software by graphical design and formal verification. *Software Engineering Journal*, January, 53-64.

Hitachi (1993), Programmable Controller H-200, Operation Manual.

Hodge, L.R. and M.T.Mock (1992) A proposed object-oriented development methodology. *Software Engineering Journal*, March, 119-129.

Joannis, R. and M. Krieger (1992) Object-oriented approach to the specification of manufacturing Systems. *Computers Integrated Manufacturing*, Vol.5 No. 2, 133-145.

Joshi, S.B., E.G.Mettala and R.A.Wysk (1984). CIMGEN- A Computer Aided Software Engineering Tool for Development of FMS Control Software. *IIE Transactions*. Vol.24, No.3, July, 84-97.

Menga, g. and M.Morisio (1989) Prototyping Discrete Part Manufacturing Systems. *Information and Software Technology*. Vol. 31, No.8, 429-437.

Mize, J.H., H.C.Bhuskute, D.B. Pratt and M.Kamath (1992). Modeling of Integrated Manufacturing Systems using an Object-Oriented Approach. *IIE Transactions*, Vol. 24, No. 3, 14-26.

Modicon 984 (1991) Programmable Controller Systems Manual.

Shaw, M. (1984) Abstraction Techniques in Modern Programming Languages. *IEEE Software* October, 10-26.

WEG (1991) Programmable Controller A080, Programming Manual.

Acknowledgments: This project is being sponsored by the Fundação de Amparo à Pesquisa do Estado de São Paulo (FAPESP) e pelo Fundo de Apóio ao Ensino e Pesquisa (FAEP) of UNICAMP.

Orlando Durán received his B.S. degree in Industrial Engineering from Universidad de Santiago (Chile), and his M. Sc. degree in Mechanical Engineering from Universidade Estadual of Campinas (Brazil). He is currently performing his Doctoral studies. Research areas of interests are Automation, Object Oriented Modelling and Artificial Intelligence.

Antonio Batocchio received his B.S. and M.Sc. in Mechanical Engineering from Universidade Estadual de São Paulo (São Carlos, Brazil), and his Doctor degree from Universidade Estadual de Campinas (Brazil) in 1991. Currently he is a Visitant professor at the Mechanical Engineering Department of the Minnesota State University (USA). Research areas are interests of Automation, Manufacturing Cell Technology, Simulation and Costs.

PART TEN

Multiagent Systems Architecture

26

Support for Concurrent Engineering in CIM-FACE

A. Luis Osório; L.M. Camarinha-Matos
Universidade Nova de Lisboa
Quinta da Torre - 2825 Monte Caparica - Portugal
Tel:+351-1-2953213 Fax:+351-1-2957786 E-mail {lo,cam}@uninova.pt

Abstract

A federated architecture for Concurrent Engineering is presented and special emphasis is put on the aspects of enterprise modelling and control knowledge to support the coordination of teams of experts. The prototype described combines both information integration and cooperation support functionalities. This paper addresses mainly the business plan interpreter and its multilevel supervision knowledge. Open questions and directions for further research in the context of networks of enterprises (extended or virtual enterprise) are summarized.

Keywords

Systems Integration, Concurrent Engineering, CIM, Federated Architecture, Modeling.

1 INTRODUCTION

The concept of lean / agile manufacturing is a result of the increasing globalization of the economy and openness of markets and the tough challenges this situation imposes to manufacturing companies. One of its manifestations is the recognition of the product, and thus product data, in its entire life cycle, as the main "focus of attention" in the CIM Information System. Product Data Management may be considered an essential set of tools for tracking products from conception / design to retirement / recycling. The concept of Concurrent Engineering (CE) is thus the result of the recognition of the need to integrate diversified expertise and to improve the flow of information among all "areas" involved in the product life cycle. Team work based on concurrent or simultaneous activities, potentially leads to a substantial reduction in the design-production cycle time, if compared to the traditional sequential "throw it over the wall" approach.

Evolving from earlier attempts, represented by the paradigms of "Design for Assembly / Design for Manufacturing", Concurrent Engineering is thus a consequence of the recognition that a product must be the result of many factors, including:

- -Marketing and sales factors
- -Design factors
- -Production factors
- -Usage factors (intended functionalities / requirements)
- -Destruction / recycling factors.

For all these areas there are hundreds of computer-aided tools (CAxx) on the market that help the human experts in their tasks. At a particular enterprise level various of these tools may be available, together with some proprietary software developments. A platform that supports the integration of such tools (information and knowledge sharing) as well as the interaction among their users (team work) is a computational requirement for CE.

Observing companies' evolution in terms of organization, a strong paradigm shift towards team-based structures is becoming evident. Team work, as a practical approach to integrate contributions from different experts, by opposition to more traditional hierarchical / rigid departmental structures, is being extended to all activities and not only to the engineering areas.

A realistic approach to design an architecture that supports Concurrent Engineering has to take into account results and tendencies emerging from various research sub fields of the advanced manufacturing area. The definition of a platform for Concurrent Engineering involves, in our opinion, three related sub-problems:

i) Definition of common models. This is a basic requirement in order to enable communications between members of the engineering team. The adoption of common modeling formalisms is a first requirement. Formalisms like IDEF0, NIAM, Express/ Express-G, Petri nets are being widely used. The consolidation of STEP (ISO 1991) may help in terms of product modeling, but many other aspects not covered by STEP have to be considered, like process and manufacturing resources modeling. MANDATE seems still far from offering usable results. Business Processes modeling, as proposed by CIMOSA (Esprit 1989), is also contributing to facilitate dialogue.

ii) Engineering Information Management. Definition of integrating infrastructures and information management systems able to cope with the distributed and heterogeneous nature of CIM, has been the subject of many research projects from which various approaches and prototypes have been proposed in last years (Camarinha-Matos 1991), (Camarinha-Matos 1993). Management of versions, a difficult problem in engineering data management, is even more complex when different versions may be produced / explored in parallel / concurrent way. Various centralized and decentralized solutions have been experimented, the concept of federated architectures developed and the issue of interoperability between different data management technologies and standards has been pursued (Camarinha-Matos 1994). The need for a more mature technology for Engineering Information Management, combining features from Object Oriented and Knowledge Based Systems, Concurrent / multi-agent systems, is becoming evident.

iii) Process supervision. To build a platform that supports concurrent engineering it is not enough to guarantee that the various computer-aided tools used by a team are able to communicate and share information. In other words, it is not enough to provide an integrating infrastructure and to normalize information models. Even though these aspects are essential, there is also the problem of coordination. It is necessary to establish a supervision architecture that controls or moderates the way and time schedule under which computer-aided tools (team members) access the infrastructure and modify shared information. In other words, it is necessary to model the various business processes and to implement a process interpreter or supervisor.

The platform for integration and concurrent engineering -- CIM-FACE: Federated Architecture for Concurrent Engineering -- being developed at the New University of Lisbon addresses these three issues.

2 THE CIM-FACE ARCHITECTURE

Figure 1 illustrates the main blocks of the CIM-FACE architecture. Application (computer-aided) tools are integrated via the integrating infrastructure, which provides access to the common Information System (IS).

One important part of the CIM-FACE prototype is the EIMS (Camarinha 1993), (Osório 1994) (Figure 2) subsystem which provides basic information management functionalities as well as an integrating infrastructure to support the connection of a federation of heterogeneous software tools.

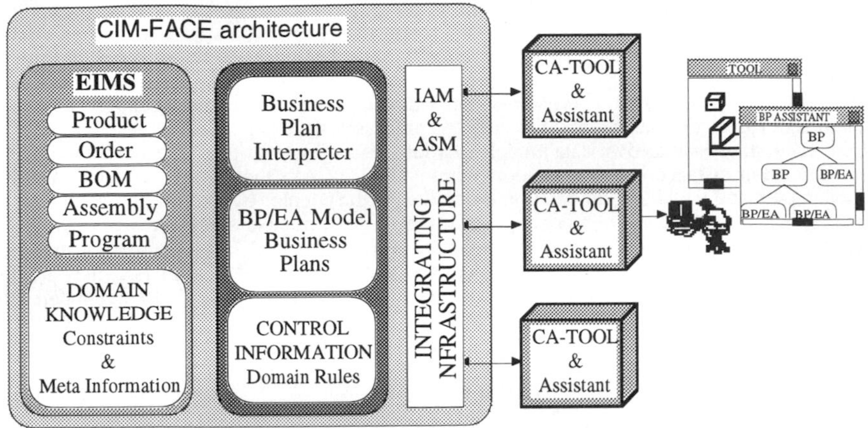

Figure 1 CIM-FACE Architecture.

The implemented EIMS prototype, inspired on the developments of STEP/ Express (Schenck 1994), is based on a hybrid and distributed programming environment supporting the connection of tools implemented in UNIX and MSDOS environments. The integrating infrastructure supports two connection modes: tight and loose connection. For tight connection a library of Information Access Methods (IAM) is linked to each tool, thus hiding the communication details (RPCs, messages format, etc.). As this part of the system was developed in an earlier stage, before the availability of the STEP SDAI specification, the IAM methods don't follow the standard, although they are quite similar, as inspired by EXPRESS concepts.

For loose connection mode, a STEP-port based on neutral files interchange is included. Reactive programming was used as an effective mechanism to implement interoperability between different data management technologies. The "visible" part of EIMS is implemented on a KBS (Knowledge Craft), but links to a RDBMS or CAD DB were established to offer object's persistency. The interoperability mechanisms play an important role in the migration of legacy systems to more advanced Engineering Information Management Systems.

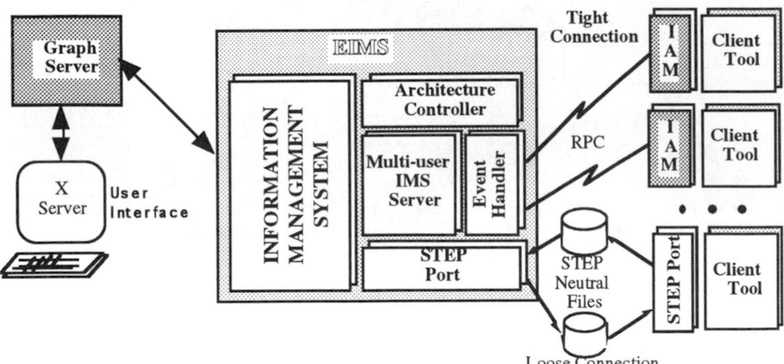

Figure 2 Engineering Information Management System.

Another major component of CIM-FACE is the Supervisor or Business Plan Interpreter (Camarinha-Matos 1995), (Osório 1994). The objective is to provide a framework that allows for a set of autonomous agents -- computer aided tools and engineers -- to cooperate in solving complex tasks.

Special tools - Business Plan Assistants - provide a front end to the human experts allowing them to take part in the execution of the business processes.

Each agent may have its own data models (partial views of the world) and its decision making capabilities. For instance, the time window within which a CAxx tool is active, interacting with the federation, and the kind of interaction is decided by the couple "CAxx tool - human user". A global supervision system -- "federal government" -- can impose some rules regarding the interaction, like refusing it if some pre-conditions are not satisfied, but it cannot consider the agents as obedient "slaves". On the other side, as agents are supposed to cooperate, they are not completely independent from each other. For instance, the actuation of a CAPP agent depends on the existence of a product model generated by a CAD agent.

3 THE ENTERPRISE FUNCTIONAL VIEW

Business plan. CIM-FACE follows the CIM-OSA approach to model the enterprise in terms of hierarchies of Business Processes and Enterprise Activities (Camarinha-Matos 1995), (Osório 1994). At each level of the hierarchy, extended Procedural Rule Sets define precedence constraints between BP/EAs of that level, as well as their starting (firing) conditions. We call this hierarchical structure business plan.

Building up business plans is a task for an enterprise modeler, Figure 3. This expert may resort to some domain knowledge, specific to the target industrial sector or to his company. Such knowledge includes a taxonomy of classes of Business Processes and a catalog of installed functionalities (Enterprise Activities, i.e., the functional model of the particular enterprise).

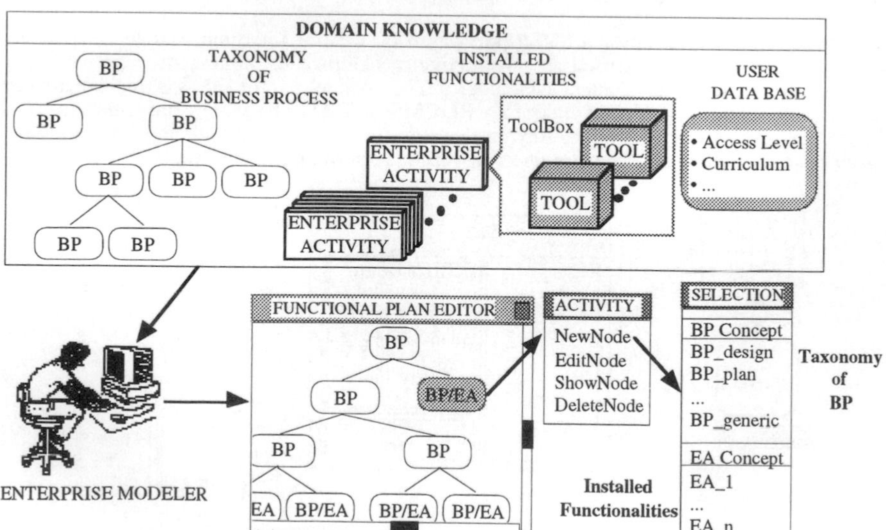

Figure 3 Edition of the enterprise's business plan.

Considering that a business plan will be performed by a team of experts (using several tools), potentially in concurrency, there are other additional aspects to be modeled. A summary of the tasks for the enterprise modeler are:

- Definition of shared information models, i.e., models common to, and that support the interactions among, team members. These models might include methods to derive particular views of the common concepts.
- Definition of business plans. This activity includes the creation of instances of the taxonomy of BP and their parametrization, definition of PRS, assignment of tools and other resources to the enterprise activities, etc..
- Characterization of the team of experts assigned to the business plan: identification and rights (namely in terms of allowed operations).

It shall be noted that these models are important for the implementation of business plan interpreter. Figure 4, shows a partial view of some concepts that support the Business Plans.

```
{{ PlanNode:
   Identifier :                              {{ BusinessProcess:
   Name :                                      is-a : PlanNode
   InputConcepts :                             SubObjectives :
   OutputConcepts :                            . . .
   PrevActivities :                          }}
   NextActivities :
   UserRules :
   StartActivity :      ; methods            {{ Enterprise-Activity:
   EndActivity :                                is-a PlanNode :
   ShowStatus :                                 ToolBox :
   ShowSubPlan :                                . . .
   SendMsg :                                 }}
   ReadMsg :
 . . . }}
```

Figure 4 Basic concepts in a business plan.

Contexts. As there might be many business plans running in parallel (supported by different teams), it is important to provide a different context to each business plan. On the other hand, even inside one team / one business plan, some members (or subgroups) might need some "privacy" while they are exploring some rough ideas and before they decide to make their results public to the team. Therefore, CIM-FACE offers the possibility to create hierarchies of contexts. All IS objects created / modified in a given context won't be visible outside this context or its children contexts. The implementation of this concept in Knowledge Craft is quite straightforward.

```
{ BusinessPlan:                  { InformationContext
    RootNode:                        BusinessPlan
    InformationContext:              ObjectList
    UserRestritions:                 SubContexts
; methods                        ; methods
    CreatePlanContext                CreateSubContext
    EnterContext                     ExportContextObjects
    ListPlanContexts                 DeleteSubContext
}                                }
```

Figure 5 Information Context concept.

When a business plan is started, a root context is associated to it. New child contexts may be created outside the current context provided the team member has enough rights (method CreateSubContext). The new context is created as a subcontext of the current one and normal inheritance rules apply, i.e., all objects visible in a parent context are available to the child context.

In this way, a team member (or subgroup) can explore his ideas without interfering with other members involved in the business plan. Once he is satisfied with his results, he can make them available to the others, i.e., export them to parent context (method ExportContextObjects).

Therefore, we can have several contexts associated (in a hierarchical way) to a business plan. The methods EnterContext and ListPlanContexts are useful to navigate through these contexts.

```
{{BP012:
Type : BP
Instance : BP_design
Name : 'Design P1'
Input :
Output : Product
PrevActivities :
NextActivities :
UserRules : all_concluded(Self),
            version(Self, less_than(5)), ...
SubObjectives : EA123, EA124, EA125
Initializer : create_instance(Product,
                   Name, NewVersion) }}
```

```
{{EA123:
Type : EA
Instance : Product_Rough_Design
Name : 'RoughDesign_P1'
Input :
Output : RoughDesign
PrevActivities :
NextActivities : EA124
UserRules : attribute_value(RoughDesign,
                 material, metal), ...
ToolBox: CAD1
Initializer : create_instance(RoughDesign,
                 Name, NewVersion) }}
```

```
{{EA124:
Type : EA
Instance : Product_Design_Analysis
Name : 'DesignAnalysis_P1'
Input : RoughDesign,
Output : DesignConstraints
PrevActivities : EA123
NextActivities : EA125
UserRules : attribute_value(RoughDesign,
                 material, metal), ...
ToolBox: PDA1
Initializer : create_instance(DesignConstraints,
                 Name, NewVersion) }}
```

```
{{EA125:
Type : EA
Instance : Product_Design
Name : 'Design_P1'
Input : RoughDesign,
        DesignConstraints
Output : Product
PrevActivities : EA124
NextActivities :
UserRules : attribute_value(Product,
                 weight, less_than(23)), ...
ToolBox: CAD1, FEA1
Initializer : }}
```

Figure 6 Examples of BP and EA concepts.

Precedence relationships among activities. The Procedural Rule Set mechanism of CIM-OSA doesn't seem quite flexible to represent the temporal interdependencies among activities in a context of Concurrent Engineering. For instance consider the activities of "Product Design" and "Process Planning". Although they can proceed with some degree of concurrency (i.e., process planning can start once a first draft of the product is made), Process Planning cannot finish before Product Design finishes. At least some details of the process plan definitely depend on the final commitment regarding the product model. In Figure 6, some rules are used to declare precedences among exemplified Enterprise Activities. Classical precedence relationships don't capture the full semantics of this temporal overlapping situations.

Therefore CIM-FACE considers rules that combine some aspects of PRS with the Allen's temporal primitives.

For example, consider the illustration of figure 7, where four activities are carried on partially in concurrency: A1- launch New Product process, A2 - Rough Design, A3 - Functional Analysis, A4 - Geometrical Design.

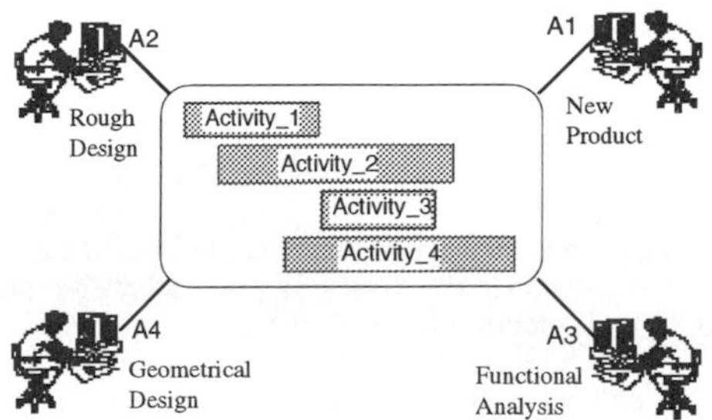

Figure 7 Relation among concurrent activities.

In this example, the following rules apply:

i) Activity A1 must start before all other activities and should finish during execution of A2
ii) Activity A2 starts after A1 and finishes after A1
iii) Activity A3 must be done during A2
iv) Activity A4 starts after A2 and finishes after A2.

The specification of this kind of constraints is very important to drive the business plan interpreter (see next chapter). Such rules are specified by the enterprise modeler.

Information views. Although the enterprise information modeling is out of the scope of this paper, for details on the CIM-FACE approaches to information integration and management in (Camarinha-Matos 1993), (Camarinha-Matos 1991) there are some aspects that are strongly related to the supervision of team - based activities.

One aspect is the concept of object view. The common IS contains the concepts / models that are shared by, or support the interaction among team members. But for some of these concepts, different team members will have different views or perspectives.

Therefore, the enterprise modeler(s) is supposed to model the different views and to associate them to the descriptors of the team members. This implementation may include methods to derive (and integrate) the various views from (into) a common model.

In some cases the views / contributions of each member / tool can be considered just as facets (slots) of a global concept (see Figure 8).

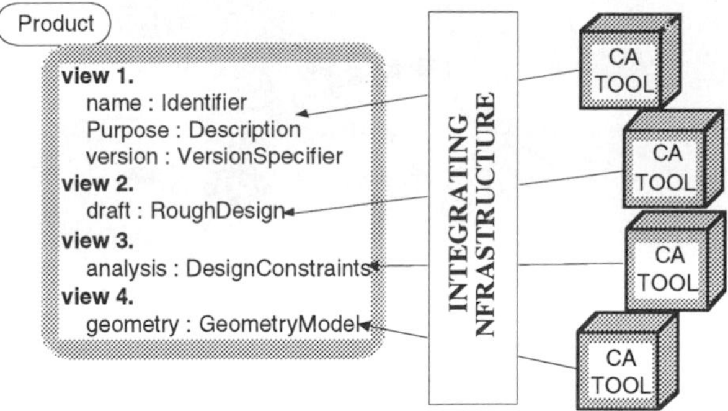

Figure 8 Partial views of an Information Object.

Therefore, optional constraints related to the facets "accessible" to each member / tool may be associated to the EA model (Input/Output slots, Figure 6) and to the user model.

4. MODEL INTERPRETER

After the enterprise modelling and, in particular, after the definition of a business plan, a model interpreter is necessary to support the operational phase.

The model interpreter is responsible for the "execution" of the various enterprise's business plans. It is responsible for:

-keeping track of the execution status of each business plan (and each node in the plan)
-supporting consistency maintenance (by verifying pre- and post-conditions for each BP/EA)
-helping or directing CA-tool selection for each EA, when more than one tool is available for an EA
-providing a common access to the enterprise information models and supporting a multicontext framework
-allowing communication among team members
-providing a platform for progressive improvement of the control structure (addition of new rules).

An important part of the execution environment is the set of CA-tools that - in cooperation with the various team experts - actually implement the EAs. A "protocol" is necessary to specify the interactions between each couple tool-user and the integrated federation. The "performer" of this protocol can be a layer separated from the applicative part of the tool. For legacy systems it is quite hard or nearly impossible to modify their control architectures. For new tools this "protocol performer" can be seen as a common script (library) that can be linked to the tool.

As a first attempt, a tandem structure (Figure 9) was implemented, separating the tool itself from the protocol performer, here called *Business Process assistant.* From the implementation point of view, this tool assistant can be a module linked to the tool or a parallel (detached) process. The second alternative is more suited to legacy systems.

With basis on this initial approach, but taking into account that:
-the "tasks" to be realized in a given business plan are hierarchically decomposed
-"real" application tools appear only associated to the leaves of this hierarchy (EAs)
-the human experts may interact with (inspect, start, finish, etc.) activities at various levels of the hierarchy.

The initial concept of tool assistant (Osório 1994) evolved to the concept of***business plan assistants*** (Camarinha-Matos 1995).

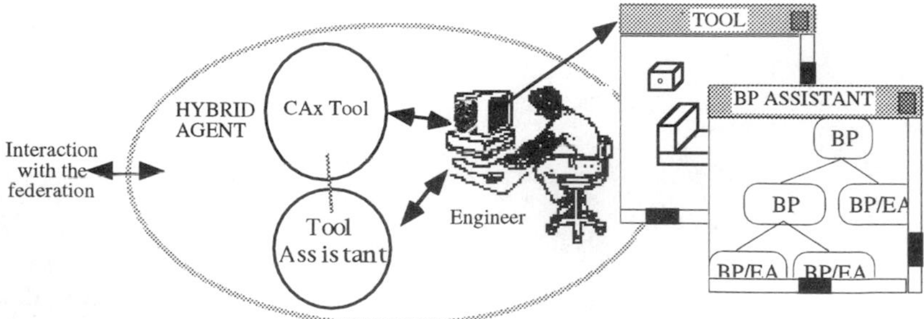

Figure 9 Business Plan Assistant.

Whenever a user "joins" the federation (logs in), a business plan assistant is launched. Through a graphical interface (Figure 10), the user has access to a specific business plan (or part of it, according to his access rights). For each node in the plan he can inspect its status, start or stop the BP/EA, create or change the default context, etc.. When a valid team member wants to enter into the CIM-FACE environment (Login) he must enter a login identification, like in classical multi-user computational platforms. This identification is used by the CIM-FACE system to condition the access to different levels of interaction and supervision rights.

As a result of login, the user is positioned in the execution context of one or more activities of the selected business plan. From that plan, the human can select, for instance, "StartActivity" for an activity not being executed (if he has enough rights). Depending on the confidence level assigned to this user, by the enterprise modeler, the interpreter can interdict the execution if some validation rule fails. When starting an EA, the associated application tools are launched.

Therefore, the human expert plays an important role in the evolution of the business plan execution.

As the control events are decided externally, by the human, how can the process interpreter be sure the decision was appropriate? For instance, lets suppose the current user of tool T_i informed the system that he finished the generation of a process plan for a given production task. Should the control system simply accept such information as a fact or should it be cautious and try to investigate the accuracy of the information? Therefore different "kinds" of control systems can be defined, ranging from a totally confident system to a cautious one.

The effectiveness of this control depends on the "quality" of the models defined in previous chapter.

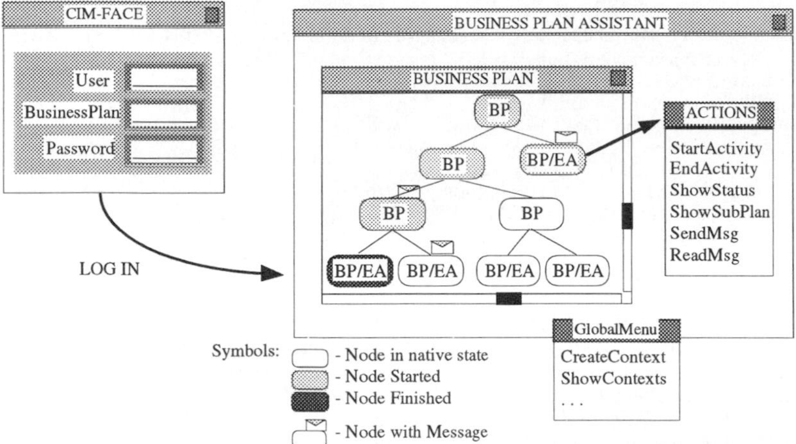

Figure 10 Business plan assistant interface.

Characterization of supervision knowledge. From the concurrent engineering point of view, the execution of an activity should be flexible enough to facilitate engineers the exploration of different alternatives or even the redefinition of a previous approach regarding a new idea received from other cooperating team member.

It is necessary however to provide the system (task for the enterprise modeler) with mechanisms to express constraints to the execution of planned activities. Taking into account the fact that the execution environment is strongly based on a distributed infrastructure, maybe even located in different enterprises, various levels of validation can be considered. Not all of these levels are direct responsibility of the plan interpreter. We think that control knowledge may be defined at three levels:

1. Information System level - constraints may be embedded into the common information models. This can be represented by local and global rules of EXPRESS. The execution of such rules is the responsibility of the Information Management System.

2. Application level. Various control aspects, due to their specificity, are better defined inside specific application software (enterprise's proprietary tools).
3. Business plan level. Rules like precedence constraints and other specific pre- and post-conditions may be associated to each BP/EA node. The business plan interpreter has mainly to take this control into account.

In some cases, and for some application domains, it will be possible to define a set of verification rules to test the validity / accuracy of each access protocol action issued by business plan assistants. In other cases that might be difficult. Therefore, our proposal is to have an architecture that can start from a level of total confidence and progress towards a more cautious system once verification rules are added to its control knowledge base.

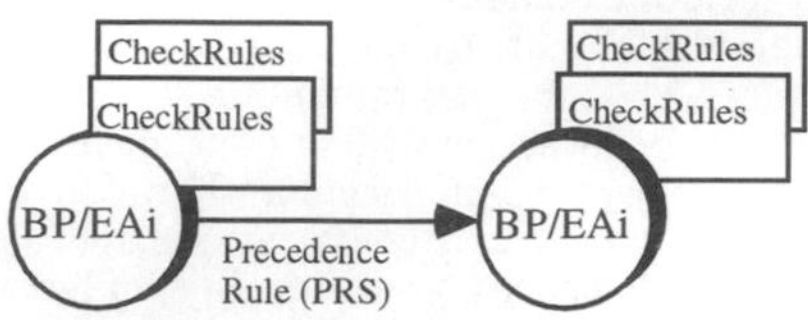

Figure 11 Check rules associated to BP/EA.

An example for a minimal checking level is a rule that implements a CIM-OSA PRS rule:

```
IF <all previous activities are finished with success>
  and <Self was requested to start>
THEN  <enable execution of Self>
```

A more specific example can be associated to the BP_Design_P1 of type BP_design:

```
IF <MarketEvaluation_p1 of type MarketEvaluation was started> and
     <BP_Design_P1 was requested to start>
THEN <enable execution of BP_Design_P1>
```

For instance the following rule could be attached to the EA124 (Figure 6):

```
IF <RoughDesign.verified of EA123 Output is true> and
     <EA123 is inactive> and <Self was requested to start>
THEN <enable its execution>.
```

It shall be noted, however, that rules associated to the nodes (BP/EA) of a business plan are related to the coordination / interaction level of the plan execution. Other kind of rule, related to information consistency (first level mentioned above), are more appropriately defined in association to the concepts present in the IS. See, for instance, the following example using Express notation. This class of rules are supposed to be checked by the Information Management System and not by the Business Plan Interpreter.

```
ENTITY manufacturing_step;
  identification : IDENTIFIER;
  manpower_time : INTEGER;
  machine_time : INTEGER;
  cycle_time : INTEGER;
  . . .
WHERE
  rule_1: cycle_time >= machine_time;
  rule_2: cycle_time >= manpower_time;
END_ENTITY;
```

Messages inter team members. As various BP/EA nodes may be active (in parallel), i.e., "operated" by different humans, possibly in different geographical locations, and a change in one may affect the progress of others, it is important to provide a message exchange mechanism. This can be seen as a particular e-mail facility associated to the plan nodes. From one BP/EA node, the human agent can send a message to another BP/EA (or to a group). See (Figure 10). At current stage, these messages are intended for communication among the human experts, but we may even think of messages to be processes by high level (autonomous) process execution. In this way, if an agent modifies the model of a part, besides making the new model available in a common context, he can notify the agent performing the process plan, which may need to revise his plan.

5 CONCLUSIONS AND FUTURE DEVELOPMENTS

CIM-FACE represents a prototype federated architecture to support systems integration and concurrent engineering activities.

The implemented parts include:

i) an infrastructure for information integration and management that was successfully evaluated in the context of an European Esprit (CIM-PLATO) project.

ii) A process interpreter that provides basic functionalities for concurrent execution and coordination of a set of business processes / enterprise activities, driven by a team of human experts. The proposed architecture supports the definition of multiple checking levels, allowing progressive degrees of robustness.

An intensive evaluation of this control strategy in a real application has still to be done. CIM-FACE is not a finished system, but an ongoing research.

A future extension of this work has to do with the concept of extended or virtual enterprise. There is a tendency to establish partnership links between companies, namely between big companies and networks of components' suppliers. Similar agreements are being established between companies and universities. Such network structures may be seen as extended or virtual enterprises. In fact, the manufacturing process is not anymore carried on by a single enterprise, but each enterprise is just a node that adds some value (a step in the manufacturing chain).

This tendency creates new scenarios and technologic challenges, specially to Small and Medium Enterprises (SMEs). Under classical scenarios, these SMEs would have big difficulties -- due to their limited human and material resources -- to have access to or to use state of the art technology. Such partnerships facilitate the access to new technologies and new work methodologies but, at the same time, impose the use of standards and new quality requirements.

The efforts being put on the implantation of high speed networks (digital high ways), supporting multimedia information, open new opportunities for team work in multi-enterprise / multi-site networks. But this new scenario also brings new requirements in terms of control: access rights to the information, scheduling of access, control of interactions, etc. (distributed information logistics infrastructure).

While in recent past the emphasis was put on enterprise integration, now the challenge is to provide tools that support an effective management, engineering and control of production in a framework where each node has a large autonomy, operates under different philosophy, and sometimes has competing objectives with respect to other nodes.

A main aspect is the flows between nodes and the monitoring of their evolution. Several aspects may be considered:

i. Inter enterprise engineering developments
iii Efficiency of orders flow
iii. Follow up of orders evolution
iv. Distributed and dynamic scheduling
v. Incomplete and imprecise orders
vi. Network-wide workload optimization

The extension of the federated approach of CIM-FACE to this scenario of network of enterprises is our next plan.

6 ACKNOWLEDGMENTS

The work here described received partial support from the European Community -- the Esprit CIM-PLATO, ECLA CIMIS.net and FlexSys projects -- and from the Portuguese Agency for Scientific and Technologic Research (JNICT) -- the CIM-CASE project.

7 REFERENCES

Camarinha-Matos, L.; Sastron , F. (1991)- Information Integration for CIM planning tools, CAPE'91 - 4th IFIP Conference on Computer Applications in Production and Engineering, Bordeaux, 10-12 Sep, 1991.

Camarinha-Matos, L.M.; Osório, A. L. (1993) CIM Information Management System: An Express-based integration platform, IFAC Workshop on CIM in Processes and Manufacturing Industries, Espoo, Finland - published by Pergamon Press.

Camarinha-Matos, L.M.; Osorio, A.L. (1994) An integrated Platform for Concurrent Engineering, Proc.s 3rd CIMIS.net Workshop on Distributed Information Systems for CIM, Florianopolis, Brazil. To appear in the Journal of the Brazilian Society of Mechanical Science.

Camarinha-Matos, L.M.; Afsarmanesh, H. (1994) Federated Information Systems in Manufacturing, Proceedings of EURISCON'94, Malaga, Spain.

Camarinha-Matos, L.M.; Osório, L. (1995) CIM-FACE: A Federated Architecture for Concurrent Engineering, CAPE'95, IFIP International Conference on Computer Applications in Production and Engineering, Beijing, China, May 1995.

Esprit Consortium AMICE (1989) Open System Architecture for CIM, Springer-Verlag.

Osório A. Luis; Camarinha-Matos, LM. (1994). Information based control architecture, Proceedings of the IFIP Intern. Conference Towards World Class Manufacturing, Phoenix, USA, Sep 93, edited by Elsevier - North Holland.

Schenck, Douglas; Wilson, Peter (1994), Information Modeling the EXPRESS Way, Oxforf University Press, New York.

STEP, ISO (1991). Reference Manual, ISO/TC 184 /SC4.

Welz, B. G. et al. (1993). A toolbox of integrated planning tools - a case study, IFIP Workshop on Interfaces in Industrial Systems for Production and Engineering, Darmstadt, Germany, 15-17 Mar 1993.

8 BIOGRAPHY

Eng. A. Luis Osório received his BSc on Electronic and Telecommunication from Polytechnic Institute of Lisbon, ISEL and Computer Engineering degree on Computer Science, from the New University of Lisbon. Currently he is adjunct professor at the Electronic and Communications Engineering Department of the ISEL. He is also researcher and PhD student in the group of Robotic Systems and CIM of the UNINOVA's Center for Intelligent Robotic. His main research interests are: Information based integration and control of CIM systems, Intelligent Manufacturing Systems, Artificial Intelligence.

Dr. Luis M. Camarinha-Matos received his Computer Engineering degree and PhD on Computer Science, topic Robotics and CIM, from the New University of Lisbon. Currently he is auxiliar professor (eq. associate professor) at the Electrical Engineering Department of the New University of Lisbon and leads the group of Robotic Systems and CIM of the Uninova's Center for Intelligent Robotics. His main research interests are: CIM systems integration, Intelligent Manufacturing Systems, Machine Learning in Robotics.

27

A Federated Cooperation Architecture for Expert Systems Involved in Layout Optimization *

H. Afsarmanesh, M. Wiedijk, L.O. Hertzberger
University of Amsterdam, The Netherlands
email: hamideh@fwi.uva.nl, wiedijk@fwi.uva.nl

F.J. Negreiros Gomes, A. Provedel, R.C. Martins, E.O.T. Salles
Universidade Federal do Espírito Santo, Vitória-ES - Brazil
email: negreiro@inf.ufes.br

Abstract

In shoe and handbag industry, there are many CIM activities that can at best be represented by expert systems. One example of such an activity is the optimization system that supports the cutting of regular/irregular shapes out of uniform/non-uniform base leather material. This paper focuses on and describes a layout optimization environment with five different kinds of agents, where every agent is an intelligent expert system. In order to perform the necessary tasks in the layout optimization activity, the involved agents need to cooperate, and share and exchange their information both with each other and with other agents involved in other CIM activities. In general, an agent representing a CIM subactivity can be represented either by an intelligent system, a simple file system, or an information management system. A natural architecture to represent such agents' cooperation and information sharing is a federated/distributed network of interrelated agents. In this paper we first address and describe the PEER federated architecture and then present a PEER implementation of the agents involved in the layout optimization system and the architecture to support their cooperation.

Keywords

Federated Databases, Integrated CIM Activities, Optimization Expert Systems

1 INTRODUCTION

To support the concurrent engineering approach, different CIM activities must cooperate and exchange information. For instance, the layout optimization in shoe and handbag industry requires to access the information from the design and production planning activities. The optimization expert systems addressed in this paper represent a Decision Support System based on the Visual Interactive Modeling that is further described in (Afsarmanesh et al., 1994c; Angehrn et al., 1990; Bell, 1991; Negreiros et al., 1993). This system aims to optimize the

*The research described here has been partially supported by the CIMIS.net project ECLA 004:76102.

layout of a set of moulds to be cut on a piece of raw material. To gain the best (optimized) layout, the system must be integrated with other CIM subsystems such as CAD, CAQC, PPC, and with a numeric control machine that executes the actual cutting. One approach that can support the representation of CIM environment activities and support both their inter-activity and intra-activity cooperations is to apply a federated/distributed information management architecture. Using this approach, every CIM activity is first divided into a number of sub activities and represented by several agents, where an agent is an intelligent system, a simple file system, or an information management system. Then, the sharing of information among the agents is established through integration (Import/Export mechanism). The PEER federated system used in this paper provides such integration architecture.

The intelligent optimization system described in this paper is an approach to cutting of layouts of different shapes out of the base leather material (Afsarmanesh et al., 1994c). The shapes can be either regular (rectangular) or irregular (geometry represented by polygons). Also, the base leather can itself be a uniform rectangular piece or a non-uniform partitioned surface with some defects. Several expert systems however can be defined to support all possible situations involved in cutting of layouts. In this paper, three expert systems, namely OS-IU, OS-IN, and OS-RU are defined to support different possible cases of layout optimizations. Two other kinds of expert systems are also defined to perform other closely related tasks in this application environment. Namely, there is a need for a surface inspection system, called SI, to determine the uniformity of the base leather material, and there is a need for user interface systems, called UI, to support the interactions among the human operators and the optimization system.

In this paper, first the PEER federated information management system is applied to this application to support the definition of these distinct expert systems as PEER agents. Then, a PEER federation of cooperative agents is defined that interrelates OSs, SI, and several UIs agents. Finally, the interrelations among agents are established using the SDDL integration language of PEER. The resulting network will support the access to the optimization system from any UI agent defined in the network, and will automatically support the distributed query processing on optimization data.

The PEER system (Afsarmanesh et al., 1993; Tuijnman and Afsarmanesh, 1993; Afsarmanesh et al., 1994a) described in this paper is an object-oriented federated/distributed database system designed and implemented at the University of Amsterdam. It primarily supports the complex information management requirements set by the industrial automation application environments. The research described in (Afsarmanesh et al., 1994b) describes some concurrent engineering requirements and the specific PEER capabilities to satisfy them. The federated architecture of PEER introduces an *integration facility* to support the cooperation and information sharing of autonomous CIM agents with heterogeneous data representation (Afsarmanesh et al., 1993). To better support users of the integration facility and for high level access to data and meta-data, two powerful and user-friendly interface tools are developed (Afsarmanesh et al., 1994a). The Schema Manipulation Tool and the Database Browsing Tool are both window-oriented and implemented using X-windows on SUN workstations. These interface tools support users with their access, retrieval and modification of both data and meta-data in PEER agents. More details and examples on the two tools are presented in (Wiedijk and Afsarmanesh, 1994; Afsarmanesh et al., 1994a). A prototype implementation of the PEER federated system is developed in the C language that runs on UNIX, on a network of SUN workstations.

The remaining of this paper is organized as follows. A brief description of the PEER's federated architecture is provided in Section 2. Section 3 first addresses the description of the realization of the optimization system agents, the user interface agent, and the surface inspection agent of the layout optimization environment. Then, the PEER implementation of these agents and their integration is presented.

2 PEER FEDERATED COOPERATION ARCHITECTURE

PEER is a federated object-oriented information management system that primarily supports the sharing and exchange of information among cooperating autonomous and heterogeneous nodes (Afsarmanesh et al., 1993; Tuijnman and Afsarmanesh, 1993; Wiedijk and Afsarmanesh, 1994; Afsarmanesh et al., 1994a). The PEER federated architecture consists of a network of tightly/loosely interrelated nodes. Both the *information* and the *control* are distributed within the network. PEER does not define a single global schema on the shared information to support the entire network of database systems, unlike many other distributed object-oriented database systems (Kim et al., 1991). The interdependencies between two nodes' information are established through the schemas defined on their information; thus there is no need to store the data redundantly in different nodes.

2.1 Distributed Schema Management

Every agent is represented by several schemas; a local (LOC) schema, several import (IMPs) schemas and several export (EXPs) schemas and an integrated (INT) schema (Afsarmanesh et al., 1993). The *local* schema is the schema that models the data stored locally. The various *import* schemas model the information that is accessible from other databases. An *export* schema models some information that this database wishes to make accessible to other databases. Usually, an agent defines several export schemas. The *integrated* schema presents a coherent view on all accessible local and remote information. The integrated schema can be interpreted as one user's global classification of objects that are classified differently by the schemas in other databases.

The 'local' schema is the private schema defined on the physical data in an agent. Derived from the local schema are 'export' schemas that each define a particular view on some local objects. Export schemas restructure and represent several related concepts in one schema. For every export schema, the exporter agent manages the information on agents who can access it, by keeping their agent-id, their access rights on the exported information, and the agreed schema modification conditions to notify the agents who use this export schema of its changes. To obtain access to data in another agent's export schema, an agent has to input it as 'import' schema. An import schema in one agent is the same as an export schema in another agent.

Originally, an agent's 'integrated' schema is derived from its local schema and various import schemas. In later stages of integration, instead of the local schema, the previous integrated schema will be used as base. At the level of the integrated schema, the physical distribution of information becomes hidden, and the contributing agents are no longer directly visible to the end user. Different agents can establish different correspondences between their own schema and other agents' schemas, and thus there is no single global schema for the network, unlike other "federated" database systems, such as in ORION2 (Kim et al., 1991), that define one global schema to support the entire network of agents.

The schemas for an agent are defined using the PEER Schema Definition and Derivation Language (SDDL) (Afsarmanesh et al., 1993). SDDL includes facilities for defining schemas, types, and maps, and for specifying the derivation among a derived schema and a number of base schemas using a set of type and map derivation primitives. An integrated type is constructed from other types by the union, restrict, and subtract primitives. Map integration is accomplished by specifying a map as either a union of other maps, or as a threaded map, or some combination of these two primitives.

2.2 SDDL Schema Integration/Derivation Primitives

The following schema derivation/integration primitives define the relationships between the types and maps in the derived (or integrated) schema and the types in the base schemas (called the base types). For example, to derive the agent's integrated schema for the first time, the base schemas are the local schema and the import schemas. Now first, the integrated types must be defined from their base types to appropriately classify all objects in the integrated schema. The next step is to use map derivation expressions to define necessary mappings for each derived type, using the base mappings originally defined on the base types.

A derived type is defined by a type derivation expression. Below, the semantics of the type derivation primitives are provided, where every 'T_i' stands for a "type-name@schema-name", 'I(T_i)' represents the set of all members of type T_i, 'a' represents a member object, and the 'restriction(a)=TRUE' defined on members of type T_i that checks if 'a' satisfies the restriction.

1. Type Rename: T = T_1
   interprets as: a $\in$ I(T) $\Leftrightarrow$ a $\in$ I(T_1), where T is the derived type and T_1 is a type in a base schema S_1.
2. Type Union: T = **union** $(T_1, \ldots, T_n)$
   interprets as: a $\in$ I(T) $\Leftrightarrow$ a $\in$ I(T_j), where 1 <= j <= n, T is the derived type and T_j is a type in a base schema S_j.
3. Type Subtract: T = **subtract** (T_1, T_2)
   interprets as: a $\in$ I(T) $\Leftrightarrow$ a $\in$ I(T_1) $\wedge$ a $\notin$ I(T_2), where T is the derived type and T_1 and T_2 are types from base schemas S_1 and S_2. Due to the support of multiple inheritance in PEER, T_2 does not have to be a subtype of T_1.
4. Type Restrict: T = **restrict** (T1, restriction)
   interprets as: a $\in$ I(T) $\Leftrightarrow$ a $\in$ I(T1) $\wedge$ restriction(a)=TRUE, where T is the derived type and T_1 is a type in a base schema S_1.

A derived mapping is defined by a map derivation expression. Semantics of the primitives defined for map integration are provided below, where the union primitive is represented by '{ , }', the treading primitive is represented by '.', every 'm_i' in the following definitions stands for a "mapping-name@schema-name" and 'a.m' stands for the range-object (value) related to object 'a' by mapping 'm'.

1. Map Rename: m = m_1
   interprets as: a.m = b $\Leftrightarrow$ $a.m_1$ = b, where m is a mapping of type T in the derived schema and m_1 is a mapping of type T_1 in a base schema S_1, and where T and T_1 are not disjoint.
2. Map Union: m = { m_1,...,m_n }
   interprets as: a.m = b $\Leftrightarrow$ $a.m_i$ = b, where m is a mapping of type T in the derived schema, 1 <= i <= n and m_i is a mapping of type T_i in schema S_i, and where for all i, 1 <= i <= n, T and T_i are not disjoint.
3. Map Threading: m = $m_1.m_2. \ldots .m_n$

interprets as: a.m = k $\Leftrightarrow a.m_1 = b \wedge b.m_2 = c \wedge ... \wedge j.m_n$ = k, where m is a mapping of type T in the derived schema, and m_i is a mapping of type T_i in a base schema S_i, and where T and T_1 may not be disjoint.

In the examples in section 3.2 the need for these primitives are illustrated. For instance, Example 1 in that section needs to apply the type and map rename and the type restrict, and Example 2 needs the type and map union primitives.

3 INTEGRATION OF OPTIMIZATION EXPERT SYSTEMS

This section describes the integration of optimization expert systems by a network of cooperating agents that together optimize the layout of moulds on leather plates. Each agent performs a specific task for which it usually needs to share and exchange information with other agents in the network. Below we describe the five kinds of agents that are involved in the optimization activity. Figure 1 shows the network of agents. A User Interface (UI) agent is the agent that

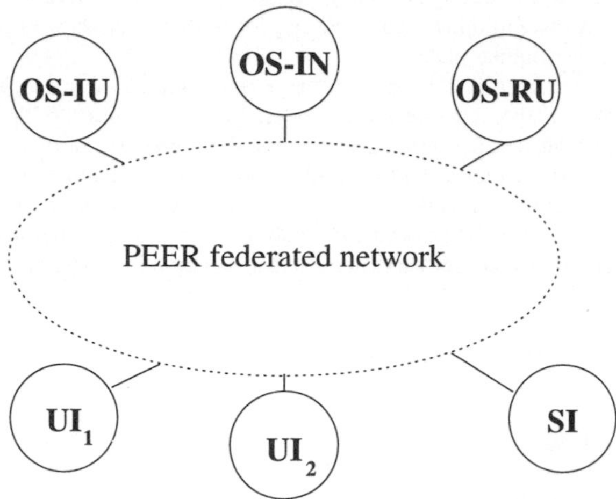

UI User Interface Agent
SI Surface Inspection Agent
OS-IU Layout optimization of irregular pieces on a uniform rectangular leather plate
OS-IN Layout optimization of irregular pieces on an irregular plate
OS-RU Layout optimization of regular pieces on an uniform rectangular plate

Figure 1 The federated network of agents involved in the optimization of layouts

performs the interaction with the human operator and interfaces to the other agents that each perform a part of the optimization. The definition of a UI agent is based on the Visual Interactive Modeling (VIM) paradigm. This subsystem is based on the idea of Modeling by Example (MbE) as a means of enhancing cooperation between user and the system. The UI supports decision making while taking into consideration four important dimensions of: End-user (decision-maker) modeling, Modeling as a concrete, visual and incremental process, Reactive system behavior, and Supplying active support (act as a consultant) (Afsarmanesh et al., 1994c). The Surface Inspection (SI) agent inspects the leather plates. Its function is the determination of the geometry of a plate and the detection and classification of defects (Salles et al., 1994). There are three specific optimization systems (OS-IU, OS-IN, and OS-RU) that optimize the layout of moulds on a leather plate. An optimization expert system consists of two processes: *problem processing* and *knowledge processing*. The problem processing performs the central control of the solution search process, while receiving the basic algorithms from the knowledge processing and coupling it with the interactive requirements of the user. The OS agents contain both the knowledge to perform the heuristic search, and the production rules to represent the problem (Afsarmanesh et al., 1994c). Each optimization system is specialized in the optimization of a specific combination of a mould category and plate category. The moulds can be classified as regular or irregular. The plates can be classified as uniform rectangular or irregular. Optimizer Agent OS-IU optimizes the irregular cutting on a uniform surface, it determines the allocation of irregular pieces on a uniform rectangular plate. Optimizer Agent OS-IN optimizes the irregular cutting for non-uniform surfaces by allocating irregular pieces on an irregular plate. Optimizer Agent OS-RU optimizes guillotine regular cutting for uniform surfaces by allocating regular pieces on a uniform rectangular plate.

Although it may seem that there is only one agent of each kind, there may be more than one instance of a kind of agent. For instance, the network in Figure 1 defines two User Interface agents (UI1, UI2) that support different users. Their functionalities and the information representation are in principle the same, but they differ for instance in the layout instantiations that they manage.

Section 3.1 describes the information that is managed by each agent in this network and contains a PEER schema that defines that information. In section 3.2 the integration of these agents and the information that they share and exchange is described using the PEER integration architecture.

3.1 Agent information definitions

This section describes the schema definitions for the agents in the layout optimization environment network. Please notice that in the following definitions // represents comments. The following base entities are defined in the local schema of every agent in the network. For simplicity reasons we define these base entities here once, so that they do not need to be repeated in the schema definition of every agent.

```
type MEASURE_UNIT subtype_of INTEGERS
type DEGREE_UNIT subtype_of REALS
type CLASSIFICATION subtype_of STRINGS
type WASTE subtype of REALS
type POLYGON_COMPONENT
  element: POINT
  next: POLYGON
type POLYGON subtype_of POLYGON_COMPONENT
type POINT
  x: MEASURE_UNIT
  y: MEASURE_UNIT
```

A measure unit represents a length as an integer value in a specified measure. A degree unit represents a rotation as a real value. Waste is an integer value that specifies the percentage of leather plate material that is not used for one of the pieces. The geometry of plates, mould and defects are represented by polygons. These polygons are modeled as a closed circular linked list in counter clockwise orientation. PEER supports lists through the generic recursion abstraction as described in (Afsarmanesh and McLeod, 1989). Within th recursive definition the element mapping represents a point of the polygon and the next mapping represents the next element in the list. Points are represented by their x and y coordinates.

User Interface agent UI

The User Interface (UI) agent has four main tasks. The first task is to collect the information about leather plates that are used for the cutting of the leather using moulds. The inspection of these plates is done by the Surface Inspection agent (SI). The second task is the interaction with the user that indicates which moulds and how many of each mould have to be cut from a particular leather plate. The third task is to determine which Optimization System (OS-IU, OS-IN, or OS-RU) has to be used to optimize the layout. The decision on the kind of OS is made depending on the kind of plate and kind of moulds that are used for the layout. The fourth task is to display the resulting layouts that are generated by the OSs. The UI locally manages the information about leather plates and moulds. The local schema (LOC) definition is shown below. Section 3.2 describes how the UI agent integrates the inspection information from the SI agent and the resulting layouts from the OSs with its local information.

```
define_schema LOC
  type PLATE
  type UNIFORM_RECTANGULAR_PLATE subtype_of PLATE
    plength: MEASURE_UNIT
    pwidth: MEASURE_UNIT
  type IRREGULAR_PLATE subtype_of PLATE
    partitions: IRREGULAR_PLATE_PARTITION
  type IRREGULAR_PLATE_PARTITION
    partition_geometry: POLYGON
    associated_defects: POLYGON
    partition_classification: CLASSIFICATION
  type MOULD
    mdemand: INTEGERS
  type REGULAR_MOULD subtype_of MOULD
    mlength: MEASURE_UNIT
    mwidth: MEASURE_UNIT
    reference_point: POINT
    permitted_rotation: DEGREE_UNIT
  type IRREGULAR_MOULD subtype_of MOULD
    mgeometry: POLYGON
    permitted_rotations: DEGREE_UNIT
    mclassification: CLASSIFICATION
end_schema LOC
```

The UI agent classifies the leather plates as uniform rectangular and irregular. The uniform rectangular plate is represented by its length and width. Irregular plates are represented by partitions, where each partition is represented by the geometry of the partition (a polygon), the associated defects (a set of polygons) and a classification of the partition. The moulds are classified as regular or irregular. A regular mould is represented by its length and width and the number of times that the piece (mould) must be cut (demand). The irregular mould is represented by its geometry (polygon), demand, permitted rotations, a reference point that is defined by the lowest left corner of the mould (usually the origin), and a classification. For instance a specific irregular mould instance imould23 is represented as imould23(mdemand=3, mgeometry=polygon45,permittedrotations=30,210, mclassification="good") with polygon instance polygon45 represents the geometry of that mould.

Surface Inspection agent SI

The Surface Inspection (SI) agent inspects the leather plates. Within the SI agent however, the leather plates are referred to as surface entities. The task of the SI agent is to determine of the surface geometry, and the detection and classification of defects. The local schema LOC of the SI agent is defined as follows.

```
define_schema LOC
  type INSPECTED_SURFACE
    surface_geometry: POLYGON
    associated_defects: DEFECT
  type DEFECT
    defect_geometry: POLYGON
    defect_classification: STRINGS
end_schema LOC
```

The inspected surfaces are represented by their geometry and the associated defects, where the defects themselves are represented by their geometry and classification.

The Optimization Systems OSs

The Optimization Systems generate an optimized layout for locating the pieces (moulds) on a plate. Following are the integrated schema definitions of the information used by each OS agent. Section 3.2 describes the integration among the UI agent and the OS agents to support the sharing and exchange of information in detail. Here we give a brief description of the integration. Each OS agent imports the plate and mould information from the UI agent. After the layout optimization is completed by the OS agent, it stores the resulting layouts with the allocation of the moulds on the plate locally and exports them to the UI agent. In the remaining of this section the three optimization systems are defined by their integrated schemas. For simplicity reasons, the local schemas of the OS agents are not presented since they all only include the layout and mould-allocation types similar to what is included in their respective INT schema.

Optimization System OS-IU: Irregular Cutting for Uniform Surface

The Optimization System OS-IU generates an optimized layout for a uniform rectangular plate and a set of irregular pieces. Following is the integrated schema of the OS-IU agent.

```
derive_schema INT from LOC, IMP-from-UI
  type PLATE
    plate_length: MEASURE_UNIT
    plate_width: MEASURE_UNIT
  type MOULD
    mould_demand: INTEGERS
    mould_geometry: POLYGON
    permitted_rotations: DEGREE_UNIT
  type LAYOUT     // irregular pieces, rectangular plate
    allocation: MOULD_ALLOCATION
    total_waste: WASTE
  type MOULD_ALLOCATION
    mould_id: MOULD
    dx: MEASURE_UNIT
    dy: MEASURE_UNIT
    theta: DEGREE_UNIT
end_schema INT
```

A layout on a rectangular plate is represented by the allocation of the moulds on the plate and the total waste of material. A mould allocation is represented by a X-axis shift (dx) and Y-axis shift (dy) of the mould (identified by mould_id) and the rotation (theta).

Optimization System OS-IN: Irregular Cutting for Non Uniform Surface

The Optimization System OS-IN generates an optimized layout for an irregular plate and a set of irregular pieces. Following is the integrated schema of the OS-IN agent.

```
derive_schema INT from LOC, IMP-from-UI
  type PLATE
    partitions: PLATE_PARTITION
  type PLATE_PARTITION
    partition_geometry: POLYGON
    associated_defects: POLYGON
    partition_classification: CLASSIFICATION
  type MOULD
    mould_demand: INTEGERS
    mould_geometry: POLYGON
    permitted_rotations: DEGREE_UNIT
    mould_classification: CLASSIFICATION
  type LAYOUT    // irregular pieces, irregular plate
    partition_allocation: PARTITION_ALLOCATION
    total_waste: WASTE
  type PARTITION_ALLOCATION
    partition_id: PLATE_PARTITION
    allocation_on_partition: MOULD_ALLOCATION
    partition_waste: WASTE
  type MOULD_ALLOCATION
    mould_id: MOULD
    dx: MEASURE_UNIT
    dy: MEASURE_UNIT
    theta: DEGREE_UNIT
end_schema INT
```

A layout on an irregular plate is represented by partition allocations and the total waste of material. A partition allocation is similar to the mould allocation on a rectangular plate, except that the waste is represented for the partition only.

Optimization System OS-RU: Guillotine Regular Cutting for Uniform Surface

The Optimization System OS-RU generates an optimized layout for a uniform rectangular plate and a set of regular pieces. Following is the integrated schema of the OS-RU agent.

```
derive_schema INT from LOC, IMP-from-UI
  type PLATE
    plate_length: MEASURE_UNIT
    plate_width: MEASURE_UNIT
  type MOULD
    mould_demand: INTEGERS
    mould_length: MEASURE_UNIT
    mould_width: MEASURE_UNIT
    permitted_rotation: DEGREE_UNIT
  type LAYOUT    // regular pieces, rectangular plate
    allocation: MOULD_ALLOCATION
    total_waste: WASTE
  type MOULD_ALLOCATION
    mould_id: MOULD
    dx: MEASURE_UNIT
    dy: MEASURE_UNIT
```

```
    theta: DEGREE_UNIT
end_schema INT
```

The layout representation here is the same as described for the OS-IU agent

3.2 Agent Integration

This section describes two examples of the integration among the UI agent and the OS agents to support the sharing and exchange of information among these agents.

Example 1: OS-IU agent integration with UI information

The first example describes the integration of the mould and plate information of the UI agent within the OS-IU agent. The OS-IU agent needs to import the irregular mould and uniform regular plate information from the UI agent. Thus, the UI agent defines the EXP-for-OS-IU export schema for this purpose as follows:

```
derive_schema EXP-for-OS-IU from_schema LOC
  type UNIFORM_RECTANGULAR_PLATE
    plength: MEASURE_UNIT
    pwidth: MEASURE_UNIT
  type MOULD
    mdemand: INTEGERS
  type IRREGULAR_MOULD subtype_of MOULD
    mgeometry: POLYGON
    permitted_rotations: DEGREE_UNIT
    mclassification: CLASSIFICATION
derivation_specification
  UNIFORM_RECTANGULAR_PLATE = UNIFORM_RECTANGULAR_PLATE@LOC
    plength = plength@LOC
    pwidth = pwidth@LOC
  IRREGULAR_MOULD = IRREGULAR_MOULD
    mdemand = mdemand@LOC
    mgeometry = mgeometry@LOC
    permitted_rotations = permitted_rotations@LOC
    mclassification = mclassification@LOC
end_schema EXP-for-OS-IU
```

Export schemas can only specify a subset of the local information. Here, the only difference between this export schema and the local schema LOC of the UI agent is omitting the generalization hierarchy for the irregular mould. The reason for this omission is that there is only one specific kind of mould defined here. The derivation specification simply specifies that the instances of uniform rectangular plate and the irregular mould are the same as in the local schema.

The OS-IU agent imports this export schema EXP-for-OS-IU as its own import schema IMP-from-UI. In its integrated schema the OS-IU agent defines the following derivation specifications. The UI agent classifies the plate and mould information as described in the derivation specification above. The OS-IU agent only knows about one kind of plate, namely the uniform rectangular, so it renames it to "plate" and similarly renames the irregular mould into "mould". The OS-IU agent defines the length and width of plates with prefix plate_ instead of prefix p, and similarly with prefix mould instead of prefix m_. The derivation renames these mappings accordingly. The value of the mclassification mapping, that is defined in the import schema IMP-from-UI, is used in the integrated schema of OS-IU as a restriction to define the MOULD type. Namely, only the irregular-moulds for which the mclassification is "good" is of interest to this agent. Furthermore, mclassification is of no importance to the OS-IU agent as a mapping, and so it is simply discarded. Since it is not defined as a mapping in the INT schema of the OS-IU, no derivation specification is necessary to define it here. Following is a part of the derivation specification for the integrated schema (INT) of the OS-IU agent:

```
PLATE = UNIFORM_RECTANGULAR_PLATE@IMP-from-UI
  plate_length = plength@IMP-from-UI
  plate_width = pwidth@IMP-from-UI
MOULD = restrict(IRREGULAR_MOULD@IMP-from-UI,[mclassification@IMP-from-UI="good"])
  mould_demand = mdemand@IMP-from-UI
  mould_geometry = mgeometry@IMP-from-UI
  permitted_rotations = permitted_rotations@IMP-from-UI
LAYOUT = LAYOUT@LOC
  ...
MOULD_ALLOCATION = MOULD_ALLOCATION@LOC
  ...
```

Example 2: UI agent integration with OSs information

The UI agent needs to import the optimized layouts information from the OS agents. Every OS agent defines an export schema for its layout information. These schemas that are imported by the UI agent are defined as follows. Please notice that for simplicity reasons the following schemas are not fully defined. Also, the mould allocation entity is not fully represented here with its details, since it is exactly the same in each OS agent and its definition can be found in the OS agent definitions.

OS-IU

```
define_schema IMP-from-OS-IU same_as_schema EXP-for-UI from_agent OS-IU
  type LAYOUT    // irregular pieces, rectangular plate
    allocation: MOULD_ALLOCATION
    total_waste: WASTE
  type MOULD_ALLOCATION
    ...
end_schema IMP-from-OS-IU
```

OS-IN

```
define_schema IMP-from-OS-IN same_as_schema EXP-for-UI from_agent OS-IN
  type LAYOUT    // irregular pieces, irregular plate
    partition_allocation: PARTITION_ALLOCATION
    total_waste: WASTE
  type PARTITION_ALLOCATION
    partition_id: PLATE_PARTITION
    allocation_on_partition: MOULD_ALLOCATION
    partition_waste: WASTE
  type MOULD_ALLOCATION
    ...
end_schema IMP-from-OS-IN
```

OS-RU

```
define_schema IMP-from-OS-RU same_as_schema EXP-for-UI from_agent OS-RU
  type LAYOUT    // regular pieces, rectangular plate
    allocation: MOULD_ALLOCATION
    total_waste: WASTE
  type MOULD_ALLOCATION
```

```
   ...
end_schema IMP-from-OS-RU
```

The following (extended) integrated schema of the UI agent shows how the layout information imported from the OS agents is integrated by the UI agent. Every OS agent only knows about one kind of layout, that it is the one that it generates. So, it is simply named as LAYOUT. The UI agent however, has to manage three kinds of layouts, one for each of the OS agents. Therefore, it defines a generalization hierarchy to manage these layouts. The different layout entities are extended with RM or IM for regular-moulds or irregular-moulds, and with RP or IP for rectangular-plate or irregular-plate. In the derivation specification below the layouts are specified according to their origin. For instance, the LAYOUT_IM_IP specifies the instances of the LAYOUT entity of agent OS-IN, while the layouts on the rectangular plates are specified as the union of the layouts of the OS-IU and OS-RU agents.

```
derive_schema INT from_schema LOC, IMP-from-OS-IU, IMP-from-OS-IN, IMP-from-OS-RU
  // local type definitions of moulds, plates, etc. will appear here ...

  type LAYOUT
    total_waste: WASTE
  type LAYOUT_RP subtype_of LAYOUT
    allocation: MOULD_ALLOCATION
  type LAYOUT_IM_RP subtype_of LAYOUT_RP  // irregular pieces, rectangular plate
  type LAYOUT_RM_RP subtype_of LAYOUT_RP  // regular pieces, rectangular plate
  type LAYOUT_IM_IP subtype_of LAYOUT     // irregular pieces, irregular plate
    partition_allocation: PARTITION_ALLOCATION
  type PARTITION_ALLOCATION
    ...
  type MOULD_ALLOCATION
    ...
derivation_specification
  LAYOUT = union(LAYOUT@IMP-from-OS-IU, LAYOUT@IMP-from-OS-IN,
               LAYOUT@IMP-from-OS-RU)
    total_waste = {total_waste@IMP-from-OS-IU, total_waste@IMP-from-OS-IN,
                 total_waste@IMP-from-OS-RU}
  LAYOUT_RP = union(LAYOUT@IMP-from-OS-IU, LAYOUT@IMP-from-OS-RU)
    allocation = {allocation@IMP-from-OS-IU, allocation@IMP-from-OS-RU}
  LAYOUT_IM_RP = LAYOUT@IMP-from-OS-IU
    partition_allocation = partition_allocation@IMP-from-OS-IN
  LAYOUT_RM_RP = LAYOUT@IMP-from-OS-RU
  LAYOUT_IM_IP = LAYOUT@IMP-from-OS-IN
  PARTITION_ALLOCATION = PARTITION_ALLOCATION@IMP-from-OS-IN
    ...
  MOULD_ALLOCATION = union(MOULD_ALLOCATION@IMP-from-OS-IU, MOULD_ALLOCATION@IMP-from-OS-IN,
                         MOULD_ALLOCATION@IMP-from-OS-RU)
    ...
```

The inspected surface information can be integrated by the UI similar to the layout information described and represented above, except that it is a matter of 'renaming' the import types instead of taking the union of types from several import schemas

3.3 Example application

Once the agents involved in this CIM activity are defined and their integration relations are established, agents can cooperate and simply share and exchange their information. This section describes the typical use of the resulting environment. First, we briefly summarize the needs for information sharing and integration among the involved agents. The UI agent needs to import the surface information (which is exported) by the SI agent. The OS agents needs to import the plate and mould information which is managed locally and exported by the UI agent. The UI agent also needs to import the optimized layouts which are managed and exported by the OS agents. The UI agent then integrates the imported information from SI and OS with its local information and creates its integrated schema. Through the UI's integrated schema, the human operator can simply (and transparently) access all the information that he/she needs and thus optimize the layouts. First, the operator chooses a surface that can be used as the base for the allocation of moulds, using the techniques described earlier in this paper. When the layout is defined, the operator selects an optimization agent that can optimize the designed layout. Later, when the optimization is accomplished successfully the operator can access the resulting layout.

4 CONCLUSIONS

In this paper we have described the implementation of an integration architecture for expert systems involved in a layout optimization environment using the PEER federated information management system. First, the detailed description of PEER agents for five specific expert systems involved in this application are provided. Then the step by step integration of the shared information among these agents are described. Further, it was shown that a human operator using a User Interface agent can transparently share and exchange information with the Surface Inspection agent and the three distinct Optimization Systems, while the agents could still retain their own local information representations. The complexity of such a federated architecture is an example of a balanced automated system in which the human interaction plays an important role in the leather cutting layout design.

5 REFERENCES

Afsarmanesh, H. and McLeod, D. (1989). The 3DIS: An Extensible Object-Oriented Information Management Environment. *ACM Transaction on Information Systems*, 7:339–377.

Afsarmanesh, H., Tuijnman, F., Wiedijk, M., and Hertzberger, L. (1993). Distributed Schema Management in a Cooperation Network of Autonomous Agents. In *Proceedings of the 4th IEEE International Conference on "Database and Expert Systems Applications DEXA'93"*, Lecture Notes in Computer Science (LNCS) 720, pages 565–576. Springer Verlag.

Afsarmanesh, H., Wiedijk, M., and Hertzberger, L. (1994a). Flexible and Dynamic Integration of Multiple Information Bases. In *Proceedings of the 5th IEEE International Conference on "Database and Expert Systems Applications DEXA'94", Athens, Greece*, Lecture Notes in Computer Science (LNCS) 856, pages 744–753. Springer Verlag.

Afsarmanesh, H., Wiedijk, M., Moreira, N., and Ferreira, A. (1994b). Design of a Distributed Database for a Concurrent Engineering Environment. In *Proceedings of the ECLA.CIM workshop, Florianopolis, Brazil*, pages 35–43. (to appear in the "Journal of the Brazilian Society of Mechanical Sciences", ISSN 0100-7386).

Afsarmanesh, H., Wiedijk, M., Negreiros, F., Lopes, R., and Martins, R. (1994c). Integration of Optimization Expert Systems with a CIM Distributed Database System. In *Proceedings of the ECLA.CIM workshop, Florianopolis, Brazil.* (to appear in the "Journal of the Brazilian Society of Mechanical Sciences", ISSN 0100-7386).

Alvarenga, A., Negreiros, F., Provedel, A., Sastron, F., and Arnalte, S. (1993). Integration of an irregular cutting system into cim – part i; information flows. In *Proceedings of ECLA-CIM'93*.

Angehrn, A., Arnoldi, M., Löthi, H.-J., and Ackermann, D. (1990). *A Context-oriented Approach for Decision Support*. Lecture Notes in Computer Science (LNCS), 439. Springer-Verlag.

Bell, P. (1991). Visual interactive modelling: The past, the present, and the prospects. *European Journal of Operational Research*, 54.

Kim, W., Ballou, N., Garza, J., and Woelk, D. (1991). A Distributed Object-Oriented Database System Supporting Shared and Private Databases. *ACM Transaction on Information Systems*, 9(1):31–51.

Negreiros, F., Alvarenga, A., Lorenzoni, L., Camarinha-Matos, L., and Pinheiro-Pita, H. (1993). Object dynamic behavior for graphical interfaces in cim. In *Proceedings of ECLA-CIM'93*.

Nilsson, N. (1984). *Principles of Artificial Intelligence*. Springer-Verlag.

Pinheiro-Pita, H. and Camarinha-Matos, L. (1993). Comportamento de objetos activos na interface gráfica cim-case. In *4º PPP/AC*.

Salles, E., F.J.Negreiros-Gomes, and G.H.Brasil (1994). Segmentação de Texturas utilizando Operadores de Convoluçãoe Redes Neurais. In *Anais do I Congresso Brasileiro de Redes Neurais, Itajubá-MG Brasil*, pages 189–194.

Tuijnman, F. and Afsarmanesh, H. (1993). Management of shared data in federated cooperative PEER environment. *International Journal of Intelligent and Cooperative Information Systems (IJICIS)*, 2(4):451–473.

Wiedijk, M. and Afsarmanesh, H. (1994). The PEER User Interface Tools Manual. Technical Report CS-94-15, Dept. of Computer Systems, University of Amsterdam.

28

A holistic approach to intelligent automated control

I. Alarcon, P. Gomez, M. Campos, J.A. Aguilar, S. Romero and P. Serrahíma
Instituto de Ingenieria del Conocimiento. Mod. C-XVI, p.4, UAM, Cantoblanco, E-28049 Madrid (Spain). Tel: +34 1 397 3973. Fax: +34 1 397 3972.
e-mail: idoia@helena.iic.uam.es

P. T. Breuer
Departamento de Ingenieria de Sistemas Telematicos, E.T.S.I.T., Universidad Politecnica de Madrid, Ciudad Universitaria, E-28040 Madrid (Spain)
e-mail: ptb@dit.upm.es

Abstract

Our belief is that industrial problems require holistic solutions if AI based control is to become an everyday reality. We have designed and implemented a software architecture which supports heterogenous AI subsystems and which is now controlling a working plant. The support architecture is associated with (1) a design methodology and (2) an underlying conceptual model of an integrated system. In the latter, a reactive system is seen as consisting of layers consisting of interacting elements called 'basic control tasks'. Designing a control system to this model requires the analyst to consider the plant, its environment and it components as a whole.

Keywords

Integration, Artificial Intelligence, industrial control.

1. INTRODUCTION

Better automation might be the key to the improvement of industrial competitiveness, but, in most industrial plants, full automation is not considered possible because intelligent decisions in a larger setting play an essential part in the managerial and control process.

The favoured approach is to rely on human overseers for intelligent input that can take account of more broadly-based considerations than may be programmed into conventional automated control designs. Although some instances of installations with AI based control subsystems do exist (Dubas, 1990), (Larsen, 1980), (Sofge and White, 1990), our belief is that industrial control problems require holistic solutions. They should involve the cooperative application of at least several AI techniques to be successful, and different techniques are appropriate in different parts of the management and control process. The problem of optimizing the behaviour of an industrial plant and its environment as a combined entity is inherently complex and multi-faceted. Seen from this viewpoint, the lack of a common framework and a software architecture for the support of AI-based industrial control designs must have contributed to their lack of penetration in the industrial arena.

This paper describes a framework that has being developed and is being used at the Knowledge Engineering Institute (IIC) for the support of heterogeneous intelligent systems applied to the management and control of whole industrial installations. Section 2 describes the underlying conceptual model. Section 3 describes the basic repeating unit that comprises instances of the model and Section 4 sets out a methodology for design under the framework. Section 5 gives some details of the software used to provide the supporting architecture for the installed systems.

2. CONCEPTUAL FRAMEWORK

The approach presented in this paper is founded on the framework provided by a general conceptual model of intelligent control (Alarcon et al,, 1993). The emphasis here is on the optimization of production taking into account a holistic measure of plant performance (including, for example, considerations of environmental safety, maintainability, etc.).

The framework formalizes the model. It provides a language in which the tasks performed by the control and production team of an industrial plant may be described. Like the underlying model, it is grounded in real daily problems and situations, and hooks for the actions required to solve or deal with them are part of the framework.

The conceptual model represents a problem oriented approach. Problems, together with their dynamics and related control actions are at the heart of the model. Problems are associated in the model with a set of activities and control actions. The latter have to be performed in a specific order and under tightly determined conditions in order to solve the problem. Here, problem has a general sense. It means any situation that can be modified (normally enhanced) by the actions performed by the plant staff. Some examples are: detection of failures in the process, improvement of the quality of the final product, treatment of the environmental impact, etc.

(In the approach proposed here, an entire group of such problem-oriented activities is analyzed and represented during the design of an intelligent control system.)

Our approach assumes that one isolated technique cannot solve the entire control problem and that more than one technique must be applied depending on the type of problem and the nature of the data/knowledge/information available. Because it is presently lacking, we have proposed a conceptual framework that facilitates cooperation among techniques in a natural way.

We have taken into account a particular aspect of process and management control in devising the model: namely, the hierarchical organization of the human staff and their industrial role. The conceptual model and the framework it supports are divided into three layers wherein all problems and tasks find a niche (see figure 1). These layers are, in order of lowest to highest:

Operational layer

This covers all reactive tasks, i.e. those problems or situations normally carried out by one single operator that require a quick action. A task in this layer usually concerns a single piece of equipment, a single input and a single output.

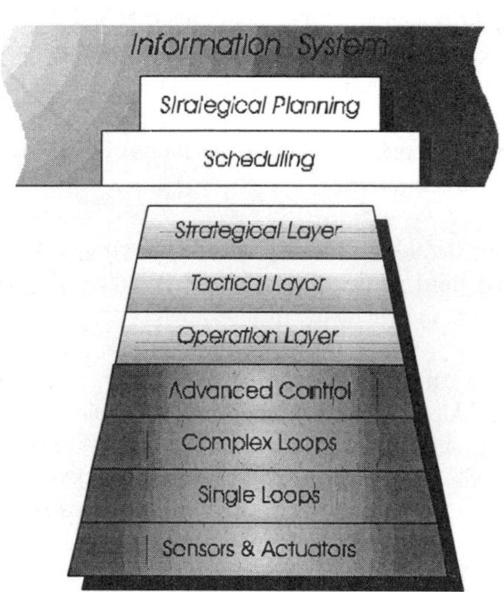

Figure 1 Control hierarchy.

Tactical Layer

This covers more complex problems that need to be followed up until successful resolved and more than one control action might have to be carried out. Generally, however, the events treated at this level are local in extent, i.e. they affect a specific subsystem. A subsystem is defined as the set of one or more pieces of equipment and associated transfer lines. A piece of equipment is a basic unit of the plant. Given a specific problem or situation, a tactic is defined as a set of control actions that can be performed sequentially or simultaneously and which are aimed at resolving the situation. The intelligent system dynamically creates a tactic tailored to the problem and the current situation and starts recommending or performing it.

The actions that compose the tactic will be those that are defined as possible to solve this particular problem and which are not being performed or already recommended. After a certain time - needed to reconnoitre the effect of the action - it is decided whether to change the tactic or to continue with it. Account is taken of the evolving situation and the alternative sequences of possibly applicable actions in order to arrive at a prediction. This procedure is iterated until the problem is solved.

Strategic Layer

This layer has a global view. It informs lower levels about the overall strategy that is being followed or the one to be followed next. It also decides when it is necessary to change the overall strategy, taking into account the state of the plant and the global objectives. Recall that the global objectives are minimization of a chosen measure of performance subject to constraints on ancillary aspects. It may also decide when to change particular tactics aimed at particular subsystems.

The responsibility of this layer is to optimize the whole production task, in particular those objectives related to production, maintenance, quality, efficiency and safety. The categories of measured quantities that are considered by this layer are detailed below:

• production quantity;
• quality (this category covers both internal and external criteria of quality and, in particular, external constraints related to contracts and sales);
• efficiency (structural and staff costs are not considered, but process costs such as energy and materials are);
• maintenance costs (including evaluations of ongoing wear and predicted downtime);
• safety considerations.

The latter includes all safety criteria affecting equipment, the whole installation, the plant staff, and third parties that might be affected. All aspects related to the pollution and the environment in general are included here.

These five categories are considered to form a network of constrained variables, as seen in figure 2. All or most of the constraints relate only variables in different categories. In optimization, the focus is on one or two of them only, but all the constraints must be considered. The network of constraints is solved by means of a rule based system and constrained optimization techniques.

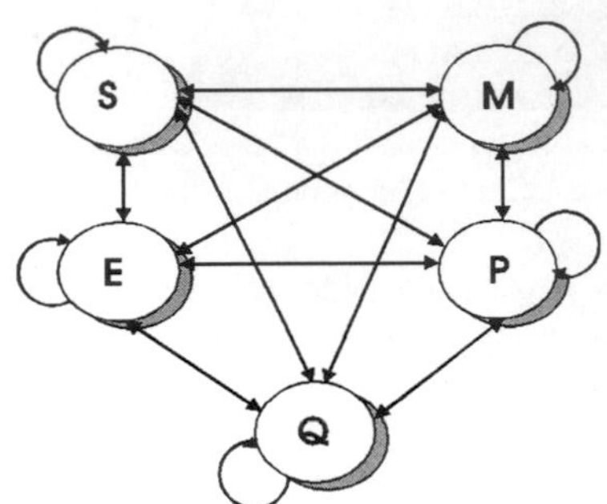

Figure 2 Plant global objectives

Also, because global strategic objectives may be inherently difficult to define, express or measure accurately, a computationally tractable representation does duty for them. Each objective is represented by means of a so-called logistic table. The table assigns quantities (which may be discrete classifications such as 'good', or 'warm') to the variables which comprise the measure of the objective. For example, the elements of the logistic table related to production are: assessments of availability of different products and materials, hard statistics, financial measures and production rates, and external, physical and time constraints.

A logistic table is built for each one of the strategic objectives considered. These tables are related in the sense that the same underlying variable may be reflected in more than one table. Entries in the table may be constructed by fuzzy reasoning from contributing quantities or qualities. Note that it is always the case that each overall strategic objective will be represented in the table. Both current values and boundary value (the limit that must never be surpassed) appear.

A fuzzy value is assigned to the level of fulfillment of the strategic objectives and a rule-based method reasons with these fuzzy values in order to suggest a possible action.

3. BASIC ELEMENTS IN THE CONCEPTUAL MODEL

Each layer of the conceptual model is constructed from the same generic classes of object, specialized and organized in different ways. The objects are called basic control tasks.

A basic control task is oriented towards solving a unique goal with a particular set of activities and persists in a specific layer of the process control hierarchy for a specific time.

The basic control task has a consistent structure through each of the layers in the control hierarchy. The concept lends a problem oriented flavour to our approach - allowing ourselves to give to the word 'problem' a broader meaning than normal.

Another important aspect of the concept is that it serves, in the software architecture built upon the conceptual model, as the vehicle of cooperation among differing technologies. A basic control task is composed of many (so-called) slots. Each of them represents a stage in the evolution of a problem and a place where one or more technologies can make their contributions to its solution, perhaps using results derived earlier in the lifetime of the task.

Basic control tasks constitute a common paradigm for knowledge acquisition, conceptualization, design and actuation. The same conceptual tools thus cover all these phases and this can facilitate development of the application.

4. DESIGN METHOD

Based on the conceptual framework described above, a methodology has been defined. It aims at providing the guidelines and support necessary for the design of applications to solve a large range of process control problems by means of the integration of heterogeneous AI techniques (Alarcon et al., 1994).

The general procedure advised by the methodology is a vertical approach. It starts at the top of the hierarchy described in the previous section and proceeds downwards through (1) an analysis and (2) a design sketch for the different layers. Then the procedure is to (3) make a pass bottom-up, defining in detail the tasks and data objects of the operational, tactical and strategic layers. A layer can start to be designed in detail once the underlying layer has been analyzed. In this way, the proper functioning of a layer can be seen to be independent of higher layers.

The first step in the analysis of the process is top-down knowledge acquisition. Identification and characterization of the problems to be tackled is carried out here. Knowledge acquisition starts with a general idea of the global objectives to be included in the strategic layer. The recommended acquisition technique for this step is the open and structured interview. Once the strategic goals of the industrial installation and their priority are known in outline, the situations and activities corresponding to the tactical layer are studied to better understand the process and its problems. Then, the problems related to the operational layer are identified and analyzed.

The second step is the sketch of the intelligent layers, that is the identification of local control cycles according to the dynamics of the process, the nature of the control tasks and the response time. The third phase, the core of the procedure, is carried out for each layer following a bottom-up approach, and consists of the following main steps:

Exhaustive knowledge acquisition and identification of problems.
The goal is the definition of the problems to be handled by the intelligent system that will by constituted by the basic control task objects. Amongst the generally applicable acquisition techniques (Hart, 1992), the ones favoured by our methodology are interviews, questionnaires, thinking-aloud and critical incident elucidation.

Assignment of identified problems to the correct layer of the model.
This takes into account the level of complexity of the related tasks, actors involved and required response time.

Definition of simple data objects.
The parameters and both measured and calculated variables are defined.

Definition and design of basic control task objects
Our methodology sets out several procedural guidelines with the aim of achieving the following goals in sequence:

• Definition of the basic control task's objective according to the situation or problem to be solved.
• Identification of the basic control task's type. We advise classifying the basic control tasks into one of four predefined types: preventive, corrective, optimizing and modal.
• Identification of the basic control task's class. A list of the most common tasks in process control and manufacturing is predefined and available on a per control layer and basic control task type basis, although new classes may be created by the knowledge engineer carrying through this method.
• Definition of the basic control task's slots. Once the basic control task has been defined and its layer, type and class are known, the set of its slots are defined on a per control layer and type basis. This step also defines the start and end condition for the task.
• Identification of which AI techniques are needed by the basic control task in question. Within each layer and basic control task type, for each one of the established classes, the technique that is appropriate to each slot is predefined. If more than one technique may be applied, the selection criteria depends on the type of knowledge/data/information available. The techniques generally considered are rule based systems, neural networks, fuzzy logic, model based reasoning and genetic algorithms.

Once the basic control tasks are defined and designed for the three layers, the identification of interactions among basic control tasks has to be investigated. Following upon this analysis, new basic control tasks may be created. The basic control tasks to be defined at this stage include (a) those modifying the status of basic control tasks allocated in lower layers; (b) those constituting a basic control task's previously defined slot and (c) basic control tasks generating a fire condition of other basic control task.

The next step is the implementation of the system.

5. IMPLEMENTATION

A software architecture has been implemented that supports the concepts presented in previous sections. It allows the interaction and cooperation of several heterogeneous techniques in order to solve industrial control problems (de Pablo et al., 1992).

This architecture follows the blackboard paradigm (Engelmore et al., 1988), (Jagannathan et al., 1989), (Carver and Lesser, 1992), in which a centralized and active data structure is the only means of communication among modules (knowledge sources). It serves as the vehicle for cooperative problem solving, and is responsible for data coherence inside the whole system. This architecture presents a control strategy that, without losing flexibility at the design phase, is efficient.

The blackboard contains objects representing the domain information that knowledge sources will share. It includes, for example, the status of the plant, results from numerical models of parts of the plant, predictions, diagnosis of problems, etc. The blackboard may also be coerced into forcing synchronizations in order to preserve coherence. The 'basic control task' object has been integrated into the blackboard mechanism by using the dependencies expressed within the detailed definitions of its slots to generate update and data flow instructions.

Knowledge sources ultimately update and use the information in the blackboard. In particular, knowledge sources are responsible for helping basic control tasks progress by contributing information ad services to their active phases.

6. CONCLUSIONS

The approach presented in this paper has been succesfully carried through in a full-scale trial at a refinery in Cartagena (Spain). Different kinds of AI modules were developed for the different kinds of problems encountered, as shown in the architecture presented in figure 3. In particular, (a) neural networks, which are able to carry out the tasks of sensor validation, forecasting, and subsystem modelling, being able to give an alternative value when the sensor credibility is very low; (b) fuzzy logic, focussing on the sensor validation and signal filtering activities; (c) model-based reasoning, dealing with diagnosis and data sensor consistency checking activities; and (d) expert systems, applied when heuristic knowledge is available and, at present, covering corrective, preventive and optimizing tasks.

Because of the generality of the approach, it is now being applied to targets within manufacturing industry with very promising results.

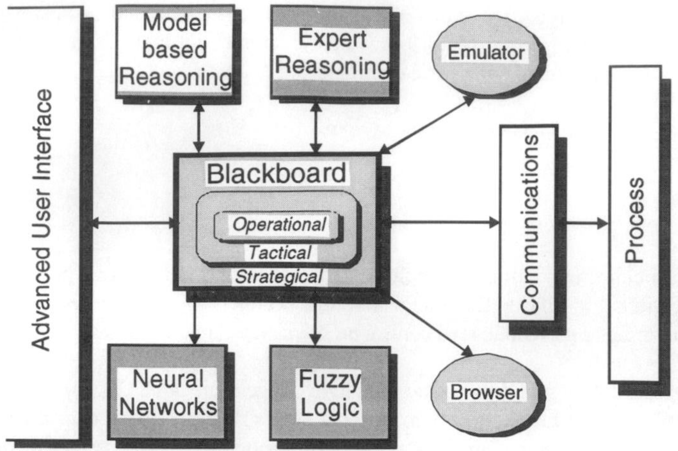

Figure 3 Experimented architecture.

7. REFERENCES

Alarcón, I., Zaccagnini, J.L., Rodriguez, P., and de Pablo, E. (1993) A Conceptual Framework for the Integration in Intelligent Control Systems *Proceedings of the IRTICS workshop*, Madrid, 05/1-05/15.

Alarcón, M.I., Rodríguez, P., Almeida, L.B., Sanz, R., Fontaine, L., G∞mez, P., AlamÉn, X., Nordin, P., Bejder, H. and de Pablo, E. (1994) Heterogeneous Integration Architecture for Intelligent Control *Intelligent Systems Engineering Journal*.

Carver, N. and Lesser, V. (1994) Evolution of Blackboard Control Architectures, *Expert Systems with Applications, An International Journal, USA, Volume 7, Number 1*, 1-30.

Dubas, M. (1990) Expert Sytems in Industrial Practice: Advantages and Drawbacks *Expert Systems, Vol. 7, No. 3,* 150-156.

Engelmore, R. and Morgan, T. (1988) *Blackboard Systems* Addison-Wesley, California.

Hart, A. (1992) *Knowledge Acquisition for Expert Systems* McGraw-Hill, Inc.

Hayes-Roth, B. (1994) Integrating Real-Time AI Techniques in Adaptative Intelligent Agents *Preprints of Symposium on Artificial Intelligence in Real Time Control (AIRTC'94)*, Ed. A. Crespo, Valencia, Spain, October, 3-14.

Jagannathan, V., Dodhiwala, R. and Baum, L.S. (1989) *Blackboard Architectures and Applications* Academic Press, New York.

Larsen, P.M. (1980) Industrial Applications of Fuzzy Logic Control *International Journal Man-Machine Studies, Vol. 12*, 3-10.

de Pablo, E., Rodriguez, P., Alarcón, I. and Alamán X. (1992) Integrating Multiple Technologies in Real-Time Intelligent Systems *ECAI'92 Workshop on Advances in Real-Time Expert System Technologies*, Vienna.

Rao M. (1992) *Integrated Systems for Intelligent Control* Springer-Verlag, Berlin.

Sofge D. and White D. (1990) Neural Network based Process Optimization and Control *Proceedings of the 29th. IEEE Conference on Decision and Control*, Honolulu.

PART ELEVEN

Enterprise Modeling and Organization II

29

Reference Models for an Object Oriented Design of Production Activity Control Systems

B. Grabot, P. Huguet°**
**Laboratoire Génie de Production, ENIT,*
Avenue d'Azereix-BP 1629, 65016 Tarbes Cedex FRANCE
Phone: (33) 62 44 27 21 Fax: (33) 62 44 27 08
°AEROSPATIALE, Centre de Recherche Louis Blériot
12, rue Pasteur-BP 76, 92152 Suresnes Cedex FRANCE
Phone: (33) 1 46 97 35 22 Fax: (33) 1 46 97 32 59

Abstract

The reference models of high level analysis and design methods for manufacturing systems, such as CIM-OSA or GRAI, provide a good support at the conceptual and organisational levels, but do not help in to integrate existing software, or to develop and reuse code. Object Oriented methods are good candidate for the improvement or integration of the analysis, specification and development phases of manufacturing systems: they are recognised as providing an efficient support in order to shorten the code development phases, and to develop modular and reusable software. Nevertheless, most of the applications developped in the past years focus on very low level control structure design, and there is a lack in supporting tools for complex production activity control architectures. We suggest in this paper to define modular reference models of PAC modules in order to provide a framework for adaptable PAC systems design. The prototype of support system which development is now in progress intends to be a real "workstation" for the PAC systems designers, in order to better take into account important factors such as reuse of models and code, and integration of existing software.

Key words

Production Activity Control, reference models, Object Oriented Analysis

1 INTRODUCTION

The Production Activity Control (PAC) system of a manufacturing system has to manage the short term planning, execution and monitoring activities required to process manufacturing orders on the physical resources of the workshop. Manufacturing systems have to react more and more quickly, efficiently and at low cost to changes in the production context, either internal (e.g. machine failure) or external (e.g. change in orders).

In order to be efficient, the PAC system must be perfectly adapted to the workshop it controls. This sets the problem of the integration of already existing software, and leads most of the time to the development of costly additional software. Reference models of manufacturing systems have been proposed for quite a long time in analysis or design methods, in order to provide a framework that should facilitate and improve this design phase. The better known of these methods (CIM-OSA and GRAI-GIM), emphasizing the conceptual and organisational levels of design, are shortly described in section 3. On the other hand, Object Oriented methods have had a great success these last years for numerous reasons, one of them being that they allow both system modularity and code reuse. In order to improve their efficiency, we suggest in this paper reference models of PAC modules (operationaly organised as clusters of classes) that can be organised in order to design versatile control architectures. The implementation of these classes in a CASE tool supporting the HP FUSION method is in progress, in order to provide a support for the system designer: the suggested design methodology of the PAC system is detailed in section 4.3.

2 THE PRODUCTION ACTIVITY CONTROL CONTEXT

Production activity control is now usually defined as the set of activities permitting short-term production and inventory control of a shopfloor as well as adapting production to the various disturbances that may occur within the shopfloor or its environment.

A lot of research works has been performed on different types of production activity control systems (see for instance (Jones and Saleh, 90), (Lyons et al., 90), (Van der Pluym, 90) or (Hynynen, 88)), most of them oriented towards the definition of a framework in order to develop a particular solution. Most of the small or medium enterprises can not afford the development of complex specific software for their PAC systems, but even industrial groups like AEROSPATIALE prefer to integrate existing software in their PAC systems, since very good tools are now available on the marketplace. Since no complete solution is available, the PAC system has then to be built using existing software (e.g. schedulers, follow-up software, controllers) integrated with specific developments. A study has then been launched in AEROSPATIALE in order to provide a tool that could facilitate the integration, maintenance and evolution of the PAC system and support the specification and design of PAC systems.

In order to satisfy these requirements, we need reference PAC models that can be easily particularised for a given application, i.e.:

- models that cover the whole abstraction cycle of the PAC system design, i.e. from conceptual aspects to code development,
- modular models (in order to facilitate integration),
- reusable models, in order to reduce development time and cost for different PAC designs.

In order to define these models, we have first studied the existing analysis and design methods for manufacturing systems.

3 METHODS AND MODELS FOR MANUFACTURING SYSTEM DESIGN

3.1 The CIM-OSA methodology

The first goal of the CIM-OSA project (AMICE, 89) is to provide an architecture of the manufacturing system that helps the integration of computer applications. This reference architecture has been widely spread through the well-known "CIM-OSA cube" that models the various aspects of the life cycle of a CIM system along:

- a "derivation axis", describing life cycle steps: requirements (definition), design (specification) and implementation (description),
- a "generation axis", composed of various views of the system: function, information, resource and organisation,
- an "instantiation axis", describing levels of implementation: generic, partial and particular.

The methodology uses a top-down approach in order to describe the three modelling levels of the derivation axis with the different views of the generation axis, passing then through the various instantiation levels. Modelling tools adapted to each step are suggested, e.g. a hierarchical approach using building blocks for the requirement definition model (which have some analogies with SADT), or the entity/relationship model for the specification of the information view.

A major interest of CIM-OSA is to provide a comprehensive framework for the life cycle of a manufacturing system. In comparison with our own particular objectives, this framework is nevertheless too high level to provide support for PAC system integration, since PAC has many specificities compare to the rest of the decisional sub-system of the manufacturing system.

3.2 The GRAI and GRAI-related methods

The original GRAI method (Doumeingts, 84) aims at the analysis, diagnosis and improvement of production management systems. The method uses three different models:

- a conceptual model, that splits the manufacturing system into physical, decisional and information systems,
- a structural model (GRAI grid), that describes the decisional system in charge of production management along two axis: a functional axis (the columns) and a temporal axis (the lines) composed of the various decision making horizons. Each cell of the grid defines a decision centre, that can be precisely described through the GRAI-nets.

The final models result from a comparison between a top-down approach (the grid) and a bottom-up approach (using the GRAI-nets). Rules are then applied in order to find the dysfunctions of the system described.

The GRAI method was extended under the name of GIM (for Grai Integrated Method) (Roboam, 88). The modelling phases are roughly similar to those of the Merise Method, dedicated to the information system development (Tardieu et al., 83). Although the GRAI method is closer to the PAC system than CIM-OSA concepts, it is more dedicated to the highest levels of production management than to the operational levels of the manufacturing system.

Another extension of the GRAI method has been made to take into account PAC problems in manufacturing systems, called GRAICO (Fénié, 94). GRAICO uses an original method based on graph theory in order to analyse the flows in the physical systems (and not only in order to hierarchise the resources like in most of the other methods). The PAC-module structure suggested remains close to the GRAI grid, and consistency rules between physical and PAC systems are suggested.

In spite of their efficiency for the improvement of the production management system, methods such as GRAI, GIM or even GRAICO do not match our objectives perfectly, since they stand quite far from software development and do not sufficiently take into account the existing experience in PAC system design.

4 OBJECT ORIENTED DESIGN METHOD FOR PAC DESIGN

Object Orientation is now widely recognised as providing reusable models and code, and improving the process of software development. One of its interests is that it provides a natural implementation of the physical entities of the workshop using objects (Nof, 94), (Rogers, 92).

4.1 Use of OO methods for manufacturing system design

A lot of work has been performed on manufacturing systems modelling using object oriented methods. Most of these works describe classes based on the manufacturing entities, that can be instantiated for the design of control systems (see for instance (Smith and Joshi, 92), (Hinde et al., 92), (Rogers, 92), (Nof, 94), (Hopkins et al., 94)) but very few are interested in the control structure. Nevertheless, it seems clear that "flat" structures (i.e. structures with a single level of control) are not sufficient in all the cases (Rogers, 92). The problem of the control structure is clearly correlated to object oriented analysis in (Elia and Menga, 94), and in this work the chosen structure is the NIST model (Jones and Saleh, 90) based on successive aggregations of the physical resources through facility, shop, cell, workstation and equipment levels. This work is oriented on the problem of communication between modules, and not on the precise content of the control modules, this last point being very important in order to prepare software integration.

4.2 Reference models of a PAC module

Since one of our goals is to provide support for different types of workshops, we do not want neither to prescribe a particular control structure, nor to define a unique PAC module. Based on the works referenced in section 2, a PAC module may be defined through four main functions:

- to plan, that considers the production plan and the objectives transmitted by the upper decisional level, and details it,
- to react, that allows real-time adaptation of the decisions taken at different management levels each time it is necessary,
- to launch, that breaks down, synchronises and executes decisions taken within a PAC module,
- to follow-up, that collects all the results provided by the sensors or the lower decisional levels as well as the requests coming from outside the shopfloor, processes them according to their type and aggregation level and transmits them to other functions and other PAC modules.

Based on an analysis of representative workshops within AEROSPATIALE, we identified seven main types of modules which differ by the functions involved and their relationships (Huguet and Grabot, 94). For instance, a PAC module can be purely reactive at the lowest level (i.e. it has no planning facility) whereas a high level module may have no other reaction than performing a new planning: this is entirely compatible with the GRAICO approach, where high level decision are triggered by a time period whereas low level decisions can be triggered by events. A typical PAC module is shown in figure 1: function "plan" transmits a set of orders to execute to function "dispatch", which is in charge of resource requisition and synchronisation. If an unexpected event occur, the "react" function may modify the sequence of orders transmitted to "dispatch", trigger the "plan" function, or ask for reaction to an upper level module if the event sets drastically into question the general planning that comes from the upper level.

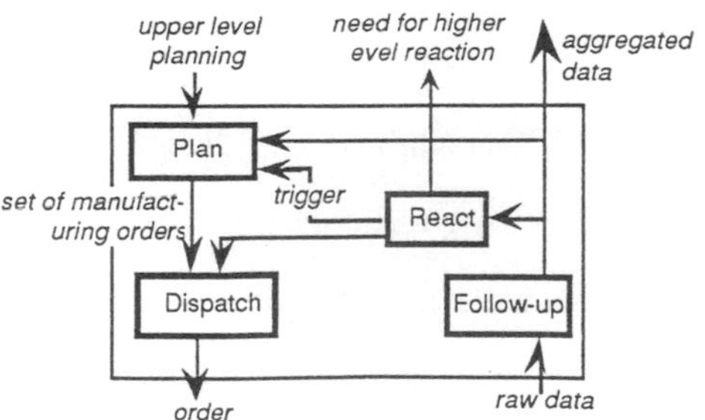

Figure 1 Example of PAC module.

These PAC modules permit to define an integration frame for existing software, but software development remains most of the time necessary, e.g. to support the "react" function which as to be closely adapted to the workshop characteristics. In order to facilitate this development and its reuse, these reference models have been implemented in an object oriented methodology: HP FUSION™.

4.3 Design methodology

Although most of the object oriented analysis methods are based on the same concepts, and use similar modelling tools, HP FUSION (Coleman et al., 94) has been chosen for the following reasons:

- it supports all the object orientation concepts (i.e. classification, polymorphism, inheritance...),
- it supports seamless development: same concepts are used from analysis to implementation, which allows iterative and incremental development,
- it covers broadly enough the different perspectives of the system development (e.g. static and dynamic aspects),
- efficient support tools can be found on the market (e.g. PARADIGM Plus™ and FUSION softcase™).

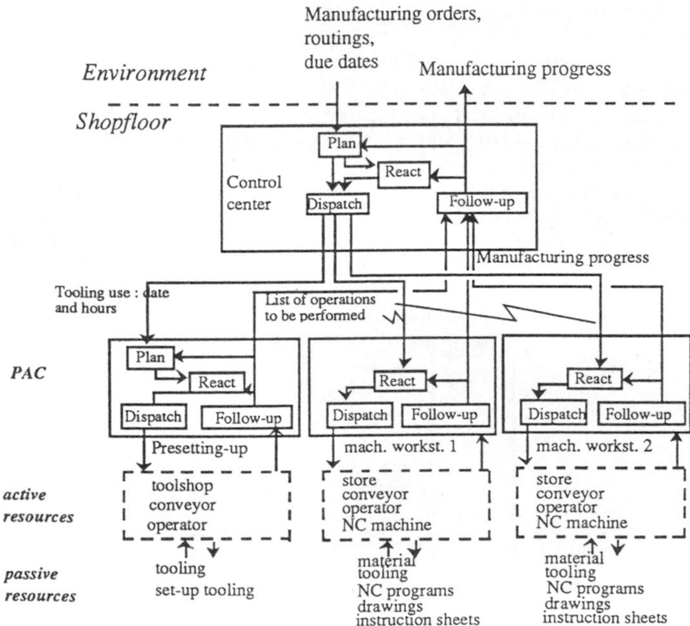

Figure 2 PAC model of an AEROSPATIALE workshop.

The main tools of FUSION are:

- the Interface model (IM). This model shows the main exchanges of information between the system and the agents acting in its environment.
- the Object Model (OM). This diagram adds the object formalism (instantiation, decomposition, inheritance) to the entity-relationship diagrams.

- the Object Interaction Graph (OIG). This graph shows the exchanges of messages between the objects described in the OM.

The PAC system design methodology defined at the moment follows the FUSION methodology:

- modelling of the relationships of the workshop with its environment, then the relationships of the PAC system with the physical resources through the Interface Model,
- modelling of the main resources of the workshop through instantiation of pre-defined classes in the Object Model,
- modelling of the PAC system through instantiation of classes deriving from the reference models in the Object Model,
- checking of the system behaviour through the Object Interaction Graph.

An analysis of the consistency between workshop and PAC models is performed through different sets of rules. Forty to fifty rules have been selected on the base of the first four models describing typical workshops of AEROSPATIALE. These rules are not definitive at the moment, and the knowledge base is subject to constant improvement on the base of further applications (Huguet and Grabot, 94).

4.4 Examples of models

Figure 2 shows the functional model of one of the studied workshops. A scheduling is first performed using a scheduler called SIPAPLUS (PRIOS). This schedule is then transmitted to the toolshop, where a planning of tool utilisation is performed on the base of the global schedule. Two machining workstations are controlled by purely reactive PAC modules, capable to manage small disturbances.

Example of reference models (or patterns) are shown in figures 3 to 5. These models provide a standard description of classical manufacturing operations, and can be modified for adaptation to specific cases if needed. An example of Interface Model is provided in figure 3, describing the passage from an operation to another on a machine.

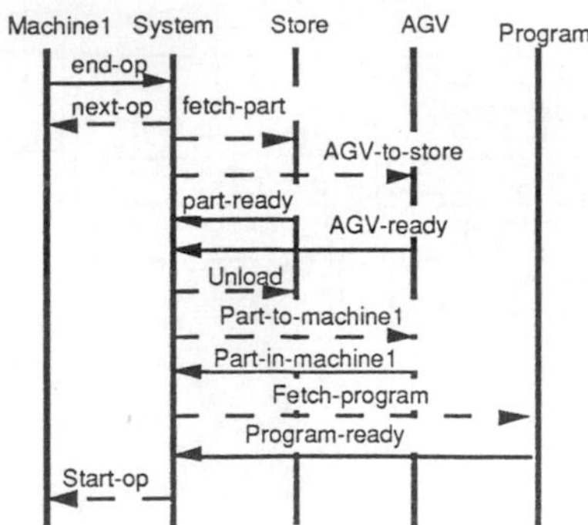

Figure 3 Example of Interface Model.

A partial object model is shown as an example in figure 4: this model shows the implementation a PAC module through objects: its components (planning system, dispatcher, follow-up system and reaction system) will then be split if the software that will support each of them has not already be chosen.

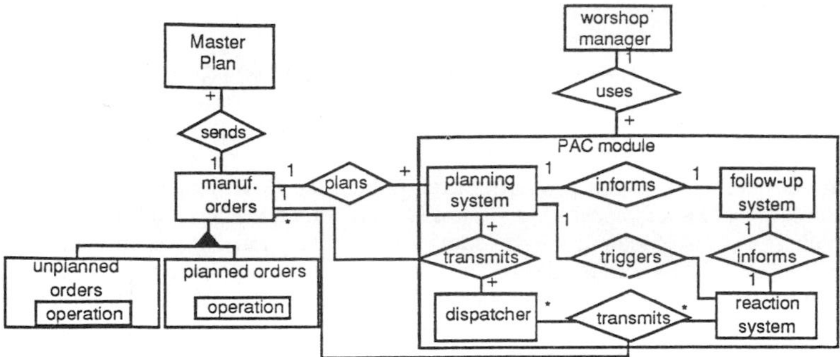

Figure 4 First level object model of a PAC module

Figure 5 shows an example of Object Interaction Model corresponding to the interface model of figure 3.

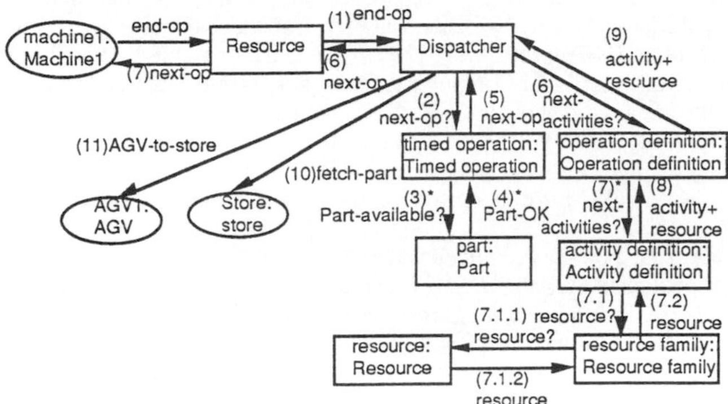

Figure 5 Object Interaction model

5 CONCLUSION

The use of Object Oriented analysis, design and programming methods is one of the elements that should help in order to design more easily and more quickly more efficient and better adapted PAC systems. This goal can only be achieved through reuse, which require reference models in order to provide a framework for the integration of existing software and specific development.

The first experiments performed with the reference models of the PAC modules that we suggest have shown that these models can be applied to a great variety of workshops. Furthermore, the implementation of these models through libraries of pre-defined objects facilitates and shortens greatly the design phase. Major improvements of the models and of the knowledge base that checks the workshop/PAC consistency are expected after other implementations, but it is clear that the support tool which development is in progress must be considered as a way to capitalise expertise, rather than as an automated way to design PAC systems.

6 REFERENCES

AMICE Consortium (1989) *Open System Architecture for CIM*, ESPRIT Project 688, volume 1, Springer Verlag, Berlin.

Codd, E.F. (1970) A Relational Model of Data for Large Shared Data Banks, *Communications of the ACM*, 13, **6**, 377-387.

Coleman, D. Arnold, P. Bodoff, S. Dollin, C. Gildchrist, F. Hayes, F. Jeremaes, P. (1994) *Object-Oriented Development - The Fusion Method*, Prentice Hall, Englewood Cliffs.

Doumeingts, G. (1984) Méthode GRAI : Méthode de conception des systèmes en productique, State Thesis, University of Bordeaux I.

Elia, G. Menga, G. (1994) Object-oriented design of flexible manufacturing systems, in Computer Control of Flexible Manufacturing Systems, Joshi S.B. and Smith J.S. Eds, Chapman & Hall, 315-342.

Fénié, P. (1994) GRAICO : méthode de modélisation et de conception des systèmes d'exploitation de systèmes de production, PhD thesis, University of Bordeaux I.

Fox, M. (1981) An Organizational View of Distributed Systems, IEEE Transactions on Systems, Man and Cybernetics, vol. SMC 11, **1**.

Hinde, C. West, A. Williams, D. (1992) The use of object orientation for the design and implementation of manufacturing process control systems, International Conference on Object-Oriented Manufacturing Systems, Calgary, Canada, 3-6 may.

Hopkins, J.M. King, R.E. Culbreth, C.T. (1994) An object-oriented control architecture for flexible manufacturing cells, in Computer Control of Flexible Manufacturing Systems, Joshi S.B. and Smith J.S. Eds, Chapman & Hall, 427-466.

Huguet, P. Grabot, B. (1994) Production Activity Control system specification : object oriented analysis and knowledge based system approach, ISPE/IFAC International Conference on CAD/CAM, Robotics and Factory of the future, Ottawa, Canada 21-24 Août 1994.

Huguet, P. Grabot, B. (1995) A conceptual framework for shopfloor production activity control", to be published in *International Journal of Computer Integrated Manufacturing*.

Hynynen, J. (1988) A Framework for Coordination in Distributed Production Management, PhD Thesis, Helsinki University of Technology.

Jones, A. Saleh, A. (1990) A multi-level/multi-layer architecture for intelligent shopfloor control. *International Journal of Computer Integrated Manufacturing,* **3**, 60-70.

Lyons, G. J. Duggan, J. Bowden, R. (1990) Project 477: Pilot implementation of a Production Activity Control (PAC) system in an electronics assembly environment. *International Journal of Computer Integrated Manufacturing,* **3**, 196-206.

Nof, S.Y. (1994) Critiquing the potential of object orientation in manufacturing, International Journal of Computer Integrated Manufacturing, vol. 7, **1**, 3-16.

Roboam, M.(1988) Modèles de référence et intégration des méthodes d'analyse pour la conception des systèmes de gestion de production, PhD thesis, University of Bordeaux I.

Rogers, P. (1992) Object-Oriented Design and Control of Intelligent Manufacturing Systems, International Conference on Object-Oriented Manufacturing Systems, Calgary, Canada, 3-6 may.

Smith, J.S. Joshi, S.B. (1992) Object-Oriented Developement of shop floor control systems for computer integrated manufacturing, International Conference on Object-Oriented Manufacturing Systems, Calgary, Canada, 3-6 may.

Tardieu, H. Rochfeld, A. Coletti, R. (1983) La méthode Merise, principes et outils, Les Editions d'organisation, Paris.

Van Der Pluym, B. (1990) Knowledge-based decision making for job-shop scheduling, International Journal of Computer Integrated Manufacturing, vol 3, **6**, 354-363.

30

Organizational behaviour analysis and information technology fitness in manufacturing: Analysis through modelling and simulation

A. Lucas Soares, J.J. Pinto Ferreira, J. M. Mendonça
INESC/Porto, R. José Falcão, 110 - 4000 Porto, Portugal
FEUP/DEEC, Univ. of Porto, R. Bragas, 4099 Porto Codex, Portugal
email: asoares, jjpf, jmm@tecno.inescn.pt

Abstract

The analysis and design of integrated manufacturing systems, under a techno-organizational approach, is a complex and time consuming process. It involves people from different knowledge areas that must cooperate in order to build a common and complementary understanding of the enterprise's organization, people and technology. Particularly the SME's, although subjected to the same pressures than their bigger counterparts, have scarcer time, people and technical resources. This paper presents a novel approach to the modelling and simulation of organizational structures and information technology resources in integrated manufacturing systems. The tool resulted from this research work enables the analysis and evaluation of alternative organizational designs and different combinations of group/software task assignment. This approach goes beyond the operational quantitative and logical analysis, enabling the assessment of socio-technical evaluation criteria. Moreover, this tool enhance the cooperation and communication within the multidisciplinary team developing the techno-organizational system: users, technical developers and organization/social specialists.

Keywords

Organization modelling, simulation, techno-organizational development, socio-technical systems.

1 INTRODUCTION

Complex restructuring processes are needed to change organization structure and culture in order to accommodate technological innovation in manufacturing. Empirical studies put into focus the influences of information technologies on the organization and people in production systems (Soares, 1994) and the difficulties to put in practice the increasingly accepted "joint development" in all of its dimensions. Anyway, it is generally accepted that the technocentric approach to the integrated manufacturing systems (IMS) development is no longer appropriate (or never has been...), and is given place to anthropocentric approaches.

INESC/Porto (Institute of Systems and Computers Engineering at Porto), besides its R&D work on advanced information technology systems for manufacturing management, has been undertaking research work in modelling tools supporting IMS development. Two main streamlines emerge from this work: the support to the development, configuration and fine-tuning of industrial IT infrastructures and applications and the support to techno-organizational development. The later is the subject of this paper and is achieved through the joint modelling and simulation of organizational structures and the IT shop floor management applications.

2 INTERDISCIPLINARY TECHNO-ORGANIZATIONAL ANALYSIS

Advanced information management systems for manufacturing applications have profound implications in the organization where are introduced. The "constructive character" (Bachman, 1994) of these technologies will only be fully exploited if they are understood in a theoretical and conceptual framework that rejects the technocentric view in favour of a vision that contemplates the interaction between the organizational subsystems which technology is part of. The joint development of information technology and organization is only possible if engineering and organizational/social knowledge areas can be applied in a cooperative way. For that, tools have to be developed in order to enhance this cooperation. Techno-organizational analysis is a development phase were a strong cooperation within the development team is needed. Modelling tools are collaboration instruments (Paul, 1994) and must be thoroughly used for an effective analysis.

2.1 Models and modelling purposes

Modelling is an increasingly important object of R&D in IMS analysis and design. To clarify the modelling purpose in the context of techno-organizational analysis and design a simple classification of models is presented below.

Descriptive models

Descriptive models capture two main properties of a system: structure and behaviour. Structural models represent function and information arrangements within a technical system or an organization. Behaviour models describe the dynamics of those structures. Basic models in this category make use of data flow diagrams, SADT, Entity-Relationship diagrams, object-oriented models, data transition diagrams, and organization charts. Composite models, aggregating or adapting the basic models, are the ones used in modelling frameworks and/or enterprise reference architectures like CIM-OSA, GRAI-GIM and PERA (Williams, 1993).

Descriptive models have a computer science root and have been used mainly for the specification and design of computerised systems. Therefore they are built on a machine metaphor resulting in major drawbacks when it comes to model the real-world of the manufacturing systems (Bansler, 1993, Hoydalsvik, 1993, Kensing, 1993). The first difficulties arose when the increasingly interactivity of those systems called for the systems environment modelling, particularly the human environment. More recently, with the growing awareness of the information technology organizational impact, new difficulties have emerged in the joint modelling of the technical system and the organization, namely in the explicit representation organizational structures and the their social aspects . Besides these drawbacks, there is the problem of how to cope with the modelling complexity due to his deepness (model level of detail) or wideness (model scope). The model construction and the evaluation resulting from model analysis can be cumbersome and time consuming calling for computer based tools that help users in this task.

Evaluation models

In the analysis and design of IMS, simulation is a frequently used tool to evaluate and compare alternative scenarios for the organization of manufacturing activities and production management strategies. Simulation systems rely on computational models that describe the operational aspects of activity networks, enabling their quantitative and logical evaluation. The construction of these models depends on the simulation system. Maximum modelling flexibility is achieved with general purpose simulation languages, but these offer very limited expressiveness for problem communication and understanding. On the contrary, "data-driven" simulation (based on pre-defined models, usually queuing networks), in conjunction with graphical interfaces, provide less flexibility but better communication for the system structure and the concepts under evaluation.

Although simulation is a well established field in the analysis and evaluation of IMS at the physical operations level, there are few contributions to the simulation of organization structures and even fewer when we add the modelling and simulation of software applications supporting manufacturing activities. Organization structure simulation has been addressed from an exclusively operational viewpoint and focusing on evaluation criteria similar to those used in the physical operations level: lead-time, WIP level, system loading and delivery dependability (Zülch, 1993). Also work on shop floor control software and physical operations joint simulation have been reported in (Ferreira, 1994). Here the modelling purpose is the shop-floor software behaviour and their interaction with the physical operations, the goal being to support the technical system development.

Hybrid models

Modern simulation tools based on object-oriented concepts and equipped with powerful graphical interfaces enable, to certain extent, the integration of descriptive and evaluation modelling. This results in simulation models with structural representation power (functional and informational) in addition to the quantitative and logical description, overcoming the lack of expressiveness of evaluation models. Moreover, object-orientation provides means for the explicit inclusion of "non-functional" information, paving the way to effective organizational modelling.

3 HYBRID MODELLING FOR TECHNO-ORGANIZATIONAL ANALYSIS

The work presented here is a contribution to the interdisciplinary development of IMS. It resulted in a simulation based system for the modelling and analysis of organizational and information technology structures in manufacturing. In this section the underlying concepts and theory of the Organization Behaviour simulation system (OBsim) will be presented .

Modelling is a mean, thus it must be clearly stated what its purpose is. OBsim is intended to help the joint development of manufacturing organization and the supporting software applications. For that, evaluation criteria were found in socio-technical systems theory and in industrial psychology. Besides basic quantitative performance indicators, e.g. task/function/process lead time or organizational unit temporal occupation, socio-technical indicators were chosen to be directly or indirectly assessed through OBsim. Examples of these indicators are the completeness of realised work functions, independence of organizational units, task interdependence, boundary regulation, group/software coupling and technical linkage (integration/dependency) (Weik, 1994, Emery, 1978). An evaluation model alone would not fulfil the purpose of a techno-organizational analysis because of the poor communication capability arising from the lack of structural description power. On the other side, the quantitative and logical evaluation of organizational scenarios calls for a simulation system with the corresponding evaluation model. Hybrid modelling is thus necessary to undertake an effective organizational analysis based on socio-technical evaluation criteria.

3.1 The Organization Behaviour Simulation system

Building classes

The techno-organizational model used in OBsim is outlined in Figure 1. *Task* and *software module* are the basic building classes, from a functional viewpoint, being aggregated by *function* and *software system* classes. The class *organizational unit* is linked to tasks and functions through assignment and qualification links. A short description of the functional building classes is given below:

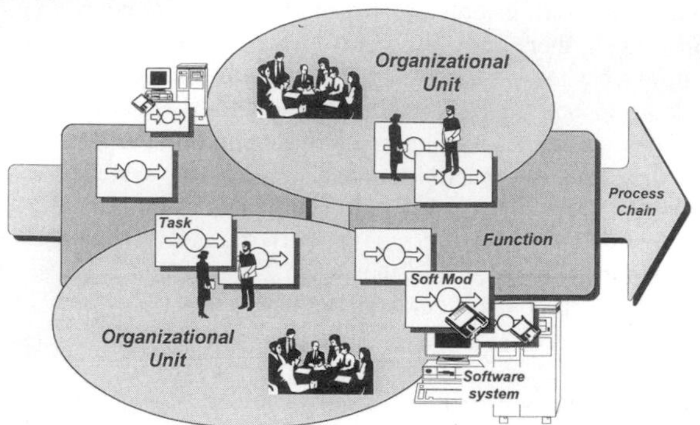

Figure 1 Techno-organizational model structure.

- *organizational unit* - models elementary units describing the task assignment, qualification structure, capacity, learning process and other relevant information to describe its behaviour; this class can model working groups, manufacturing sections, production cells, production islands, etc.,
- *task* - describes the basic behaviour of organizational units; it includes, besides the behaviour model, local information for the task fulfilment and evaluation; can be further decomposed in elementary activities,
- *function* - aggregates closely related tasks, and is the basis for the process chain modelling,
- *software module* - models elementary functions realised by a software system; the software behaviour description is made from an external (interaction) point of view; can be further decomposed in elementary software activities,
- *software system* - aggregates software modules that form a self-contained software application; the software model represents essentially the software applications specification, being one of the inputs for the software system construction.

The information network is modelled after two basic building classes:

- *operational information* - information needed to coordinate and describe the organizational tasks.
- *monitoring information* - feedback information on the degree of achievement of the task actions.

These classes are further specialized in *operational coordination* and *operational description* information flows. The former represents temporal plans that coordinate when organizational units realise their tasks. Examples are production schedules, maintenance plans, etc. The later represents how the organizational units realise their tasks, e.g. product structures, process plans, NC programs, etc.

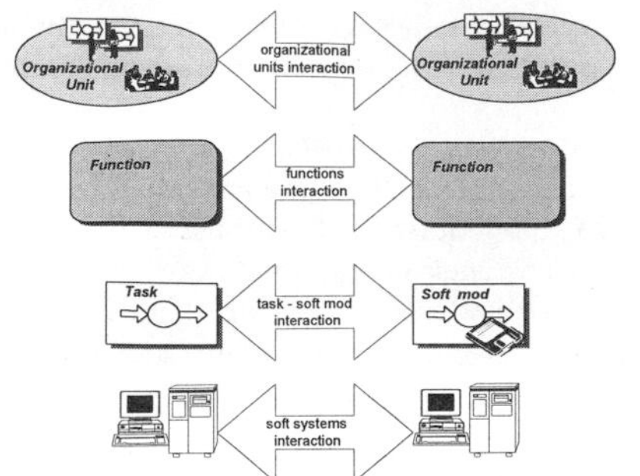

Figure 2 Information flows interaction roles.

During the model construction are given interaction roles to the information flow classes depending on the sender and the receiver of the information. This roles are shown in Figure 2.

The model building classes belong to a class library and are instantiated during the model construction. Furthermore, new classes can be obtained from the existing ones through inheritance mechanisms and added to the library, enabling an effective reuse of previously done analysis.

The modelling and simulation infrastructure

OBsim is built on a object-oriented simulation tool, SIMPLE++. This tool is primarily intended to simulate manufacturing operations at the physical level, although it incorporates mechanisms to model information control flow. Novel work has been undertaken at INESC/Porto to provide SIMPLE++ with facilities to specify and model shop floor software applications. The resulting integrated modelling environment encompasses both material and information flows, supporting R&D work in information flow and application software modelling, as well as its full integration with shop floor material flow models. OBsim used the additional facilities described to model and simulate the organizational tasks and software modules.

The behaviour of each function or software system is represented using SDL (Specification and Description Language) which is a design language standardised by ITU-TSS (International Telecommunications Union) for the design of computer based exchange systems. Besides originating from the telecommunications areas and having been used in this field for a long time, SDL is a powerful modelling language and may be applied in many other areas including the design of hybrid hardware & software systems (Ferreira, 1994) . The SDL basic structure and its mapping to the techno-organizational model is shown in Figure 3.

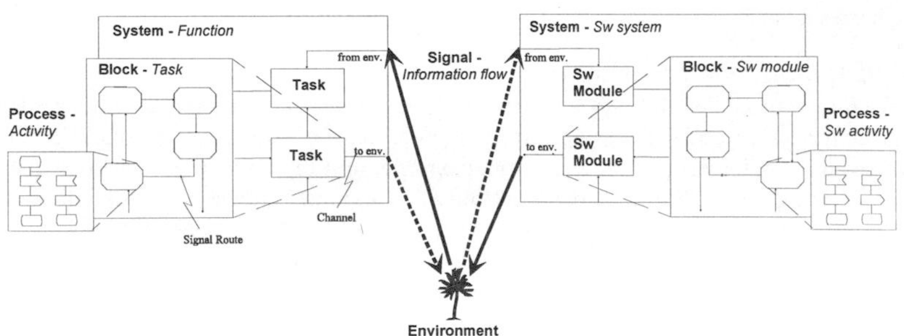

Figure 3 SDL hierarchy and its mapping to the techno-organizational model.

3.2 Socio-technical analysis and evaluation

The novel approach in OBsim brought socio-technical analysis criteria into the organizational structures simulation, thereby overcoming the "machine metaphor" which is inherent to quantitative models. Socio-technical analysis criteria can only be fully addressed if descriptive and evaluational representations are available. Hybrid models are thus suitable for this analysis. These criteria and the way OBsim can assess them will be briefly described below .

Completeness of realised functions in the organizational unit

Motivational factors lead to the need of job enrichment in the organizational unit. The processing depth of the realised tasks, i.e. vertical integration of functions, is the way of achieving this goal. In OBsim, vertical integration of functions is structurally assessed by visualising the task assignment together with the defined process chains. The higher the number of tasks belonging to the same process chain is assigned to the organizational unit the higher is the completeness of realised functions.

Independence of the organizational unit

The degree of dependence of an organizational unit on other units, being either vertical or horizontal, is an important effectiveness factor. The minimisation of this dependence is a requirement of actual organizational paradigms: production islands, autonomous working groups, manufacturing cells. OBsim assesses this criteria in a structural and quantitative way. When an information flow has an organizational units interaction role, it connects two tasks assigned to different organizational units, this representing a dependence link. The link can be further classified as operational or monitoring flow. During a simulation session the intensity of the flows between the organizational units can be recorded. The higher the intensity of flow the lower the organizational unit independence. This is also related with the following criteria.

Function and task interdependence

The form of cooperation between tasks can be classified after Thompson's categories of interdependence (Thompson, 1967). Three categories are considered:
pooled interdependence which is the lowest form of interdependence meaning that none or very few information flows link the tasks, *sequential* interdependence that exists when information flows between tasks are uni-directional, *reciprocal* interdependence is the highest form of interdependence and appears when there are bi-directional information flows between tasks.

As the interdependence increases, it makes greater demands on intra and inter organizational units coordination, communication and decision-making. Again, this criteria is evaluated in OBsim through the monitoring of the intensity and direction of the information flows between tasks and functions.

Organizational unit qualification

Qualification is modelled through a matrix where a ranking of the tasks that the organizational unit has competence and skills to perform is stored. During simulation the qualification table is used to quantify the efficiency of the organizational unit in realising the assigned tasks. One important feature of OBsim is the possibility to evaluate functional flexibility. This is achieved through a dynamic task assignment that occurs e.g. when the capacity of an organizational unit is exceeded and a task must be transferred to another unit. Another related feature is the evaluation of functional integration also related with the completeness of the realised functions. The coordination effort between organizational units depends on the interdependence of the tasks they realise. Related tasks (aggregated in functions in the OBsim model) have a strong interdependence thus it is desirable that they are realised in the same organizational unit as to minimise the coordination effort. As in the OBsim model information flows with different interaction roles are treated accordingly, alternative scenarios of functional integration levels can be comparatively evaluated. During simulation the qualification structure can be dynamically modified. With this feature organizational unit learning can be simulated using a learning function dependent on time, organizational unit composition, frequency of task assignment change or another structural factor that the user can program.

Boundary regulation

Boundary regulation is the extent in which one organizational unit monitors and coordinates another. This is easily evaluated through the intensity of the coordination and monitoring information flows with organizational units interaction roles. As a result, self-regulation, one important factor for the organizational units autonomy, can also be assessed.

Techno-organizational coupling

Techno-organizational coupling is the extent to which the software system determines the activity of the organizational unit. A first evaluation of this criteria is made through the analysis of the model structures representing the task and software module that interact. A second level of evaluation is achieved during simulation through the monitoring of the amount of time that the task waits for an information output from the software module or waits to be allowed to input information into the software module. The techno-organizational coupling evaluation facility is of utmost importance in helping the specification and design of the software system.

Technical linkage

Following the approach described in function and task interdependence, it is possible to evaluate the interdependence between software systems both through the description of the information flows with software systems interaction roles and the intensity of that information flows at simulation time.

3.3 Dificulties and limitations

Information gathering for the model construction is the biggest difficulty in utilising this tool. For example, the average lead time of certain tasks like the production of a plan or the orders rescheduling because of a machine breakdown, cannot be easily quantified. The lack of accuracy in some tasks data is partially overcomed because the system is intended to *comparatively* evaluate scenarios and not to perform an absolute and accurate simulation of reality. Techno-organizational development doesn not rely exclusively on computer supported tools to analyse and specify the organization and production support software. Even in participating contexts, difficulties and hindrances in the systems development have to be surpassed taking care of social and power relations aspects. These aspects cannot be put in a computer model. Being conscious of this, OBsim can be a valuable analysis and colaboration instrument and function also as an “awareness tool” for those small budget projects in SME’s where multidisciplinary development teams are not feasible.

4 CONCLUSION AND FURTHER WORK

Interdisciplinary techno-organizational development of IMS needs tools to shorten and improve the analysis and specification phases of the development life cycle. Moreover, trial and error methods are no longer feasible in organizational design, specially in SME’s where time and resources are scarce. With the modelling and simulation system presented here, trial and error can be done with little costs for the manufacturing enterprise and a more comprehensive approach to the analysis and specification of IMS can be undertaken. In order to validate its underlying concepts, OBsim is now starting to be applied in supporting the introduction of PROFIT, a marketed advanced shop-floor management system developed under the ESPRIT projcct 5478 Shop-Control, in SME's. Further work includes the "plug-in" of a physical operations model as to enhance the simulated scenarios, and the incremental construction of an adequate class library based on several case studies. This work is part of more extensive research that is being undertaken in the development of modelling tools for techo-organizational optimisation of IMS at INESC/Porto.

5 REFERENCES

Bachman, R., Möll, G. (1994) Participating in CIM Systems, in *Advances in Agile Manufacturing* (ed. P.T. Kidd and W. Karwowski), IOS Press.

Bansler, J.P., Bodker, K. (1993) A Reappraisal of Structured Analysis: Design in an Organizational Context. *ACM Transactions on Information Systems*, **11**, n° 2.

Emery, F.E. (1978) Characteristics of Socio-Technical Systems, in *The Emergence of a new paradigm of work* (ed. F. Emery), Australian National University, Canberra.

Ferreira, J.J., Mendonça, J.M. (1994) Supporting CIM Systems Life Cycle Through Material and Information Flow Executable Models in *Proceedings of the 10th ISPE/IFAC International Conference on CAD/CAM, Robotics and Factories of the Future*. Otawa.

Hoydalsvik, G., Sindre, G. (1993) On the Purpose of Object-Oriented Analysis, in *OOPSLA'93 Proceedings*, pp. 240-255.

Kensing, F., Munk-Madsen, A. (1993) Participatory Design: Structure in the Toolboox. *Communications of the ACM*, **36**, N°4, 78-83.

Paul, R.J, Thomas, P.J. (1994) Computer-Based Simulation Models for Problem-solving: Communicating Problem Understandings. *Electronic Journal of Virtual Culture*.

Soares, A.L., Mendonça, J.M. (1994) Interaction of Advanced Shop-Floor Management Systems with Production Systems Organization: An Exploratory Study, in *Advances in Agile Manufacturing* (ed. P.T. Kidd and W. Karwowski) - IOS Press.

Thompson, J.D. (1967) *Organizations in Action*. MacGraw-Hill, New York.

Weik, S., Grote, G., Zölch, M. (1994) KOMPASS, Complementary Analysis and Design of Production Tasks in Socio-Technical Systems, in *Advances in Agile Manufacturing*, (ed. P.T. Kidd & W. Karwowski). IOS Press.

Zülch, G., Grobel, T. (1993) Simulating Alternative Organizational Structures of Production Systems, in *Production Planning & Control*, Vol. 4, N° 2, 128-138.

Williams, T.J. et al. (1993) Architectures for Integrating Manufacturing Activities and Enterprises, in *Information Infrastructures Systems for Manufacturing* (B-14). (Eds. H. Yoshikawa and J. Goossenaerts). Elsevier Science B.V. (North-Holland).

6 BIOGRAPHY

A. Lucas Soares was born in 1964 in Porto. He received the MsC. in Industrial Automation in 1991 and is currently assistant at the University of Porto. His interest areas are: techno-organizational development, human and social aspects of information technology and photography.

J.J. Pinto Ferreira was born in 1964 in Porto. He received the MsC. in Industrial Information Systems in 1991 and is currently assistant at the University of Porto. His interest areas are: enterprise integration and modelling, formal description techniques, distributed systems and windsurfing.

J. M. Mendonça was born in 1955 in Porto. He received the PhD in Electrical Engineering in 1985 and is currently Auxiliary Professor at the University of Porto. His interest areas are: manufacturing management, enterprise modelling, shop floor control and martial arts.

An Integrated Framework for the Development of Computer Aided Engineering Systems

A. Molina, T.I.A. Ellis**, R.I.M. Young**, R. Bell***
**Centro de Sistemas Integrados de Manufactura, ITESM, Campus Monterrey, CP 64849, Monterrey, N.L. Mexico*
fax: +52 (8) 358–1209, e-mail: molina@mansun.lut.ac.uk
***Manufacturing Engineering Department, Loughborough University of Technology, Loughborough, Leicestershire, LE11 3TU, U.K.*

Abstract

The development of complex engineering environments, such as Computer Aided Simultaneous Engineering Systems (CAE System), requires that system developers and users are able to hold meaningful discussions and share a common understanding of their underlying goals and ideas. A CAE Framework that aims to aid in the achievement of this goal is presented in this paper. The CAE Framework uses elements of the CIMOSA architecture framework for enterprise integration and the Reference Model for Open Distributed Processing (RM–ODP). The experience gained whilst using these reference models to define a hybrid methodology for the definition, design and implementation of CAE systems is discussed.

Keywords

CAE System, Framework, Reference Model, Simultaneous Engineering

1 INTRODUCTION

The development of integrated information systems to support design and manufacturing activities, and hence satisfy company requirements, represents a major problem for system developers and users. The authors have recognised that an important task in the successful computer support of simultaneous engineering is the identification of the enterprise requirements, i.e. where the Computer Aided Simultaneous Engineering System (CAE system) is going to be installed, what its functionality has to be and how it is operated (Molina et al. 1995). Being able to recognise these needs enables the CAE system to be configured to a specifc manufacturing environment. Different aspects which have to be considered for the elicitation of requirements are: the manufacturing strategy, the activities that require support, flow of information, availability of resources, organization and responsibilities, etc. A key problem is to determine the method, or set of methods, to be used in order to capture all these aspects in a formal representation, and hence guide the development of an integrated information system.

The IFAC/IFIP Task Force report on Architectures for Integrating Manufacturing Activities and Enterprises (Williams et al. 1993) found the following reference models suitable for the task of describing an integrated system, its life cycle and the methodology for its application: CIMOSA, GRAI–GIM and the Purdue Enterprise Reference Architecture. This Task Force aims to develop a Generic Enterprise Reference Architecture and Methodology (GERAM) on the basis of the above architectures (Bernus and Nemes 1994). Based on the concepts of reference models the authors argue that a framework or reference model is needed to design, develop and integrate CAE systems which effectively support Simultaneous Engineering. The CAE framework should assist the users and developers in the following three tasks:

1. Identify the enterprise CAE system requirements for the support of simultaneous engineering i.e. functionality, information, resources, role in the organization, etc.

2. Guide the design and implementation of the CAE system itself.

3. Organize the people and establish the set of methods and tools to evolve the CAE system towards the desired level of integration and automation.

This paper introduces a CAE framework which aims to address these issues by combining the use of two reference models: the Open System Architecture for Computer Integrated Manufacturing–CIMOSA (ESPRIT Project 688/5288) and the Reference Model for Open Distributed Processing, named RM–ODP (ISO/IEC JTC1/SC21/WG7 N 755).

The paper is organized as follows, the next section describes in detail the issues involved in the definition of the CAE framework. The use of the CIMOSA model is presented in Section 3. Section 4 describes the utilization of the RM–ODP to define what has been called the MOSES CAE Reference Model. A discussion about the importance of using reference models for the definition, design and implementation of CAE systems is offered in section 5. Finally conclusions are proffered on the authors' experiences in the use of reference models.

2 A CAE FRAMEWORK FOR COMPUTER AIDED SIMULTANEOUS ENGINEERING SYSTEMS

The rationale of the CAE Framework is based on the idea that a formal definition of enterprise requirements can be used as the driver for the selection of alternative CAE systems. The match between the particular needs of an enterprise against different available CAE systems can be

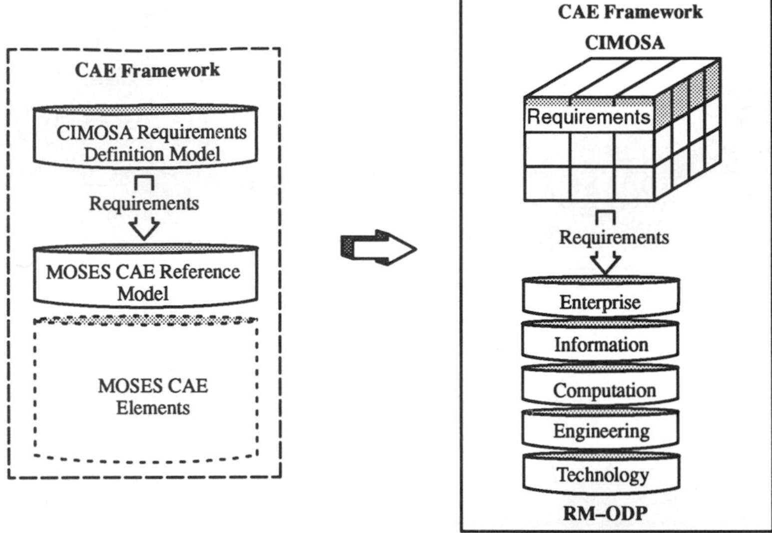

Figure 1 CAE Framework based on CIMOSA Requirements Definition Model and Reference Model for Open Distributed Processing (RM–ODP).

facilitated, if the CAE system concepts are represented in a reference model which allows the description of the CAE systems functionality and elements. This idea of using a model to define the enterprise requirements has been introduced by the authors in the MOSES research project in order to define the CAE framework. The use of a requirements model allows different enterprises to specify what functionality is required of a CAE system in order to support simultaneous engineering. To define the enterprise requirements in a formal manner, the authors decided to use the Requirements Definition Model of the CIMOSA architectural framework. The enterprise requirements are those needed to realize the design of products using simultaneous engineering principles. The main reason for this choice was that CIMOSA provides a set of methods to systematically develop the requirements model.

The CAE Framework is therefore composed of two reference models (Figure 1):

1. The CIMOSA Requirements Definition Model

2. The MOSES CAE Reference Model based on the Reference Model for Open Distributed Processing (RM–ODP)

The Reference Model for Open Distributed Processing (RM–ODP) was used to define the MOSES CAE Reference Model (CAE–RM) in order to provide a multi–viewpoint representation of the MOSES CAE system concepts (Molina et al. 1994). The CAE–RM describes the functionality, configuration and technology necessary to satisfy the requirements specified by the CIMOSA Requirements Definition Model.

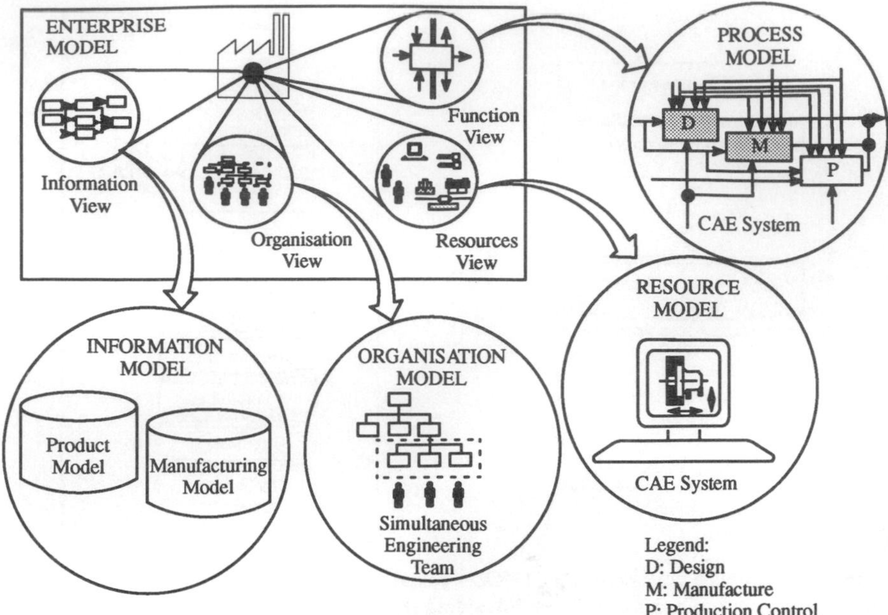

Figure 2 CIMOSA Partial Enterprise Requirements Definition Model.

The CAE Framework is therefore an hybrid methodology based on the CIMOSA Requirements Definitions Model and the multi–viewpoint MOSES CAE–RM based on RM–ODP. This CAE Framework aims to enable the identification of requirements for a CAE system, the provision of guide-lines for CAE system development and the organisation of methods and tools to evolve the CAE system towards the required level of integration and automation.

3 THE USE OF CIMOSA IN THE CAE FRAMEWORK

The CIMOSA Reference Model allows the construction of enterprise models. In order to build the CIMOSA Requirements Definition Model model the authors followed the CIMOSA modelling process which requires the enterprise model to be built in terms of four interrelated views: function, information, resources and organisation (Figure 2). By modelling the function view, the activities to design, produce and maintain products with consideration for the whole product life–cycle are identified. The model developed in this research only covers the activities relevant to the development of products using simultaneous engineering principles, and not all the activities in the enterprise. During the modelling of these activities the information required to realize these activities is determined. This information is modelled and captured in the information view by using information models. By combining the results of the information and function views the required capabilities for the CAE system to support simultaneous engineering, as a resource, can be defined. Thus, the CAE system capabilities form the resource view. The organisation view models the responsibilities and authorities of the group of people who have influence over the elements defined in the function view, information view and resource view (in this case, the Simultaneous Engineering Team). This team is responsible for the realization of the concurrent design of products.

Once the Partial CIMOSA Requirements Definition Model was defined, this model was instantiated to specify the CAE system requirements for a particular enterprise. The CIMOSA model is therefore a driver for the definition of the functions required by a CAE system. The use of CIMOSA allows the creation of a more structured and flexible CAE Framework. These characteristics enable the framework to be based upon an available, acceptable and formal terminology and methodology; and perhaps a future standard. In addition, the model can be applied to a wide range of systems within the domain of information systems to support simultaneous engineering. The use of CIMOSA in the CAE Framework allows future CAE systems, developed based on the MOSES CAE–RM, to be easily integrated within the enterprise. This integration can be more easily achieved because the system requirements are clearly defined and may be used as the drivers for the development of the CAE system.

The author's main reasons for chosing the CIMOSA reference model as part of the CAE Framework were: it is a well defined and documented reference model, it has a formal approach to system modelling, the CIMOSA concepts match the ones required for the definition of the CAE–RM at the enterprise view, it has the potential to become an international standard

4 THE USE OF RM–ODP IN THE CAE FRAMEWORK

Once all the enterprise requirements for the computer support of simultaneous engineering have been identified and defined using CIMOSA, the context for the development of the CAE system, based on the MOSES CAE Reference Model (CAE–RM), is set. The aim of the CAE–RM is to provide a description of a CAE system and its functionality that matches and satisfies those requirements, and then to guide the development of the CAE system itself. On this basis, the design, configuration and implementation of the CAE system can be undertaken by selecting the system elements, defined in the CAE–RM, which are important for that particular manufacturing environment such as: information models (e.g. Product and Manufacturing Models), decision support environments (e.g. applications for Design for Function, Design for Manufacture, etc.), and the adequate integration infrastructure (i.e. information system architecture).

The CAE–RM is based on the RM–ODP (ISO/IEC JTC1/SC21/WG7 N 755), which is a five level model intended to represent open distributed systems. To achieve this, the following five levels have been defined: Enterprise, Information, Computation, Engineering, and Technology. Although the RM–ODP allows the thorough description of a CAE system from different views, the MOSES research focused primarily in defining the first three viewpoints i.e. Enterprise, Information and Computation (Figure 3).

The purpose of using the CAE–RM is to model a real system, whether existing or planned. The enterprise may have requirements which are difficult to realise. Any gap between what can be achieved by the system and what is desired will be highlighted by a disparity in the mapping between the CIMOSA Enterprise Requirements Definition Model and the Enterprise viewpoint of the CAE–RM.

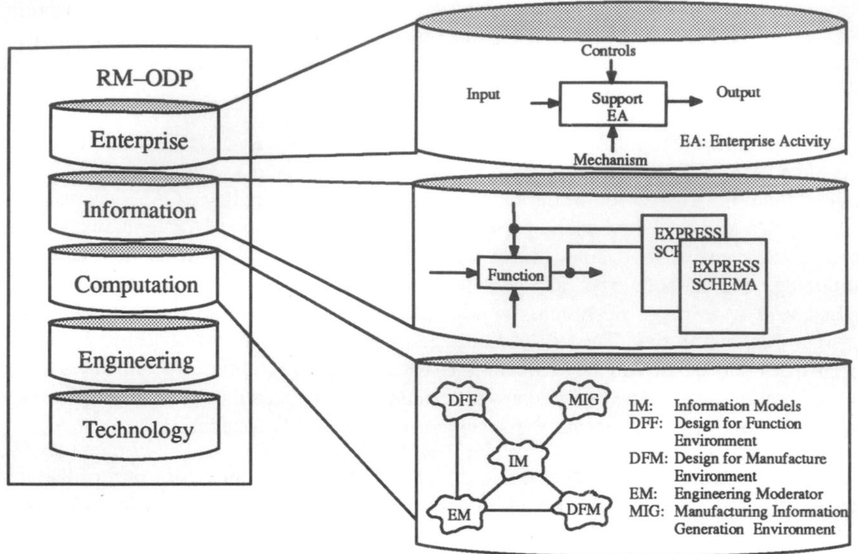

Figure 3 The MOSES CAE Reference Model.

In the MOSES CAE Reference Model the Enterprise Viewpoint is described using IDEF0 activity diagrams (ICAM 1980). The IDEF0 model describes the purpose of the CAE system, i.e. support the enterprise activities undertaken by the simultaneous engineering team. The Enterprise Viewpoint of the CAE–RM allows the establishment of a relationship between the CIMOSA Requirements Definition Model and the CAE–RM. This relationship is created between an enterprise activity defined in the CIMOSA model and an activity defined in the IDEF0 model of the Enterprise Viewpoint. Each enterprise activity is defined in terms of inputs and outputs. The CIMOSA resource inputs define the capabilities that are required from a resource in order to perform certain activities. Thus the specific CAE system capabilities which are needed to realize certain enterprise activities are defined and from this definition the CAE system requirements are derived. The IDEF0 activity diagrams of the Enterprise Viewpoint describe the functions of the system which can satisfy the enterprise requirements, and therefore the CIMOSA enterprise activities which can be supported.

The information flows, together with information structures, are represented at the Information Viewpoint. In the CAE–RM this viewpoint is defined via a combination of IDEF0 models and EXPRESS (ISO 10303–11). This combination of models allows the description of the information flows together with the structure of the information elements, their relationships and relevant attributes.

The description of the Computational Viewpoint is based on the Booch Methodology (Booch 1991). This methodology seems to fulfil the requirements needed to describe this view, i.e. define all the computational objects within the CAE system, the activities that occur within those objects and the interactions among the objects. The MOSES CAE system comprises the following objects: Information Models, Design for Function Environment, Design for Manufacture Environment, Engineering Moderator, and Manufacturing Information Generation Environment. These objects are required to design and implement MOSES type CAE systems as object oriented information systems with the characteristics of being modular, flexible and open.

The Engineering Viewpoint focuses on the infrastructure required to support distribution. This view enables the specification of the processing, storage and communication functions required to implement the system. This viewpoint, in our case, is supported by defining an Integration Environment which supports remote procedure call functions. The CIMOSA integrated infrastructure can be used as the basis for the definition of this viewpoint in order to enable the integration of the MOSES CAE System with other CIMOSA based systems. Finally, the Technological Viewpoint focuses on the selection of the necessary technology to support the system. In this research the object oriented database DEC Object/DB and the object oriented programming language C++ are being used.

5 DISCUSSION ON THE USE OF REFERENCE MODELS

Increasing the levels of automation in information systems for the support of design and manufacturing activities is a key issue that must be tackled jointly by users and developers. These information systems should provide decision support and offer significant advantages for the improvement of product development activities. The systems are intended to assist human decision makers exercise judgement. A consequence of this is that the complexity of data, information and knowledge that must be modelled and used by a CAE system increases dramatically. The management of this complexity is an onerous task, and tools and methodologies to help support system development are essential if the current pace of system evolution is to be maintained or improved.

The authors' experience in developing a computer system to support simultaneous engineering activities has allowed them to recognise two key issues in integrated system development: the importance of identifying the real enterprise requirements for information system support, and the need to design and implement systems which are easily configured, standardized and integrated. These issues have been addressed in the MOSES research project by using the CAE framework in order to achieve the required system configuration and by considering the particular requirements of each company involved in the MOSES project. Different configurations of the MOSES CAE system have been developed to tackle the specific problems faced by each of the industrial collaborators.

The CAE Framework allows assessment of the ways in which a particular CAE system, developed based on the CAE–RM, could provide support to different CIMOSA enterprise requirements models which each represent the needs of a particular firm. The Partial CIMOSA Enterprise Requirements Definition Model, developed in this research, can be instantiated to represent the requirements of a company and then the functionality of the MOSES CAE–RM is matched to this model in order to establish the important link between requirements to be satisfied, and the functions provided in the system to satisfy them. This approach has proven to be successful in the case studies carried out to date.

6 CONCLUSIONS

The combined use of CIMOSA and RM–ODP in the CAE Framework has enabled the support of two tasks: identification of requirements for CAE systems in an enterprise, and the introduction of guide-lines for CAE system design and development. However, the organization and management of the timely incorporation of people, methods and tools to evolve the CAE system environment towards the concept of an integrated enterprise requires further research.

7 ACKNOWLEDGEMENTS

The research is part of an ACME funded project, 'Exploiting Product and Manufacturing Models in Simultaneous Engineering', pursued at Loughborough University of Technology and Leeds University, and supported by a group of industrial collaborators [SERC ref. GR/H 24273 and GR/H 24266 respectively]. The research undertaken by A. Molina is funded by the Mexican Government (CONACyT), Monterrey Institute of Technology (ITESM, Mexico) and ORS Award Scheme (ORS/9226010).

8 REFERENCES

Bernus, P. and Nemes, L. (1994) A Framework to Define a Generic Enterprise Reference Architecture and Methodology. *Draft proposal, IFAC/IFIP Task Force on Architectures for Enterprise Integration*, Eight Workshop Meeting, Vienna, Austria, June 11–12.

Booch, G. (1991) *Object–oriented design with applications*, Benjamin/Cummings Inc.

ESPRIT Project 688/5288, AMICE, Volume 1, *CIMOSA: Open System Architecture for CIM*, ESPRIT Consortium AMICE (Eds.), 2nd, revised and extended edition, Springer–Verlag, 1993.

ICAM (1980) *Architects' Manual ICAM Definition Method* (IDEF0), DR–80–ATPC ol.

ISO CD 10303 – 11, *Industrial automation systems and integration – Product data representation and exchange – Part 11: Description methods: The EXPRESS language reference manual.*

ISO/IEC JTC1/SC21/WG7 N 755 (1993) "Draft Recommendation X.901: Basic Reference Model of Open Distributed Processing – Part 1: Overview and guide to use", *ISO/IEC JTC1/SC21/WG7.*

Molina, A., Ellis, T.IA., Young, R.I.M. and Bell, R. (1994) Methods and Tools for Modelling Manufacturing Information to Support Simultaneous Engineering, *Preprints, 2nd IFAC/IFIP/IFORS Workshop Intelligent Manufacturing Systems – IMS'94*, (Ed. P. Kopacek), Vienna – Austria – June 13–15, 1994, pp. 87–93.

Molina, A. et al. (1995) A Review of Computer Aided Simultaneous Engineering Systems, *Research in Engineering Design* 7:38–63.

Williams, T.J. et al. (1993) Architectures for Integrating Manufacturing Activities and Enterprises, in H. Yoshikawa and J. Goossenaerts (eds.), Information Infrastructure Systems for Manufacturing, *Proceedings, JSPE/IFIP TC5/WG5.3 Workshop on the Design of Information Infrastructure Systems for Manufacturing*, DIISM'93, Tokyo, 8–10 November, 1993, North–Holland, pp. 3–16.

9 BIOGRAPHY

Dr Arturo Molina is an Associate Professor in the Integrated Manufacturing System Center at Monterrey Institute of Technology (ITESM), Campus Monterrey, Mexico. He received his University Doctor degree in Mechanical Engineering at the Technical University of Budapest in November 1992, and his M.Sc. degree in Computer Science from ITESM, Campus Monterrey in December 1990. He is a PhD Candidate in Manufacturing Engineering at Loughborough University of Technology. His current research interests include information systems support for concurrent engineering, information modelling, enterprise integration, enterprise modelling, and design and planning of manufacturing systems.

Mr. Tim Ellis is a research associate on the MOSES project. He is primarily involved in determining how best to support design for manufacture in a data model environment. This encompases product and manufacturing modelling and expert application development. He holds a degree in Computer Aided Engineering and an MSc in Computer Integrated Manufacture. His work experience includes time in the structural mechanics and computational physics departments of a large engineering consultancy and the development of multimedia application prototypes for a large automotive manufacturer.

A Mechanical Engineer by training, Dr. R.I.M. Young is currently a lecturer in the Department of Manufacturing Engineering at Loughborough University of Technology. His research interests are in integrated design and manufacture, and software support for concurrent engineering. Prior to joining Loughborough in 1985 he worked for some 10 years in industry, being mainly concerned with the design and development of new products and processes. Dr. Young's recent research publications have addressed issues in product modelling, features technology, process planning, design for manufacture and modelling manufacturing information.

Professor Robert Bell graduated from UMIST in Electrical Engineering in 1954, and with a Masters degree in Textile Technology in 1956. Graduate trainee at Metropolitan Vickers 1956–58; research engineer at UMIST in machine tool research 1958–60. Teaching and research in the field of machine tool engineering at UMIST 1960–75. Appointed Reader in 1972; awarded DSc in 1975. Appointed Professor of Manufacturing Technology at LUT in 1978. Established research in flexible manufacturing systems with particular emphasis given to modelling methods for cell design. Contemporary research interests concerned with the role of product and manufacturing models in computer integrated engineering, concepts for factory modelling, and research into tool management systems. A major additional interest over the last three years has been the international activities in IMS, being a member of the European and International Technical Committees throughout the feasibility study. A further major interest has been support for academic development under the aegis of the ODA and UN, with assignments in Brazil, Hong Kong, India, Mexico, Singapore, Sri Lanka and Turkey.

PART TWELVE

Modeling and Design of FMS II

32

Formal and Informal in Balanced System Specifications

P. Dini, D. Ramazani, G. v. Bochmann
University of Montreal, CP 6128, Succ. Centre-Ville, Computer Science and Operational Research Department, Montreal, H3C 3J7, Canada
phone: (514) 343-6111, fax: (514) 343-5834, e-mails: {dini, ramazani, bochmann}@iro.umontreal.ca

Abstract

The paper presents the relation between balanced system specification by a formal or informal documentation. The text is a technical description of experienced variants, rather than pretentious advices on this topics. We claim that the advent of the object-oriented approach, combined with a mix, formal and informal documentation, in the context of automation, represents a possible solution for either the easiness of knowledge transfer, the use of advanced tools for the system design, or a rapid adaptation from large size enterprises to small ones because of the modularity and reuse. Our report focuses first, on existing techniques, object-oriented methodologies, and second, on lessons learned and retained with respect to the system documentation.

Keywords

Models, formalism, documentation, object-oriented, design methods, experiences

1 INTRODUCTION

The *design of complex systems* is one of the most difficult human endeavour, and sometimes it is difficult to *directly transfer experiences* from a country to another, and even from an enterprise to another. Even further, *human experience* differs from a small to a medium, or large size enterprise. And the *financial support* also differs.

On the other hand, the *automation* of industrial processes or the system design seems to be a right way to enhance the productivity, the quality of service, and the correctness of industrial systems. CAD/CAM systems are the best example of this approach. Also, in the area of telecommunication systems, the automation is inevitable because of increasing tasks and their complexity for a human operator. *Automation* is almost always accompanied by *standards* and *formalisms* for the *system representation. Formal specification* requires staff with high level abilities which is not possible in all kind of enterprise because of *expenses engendered.* How to marry our needs with the resources we have, in the context of a continuous evolution of concepts, technologies, and approaches?

Concept evolution versus industrial needs

The marriage between the field of computer science (CS) and information system (IS) creates a lot of contradictions regarding fundamental concept definitions. The advent of the *object-oriented (OO)* approach having the *object* as a basic paradigm has generated the proposal of many OO methodologies. We consider that an inadequate formalization of basic constructs reflects that the progress is not finished. Consequently, designers of manufacturing systems need continuously to make trade-offs between the requirements resulting from the mismatch of *conceptual evolutions* and the *industrial reality.*

Since *formal specification* is a necessary condition for *automation* in the area of *distributed communications systems* (DCSs), we concentrate in this paper on our experience in using formal and informal specifications for the system description into the object-oriented (OO) approach. We report our experience with respect to the specification of such complex systems. Relevant techniques and methods are briefly presented. Finally, we claim that a certain trade-off between formal versus informal specification must be adopted because the costs spent to re-convert the personnel by training, the time required for a real expertise to be achieved, and system implementations become important.

Experience framework

The present reported experience has been achieved as members of the project IGLOO (Ingénierie du Génie Logiciel Orienté Objet) which is an ongoing research collaboration between the Computer Research Institute of Montreal, three universities, and six industries. While most research in the past has been done in relation with *object-oriented programming* in the context of sequential programming languages, the objective of the *IGLOO research project* (IGLOO, 1992) is to advance the state of the art in the *earlier stages* of the development cycle, related to *requirements analysis* and *specification development.* Special attention is given to the modeling of *active objects' behaviors* and *concurrence*, aspects that are important for the *specification* of *reactive systems* and *real-time applications.*

Paper outline

The remainder of this paper covers several aspects as following. In Section 2 we consider major paradigms of the system development such as *automation, formalism,* and *OO approach.* Section 3 reports our experience with respect to several *OO methodologies* and *formal modelling techniques* used by them. Finally, Section 4 contains what we have *learned* and what we *recommend* with respect to the system documentation.

2 AUTOMATION PREMISES

2.1 Management automation, a fashion?

Usually, the DCS management is manually accomplished by a human operator at a management centre (for the case of the centralized management), or by many operators using their own management stations (for the case of a distributed platform). Evolutionary changes, usually called *dynamic reconfiguration*, can be automatically applied in response *to failures* (sometimes, for unexpected events), to *on demand events* (changes desired by the human operator, or by system components), or by *prevention* (in order to avoid undesired situations, or to prevent a rapid degradation). Methods using predictive mechanisms must be implemented in order to avoid bottlenecks, and to adapt parameters' values.

Coping with these aspects, the *automatic management* consists of three major tasks namely, (i) the acquisition and the representation of data with respect to the state of the system, (ii) the interpretation of these data having as results appropriate decisions of distributed computing systems (DCSs) state changes, and (iii) the application of these decisions (reconfiguration) and the evaluation of the quality of reconfiguration.

Many studies have identified the *automatic management* paradigm as having a major importance in DCSs, coping with the complexity and the diversity of several management aspects such as management architectures, data updating techniques, reasoning policies, reconfiguration techniques, as well as methods to evaluate the need and the quality of reconfiguration (Race, 1991)(Stalling, 1993)(Bapat, 1994).

2.2 Formalism, a need?

In the last decade, more and more authors emphasize the major role of the formalism in software development. Some consider that a correct specification is formal (Kilov, 1994). Coleman argues that if formal notations were incorporated into object-oriented methods, then we could expect coherent models, rigorous object-oriented methods, and CASE tools which can check the semantic of models (PANEL, 1991).

There are four significant formal approaches for the DCS's specification namely, *modular configuration languages, formal description techniques, axiomatic languages, and object-oriented concepts.* The standardization organizations have adopted so-called *formal description techniques* (e.g., LOTOS, Estelle, SDL) for the description of OSI protocols and services, each of them having certain advantages. *The axiomatic languages* (Z, Object-Z) are extensively analyzed in (Stepney, 1992). *The object-oriented approach* is a new modelling approach where entities's services are presented into so called «object interfaces».This approach is taken for the *information architecture* by the international organizations of standardization.

2.3 OO approach, a mode?

Current object-oriented (OO) software techniques based on the fundamental paradigm of *object* have been extensively presented in (Snyder, 1993)(Bochmann et al, 1994). The OO approach must cope with the complexity of DCSs and solve different aspects related to 1) *cooperating/coordinated/communicating systems,* (2) *reactive/interactive/reflective systems,* (3) *centralized/distributed systems,* and (4) *deterministic/nondeterministic systems.* In using OO concepts for DCSs, we distinguish two distinct goals namely, (1) the structural and functional OO models, and (2) the management OO model.

Each software or hardware DCS entity is modelled as an object which encapsulates data as *attribute values* and the behavior as methods implementing its own services *(operations).* Objects communicate via *messages* with other objects, which means that an object requests to another object to execute one or several of its own operations.

3 EXPERIENCED METHODS

We describe here several experienced solutions with respect to existing OO methodologies, frequently used formalisms, and documentation of complex systems.

3.1 Existing OO methodologies

Since detailed comparative studies have been presented in the last three years, we only illustrate major specific features of the most used object-oriented methodologies.

- *Shlaer-Mellor's method.* In this method (Shlaer and Mellor, 1988, 1992) objects are involved in relationships and constructions (subtypes/supertypes or associative objects). The behavioral view is concerned with the Object Communication Model (OCM). Each class is given a state diagram which describes its life-cycle (state model). State models are expressed in state-transition diagrams (Moore model) and tables. The dynamics of relationships captured by state-transition models, relationships without life-cycles, concurrency, and monitors must be analyzed. Action data flow diagrams (ADFDs) depict the Process Model.

- *Wirfs-Brock et al's method.* This method covers both the analysis and design phases, since no structured input is required (Wirfs et al, 1990). The analysis phase refers to factoring common responsibilities in order to build class hierarchies, streamlining the *collaboration between objects* by nesting same encapsulation levels or by creating subsystems. The design phase consists of the definitive collaboration establishing, based on interaction of kind of client/server paradigm.

- *Coleman et al's Fusion method.* The analysis phase of the Fusion methodology separate the behavior of a system into a user perspective (Coleman et al, 1994). The analysis captures different aspects within two models namely, Object Model which defines the static structure of the information, and Interface Model which specifies the input and output of the system. A System Object Model is derived from the Object Model as a subset of objects relating to the system to be built.

- *Booch's method.* Specification is build within two logical models (Class, Objects), and two physical models (Module, Process) each of them having in turn a static or dynamic semantic. This permits to separate independent types of analysis and design (Booch, 1994). Statics refers to individual objects and the relationships between classes and modules, while events occurring dynamically can be expressed for classes in the State Transitions Diagrams and for objects in Scenario Diagrams. The inter-object behavior is set out in the object scenario diagrams.

- *Rumbaugh et al's OMT method.* The methodology uses three kinds of models to describe a system (Rumbaugh et al, 1991). The Object Model defines the objects in the system and their relationships, while object interactions are specified in the Dynamic Model. Data transformations are represented in the Functional Model. Rumbaugh's OMT uses three techniques namely, Object Diagrams for the Object Modeling into a data-perspective, Harel's Statecharts for the Dynamic Model into a behavioral perspective, and Data Flow Diagrams in the Functional Model.

3.2 Formal modeling techniques used by methodologies

Several modeling techniques used in software design methods have been experimented and analyzed. Some of them are recommended by the previous OO developing methodologies.

- The *entity-relationship (E/R)* approach is a widely accepted technique for data modeling (Chen, 1976). Many OO methods apply the E/R model for identifying objects and their relationships. We consider that this model can be used to easily separate objects (subsystems) from their environments and to discriminate between system internal relations and system-environment relations. A dialect called *Object Diagrams* is frequently used by different OO methodologies.

• The *FSM* model is a well-known model for describing synchronous sequential machines (Gill, 1962). The model explicitly represents the entire state space. The *Communicating Finite State Machines* (CFSMs) could be used for describing cooperations.

• *Petri Nets* (PNs) is more general than FSMs. A Petri Net of cooperating objects formally models the information flow. The PNs (Peterson, 1977) allows to hierarchically model a system, both the hardware aspects (control units), and the software ones (resource allocation, deadlock, process coordination, etc.).

• *Statecharts* additionally permits timed transitions and nested states (Harel, 1987). Statecharts diagram is an adapted tool to describe the behavior for one object since they could accommodate temporal aspects and the state evolution.

• *Objectcharts* combine an extension of Statecharts with object-oriented analysis and design techniques (Coleman, 1992). Objects have a trace formalization of their behavior, that is, a set of service requests as a triple (a, s, b) where s is the service between the service required (a) and the service provider (b). There is no type definition, but a class is defined as a template for objects belonging to it. Object relations are described in a configuration diagram.

• *Message Sequence Charts* refer to the systems specifications with respect to the *behavior.* Z.120 by WPX/3 (Ccitt, 1993) has *standardized MSCs* that correspond to *information flow diagrams.*The *object behavior* is represented by state transition graphs, whereas *block interactions* by *interaction diagrams* accompanied by *structured English text.*

• *Traces* are a technique invented by Parnas and Bartussek (1978). An object's *externally observable behavior* is completely defined if its output values are defined for any trace. We experienced that traces are useful to build scenarios at the border between a system and its environment (Bochmann, 1994).

• *Temporal logic* (Pnueli, 1977) is suitable for specifying properties on each level of abstraction (requirements), for formal reasoning, and verification. Different non-traditional temporal logic have been proposed, in order to *produce precise and complete formal specifications* and to *improve the understanding.* Our experience demonstrates that temporal logic is a useful tool for *specifying* interactions between objects (subsystems). However, the link of specification with the implementation is more difficult to be established.

4 WHAT HAVE WE LEARNED?

4.1 Development process

Object-oriented computing is a developmental approach which encompasses a number of concepts whose goal is to produce in a relatively short time software which is more reliable and easier to maintain. It covers all the steps of software development, from analysis to programming. An important feature of a such development approach is that the analysis, design and implementation phases adopt a similar model. The approach permits to model system structures which consists of objects and their relationships, the dynamics of systems (behavioral interaction among these objects), and to impose rules of cooperations (assumptions on invariants, preconditions, or postconditions). Software development tasks are distributed in models corresponding to the appropriate phases. At the analysis phase the *analysis model* prescribes what is the system and its environment. The *design model* adds to the analysis model aspects defining how the system operations are performed by object behaviors or inter-objects behaviors. Since machines commonly recognize only programming languages, the design model is converted into an *implementation model*. Consequently, the generality of concepts across software development cycle is embedded into a particular and restrictive model expressed in a *formal* code (language primitive) and in an *informal* (structured or non-structured) English text. Ideally, similar to the OO model, an unified and runable specification is desired, in order to avoid the manual transition between the design model and the implementation model. By experience, an OO design model does not fit well into a non-OO implementation model. But, and this is a thorny aspect, even into an OO language this translation is not easy and, more important, it is difficult to re-trace all correspondences between these two models. The problem could be solved across well identified translation relations which we briefly represent in Figure 1. The programer performs an horizontal translation from informal specification to formal specification according to the power of the implementation language and its experience (at least for the structured English text). Non-structured text is hopefully formalized by code components connections. The remainder is classified as code annotations. We remark that this translation is language-dependent and programer-dependent.

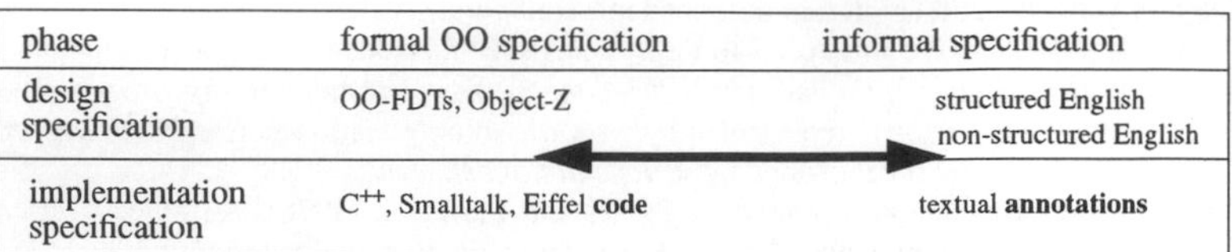

phase	formal OO specification	informal specification
design specification	OO-FDTs, Object-Z	structured English non-structured English
implementation specification	C^{++}, Smalltalk, Eiffel **code**	textual **annotations**

Figure 1 Horizontal translation between formal and informal specification

4.2 System documentation

For a good specification, the programmer plays an important role. First, there are two programmer classes namely, which easily use formal description techniques and which do not. Commonly, actual software documentation has a *formal part* and an *informal part* at any level of the development cycle. Second, part of the design informal documentation could be formalized by means of programming languages, while part of design informal documentation is identified as either a *structured* or *non-structured English* text in the implementation documentation (program annotations). We mention that the formal and informal parts at the analysis/design level are not exactly the same with respect to those at the implementation level. The balance is influenced by many *objective(o)* or *subjective(s)* factors.

Among these factors there are (1) the magnitude of modeled systems (o), (2) the methods chosen for the description of object-oriented models (o), (3) the power of expression offered by the used description techniques (o | s), (4) the kind of programming language (o), and (5) the ability of the programmer (o | s). Establishing itself these criteria and their nature is an objective or/and subjective decisions:

(1) Criteria for the magnitude depend on the real systems versus a particular specification method, i.e. PNs, FSMs, Traces, Objectcharts are suitable for different sizes of applications;

(2) Despite of some authors claim of generality, an object-oriented method is better for a particular type of systems than others;

(3) Here are some examples of the formalism expressiveness: the PNs formalism offers the possibility to describe the concurrence, while the language Z does not;

(4) Only with respect to the inheritance we present two criteria: (a) some programming languages allows the multiple inheritance (C++), while other do not (Smalltalk), and (b) commonly all OO programming languages offer the possibility of monotonic inheritance, however, Eiffel permits to specify the selective inheritance.

(5) The programmer is an important player. Its experience, ability with a particular method, language, or techniques can effectively improve the quality of the implementation documentation.

In Figure 2, we present possible translations between non-structured English, structured English or formal specification across the analysis, design and implementation phases. We recommend that at each development phase these three parts of the documentation co-exist.

Transitions between software development phases must be carefully documented, especially when the specification forms are translated (for example, cases B and C). Thus, activities attached to each arrow must be precised into an additional documentation.

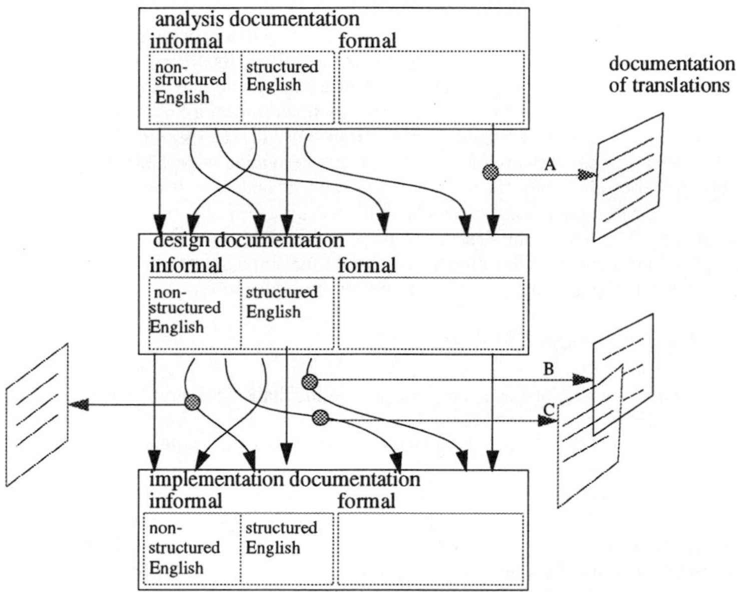

Figure 2 Documentation between phases of development lifecycle

For example, particular procedures describe the *A case* (from *formal* to *formal*) when one passes from analysis to design in the Shlaer-Mellor's method. The authors prescribe transformations rules as procedures for producing architectural objects (from the analysis to design) in the form of specialized class diagrams and class structure charts. From the design to the implementation multiple transformations could be performed, depending on the programmer or on the programming language. In the same method, for example, modules of the archetype class structure charts are expressed as archetype code (code templates) in the chosen implementation language. These templates can be filled out from the specialized class structure charts. As presented in Figure 2, this filling is not easy because of various translations (either explicitly or implicitly) when one transits from a development phase to another.

Many structured English texts can be inserted in algorithms (*case B*). Sometimes this insertion is a part of an implemented method. Sometimes it is together with formal specification, leading the identification of translation more difficult (example, IF <structured English text | formal specification> THEN <structured English text | formal specification>, ELSE <structured English text | formal specification>). Nevertheless, all it is the program code, but we can distinguish these two forms. Otherwise, part of these structured English texts could be informal annotations of the code. We experienced that it is recommended that the implementation documentation records the reasons of such distinctions. A later interpretation of these documentations could open ways for future adaptations or facilitate them.

Even further, it is possible that a non-structured English text semantics has its formal representation in the code (*case C*). Usually, is the programmer which must update the implementation documentation showing unambiguously where, how, and why this approach is used. Commonly, such of situations appear if either there is a superposition between the informal text and the formal specification (implicitly), or the programmer has identified programming language features which can express informal assumptions. Usually, an implementation from a formal specification could contain annotations in structured English derived from those design specifications which are not directly translated into a programming code. Since this derivation is possible, the reverse is not true because first, one could have many derivations, and second, it is difficult to directly isolate within the code a part which is not explicitly indicated by the programmer. Moreover, existing non-structured English text in the design documentation could be textually represented in the implementation documentation in order to ameliorate the implementation comprehension. For example, an Object-Z specification could contain English explanations.

4.3 What have we retained?

An ideal solution is not possible, and compromises are inevitable. The best issue is to choose the *right level of abstraction* in order to easily *formalize* both the system and its environment. This formalization makes possible the matching between models and their implementations, and facilitates the evolution of the systems by the *reuse* promotion. An assertional or axiomatic formalization (rules or axioms) facilitates the traceability of development assumptions across the development models, while an operational formalization (state sequences, algorithms) does not in the communicating systems. The drawback is that the former are recommended for small and middle systems, while the latter are specific for large systems. A trade-off is the best solution in order to get advantages from both two forms of formalization.

Evidently, we cannot formalize all. Our opinion is we must first distinguish between *critical aspects* which will be formalized, from the *non-critical aspects,* which could be informally done. A *structured English* for the latter help us to easily understand the code and to make updates without disturbing in an unexpected way the contract between system and its environment.

Acknowledgments

We would like to acknowledge and thank all colleagues involved in the IGLOO project. Particularly we thank Michel Barbeau, Hafid Mili, Jean-Michel Goutal, and Jean-Marc Serre for fruitful discussions, cooperations, and their support. However, the presented opinions are our own experience and do not necessarily reflect the positions of all team members.

5 REFERENCES

Bapat, S. (1994) Richer Modeling Semantics for Management Information, *Integrated Network Management,* III, eds. H.-G. Hegering and Y. Yemini, Elsevier Science Publishers B.V. (North-Holland) 1993, IFIP, pp. 15-28

Bartushek, W., Parnas, D.L. (1978) *Using Assertions about Traces to Write Abstract Specifications for Software Engineering*, Proceedings of the Second Conf. European Cooperation in Informatics, New York, Springer-Verlag,1978

Bochmann, v. G., Dini, P., Barbeau, M., Colagrosso, P., Keller, R., Ramazani, D., Somé, S. (1994) *Common Concepts for Object-Oriented Analysis and Design*, IGLOO Project, Deliverable M.a.1.2., CRIM, January 1994

Booch, G. (1994) *Object-Oriented Analysis and Design*, The Benjamin/Cummings Publishing Company, Inc.

Ccitt (1993) CCITT Recommendation Z.120, *Message Sequence Chart*, Geneva, 1993

Chen, P.P. (1976) *The Entity-Relationship Model: Toward a Unified View of Data*, ACM Transaction on DataBase Systems, vol. 1, no. 1, 1976

Coleman, D., Arnold, P., Bodoff, S., Dollin, C., Gilchrist, H., Hayes, F., Jeremaes, P. (1994) *Object-Oriented Development: The Fusion Method*, Prentice Hall, Englewood Cliffs

Gill, A. (1962) *Introduction to the Theory of Finite State Machines*, McGraw-Hill

Harel, D. (1987) *Statecharts: A visual formalism for complex systems*, Science of Computer Programming no. 8, 1987, pp. 231-274

IGLOO (1992)*** *Object-oriented Specification and Design Methodologies*, Research Proposal, CRIM, March, 1992

Kilov, H., Ross, J. (1994) *Information Modeling*, PTR Prentice Hall, Englewood Cliffs

PANEL (1991), *Formal Techniques for OO Software Development*, OOPSLA'91

Pnueli, A. (1977) *The Temporal Logic of Programs*, 19-th Annual Symposium on Foundations of Computer Science, Providence, RI (IEEE, 1977), pp. 46-57

Race (1991) CFS M200-M210, *RACE Common Functional Specifications, Document 8, Network Management Evolution,* Issue B (Research and development in Advanced Communications technologies, in Europe), 1991

Rumbaugh, J., Blaha, M., Premerlani, W., Eddy, F., and Lorensen, W. (1991) *Object-Oriented Modeling and Design*, Prentice-Hall, Inc.

Shlaer S., Mellor, J.S. (1988) *Object-Oriented Systems Analysis: Modeling the World in Data*, Yourdon Press, PTR Prentice Hall, Englewood Cliffs

Shlaer S., Mellor, J.S. (1991) *Object Life Cycles: Modeling the World in States*, Yourdon Press, PTR Prentice Hall, Englewood Cliffs

Snyder, A. (1991) *The Essence of Objects: Common Concepts and Terminology*, Software and Systems Laboratory, HPL-91-50, June

Stalling, W. (1993) *SNMP, SNMPv2, and CMIP: The Practical Guide to Network Management Standards,* Addison Wesley Publishing Company

Stepney, S. (1992) *Object Orientation in* Z, Logica Cambridge Limited, U.K.

Wirfs-Brock, R.J., Wilkerson, B., Wiener, L. (1990) *Designing Object-Oriented Software*, Prentice Hall, Englewood Cliffs

6 BIOGRAPHY

Petre DINI received the MS in Computer and Automation Science from the Polytechnical Institute of Timisoara, Romania, in 1976. Since, he has developed industrial systems related to CAD/CAM, robotics, and automatic machine tools. He has interested on knowledge representation in electronics, electrical and mechanical domains, modelled and designed distributed industrial chains, having many publications in these areas. He is currently involved in Ph.D. program at the University of Montreal. His interests are related to automatic reconfiguration in networks and open distributed systems, QoS in multimedia applications, software engineering formalization, software and hardware tests, telecommunications, and the object-oriented concepts related to these fields.

Dunia RAMAZANI received the MS in Computer Science from the University of Montreal, Canada, in 1992. He was a visiting researcher at the Bavarian research Center for Knowledge Based Systems (FORWISS), Munich, Germany, during the first quarter of 1993. There, he has worked on the design and implementation of Object-oriented database management systems. Since then, he has been working towards the Ph. D. at the University of Montreal. As part of his Ph. D. work, he is involved in the IGLOO project, a project aimed at contributing to the advancement of the object paradigm. His current research interests include formal methods, object-oriented software engineering and telecommunication systems.

Gregor v. BOCHMANN (M'82-MS'85) received the Diploma degree in physics from the University of Munich, Munich, West Germany, in 1968 and the Ph.D degree from McGill university, Montreal, P.Q., Canada, in 1971. He has worked in the area of programming languages, compiler design, communication protocols, and software engineering and has published many papers in this area. He holds the Hewlett-Packard-NSERC-CITI chair of industrial research on communication protocols in the Département d'informatique et de recherche opérationnelle, Université de Montréal, Montréal. His present work is aimed at design methods for communication protocols and distributed systems. He has be involved in the standardization of formal description techniques for OSI. From 1977 to 1978 he was a Visiting Professor at thc Ecole Polytechnique Fédérale, Lausanne, Switzerland. From 1979 to 1980 he was a Visiting Professor in the Computer Systems Laboratory, Stanford University, Stanford, C.A. From 1986 to 1987 he was a Visiting Researcher at Siemens, Munich. He is presently one of the scientific directors of the Centre de Recherche Informatique de Montreal (CRIM).

PFS/MFG: A High Level Net for the Modeling of Discrete Manufacturing Systems *

José Reinaldo Silva †
Paulo Eigi Miyagi ‡
University of São Paulo, Escola Politécnica, São Paulo, SP, Brazil, fax: +55-11-818-5471, e-mail: jorsilva@fox.cce.usp.br

Abstract

A formalism for an extended Condition/Event Petri Net, PFS/MFG (Production Flow Schema/Mark Flow Graph) is introduced in this work. The algebraic relations provided can be used in behavior and structural property analysis of discrete systems modeled in this technique. Behavior analysis is based on a simulator which specifications and some algorithms are presented here. The resulting net can be applied to the design, modeling and evaluation of configurations in Factory Automation as well as to discrete shop floor control. Briefly, we describe a realistic application of the revised PFS/MFG net in a medium size printers factory.

Keywords

Factory automation, manufacturing systems, discrete event dynamic systems, Petri Net, Production Flow Schema, Mark Flow Graph

1 INTRODUCTION

A Petri Net (PN) (for example: Reisig, 1985) is a graph representation composed of two classes of elements, normally associated with static and dynamic aspects of a system, respectively. Suitable target systems generally are non deterministic discrete event systems. Concepts such as concurrence, parallelism and synchronism are the basis over which the real model are depicted. The graphic presentation works as a design tool and plays a visual-communication role similar to flow charts, block diagrams and networks (Murata, 1989). Thus, a PN representation of a system can be used as a design schema as well as a supervising

* This work is part of the research activities of the Flexsys-Esprit 76101, UE.
† Partially supported by FAPESP, Brazil.
‡ Partially supported by CNPq, Brazil.

framework, or feedback control (including the verification of correctness where PN supports simulation algorithms).

The formal interaction between abstract specification and practical implementation makes the modeling and analysis of production systems using PN an important contribution to the formalization of the design process in engineering. However, the modeling of large and complex[1] systems in PN raises some difficulties. The problem will be even worse if we try to analyze structural properties such as invariants, synchronic distance, etc. One solution for this problem would be to search for a more synthetic representation based on the definition of high level and extended nets. A successful approach in discrete manufacturing systems is the MFG (Mark Flow Graph), which include compounded elements (for example: Hasegawa, 1987). In a MFG, each element called *box* can represent a single element or a "static composed element", that is, a subnet. Thus specific ways to store items, such as FIFO, LIFO, etc., can be represented. Assembling and disassembling processes (which affects only the item flow) can be also abstracted by a *box*.

Another view of the same problem argue that the modeling difficulties derive from the inexistence of a more structured design methodology. An hierarchical approach, called PFS/MFG (Production Flow Schema/MFG) was developed by Miyagi, 1988, based on Condition-Event PN (C/E nets) and including abstract elements called *activities.* Activities stand also for an entire sub-net and introduce the concept of "dynamic composed element". The duality of PN is then restored. In this approach, each *activity* can be substituted by a sub-net starting and ending with an event, while a *box* can be substituted by a subnet starting and ending with conditions [2].

In the present work we discuss the algebraic representation of PFS/MFG briefly referring to its structural properties. Reachability conditions will take a more important role when we discuss the specifications of an object-oriented simulator now being developed. To enrich the presentation we will show an application example based on a real medium size company that produces printers. The application is concerned with the analysis of the item flow in the factory in order to automate the transport system. The analysis was based on a software tool called AIPUSP (Intelligent Environment to the Design of Production Systems) which first version was presented in (Lucena, 1989) and refined since then until the new object-oriented model showed here.

Before discuss the structural analysis we will present some basic definitions and notations.

2 BACKGROUND DEFINITIONS

In this section we will introduce the basic notation that will be used in this work regarding C/E nets and its derivations.

---

[1]The concept of large and complex are quite imprecise since there is not a formal definition of "large system". We risk a heuristic concept of "large and complex"(based on the experience with Petri Nets Design) as a system with more than 40 conditions and more then 40 events, and where the respective incidence matrix is not sparse.

[2]We are in fact admiting the existence of two kind of subnets: those which are ressamble active elements - as milling machines or any kind of transformation of material - and pasive subnets like assembly process or storing, where no transformation is performed.

[*Def 2.1*] A *C/E net* N is a triple, $N = (B,E,F)$, where:

i. B and E are disjoint sets;

ii. $F \subseteq (B \times E) \cup (E \times B)$ is a binary flow relation in N;

iii. A subset $c \subseteq B$ is called a *case*.

iv. If $e \in E$ and $c \subseteq B$, e is enabled by c iff $^{\bullet}e \subseteq c \wedge e^{\bullet} \cap c = \varnothing$.

A function $M{:}B \to \mathbb{N}$ is called a *marking*. A marked C/E net $<N,M_0>$, is a C/E net N with an initial marking M_0.

Generally, we define the *enabling vector* as the mapping $\sigma : E \to \{0, 1\}$, indicating which events are currently enabled.

If $e \in E$ is enabled at a marking M, then e may be fired yielding a new marking M' given by the equation :

$$M' = M + A^T\sigma \qquad (2.1)$$

for $\forall c \in B$, $M[e> M'$ denote that M' is reachable from M by firing the event e.

The *incidence matrix* $A=[a_{ij}]$ is an $n \times m$ matrix of integers $\{-1,0,1\}$ given by: $a_{ij} = a_{ij}^{+} - a_{ij}^{-}$ where, a_{ij}^{+} is the arc from event i to its output box (post-condition) j, and a_{ij}^{-} is the arc from the input box j to event i.

A finite sequence of transitions $\sigma=\{e_1,e_2, \ldots\ e_n\}$ is a finite firing sequence of $<N,M_0>$ iff there exist a sequence: $M_0[e_1>M_1[e_2>\ldots\ [e_n>M_n$ such that, for $\forall i,\ 1 \le i \le n,\ M_{i-1}[e>M_i$. It's said that the marking M_n is reachable from M_0 by firing sequence σ, i.e., $M_0[\sigma>M_n$.

C/E nets are very useful in the analysis of the qualitative behavior of discrete systems. The drawback in applying this technique to the modeling and analysis of manufacturing systems is its expressiveness, principaly when a big number of conditions is involved. Unfortunately, modeling real systems frequently implies in the representation of a great number of conditions and leads to complex net structures. This is the principal reason to study extensions to the basic formalism even when they are amenable only to some specific application domains such as manufacturing systems.

MFG is one PN extensions attached to the modeling of discrete manufacturing systems. The main idea was to base the design of factory systems in suitable macro-elements such as assembling stations, intermediary buffers, etc. When dynamic parts are not important in the modeling, assembling stations and storage mechanisms can be considered static elements and reduced to conditions such as "if a number of pieces k is ready then a product is assembled".

PFS, on the other hand is a representation of the specifications (or design) in a top-down refinement methodology. It introduces abstract elements which are supposed to be filled later with a proper detailed version. The final result would be a MFG, and that is the reason the whole methodology is called PFS/MFG. However the abstract model (PFS) and the final representation of the system (MFG) are kept apart, that is, the designer should make a PFS representation first and then refine it until it reaches a MFG form. Also, the verification of the model (by simulation for instance) could be done only when the MFG form is reached.

Following we resume a revision of PFS/MFG to unify the abstract representation of PFS and the macro-objects in MFG by interpreting it as a bipartite graph where the nodes belong to two different classes of sub-graphs (similarly to the conventional PN).

2.1 PFS/MFG Basic Elements

A PFS/MFG would be defined as a triple (*B*,*A*,*G*), where *B* and *A* are disjoint sets of composed elements called boxes and activities respectively. Each element in *B* or *A* is represented by an object characterized by a set of attributes. For simplicity let us consider that a generic class box can be characterized by *box* :: <*#name*,*#mark*>, where *#mark* is a boolean variable to indicate when a box is marked. A condition, such as in conventional C/E net is an object instance of the class box.

In PFS/MFG there is also the following sub-classes of the class *box*: {*time_box*, *capacity_box*} where, *time_box* :: <*box* | *estimated_time*>, that is, it inherits all the attributes of box and have an extra parameter with the estimated time to enable the firing. Similarly *capacity_box* :: <*box* | *capacity*> where capacity is the maximum number of marks it could accept. There are also other sub-classes for *capacity_box* defined as: *static_subnet* :: <*capacity_box* | *#io_relation*, *#pt_subnet*> where, (to be consistent with the original net) we would call *distributed_box* to an object of the class *static-subnet* that has *io_relation*=(1 : *n*) and *assembling_box* to one with *io_relation*=(*n* : 1). The attribute *#pt_subnet* is a pointer to a static subnet. We could also go further, defining elements with other *io_relation* values but what we have done so far is enough to establish the relationship between conceptual elements and extended ones.

The classs *activity* would be defined as the aggregate: *activity* :: <*#name*> with sub-classes:

- *single_activity* :: <*#name*, *#list_of_con*> and
- *compound_activity* :: < *activity* | & {[2, *single_activity*],[1,*time_box*]}>

where, the last definition includes a part-of aggregation (Embley, 1992),

Similarly to conventional nets, the *G* elements (*gates*) are relations defined in the set $(B \times A) \cup (A \times B)$ and will not be represented as another class of objects. However we can classify *G*-elements as: {*$flux*, *$int_gate*, *$ext_gate*} acoording to the following criteria: for a generic flux relation $g=(a,b)$, *g* will be a single flux relation if there is no persistent marked element in the pair (*a*, *b*). If *a* or *b* has a persistent mark, then the relation will be an internal or external gate if the element with the persistent mark belongs or not to the system (it is not a pseudo-box) respectively. Then, an "external condition" is a rule that is not part of the system in the sense that its satisfaction is not under its control. Reset buttons, power switches, quality assurance, and other anthropomorphic control actuators are examples of "external conditions". Internal and external gates can still be classified as enabling or disabling according to whether they can enable an activity when it is marked or when it is unmarked.

We can define our extended net as a bipartite graph, similarly to conventional C/E nets. Activities will play the role of dynamic elements and boxes the static counterpart. Conventionally, activities will start and end with input and output events, while macro boxes will start and end with input and output conditions respectively. Both can stand for sub-graphs. This full complementary relationship between our definitions and those of conventional PN make it possible to mix abstraction levels in the same representation, that is, composed elements such as macro boxes or activities can be connected to single conditions and events.

A normal flux relation can be compared with internal and external gates. An external gate is a flux relation between an element within the component or object being considered and another one outside the component. We associate the source (destination) of the relation as a pseudo-condition, in the sense that it will appear to the component as an unique condition which could not be explicitly represented with its base set of conditions. For instance, if we consider a manufacturing cell controlled by a central station, a control sign sent to a NC-machine in a cell would be represented by an external gate in the sense that the rationale of the sign is in the control algorithm located in the station. If the design (model) should be modular the machine cell should not be concerned with the functionality of the controller. An enabling external condition can then be interpreted as a relation between an external pre-condition and a dynamic element in the component. On the other hand, a disabling external condition would stand for a relation between a dynamic element in the component and an external post-condition.

The definition of enabling activity is exactly the same as in [*Def 2.1.iv*], except by the requirement to verify the local time. Thus, an activity with estimated time τ will be enabled if when its local time is zero (an activity has local time zero if it is not operating) all its pre conditions and input macro boxes have at least one mark, and all its post-conditions and output macro boxes have at least an empty space by that time. When thw firing occurs, the time box will retain the mark, the local time will start counting and the process represented by that activity will be unavailable. If the post-condition is not enabled when the estimated time is reached the activity would hold the mark. Thus τ is in fact a lown bound.

In the next sub-section a state equation for this firing process will be proposed.

2.2 PFS/MFG State Equation

As we have seen in eq. (2.1), a new state in a conventional C/E net depends linearly from the previous events and from the events currently enabled. We can show that in PFS/MFG the state equation is similar to the conventional if we follow the definitions above. Actually, we intend to show that PFS/MFG preserves most of the formalism and properties of PN and also improves its expressiveness.

First of all we will establish that the incidence matrix will be built including all the pseudo-conditions and consequently all the gates will be normally included. Since pseudo-conditions have a persistent marking it is imperative that the marks be restored in the firing process. A term $\Delta^T\sigma_k$ replaces the marks for a given enabling vector. Thus modified state equation is, $M_{k+1} = M_k + (A^T - \Delta^T)\sigma_k$ where, A is the new incidence matrix and σ_k is the enabling vector in a state of M_k. The matrix Δ has the same order that the incidence matrix and is given by: $[\ 0 \ \ \Pi\ ]$ where, Π is the sub matrix of the pseudo-conditions and 0 is a matrix $(n\text{-}l) \times (m\text{-}k)$ composed of zeros if the order of Π is $(l \times k)$.

As an example, let us take the PFS/MFG of Fig.1, which incidence matrix (each column describe the link relations to each passive element in the order of the indexes in the net) and Δ matrix (we prefer to keep this term - sub matrix of the pseudo-conditions - separately from the incidence matrix because matrix A has all the property relations of the net, what will be used to evaluate properties as well as the firing vector) are:

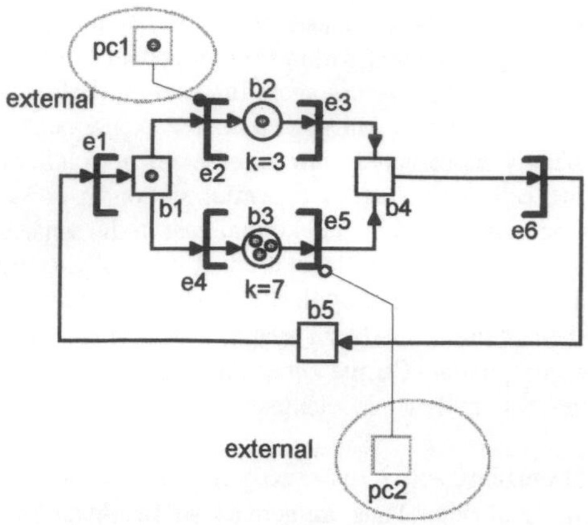

Figure 1 A PFS/MFG example.

$$A = \begin{pmatrix} 1 & 0 & 0 & 0 & -1 & 0 & 0 \\ -1 & 1 & 0 & 0 & 0 & -1 & 0 \\ 0 & -1 & 0 & 1 & 0 & 0 & 0 \\ -1 & 0 & 1 & 0 & 0 & 0 & 0 \\ 0 & 0 & -1 & 1 & 0 & 0 & 1 \\ 0 & 0 & 0 & -1 & 1 & 0 & 0 \end{pmatrix} \qquad \Delta = \begin{pmatrix} 0 & 0 & 0 & 0 & 0 & 0 & 0 \\ 0 & 0 & 0 & 0 & 0 & -1 & 0 \\ 0 & 0 & 0 & 0 & 0 & 0 & 0 \\ 0 & 0 & 0 & 0 & 0 & 0 & 0 \\ 0 & 0 & 0 & 0 & 0 & 0 & 1 \\ 0 & 0 & 0 & 0 & 0 & 0 & 0 \end{pmatrix}$$

A Direct Algorithm to Evaluate the Enabled States

Although the firing vector cannot be built by simple numerical calculations it is important to draw a line between what could be done by direct analysis and what should be partially automated by using knowledge systems (Lucena, 1989). In order to specify formaly the algorithm let us first define a direct product of vectors:

[*Def.* 2.2] The direct product of two column vectors A and B is a column vector C with the same order, such as $A \otimes B = C = [c_{ij}] = [a_{ij} .b_{ij}]$.

If we notice that each line v_k in the matrix A describe the pre and post-conditions of an event e_k the proposition follows,

[*Proposition 1*] An event e_k is enabled in a marking state M, if and only if $M_i \otimes v_k = [{}^{\bullet}e_k]$ (where $[{}^{\bullet}e_k]_j = 1$ if b_j is a pre-condition of e_k and zero otherwise).

This proposition is a PFS/MFG rephrasing of the conventional firing conditions from which the demonstration can be derived straightforwardly. Therefore, the firing vector is build according to the rule:

$(M_l \otimes v_k = [\bullet e_k]) \Rightarrow \sigma_l|_k = 1$
otherwise $\sigma_l|_k = 0$ (2.2)

Taken in account that the direct product is associative, commutative and distributive, the antecedent of the previous equation can be rewritten as:

$$[(2.M_l - 1 + v_k) \otimes v_k]^2 = 0 \quad (2.3)$$

where, the square means the scalar product . This equation can be evaluated directly from the terms in the state equation, dispensing the need to distinguish the input relations from the output one as implied in the original construction (equation 2.2). However those are only necessary conditions. Enabled events can be in conflict or contact what demands a further analysis of the enabling vector derived from the rule above. Two generic events e_k and e_r can be:

- independent if, $\forall j,\ (v_k \otimes v_r)|_j = 0$
- in contact if, $\exists j,\ (v_k \otimes v_r)|_j = -1$
- in conflict if, $\exists j,\ (v_k \otimes v_r)|_j = 1$

Normally we should allow only orthogonal events to be enabled simultaneously (independent events). This is a very strict rule which make models less flexible. We can now allow some of the conditions in the second group to be fired even because contact do not have the same meaning in a safe net that has capacity boxes or because we would like to model explicitly the synchronized start of two parallel events in distributed systems. Such decision can be taken according a policy previously set in knowledge systems or taken with the assistance of expert systems. This work is not concerned with such decision support systems.

Managing Multiple Marks in Capacity Boxes

One of the original features of PFS/MFG is the capability to represent the flux of control as well as numerable items or conditions. The capacity box is the basic element capable to hold multiple marks. We will deal only with this element here, since any other boxes can represented by a static sub-net (starting and ending with a box or condition) of the basic elements.

As it was mentioned before, PFS/MFG is a safe net, that is, even holding multiple marks in capacity boxes, each connection admit the flux of one mark at a time. Thus, the multiplicity of marks should be taken in account only to build the firing vector.

According the strict firing rule represented in Proposition 1, we need to know if there is at least one mark in each pre-condition and if there is capacity (space) for at least one more mark in each post-condition. Therefore, the marking vector (which is no longer unitary) has to be modularized to a unitary one. The modularization function is defined as follows:

[*Def. 2.3*] Let C be the capacity vector of a PFS/MFG net, that is, a vector where each c_{lj} represents the capacity of the static element j. Let be the marking vector M, and the event vector e_r (a line r in the incidence matrix). The modularization is defined as: $\mu(M_l, e_r) = \overline{M_l}^{e_r}$ where:

$$\mu(M_l, e_r)|_k = \overline{M_l^{e_r}}|_k = \begin{cases} 1 & \text{if } (M_l \otimes e_r)|_k < 0 \\ 0 & \text{if } (M_l \otimes e_r)|_k > 0 \text{ and } C|_k - (M_l \otimes e_r)|_k > 0 \\ M_l|_k & \text{in all the cases remaining} \end{cases}$$

Returning to the example in Fig.1, let us build the enabling vector for the initial marked state: $M_0^T = (\ 1\ 1\ 3\ 0\ 0\ 1\ 0\ )$ The capacity vector for this net is: $C^T = (\ 1\ 3\ 7\ 1\ 1\ 1\ 1\ )$

The modularization function for the event e_1 returns the same initial marks, that is, $\mu(M_0, e_1) = M_0$, and if we substitute in equation (2.3), result in $[(2.M_0 - 1 + v_1) \otimes v_1]^2 = 8$ meaning that event e_1 is not enabled. We can repeat the process to event e_2 : $\mu(M_0, e_2) = (\ 1\ 0\ 3\ 0\ 0\ 1\ 0\ )^T$, and $[(\ 2.M_0 - 1 + v_2) \otimes v_2]^2 = 0$ means that the event e_2 is enabled. Continuing the process to the other events we can conclude that the events e_2, e_3, e_4 e e_5 are enabled, and that the pairs $\{(e_2, e_4), (e_3, e_5)\}$are in conflict, $\{(e_2, e_3), (e_4, e_5)\}$ are in contact, and $\{(e_2, e_5), (e_3, e_4)\}$ are independent. If the events are allowed to fire in parallel the following enabling vectors are equally possible[3]: $(\ 0\ 1\ 1\ 1\ 0\ 0)^T$ or $(\ 0\ 1\ 1\ 0\ 1\ 0\ )^T$

Introducing Activities

Activities are essential to encapsulate the concept of process and sub-system in a large model. Originally activities were proposed by (Miyagi, 1988) to extend the MFG formalism to the modeling and design phases. The extended model could then play the important role of systematizing and structuring the modeling until it results in a plain MFG. However, in this approach, the verification of correctness by simulation could be done only in the final steps.

To solve this problem, we present another view of activities as a compounded element including an input and an output event separated by a timed box. Thus, partial models are complete, in the sense that they could be simulated once an estimated time argument is provided. The encapsulation would be provided by representing all elements as objects, and activities as an aggregate of at least three objects. Once the model is refined a pointer to another subnet should be added to the original object representation and internal gates would connect and synchronize the aggregated subnet.

The current approach can suit better those applications in manufacturing systems where models are revised several times, and can also improve the reusability of models. Also, decomposition of large nets in several sub-nets can produce a representation more appropriate to a distributed implementation of the PFS/MFG simulator.

As in the case of the multiplicity of marks, introducing activities will have some impact in the evaluation of the enabling vector, since some boxes will now have a time constraint to be considered as a pre or post-condition for some events. Similarly to the previous case we will introduce the time vector as a vector with the same order of the marking vector and which elements are the expected time to preserve its marks.

If we adopt the policy that conditions will be prepared to dispose its marks as soon as possible, conventional conditions would have expected delay equal to zero. Conversely, pseudo-conditions whose marks are not under the control of the modeled system would have

[3] We assume that the simulation policy is to fire as many events as possible.

an infinite expected delay[4]. Time-boxes would have an integer expected time (the delay assumed by the simulator).

Thus, one more rule should be added to our algorithm which is a new modularization to the marking vector M_l: the local time for each box (stored in its object data model) must be greater then or equal to the corresponding value in the time vector. We define this second modularization as:

[*Def. 2.4*] Let $\overline{M_l}^{e_r}$ be the modularized vector of marks, the global (simulation) time vector T and the local time vector t, that is, where each element is the current value of the simulation time. The second modularization is defined as: $\eta(\overline{M_l}^{e_r}, t_l) = \overline{\overline{M_l}}^{e_r}$ where,

$$\eta\left(\overline{M_l}^{e_r}, t_l\right)\Big|_k = \overline{\overline{M_l}}^{e_r}\Big|_k = \begin{cases} 0 & \text{if } t_l|_k \text{ is finite, greater then 0 and } T|_k > t \\ \overline{M_l}^{e_r}\Big|_k & \text{in all the remaining cases including } t_l|_k \text{ infinite or } 0 \end{cases}$$

To clarify the ideas, let us take a slightly modified version of the example in Fig. 1 (illustred in Fig.2), which has the following incidence matrix, capacity and time vectors:

$$A = \begin{pmatrix} 1 & 0 & 0 & 0 & -1 & 0 & 0 & 0 \\ -1 & 1 & 0 & 0 & 0 & 0 & -1 & 0 \\ 0 & -1 & 0 & 1 & 0 & 0 & 0 & 0 \\ -1 & 0 & 1 & 0 & 0 & 0 & 0 & 0 \\ 0 & 0 & -1 & 1 & 0 & 0 & 0 & 1 \\ 0 & 0 & 0 & -1 & 0 & 1 & 0 & 0 \\ 0 & 0 & 0 & 0 & 1 & -1 & 0 & 0 \end{pmatrix} \quad C = \begin{pmatrix} 1 \\ 3 \\ 7 \\ 1 \\ 1 \\ 1 \\ 1 \\ 1 \end{pmatrix} \quad T = \begin{pmatrix} 0 \\ 0 \\ 0 \\ 0 \\ 0 \\ \tau \\ \infty \\ \infty \end{pmatrix}$$

The state shown in Fig.2 corresponds to the vector: $M_3^T = (0\ 1\ 3\ 1\ 0\ 1\ 1\ 0)$

Assuming that τ=3, let us now evaluate the enabling vector in this state to reach the next state. Once we increment the simulation clock the current time vector will be, $t^T = (0\ 0\ 0\ 0\ 0\ 1\ \infty\ \infty)$ and using the definition for the first and second modularization above, we have that,

$$\overline{M_3}^{e_7} = M_3, \text{ and } \overline{\overline{M_3}}^{e_7} = (0 \quad 1 \quad 3 \quad 1 \quad 0 \quad 0 \quad 1 \quad 0)^T$$

and, because of the time bound of the activity,

[4]From the point of view of the model these marks would persist forever if an external agent do not change them.

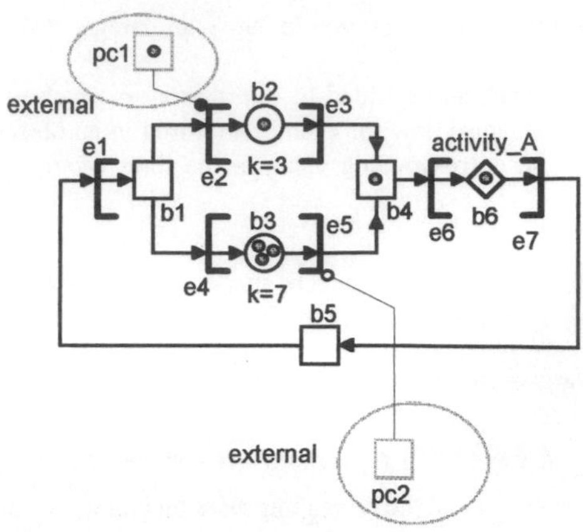

Figure 2 Modified version of Fig.1 after the third firing.

$M_4 = M_3$ and $\overline{\overline{M}}{}_4^{e_7} = \overline{\overline{M}}{}_3^{e_7}$

$M_6 = M_5 = M_4$ and $\overline{\overline{M}}{}_6^{e_7} = M_6$, thus

$M_7 = \begin{pmatrix} 0 & 1 & 3 & 1 & 1 & 0 & 1 & 0 \end{pmatrix}^T$

Suppose now that the *activity_A* has to be refined so that, instead of only the previous aggregate of elements we also have a subnet associated with that. Consider for instance, the refinement shown in Fig.3.

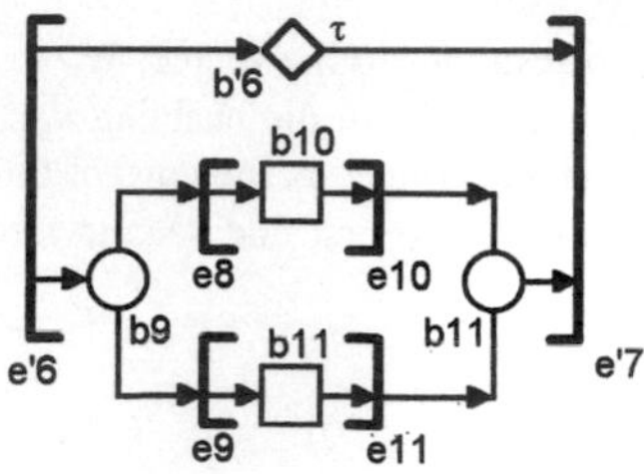

Figure 3 Refinement of *activity_A*.

A interesting thing is to assure that by using abstraction we do not have to reconsider all the representation of the models each time we want to add one more element or one more subnet. However, each time a new single element is included the incidence matrix and all other relations are modified.

Therefore, our next goal is to minimize this re-engineering and maximize the reuse of the representation already done by keeping the subnets apart. For instance our refinement of the net in Fig.2 is the subnet between the input and output events of the activity. They could be two different nets connected by external gates and having two events in common as depicted in Fig.4.

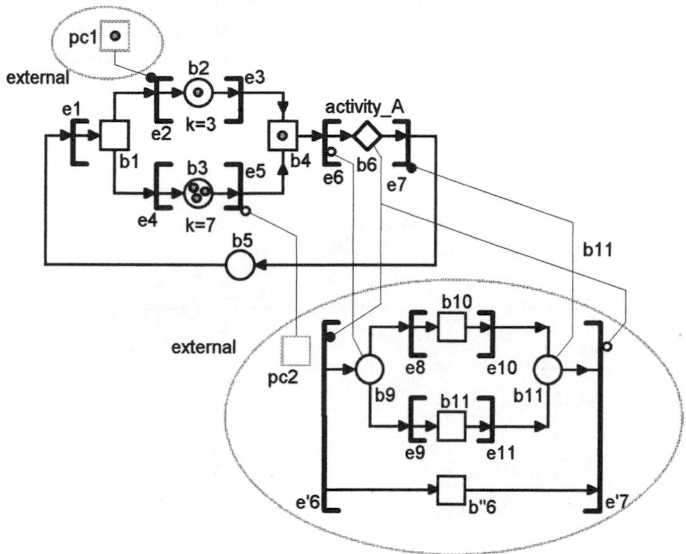

Figure 4 Other form to represent the system of Fig.2 and 3.

Thus, in our previous example only one column would be added to the incidence matrix.

3 A PRACTICAL EXAMPLE

A simulator was implemented in C++ using a personal computer. The objective was to have a feeling about the application of the ideas presented here in a more realistic environment where real requirements were present.

As a target application we analyzed the feasibility of automating the transport service and the flux of manufactured pieces in a printer's factory. All the work was concentrated in a simple path from the burning test to the packing station which had to cross several other sections and paths in the factory environment, due to present disposition of the site.

The proposed problem was to analyze the possibility of using AGV's (Autonomous Guided Vehicles) to transport the pieces from the burning test cell to the packing center and evaluate the performance of such design compared to the previous approach based on human work. Since this is a feasibility analysis, all the work were more qualitative using time estimates based on the mean performance of the equipment (there were no real selection of model or kind of robots or vehicles).

Our analysis was directed to verify if the new automated system could do better then the old system. The parameter of analysis were not only the number of artifacts transported but also the minimization of the number of AGV's in order to reduce the cost with equipment and also avoid traffic jam in the site. Another important point were the optimization of the maintenance pit stops to recharge batteries and the flexibility to increase or reduce the workload according to the necessities of the company. Fig. 5 shows the PFS/MFG model for the system.

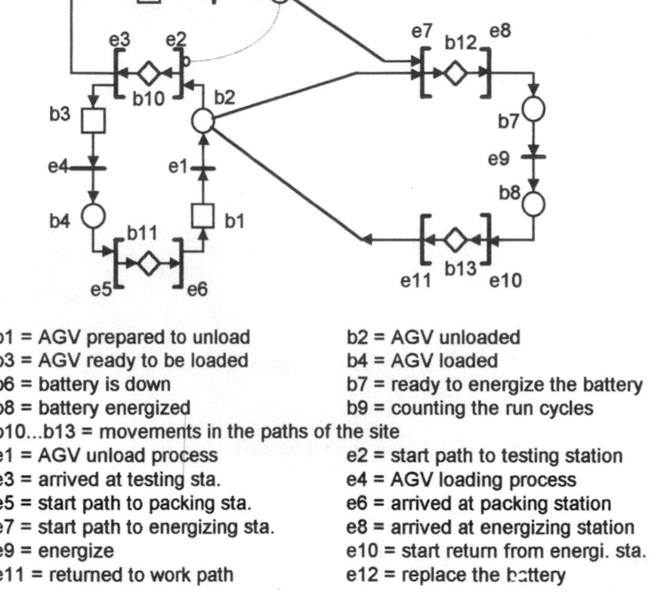

Figure 5 PFS/MFG of a printer's factory (marks corresponding the AGV's are not represented).

To this particular problem we supposed that the transport vehicles had an autonomy of 4 hours, and a workload capacity of 8 printers. We consider two possibilities to recharge the batteries: a) making a pit stop (e_9) or b) replacing the battery at once (e_{12}). We worked with the pessimist estimate that the pit stop process would take twice the time of the battery replacing (in fact we expect the complete replace be even faster).

We also use some structural properties (not mentioned in this work) by analyzing the s-invariants of the system.

Based on the hypotheses above we simulated the net with different numbers of AGV's and conclude that the best solution is achieved with two AGV's. We can increase the working capacity of 30% over the intended in the maximum working cycle. This will give some flexibility in the production schedule. The model presents a deadlock when we use three AGV's. Even if that can be solved with a proper adjust of the timing in the processes it do not present any other advantage besides a bigger lazing time. In other words using three AGV's would no increase significantly the performance already achieved.

4 FURTHER WORK

In this work we tried to establish a more strong link between the concept in PFS/MFG and conventional C/E nets in order to set the basis for a new simulator. The main idea is to consider each basic element (event, condition and relation) as a more complex element with some associated parameter as the capacity, the kind of relation, the kind of marking and the estimated time for an activity. The number of parameters can increase since the relation between them remain the same, what was represented by the Proposition 1 and by the state equation.

We are now implementing a new version C++ of the PFS/MFG simulator based in the algorithms mentioned here and using the concept of objects to encapsulate the representation and properties of each element. Aggregates are being used to define activities as compound elements and to allow more reusability in the process of modeling.

The new simulator is the core for a new version of the AIPUSP system (Lucena, 1989), renamed as AIPUSP II. All numerical algorithms for qualitative analysis will be connected in the future with AI tools to arbitrate conflicts and to help the designer to perform behavioral and structural analysis.

Two more improvements are being prepared: the association of the representation of PFS/MFG with new object-oriented design methodologies and its use in distributed control environments.

5 REFERENCES

Embley, D.W., et al. (1992) *Object-Oriented Systems Analysis: A Model-Driven Approach*; Prentice-Hall, Englewood Cliffs.

Hasegawa, K. et al. (1987) Simulatrion of Discrete Production Systems Based on Marl Flow Graph; *Systems Science*, vol.13, no.1-2, pp.105-121, Poland.

Lucena, C.J.P., et al. (1989) The Specification of a Knowledge Based Environment for the Design of Production Systems; *Proceedings of INCOM'89, 6th Sym. Inf. Control Problems in Manufacturing Technology*, Madrid.

Miyagi, P.E. (1988) *Control System Design, Analysis and Implementation of Discrete Event Production Systems by Using Mark Flow Graph*; PhD Thesis, Tokyo Institute of Technology, Tokyo.

Murata, T.(1989) Petri Nets: Properties, Analysis and Applications; *Proceedings of The IEEE*, vol.77, no.4.

Reisig, W.(1985) *Petri Nets an Intorduction*; Springer-Verlag, Berlin Heidelberg.

Silva, J.R., Pessoa, F.J.B. (1992) Análise Semi-Automática de Mark Flow Graphs; (in Portuguese) *Ibero-American Workshop in Autonomous Systems Robotics and CIM*, Lisbon.

6 BIOGRAPHY

José Reinaldo Silva received the Dr.Eng. degree in 1992 from University of São Paulo (USP), Brazil. From 1989 to 1992 he was Assistant Faculty of USP. Since 1992 he is Assistant Professor of USP. His research interests are in design theory, software engineering, intelliget CAD.

Paulo Eigi Miyagi received the Dr.Eng. degree in 1988 from Tokyo Institute of Technology, Japan. From 1988 to 1993 he was Assistant Professor of University of São Paulo (USP), Brazil. In 1993, received the L.Doc degree from USP and since then is Associate Professor. His rsearch interests are in discrete event dynamic systems, design of control systems, mechatronic systems.

34

Discrete event and motion-oriented simulation for FMS

R. Carelli, W. Colombo
Instituto de Automatica - Universidad Nacional de San Juan (UNSJ)
San Martin 1109 (o), 5400 - San Juan, Argentina
Tel +54-64-213303, Fax +54-64-213672
E-mail: rcarelli@inaut.edu.ar

R. Bernhardt, G. Schreck
Fraunhofer Institute for Production Systems and Design Technology (IPK)
Director: Prof. Dr.h.c.mult. Dr.-Ing. G. Spur
Pascalstraße 8-9, D - 10587 Berlin, Germany
Tel +49-30-39006-0, Fax +49-30-3911037
E-mail: Gerhard Schreck@ipk.fhg.d400.de

Abstract

In this work, an off-line simulation approach for simultaneously handling the discrete event and motion orient aspects of flexible manufacturing systems (FMS) is presented. A concept for signal handling in a discrete-event simulation based on Petri Nets and a proposed interface to motion simulation is outlined. The description of motion simulation covers the introduction of a unique interface concept for all FMS components. As an example for this approach, the RRS-Interface developed for robot simulation is used.

Keywords

Flexible manufacturing systems, discrete event simulation, Petri net, motion simulation

1 INTRODUCTION

Simulation techniques aid in optimizing flexible manufacturing system (FMS) design. A simulation program is produced in two stages (Figure 1). The first stage involves creating a formal model of the FMS. This entails abstracting physical reality as far as problem parameters allow.

The second stage converts this formal model into a simulation program with the aid of a simulation language. Using analysis of experiments with the simulation model (Figure 2), designers now seek evidence that the chosen formal model closely represents the actual system (validation) and that the simulation program satisfies the formal model (verification) [1].

Discrete simulation refers to the jumping by the simulation model time parameter from one event to the next; simulations without jumps are called continuous. Most FMS involve discrete simulation programs, although mixed forms are possible. A continuous process may activate a discrete event, for example: the end of a continuous heating process may trigger a request for a transfer trolley.

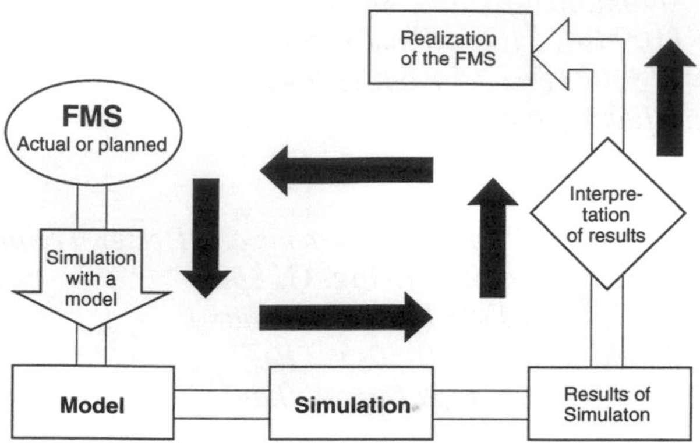

Figure 1 A formal model precedes each simulation program.

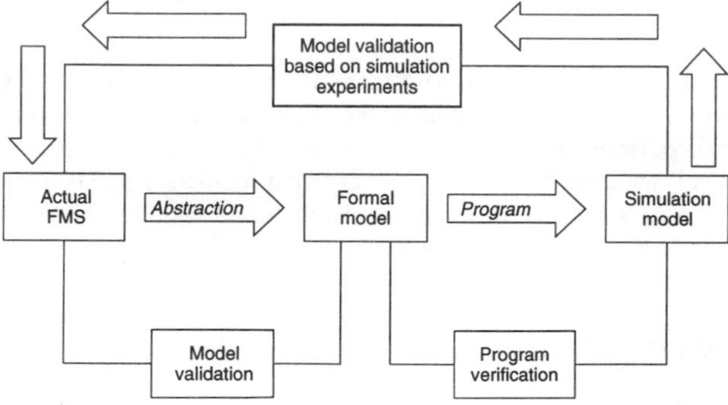

Figure 2 Designers realize a mature simulation only after several model-building stages.

An off-line simulation is used during the production planning phase and operates in isolation from the actual process. Numerous simulations, with widely varied data, must be run to obtain reliable information on future process behavior. It is also desirable, however, to optimize actual process control by coupling a simulation to an actual system in operation. This immediate integration of actual data is known as on-line simulation.

In this work an off-line simulation approach is proposed for handling the discrete and the motion oriented (continuous) aspects of the FMS operation simultaneously. This off-line simulation allows to dimension the systems under study and to determine process control parameters for satisfying -insofar as possible- the following objectives [2]:
• Maximum utilization of machines and personnel;
• minimum set-up times;
• short lead times for orders;
• minimum workshop and warehouse stockage levels;
• low transportation costs;
• investment optimization;
• other objectives may be added, depending on the applications.

Furthermore output variables are influenced by
• the work (machine sequence);
• the capacity of machines and personnel available;
• the batch size;
• the sequencing of incoming orders;
• the number and type of transportation facilities;
• the production process;
• future trends in the volume and range of orders.

At this stage, the simulation aims to investigate the effects of changing input parameters on the overall system, as well as to define the functional relationships between input parameters and output variables. By including investment considerations and other criteria, like flexibility and delivery times, analysts can assess -using this proposed methodology- the overall system and describe the best approach.

At present discrete event and motion oriented simulations are performed by using different modelling techniques implemented in separated simulation tools. The benefits of an integrated simulation environment, covering both aspects have been investigated in prototypical implementations at IPK. For a robotized FMS system a motion and process related simulation module has been connected to the high level cell control system. The development and test of cell programs (event level) has been supported by simulation and led to a reduction of set-up times at the physical system. However this experiments have been performed in a specific implementation dedicated to the available FMS system and the related system components (e.g. cell controller). In the frame of the cooperative work between INAUT and IPK a more generalized approach, independent of specific components is investigated. For this reason a discrete event simulation module (based on Petri Nets), under development at INAUT, will be used to cover the high level control and to identify the required interfaces to motion simulation modules.

2 SIGNAL HANDLING IN THE DISCRETE-EVENT SIMULATION MODEL

A manufacturing system -whether simple or complex- includes raw materials, tools, operators, and control policies for the operation of the equipments. At the sector level, it also includes flow of material and information through a collection of tools. It can even extend to the procurement, production and distribution networks of suppliers, plants, and distributors. Using computer simulation for modelling and analyzing such systems helps the designers to predict and improve a system's performance, as measured by such elements as capacity, cycle time, inventory, utilization, service level, and costs [3].

Information about the dynamics and kinematics of the involved equipment of the FMS, the product routing, and the production and parts-availability schedules must be the inputs to the discrete event simulator. Using the graphical user interface of the simulator, these inputs could be described and modified easily without programming. Information such as the production capacity of machines on the FMS, the duration of various operations performed by different machines on different products, buffer sizes, and the staffing must be processed before they entry to the simulator, in order to allow their adaptability to the signal handling of the discrete-event simulator.

Key outputs of the discrete-event simulator are the system throughput (the quantity of products produced in each time period), a work-in-process inventory (the quantity of products being worked on in the system), the cycle time (the time from release of jobs into the system to completion), and machine utilization (the proportion of time the machines work on products). These results can be displayed as bar or line charts. By examining the above named results, the designers can identify easily the bottlenecks of the systems under study, and suggest on how to alleviate the problems that could be generated. Figure 3 shows a typical scheme about the functioning of a manufacturing simulator, and its more common input and output information [4].

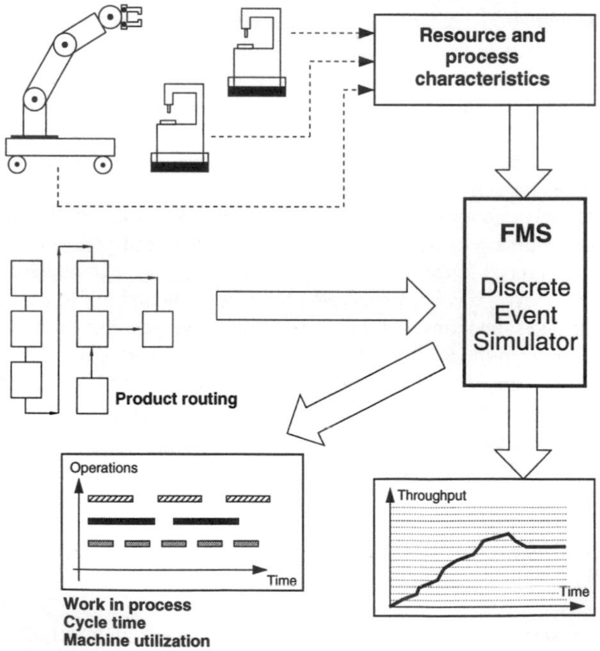

Figure 3 Input and output information related to a FMS discrete event simulator.

Since a simulation imitates the behavior of a system as it evolves over time, basic to the approach of this simulation is the building of a model that highlights the vital characteristics of the system [5]. Good models are needed to capture the characteristics of the discrete event systems [6], namely: concurrency or parallelism, asynchronous operations, deadlock, conflicts and event driven. In this work it is proposed the use of Petri Nets (PNs) as a discrete event specification tool for the modelling of the manufacturing systems under study. Petri Nets have been designed specifically to model FMSs [6], [7], [8]. They provide suitable models for the following reasons:

- PNs capture the precedence relations and structural interactions of unpredictable, concurrent, and asynchronous events. In addition, their graphical nature helps to visualize such complex FMSs.
- Deadlock, conflicts, and buffer sizes can be modelled easily and efficiently.
- PNs models represent a hierarchical modelling tool with a well-developed mathematical and graphical foundation.
- PNs allow the use of different abstraction level models and the use of powerful modelling methodologies: refinement and modular composition, each of them with its corresponding validation analysis.

It is very important to take into account that the main function of FMSs is to input rough and separate material parts, to achieve a certain amount of transformations, assemblies, disassemblies on the initial parts, and finally to output finished parts. Therefore, the designer has to identify the input and output sequences of material parts, at the same time all the elementary operations which have to be applied on these parts, and the main characteristics of the resources (turning, milling machines, manipulators, AGVs, etc.) involved on each of these operations. According to the above considerations, the Petri Net based discrete event simulator has a logical model where [9]:

- One place is associated to one state of the produced part in its operating sequence, or the state of a resource (idle or busy by a part).
- One transition is associated to only one operation on the produced part, or to the assembly operation of several parts.
- One token is associated to one material part, or information about the resources of the FMS.
- To each place of the net, a capacity place or capacity monitor is associated to indicate the maximum number of tokens this place may content. This capacity is introduced to indicate that a buffer, a storage area, or another resource can only contain a limited number of parts.

In addition, all the resources and operations verify the following set of constraints:

- The operations are completed in finite time.
- One part is submitted to only one operation at a time, and each resource can perform only one task at a time.
- One separate part can be submitted only to transformational or informational functions. A transformational function consists in modifying the physical attributes of the part (shape, constitution, surface, etc.). They are the machining functions: turning, milling, manipulation, and conditioning functions: termical treatment, painting, washing, etc.. An informational function consists in verifying that the operations have been accomplished correctly.

At the moment, a delay on transitions are usually used to model the dynamic behavior of the PN model. This temporization must be compatible with the capacity mechanism and with the control processes which are modelled. The discrete event simulator presented here has the parameter time associated to each temporized transition.

Petri Net models

Petri nets are now defined so that Petri nets can be discussed in a formal manner. The notation introduced in this section will be used throughout the paper. More details about this tool are in the following references [10], [11], [12].

Formal Definition

A Petri net (PN) is a directed graph consisting of two types of nodes, places and transitions, in which places are always followed by transitions, and vice versa. A Petri net is defined by

$$PN = \{P, T, E, S, m_0\}. \tag{1}$$

Where:

P : Set of the PN places. This set has cardinality m.

T : Set of the PN transitions. This set has cardinality n.

E : Input matrix. It can be seen like a function $E{:}P \times T \rightarrow \mathrm{N}$
Each element of E represents the weight of the input arc from place p_i to transition t_i.

S : Output matrix. It can be seen like a function $S{:}T \times P \rightarrow \mathrm{N}$.
Each element of S represents the weight of the output arc from transition t_i to place p_i.
m_0 is the initial marking, with dim $(m_0) = m$.

A Temporized Petri Net is now defined by:

$$TPN = \{P, T, E, S, m_0, \mathbf{q}\}. \tag{2}$$

Where each element has been previously defined, and $\mathbf{q}$ represents a time function. It attaches a time to each temporized transition of the net, that is $\mathbf{q} : T \rightarrow \mathrm{R}^+$.

The token-game of the ***TPN*** is performed as described in Figure 4. According to the adopted modelation methodology, a transition models an operation of T time units in the modelled system. The editor of the discrete event simulator represents the transitions with blue boxes. When a transition is enabled, the corresponding box becomes light blue color during a time T period (the modelled operation is actually performed). After the time T, the transition fires (the modelled operation finished). The last condition is showed by a green color box during a unit graphical time in the simulator.

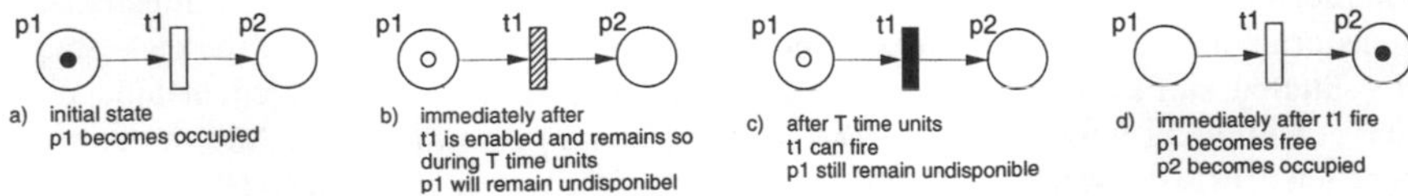
p1
t1
p2
a) initial state
p1 becomes occupied
b) immediately after
t1 is enabled and remains so
during T time units
p1 will remain undisponibel
c) after T time units
t1 can fire
p1 still remain undisponible
d) immediately after t1 fire
p1 becomes free
p2 becomes occupied

Figure 4 Token-game for a TPN model in the proposed discrete event simulator.

3 MOTION ORIENTED SIMULATION OF FMS

An abstract simulation of FMS can be done at the event level. A more precise simulation requires the consideration of causes of events. At the physical system, events are defined related to states of technological processes or motion states of FMS components. For the second aspect, motion models are required for all components which can generate a motion (e.g. industrial robot, conveyor, turn table). These components can be characterized as active components. Their drives can be controlled by numerical controllers including feedback loops or by simple on/off switches using a PLC. In the case of an industrial robot the motion model is represented by a model of the behavior of the robot controller (control model) [2]. This example will be used for further investigations.

Robot control models can be used to simulate the motion path, but also to calculate the motion time which is an important value for the estimation of cycle times. The problem arises in getting a control model which represents the real system behavior and which leads to precise simulation results. For this reason, a consortium of European automotive companies initiated the project Realistic Robot Simulation (RRS) [13]. The goal of the project was to overcome the inconvenience of accuracy by defining a simulation system interface, which enables the integration of control models delivered by the robot manufacturer. Furthermore, the interface concept had to allow the integration of original controller software. In the meantime the interface specification is available [14] and the RRS-Interface has been implemented in off-line-programming and simulation systems on the market. First robot control models are integrated and in industrial application.

The RRS-Interface had to be uniform for all controllers, but also had to cover the wide range of different concepts for motion handling. In the result, the interface provides a set of principle services. Without these services, no reasonable operation is possible. In addition, several groups of further services are defined (e.g. services for motion related event handling on RC level). The basic idea of the RRS approach is to use precise simulation models delivered by the control / robot manufacturer. For motion simulation, this concept could also be applied for other FMS components, if a suitable and accepted interface is available, which includes the definition of services to be provided by simulation modules. The usability of the RRS-Interface concept as a basic approach will be subject of further investigations.

4 INTEGRATION OF MOTION ORIENTED SIMULATION

From the control point of view, the discrete event simulator communicates with the real FMS or the motion oriented simulator through a set of signals such as: orders for the execution of activities, activity-end messages, sensor signals, alarm and error signals. In this work, the motion oriented simulator has replaced the real resources of the FMS and simulates the signal interchange with the discrete event simulator. It is necessary to develop a special system called event monitor, which is responsible for the interfacing between the PN model and the motion oriented simulator.

The interactions from the motion oriented simulator to the discrete event simulator are called events, and the interactions from the discrete event simulator to the motion oriented simulator are termed actions. A transition enabling and firing (according to Figure 4) may suppose the execution of an associated action. If this action implies an external action, a message is sent to the event monitor. Events are associated to the incoming signals from sensors or alarm signals (such as machine and part breakdowns, abnormal stoppages, etc.). Events produce additional constraints for the transition firing rule of the PN based discrete event simulator.

During simulation time, the discrete event simulator makes the interpretation of the dynamic behavior specifications represented in the system model by the underlying Petri Net. The discrete event simulator performs the token-game in the net, which generates actions sent to the appropriate module of the motion oriented simulator (Figure 5). At the occurrence of an event, a message is ent back to the discrete simulator producing a new marking which enables new transitions.

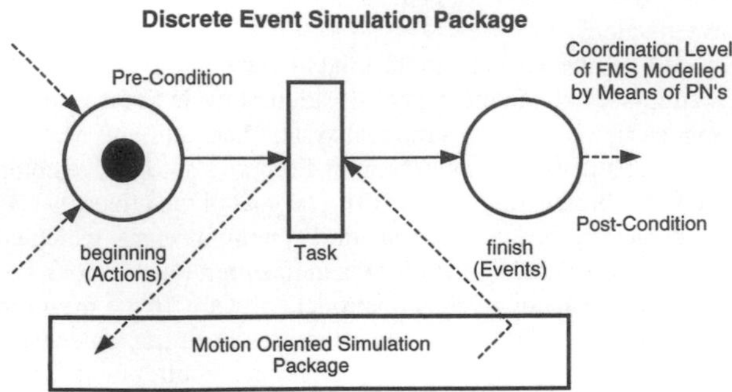

Figure 5 Interface for Communication between both Software Packages.

5 CONCLUSIONS

The present paper gives an idea of the main research goals of the joint project between the Instituto de Automática (INAUT) -Argentina, and the Fraunhofer Institut (IPK) -Germany, and the way they are associated to develop a whole simulation based methodology for designing and analyzing flexible manufacturing systems. Both research groups have decided to develop the concept of an hybrid software tool based on the integration of discrete event and motion oriented simulation concepts. The INAUT provides a recently developed Petri Net based discrete event simulator [4], and the IPK a motion oriented simulator. In this work, a basic interface architecture for connecting the above software tools is proposed. In a first step the FMS system from IPK has been modelled and simulated by using the Petri Net simulator. By associating a firing delay time to each operation modelled by means of the transitions of the Temporized Petri Nets, a Gantt diagram was obtained. When both simulators (discete event and motion oriented) work together, the firing delay time will be obtained from the continues simulation.

The obtained results are promising and shows that the proposed methodology allows to test different FMS configurations and to attain:

- quantitative information about the behavior of the simulated system;
- a fast prototype of the system under study;
- qualitative information on the system evolution.

6 ACKNOWLEDGEMENT

The work presented in this paper is the result of the cooperative action in the frame of the ECLA FlexSys project (partly funded by the European Commission) between the Instituto de Automatica San Juan, Argentina and the IPK Berlin, Germany.

7 REFERENCES

[1] Willi Hardeck. "Simulation optimizes planning and design". Siemens Review 2/85. Vol. LII n° 2, p.p. 28-29. March/April 1985.

[2] R. Bernhardt, G. Schreck, A. W. Colombo, R. Carelli. "Integrated Discrete Event and Motion Oriented Simulation for Flexible Manufacturing Systems". Journal on Studies in Informatics and Control. Vol. 3, n° 2-3, p.p. 209-217. Ed. Romanian Academy. September 1994.

[3] Rangarajan Jayaraman (IBM Research Center). "Manufacturing Systems Simulated". IEEE Spectrum 9/93, p.p. 60-62. September 1993.

[4] A. W. Colombo, J. Pellicer, M. Martín. B. Kuchen. "Simulador de Sistemas Flexibles de Manufactura usando Redes de Petri Temporizadas". Proceedings of 6° Congreso Latino Americano de Control Automático, vol. 1, p.p. 100-102. Río de Janeiro. Brasil. 19 al 23 de Septiembre de 1994.

[5] J. Ezpeleta, J. Martínez. "Petri Nets as a Specification Language for Manufacturing Systems". 13th. IMACS World Congress on Computation and Applied Mathematics. Dublin. July 1991.

[6] M. Silva, R. Valette. "Petri Nets and Flexible Manufacturing Systems". Advances in Petri Nets. Lecture Notes in Computer Sciences, vol. 424, p.p. 374-417. 1989.

[7] Invited Sessions on Petri Nets and Flexible Manufacturing. Proceedings of the International IEEE Conference on Robotics and Automation. Vol. 2, p.p. 999-1186. Raleigh, North Carolina. EEUU. March/April 1987.

[8] Proceedings of the 3rd International IEEE Conference on Computer Integrated Manufacturing. Rensselaer Polytechnic Institute. Troy, New York. EEUU. May 1992.

[9] A. W. Colombo. " Modelling and Analysis of Flexible Production Systems". MsC Thesis. Universidad Nacional de San Juan. Argentina. November 1994.

[10] J. L. Peterson. "Petri Net Theory and the Modelling of Systems". Prentice Hall. Inc. Englewood Cliffs. N.Y. 1981.

[11] M. Silva. "Las Redes de Petri en la Automática y las Informática". Ed. AC, Madrid. España. 1985.

[12] T. Murata. "Petri Nets: Properties, Analysis and Applications". Proceedings of the IEEE, vol. 77, n° 4, p.p. 541-580. April 1989.

[13] R. Bernhardt, G. Schreck, C. Willnow. "The Realistic Robot Simulation (RRS) Interface". Proceedings of 2nd IFAC/IFIP/IFORS Workshop on Intelligent Manufacturing Systems (IMS ´94). June 13-15, 1994, Vienna, Austria.

[14] N.N. (1994) RRS-Interface Specification (Version 1.0), distributed via IPK-Berlin.

8 BIOGRAPHY

Ricardo Carelli received the Electrical Engineering degree (with honours) form the National University of San Juan (Argentina) in 1976 and the PhD degree in Electrical Engineering from the National University of Mexico (UNAM) in 1989. He is presently professor of Automatic Control at the National University of San Juan and research-worker of the National Council for Scientific and Technical Research (CONICET). His current research interests are in the fields of robot control, manufacturing systems and artificial intelligence applied to systems control. Dr. Carelli is a member of IEEE and AADECA, The Argentine NMO of IFAC.

Armando Walter Colombo war born in Mendoza, Argentina in 1960. He graduated as an Electronic Engineer from the National Polytechnic University of Mendoza in 1990. He received the Master Degree in Control Engineering at the National University of San Juan in 1994. He is currently pursuing his Doctoral Degree on Design of Flexible Production Systems and is developing a DAAD fellowship at the University of Erlangen-Nürnberg, Germany. His interests are in the fields of automatic control, Petri Nets and flexible manufacturing systems.

Rolf Bernhardt studied Electrotechnique and Communication Technique at the Fachhochschule Coburg and at the Technical University Berlin (TUB) at which he also received a Doctoral Degree in the field of navigation systems. Since 1984 he is working with the Fraunhofer Gesellschaft, Institute for Production Systems and Design Technology (IPK) in Berlin. He is and was involved in many national and international industrial and public funded R&TD projects in the area of planning, simulation and programming of manufacturing systems with robots. At present he is head of department in the division Control Techniques for Production Systems and provisional director for R&TD in robotics. He is lecturer for Automation at the TUB.

Gerhard Schreck studied Mechanical Engineering at the Technical University Berlin. Since 1985 he is a researcher at the Fraunhofer Gesellschaft, Institute for Production Systems and Design Technology (IPK, Berlin) in the area of off-line programming and simulation for industrial robots and was involved in many national and international projects for robot application in manufacturing, off-shore, and space. At present he is head of department in the division Control Techniques for Production Systems at IPK.

PART THIRTEEN

Balanced Flexibility

35

·roup technology considerations for ıanufacturing systems with balanced exibility

·. J. Zelený, Dr. P. Grasa S.
·ofessors, Dept. Sistemas de Manufactura,
ESM - CEM Instituto Tecnologico y de Estudios Superiores de ·onterrey, campus Estado de Mexico, apdo postal 50, 52926 Atizapan , ·tado de Mexico, Mexico. Tel. (52 5) 326 55 55, fax (52 5) ·65789, e-mail 148.241.44.81

Abstract

e paper is devoted to a special class of manufacturing systems with limited and balanced ·hnological flexibility. The technologic limitation consists in selection of technologically similar ·sses of workpieces for machining in a particular manufacturing configuration and in elimination special or "difficult " parts from the automated part of the process. The special approach, ·scribed in this paper looks for classes of workpieces with very similar or even uniform handling ·rfaces. Unification of external handling surfaces simplifies considerably the function of handling d transport devices within the manufacturing system and avoids the use of interchangeable ppers. This makes the system more robust, reliable, practical and economical.

Within the paper, three particular flexible manufacturing systems with unified handling surfaces workpieces, tools and fixtures are described as examples .

Keywords

·mputer aided design / manufacture, flexible manufacturing cell, flexible manufacturing line, flexible ·nufacturing system, group technology

INTRODUCTION

·exible manufacturing systems

·exible manufacturing systems, together with highly efficient CAD-CAM methods for generation geometric and technologic data for automated production of complex mechanical parts belong ·esently to the most progressive industrial applications of computer technologies. Flexibility of ·nufacturing processes has many aspects and limits. It is often interpreted as the ability of the ·oduction system to readapt its parameters and structure and change easily to manufacturing of a ·w class of products. Another, probably more practical, interpretation could be the ability of the ·stem to produce in a random sequence different types of parts without changing its parameters and ·ucture.

Group technology

Group technology analyses the spectrum of geometrically and technologically similar parts whi can be economically produced in particular manufacturing configurations. Geometric similarity c be very important for machining of parts on conventional, hand controlled machines. Neverthele: with the introduction of CNC machines and CAD/CAM programming methods, the generation geometric data even for very different and very sophisticated parts, has been greatly simplified, a the close geometric similarity within the part "families" has lost its previous significance. Ne concepts in the area of group technology analyze and evaluate the width of technology of comm CNC machines, flexible cells, lines or systems and select the families of parts for maximu utilization of inherent flexibility of numerical control, making simultaneously full use of the availat width of technology of CNC machines. General filosophy of group technology approach has be published by Bednarek (1994) in the contribution 2. Special group technology approaches for CN machining centers and flexible manufacturing systems have been published by Zeleny (1984) in t reference 4.

Automation

Automation seeks for production means and methods of elimination or substantial reduction manpower from the production process and maximization of productivity by introduction of ne technologies and intensification of production processes. Regularity, homogeneity and controllabili of automated processes contribute significantly to their long-term stability and to the quality of fin products.

Batch organization

In conventional metalworking production, parts are produced in batches and stored for lat assembly of final products. Any conventional machine producing a batch of identical workpiec performs a regular operation for a certain "homogenous" period of time and this is probably t main advantage of the batch - type of organization. Any "switch-over" to another type workpiece means interruption of the homogenous period, loss of time and organization irregularity. From this point of view, the homogenous periods should be kept as long as possibl with the highest desirable numbers of identical workpieces in every batch. This contradicts to t time and costs of storing the parts for later assembly. Moreover , the production has to be planne months ahead and adaptation to changing market demands is difficult.

Flexible automation

In the area of CNC machining, we often speak about flexible automation, which tries to app automation principles even for processes which actually are very irregular, heterogenous and near incontrollable. Flexible automation tries to reach new, higher levels of homogeneity in the mic volume production of parts. CNC machining centers can be equipped with high- capacity to magazines which makes it possible to machine any time and in any sequence different types technologically similar workpieces. Machines can change fully automatically to another type workpieces within quite a broad technologic "family", without attendance or intervention operators, making use of the stored tools and stored or communicable program data. Worn too can be monitored and periodically replaced without interruption of the automatic an homogenous regime of the machine. With this achievement of flexible automation, the batch type organization of mid- volume production can be overcome and mechanical parts for th assembly of final products can be manufactured in a "just-in- time" regime, with random-typ flow of parts, following the needs of the assembly.

ystems with balanced flexibility

he just described new philosophy of flexible automation can be directly applied to autonomous NC machining centers for prismatic workpieces or autonomously operating CNC turning centers r rotary parts with additional "prismatic" operations. Nevertheless, there exists a serious problem nnected with the manipulation of parts. Parts within the technologic family may have very fferent geometric forms, dimensions and shapes. Moreover, parts change their shape during the achining process and the final shape may differ significantly from the shape of the input aterial. If the parts are handled by a human operator, the machine keeps full "random" xibility within the whole technologic family of parts due to the intelligence and adaptability of the erator. Automation of handling operations is highly desirable for "round the clock" operation machines, but no robot or manipulator has enough adaptability to replace a human operator in e permanently changing conditions of random- type flexible production. Among other possibilities, e development of "balanced" manufacturing systems with limited flexibility may represent an lequate and practical approach to the implementation of flexible manufacturing systems with ndom-type operation into the mid-volume production of parts. Within this paper, some special proaches will be discussed and demonstrated in this area.

GROUP TECHNOLOGY CONSIDERATIONS

utomation of handling processes in mid volume, flexible production of parts is a very difficult oblem which cannot be solved in general. Nevertheless , there are solutions for some particular asses of parts. Let us mention at least some of them:

For small rotary parts with additional prismatic surfaces, it is possible to use bar-type CNC turning centers with contra- spindles and perform all the turning and "prismatic " operations from both sides within one highly integrated machining process. The bar can serve as input material for all parts of the technologic family which is in this case limited mainly by the diameter of the bar. Finished parts are mixed, have different shapes and cannot be automatically manipulated. They have to be sorted and handled manually for further operations or assembly.

For medium-size rotary parts with no additional prismatic surfaces, it is possible to develop and use self- centering or automatically exchangeable grippers for manipulation of both the input materials and finished parts. Batch-type regime with adequate batch sizes is here recommendable rather than a pure random operation. Further automatic handling of pieces, stored and sorted in transport pallets, is here possible.

For smaller prismatic parts in bigger- volume production, the batch-type regime with adequate batch sizes is recommendable. A robot or manipulator with an automatically exchangeable batch-dedicated gripper may be used for handling the parts to and from batch-dedicated fixtures placed permanently on the machine table or fixed to technologic pallets and exchanged automatically by pallet changing equipment. Production configurations of this type are recommendable only for a very limited variety of parts because of high costs of batch - dedicated fixtures and grippers.

Medium- or bigger-size box-like workpieces with a high variety of technologic operations also need a high variety of tools. For this class of workpieces, the system should be equipped with automatic handling and transport of tools between the tool room and tool magazines of installed machining centers. Homogenous regime of this flexible manufacturing system with random flow of parts and tools may include the automatic reconfiguration of stored tools.

Flexibility of these systems has to be "balanced" and optimized between the problematic ne of ideal flexibility and costs of the stored tools and the tool transport system.

- In some special cases, some technologic families of workpieces with inherent uniformity handling and clamping surfaces can be found and selected for automatic production, avoidi the necessity of technologic pallets. Important condition for realization of such systems is t invariability of uniform handling (and if possi ble also clamping) surfaces throughout t whole process of automatic machining and manipulation. Manufacturing systems with unifor handling and clamping surfaces can reach the highest level of flexibility and effectivity and mal possible a random flow of free, unclamped parts throughout the manufacturing area. Numb of fixtures and technologic pallets is reduced to a minimum. Piece- or batch- dedicated tooling fully eliminated.

3 RESEARCH AND REALIZATION

Three examples of systems described above last two point will be shown within this paper, the being:

- *System A*: The flexible manufacturing cell for automized production of a broad family of IS 40 - cone tools and tooling parts.
- *System B*: The flexible manufacturing line for flexible production of multicolored PVC par with CNC machining of molds.
- *System C*: The flexible manufacturing system with random flow of parts and tools with th highest degree of technological flexibility in production of medium size, box-like , cast iron parts.

Realization

- Research and realization of the system A is being performed at the ITESM-CEM Institut Tecnologico y de Estudios Superiores de Monterrey - Campus Estado de Mexico , Mexic Realization is being prepared within the laboratories of the department "Sistemas d manufactura" of the ITESM - CEM. The basic configuration of the system has been described b Grasa ,Chavoya (1993) and Zeleny (1994) as quoted in references 3 and 6.
- Research of the system B has been performed by the INSTITUTO DE INGENIERIA UNAM in Mexico, D.F. The system has been described in more detail by Zeleny (1995) in th reference 7. General state of art in production of molds has been published by Altan and co authors (1993) in the reference 1.
- Research and development of the manufacturing system C has been performed in th VUOSO Research institute of machine tools and machining in Praha (Czech Republic). Th system had been repeatedly realized in four Czech plants these being the TOS - OLOMOU machine tool plant, the TOS - CELAKOVICE machine tool plant , the OSTROJ - OPAVA min machine plant and the TRANSPORTA - CHRUDIM transport machine plant. A mor detailed description and realization of these system has been published by Zeleny (1993) a quoted in references 4 and 5.

4 FLEXIBLE MANUFACTURING CELL - SYSTEM A

Technology

Recent revolutionary development in tools, tool holders and tool inserts for milling, drilling boring and turning operations opened new application areas for most progressive CN technologies. Technology of the system A is based on an integration of a CNC lathe with a multi is "prismatic" machining center, both installed in the laboratories of the department "sistemas de anufactura" in the ITESM.

echnologic family of workpieces:
ools, tool holders and tool bodies with the ISO 40 cone.

Main ideas

Using the handling grooves of the future tools as the uniform handling surfaces during their production.
Using the clamping cones of the future tools as the uniform clamping surfaces for performing of prismatic operations.
Preparing in advance the interface surfaces on tool bodies, assembling composed (semi - finished) tools and letting them enter as family members for automatic machining of tool bodies.
Preparing in advance only the clamping surface for the first (turning) operation and machining other clamping surfaces within the automatic function of the cell.

lock diagram of technologic operations
igure 1 shows the sequence of main operations within the flexible manufacturing cell. Bold lines ow the "hard core" of the cell with fully automatic function. Composed tools contain only tool olders and tool bodies, whereas the cutting inserts are assembled manually after machining perations.

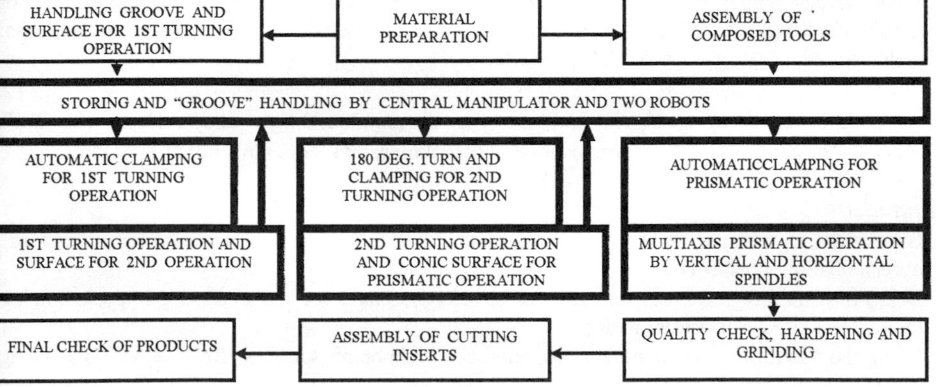

Figure 1 Main technologic operations in the system A.

ay-out diagram
igure 2 shows the system lay-out. Here again, the accented part shows the "hard core" of the ystem. Central manipulator (type CINCINNATI T3-373) has access to the working space of the athe, to the fixtures placed on technologic pallets of the machining center, to the cleaning station and o the re-circulating store with 12 recirculating system-pallets. Pallets have multipurpose nests which allow to store free pieces with both cylindrical (for turning) and conic clamping surfaces. Parts can be stored during all phases of the process. Two robots (type IBM7576 "SCARA" industrial obot) serve as an interface to the external transport of parts on transport pallets.

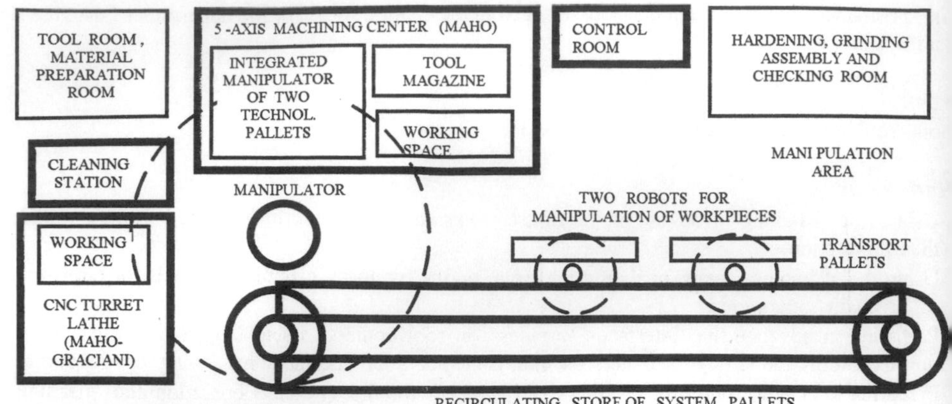

Figure 2 Lay-out of the system A.

5 FLEXIBLE MANUFACTURING LINE - SYSTEM B.

Technology

The rapidly changing market of PVC products needs an unbelievable variety of molds for inject machines and production processes using gravitation technology rather than high-pressure injecting., applications of the gravitation technology, mono or multicolored fluid plasts are filled into pockets hollows of flat- shaped molds under atmospheric pressure conditions. Solidification of the flu originally cool plastisol material is performed by thermal cycles which include heating and cool of molds together with the future PVC products.

Technologic family of workpieces: Flat - type aluminum and light alloy molds with unifo external dimensions 500 x 12.7 mm. Longitudinal dimensions 500, 600, 700 or 800 mm.

Main ideas

- Elimination of transport pallets and circulation of free molds with uniform external dimensions a recirculating flexible line .
- Using the uniform external surfaces as a reference base for the CAD/CAM programming internal functional surfaces of molds.
- Using the uniform external surfaces as reference clamping surfaces for CNC machining molds.
- Introduction of protective layers at the beginning of the production process for significa improvement of quality and productivity.

Block diagram of technologic operations

Figure 3 shows the main technologic operations within the flexible manufacturing line (system B The main part of the system, ("hard core") shown by bold lines, includes in this case some statior with a lower degree of automation. Typically, the station for colored layers cannot be full automated for the high degree of flexibility needed for "random" flow of molds with a high variet of internal functional surfaces . A protection layer provides a mono-colored protection of all to rfaces of molds . It doesn't enter into hollows reserved for colored layers and protects perfectly boundaries between mono - and multicolored parts of the final PVC products.

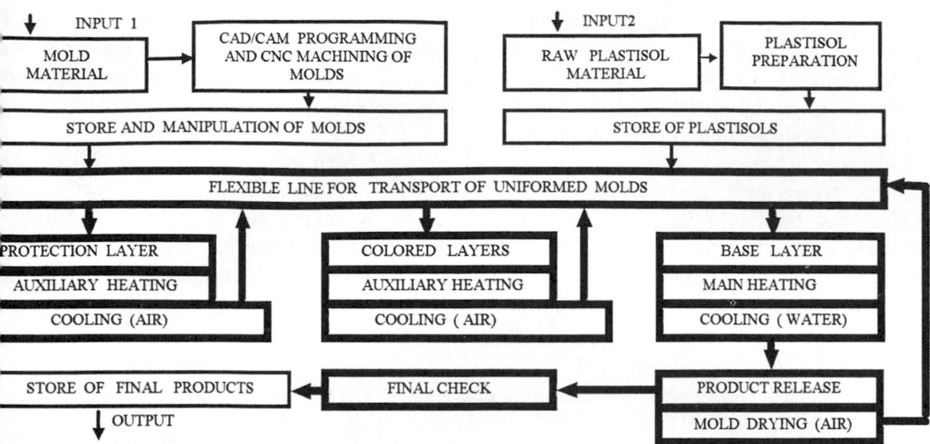

igure 3 Main technologic operations in the system B.

ay-out diagram

igure 4 shows the layout of the flexible line (system B). CNC machining of molds represents an f-line part of the production system . Also a very important off-line part of the process is the quality ieck of the plastisol materials at the beginning of the process. Manual manipulation of molds etween the store of molds and the protection layer station can change anytime the types of molds rculating within the flexible line.

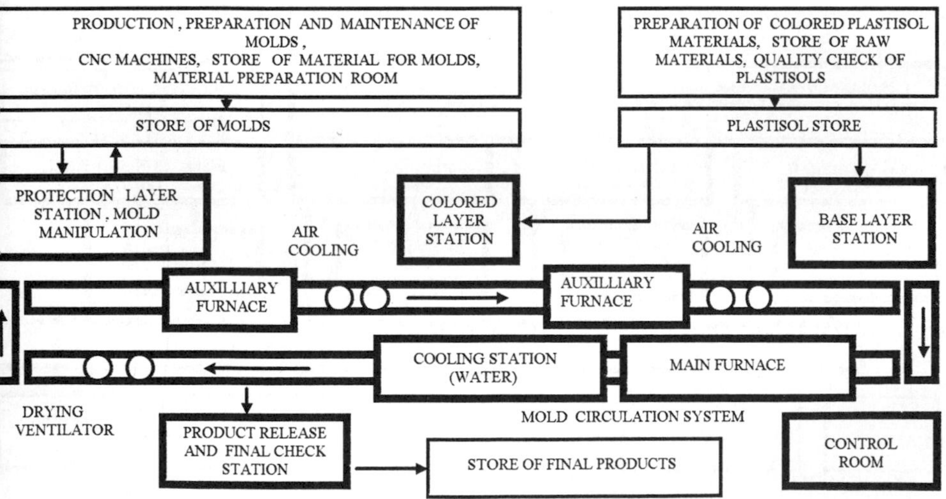

Figure 4 Lay-out of the system B.

6 FMS WITH RANDOM FLOW OF PARTS AND TOOLS - SYSTEM C

Technology.

Unmanned, "round- the- clock" machining of box-like , cast iron workpieces in flexib manufacturing systems already proved its vitality in many areas of metalworking industries yea ago. Machining technology is based on application of CNC machining centers with horizon spindles and pallet manipulators. Workpieces are clamped by means of piece-dedicated fixtures technologic pallets with uniform handling surfaces.

Within the described system, 8 identical machines have been used of the type MCFHD machining centers with horizontal spindles , high-capacity tool magazines and pallet changers. Ea machine for this particular FMS application includes a module for in-process measuring workpieces and tools, a module for automatic in-process control of boring bar diameters and a hig capacity tool magazine for 144 tools. Both the pallet changer and the tool magazine are linked to t external automatic transport of pallets and tools.

Technologic family of workpieces

Box-like cast- iron workpieces to maximum dimensions of 400 x 400 x 400 mm.

Main ideas

- FMS machining module with "random - type " flexibility in both pallet and tool flows.
- High- capacity tool magazines at the machines (144 tools) for increasing the variety workpieces within the group technology of the system.
- Automized transport link between the tool magazines and the tool room.
- In- process measuring of tool dimensions and automatic correction of boring bar diamete directly in the machine spindleSoftware for tool life monitoring within the whole tool syste (1440 tools) and generation of messages for the tool room .

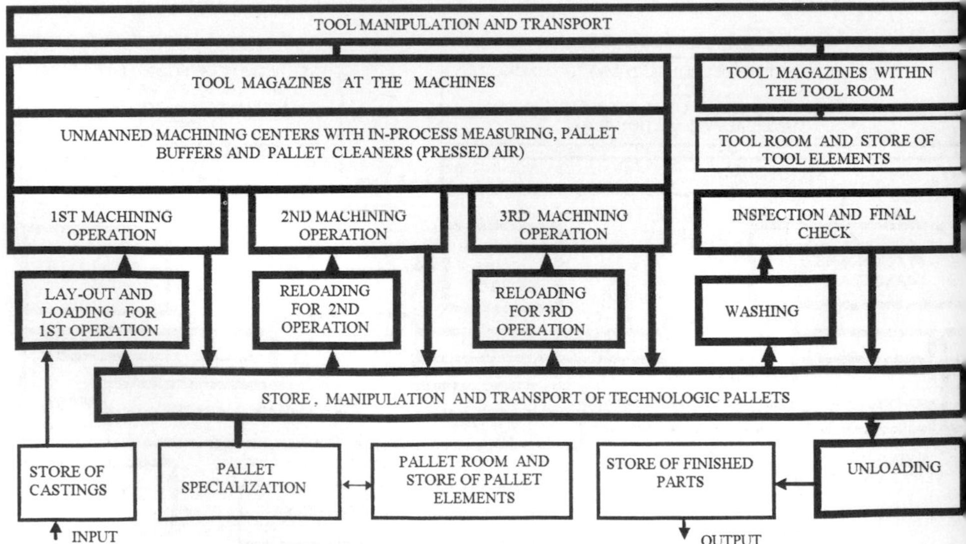

Figure 5 Main technologic operations in the system C.

ock diagram of technologic operations

;ure 5 shows the typical sequence of technological operations within the system C. The :ented part relates to the area of automatic, "round-the-clock" operation . Machining of parts :urs typically in three operations, with possible exceptions. The system can handle any time any 200 types of operations, with a reserve of 50 operations within the pallet store capacity (250 llets). This corresponds to about 70 different workpieces.. Total variety of tools is about 350 ol types for all ttechnologic operations. The rest of the tool system capacity (1440 tools) is erved for spare- and multiple tools.

y-out diagram.

gure 6 shows the lay-out of the system C as it has been realized in the TOS-OLOMOUC plant. calization of individual stations and rooms minimizes the transport routes and considers the fety regulations. In fact, every machine enables access to its working space for inspection and intenance. Manual manipulation with tools is not practiced with possible exception when rifying new workpieces. Coordinate measuring machines are placed in a special room with bilized temperature. Palletized workpieces are introduced to this room inmediately after washing the automatic storage- and retrieval system.

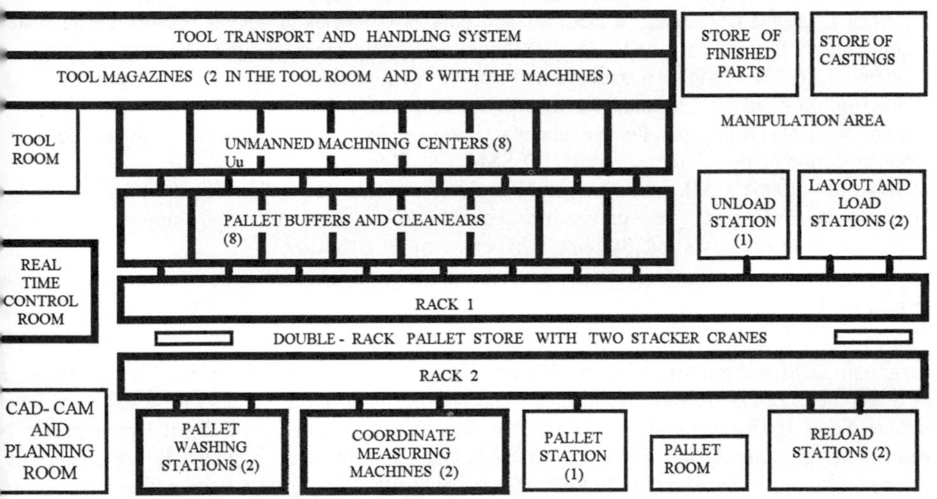

igure 6 Lay-out of the system C.

CONCLUSIONS

roup technology considerations are extremely important in conceptual design of flexible anufacturing systems . The paper is devoted to flexibly automated manufacturing constellations ith uniform handling and clamping surfaces and random flexibility of operations within the osen range of technology. This represents certain limitation in the selection of part families and onceptional design of the system. On the other hand, systems of this class make full use of the omputer-flexibility of information processing and real time control of CNC machines at multaneous robustness and relative simplicity of handling and transport operations.

Shown examples from the area of tooling parts (system A), molds for PVC parts (system : and box-like cast iron parts (system C) demonstrate cases where hundreds of geometric different workpieces have uniform handling and clamping surfaces and can be manufactu and handled with considerably simplified and robust handling and tooling devices.

Also, it was the intention of the authors to show that certain engineering limitations a optimizing "balance" in the conceptual design of manufacturing systems create much bet chances for highest, random-type flexible automation of quite difficult production processes a for realization of most progressive production philosophies.

8 REFERENCES

1 Altan T.,Lilly B. V., Kruth J.P., Koenig W., Toenshoff H.K., Lutterwelt V.A., Khairy A.B. (19 Advanced technique for die and mold manufacturing . CIRP Annals " Manufactur Technology" , (with other 103 bibliography items).

2 Bednarek M. (1994) Tecnologia de grupos, la clave para la implementación del CIM. AC Editor, Mexico.

3 Grasa P. S., Chavoya O. (1993) Integración de una celda flexible de manufactura. Revista Ingenieria de la Universidad de Costa Rica, vol. III, No. 1.

4 Zeleny J. (1984) Unmanned technology and control strategies in flexible manufacturing syste with random flow of parts and tools. CIRP conference on manufacturing technology, Palermo.

5 Zeleny J. (1993) Flexible manufacturing systems with random flow of parts and tools. XI Congreso de Academia Nacional de Ingeniería, Acapulco.

6 Zeleny J. (1994) Robotized flexible manufacturing cell for 5-dimensionally complex parts in tooling technologies " Internal report, ITESM-CEM, Mexico.

7 Zeleny J. (1995) CAD / CAM machining of molds and intelligent manufacturing multolored PVC parts by gravitation technology", First World Congress of Intellige Manufacturing Processes and Systems, University of Puerto Rico.

BIOGRAPHY

Dr. Jaromir ZELENY got his MSc in 1953 and his PhD in 1961, both from the CVUT Prague. I was researcher, head of research departments and vicepresident of the VUOSO Research Institute Machine Tools in the period 1952 - 1991. He directed 8 succesful governmental projects in N machines, CNC controls and FMS and was coordinator of research activities in the czechoslov machine tool industry. He directed 20 PhD theses, published 80 papers/articles and got 78 industr patents. He is active member of CIRP since 1974. He was lecturer, Professor and researcher at t CVUT, Michigan State University and UNAM and currently is Professor at the ITESM - CEM Mexico.

Dr. Pedro GRASA Soler got his Bsc from the UNAM, Msc from theUniversité de Bordeaux and Ph from the Université de Nancy. He was Professor and Chief of Department at the UAM a participated in 7 research projects in the UAM and in the Automatics Research Center in Nancy. I has also been director of the Manufacturing Research Center and Technology and Productivity Cent in the ITESM-CEM. He published 20 research reports and directed 19 Msc theses. He received t Golden Ram award in the ITESM-CEM and candidates for National Researcher. He is now Profess and director of theDivision for Graduate Studies and Research at the ITESM - CEM.

36

Product differentiation and process flexibility as base for competitive strategy

H. A. Lepikson
Univ. Federal de Santa Catarina, Dept. of Mechanical Engineering
88040-100 Florianópolis, Brazil
Phone: +55 (048) 231-9387, Fax: 234-1519, E-mail: hal@grucon.ufsc.br

Abstract

One of the key problems faced by the companies nowadays is the management of their competitive positions is an increasingly more complex environment. This paper places a discussion in the strategic position of the companies and analyses the consequences originating from this position, mainly the imposed need to conciliate two conflicting elements of one manufacturing company: the profusion of new and differentiated products, in detriment of manufacturing economy, which demands large batches and slight changes in its planning in order to get the best performance. After a brief context analysis, an approach to this conciliation is evaluated, based on flexible product design and manufacturing, oriented by technological patterns and organizational culture to achieve this goal.

Keywords

Strategic planning, product design, flexible design and manufacturing, core competence

1 INTRODUCTION

The present market condition for industrial goods faces rapid and deep changes. The demand for more specialized products and its premature obsolescence impel companies to take a more flexible and dynamic attitude to face competition. In the companies' highest concerns the customers are no longer the same, nor are the workers or competitors. As markets expand worldwide, technological advancements exert a permanent pressure on products and their production process. The relations of the company to its environment have turned out to be very complex.

As a consequence, the traditional mass production concept is overshadowed by concerns with flexibility and competitiveness. Companies are giving more emphasis to quality and automation approaches to face the challenges set by these changing consumption patterns. But this is not enough, since these approaches are oriented to enhance existing stabilized systems. In order to cope with these new paradigms, companies need to deeply transform their product development structure and this must be also accompanied by transformations in the production system.

This paper deals with these transformation processes and suggests an integrated approach to face the challenge of assuring the survival of companies in a competitive environment on a long term basis. Rather than offering a "revolutionary approach", it shows how to put together known concepts in an effective way to generate real competitive advantage.

2 PRODUCT DIFFERENTIATION AS STRATEGY

There are two main generic strategies that are successful: total cost and differentiation leaderships (Porter, 1986). The strategy of total cost leadership is achieved by a number of functional policies set to this objective, like facilities to manufacture on an efficient scale, a strong effort towards cost reduction and control and minimum investments in areas such as R&D, design, etc. This strategy, pursued by most companies, is characteristic of mass production industries, which utilize scale economy as their major weapon for competition.

The second strategic option sets out to differentiate the product the company offers, creating something considered unique in the industry, in proportions appreciated by the buyers. The differentiation may be done in product design, technology, peculiarities, ordered services or other dimensions. It allows a above average return, as it is paid off by its singularity, called premium price. Responding to these dimensions causes the products to be adapted to customer needs, with consequences in the production structure, as it favors the flexibility and the knowledge base on which it is sustained. A third option, focusing strategy, may be considered as a subset of one of the other two.

These strategies are not necessarily contradictory and could be used together (Day, 1989). Product quality may reduce costs as it increases the market share. As a consequence, gains are obtained in production scale. Cost is always an important factor: no matter how much the consumer may wish for a particular product, if cost is beyond his means, he will buy a cheaper one. However, as the company creates superior value for its clients, it can choose between obtaining a premium price for its advantages or have them transferred to a more competitive price, which will increase its market share (Clark 1993).

The two generic competition strategies take very different directions, but it may be concluded that they lead to only one result: what the customer really wishes for and tries to obtain is a differentiated product. Therefore, only product differentiation actually guarantees real competitive advantage in the long run. Cost strategy will always be a transitory advantage as it is easily followed by competitors. This paper seeks to analyse and suggest an approach for reorientation of strategies towards differentiation, with its implications on organizational flexibility.

2.1 Flexibility and competitiveness

Flexibility is here understood as the ability of making fast, representative, changes in company's functions as result of the emerging needs. In order to develop their capacities for the differentiation strategy, the companies should direct their efforts to product development and manufacturing, with functional areas articulated towards client needs. From this "looking outside" it will be possible to develop products able to meet the real market demands and adjust the production structure to such a purpose. As a consequence, there comes an unavoidable need to turn to flexible production, according to reduced lot sizes, due to the variety of product options offered.

This conception is supported by the evidence that environmental uncertainty generates needs for flexibility and this induces developments in technology (Chen, 1992). As a consequence, this tends to increase the range of processes attended by flexible systems due to the breaking up of technological restrictions, accomplished by innovation. Companies that invest in product and process flexibility may expect to have a better response to environmental changes.

The question to be answered is how to introduce this differentiation strategy based on flexible production. In this introduction two concerns must be considered. First, it is essential to invest in advanced technology. But this alone is not enough and the frustrating efforts in CIM (Computer Integrated Manufacturing) in the last decade have proved that. In addition, it is necessary to have an organizational culture able to handle these technologies. The integration of both aspects is the key to the competitive leverage in the approach defended and discussed here.

2.2 Technological aspects: flexible design and manufacturing

Three technologies of advanced production systems are the core of the evolving process towards a differentiation strategy and they must be made available and integrated: CAD- Computer-Aided Design, GT- Group Technology, and FMC- Flexible Manufacturing Cells. The other areas will follow the orientation given by this three computer-aided technologies in a natural way.

CAD guides product development. But its potential will only be fully employed and will only be technically and economically justified if it is integrated in the productive process as a whole and in an integrated structure (Lepikson 1990a).

Among the options of automatic production, FMCs are the best adapted to the differentiation needs and recent technological advances are making FMC economically viable for many traditional processes (Chen, 1992). Its characteristics, if properly applied, allow fast adaptation to market conditions and possibility of gradual and modular implementation.

The GT is an already old concept that sets out to put together similar parts into families in order to optimize the many production stages. It is based on the recognition of similarities (in design or in process) and the memory of past problems with its solutions. However, GT has been traditionally used only as a supporting tool to process planning or to design. To make viable the approach here defended, GT has to be integrated first (and this is an internal question for every company). In this condition, it starts as the link joining the different views of integration. Figure 1 shows the role of GT and how the main different functions are interrelated. In this conception, GT disciplines and imposes restrictions on the many areas involved (Lepikson 1990a).

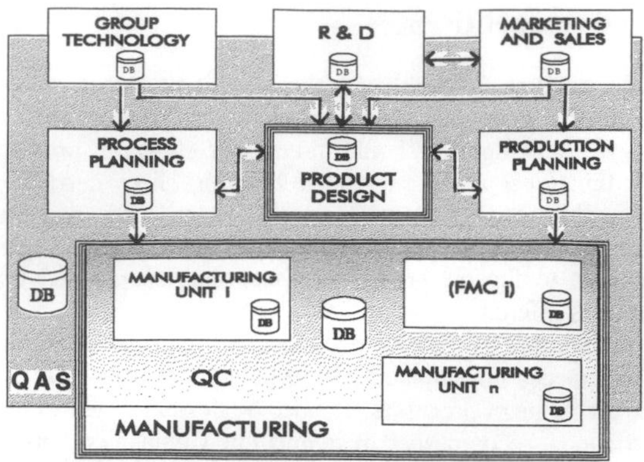

Figure 1 Basic interrelationship among areas directly related to product and its production.

In this sense the enclosing and feed-back role of the QAS (Quality Assurance System) is important to the whole production system and to the manufacturing function. QA shall be understood as a global function in the company, oriented towards the client. In this sense, every action in the company must be planned, controlled and audited, taking in account the QA objectives. QAS links the organizational culture to the technological frame in order to assure integration of the systems and to give direction to the technological demands.

The three technologies are the pillars and beams which build the system and QAS is a cement which plays a catalyst role. Nevertheless, in order to cope with their differentiated strategy demands, they need a new foundation to direct synergy towards technology. These are the company's core competencies (Prahalad, 1990). Core competencies constitute the set of resources and capacities absorbed by company's culture over a period of time and hold an important advantage for being unique, hard to imitate. Successful companies today must identify, develop and manage their core competencies to build their long term strategies. There are four outstanding aspects in this foundation on core competencies:

- it provides potential access to a large variety of markets with the same technological base;
- it favors technology flow and organizes work to obtain the optimum;
- it provides a significant contribution to the benefits noted by clients;
- it is difficult for competitors to imitate, due to the complex harmonization of individual technologies and competence generated in design and manufacturing.

A question to be solved is how to establish this unique competence, tuned in to the highly dynamic and turbulent environment in which the company is to survive and also how to take the fullest advantage of the situation, associating technological structure with organizational culture. This is done by the identification and development of core competencies based on technology management. In this case, technology is used to empower the company to face the rapid changes that its very policies have brought about (as it promotes innovation which competitors somehow try to follow). Technology adapts the company to changes and, at the same time, prepares for the necessary leap towards the future, strengthening its core competencies in a closed loop way. The company which has the best condition to promote the most frequent, and the best oriented technological advances in its production line (with the consequences in increased value and cost reduction that each advance means) will draw away, as the time goes by, from its competitors. The advantage obtained will be represented by higher profitability, higher market share, or both.

2.3 The mean term: standardizing for the flexibility

Although simple in conception, it is a nevertheless complex and wearing task to put this structure to work with positive synergy, specially considering the very different views of the same reality held by marketing, engineering and design. This approach allows greater holistic perception of the problems but also brings unavoidable conflicts.

Thinking first of operational basis, to have the system working effectively, it is important to associate it to standardization. Standards are essential to set processes in order, through an organized base on which the relationships are built. They are a strong point in the cost strategy and can be also useful for the differentiation strategy. In this way, GT offers, as shown, many advantages in disciplining the relations in the production system. It allows, from the very beginning, a unique language for communication and discipline in the relations between suppliers and customers.

Advantages are obtained through standardization of the product development philosophy itself. The development process of new products and its parts can be done by recovering the products and process histories, in a systematic way, taking advantage of similarity and modularity concepts (Pahl, 1988). This last item brings a number of implications (Logar, 1991). Among them, certainly the most important is the one regarding the power of features to support products by pre-configured design macros, with interfaces based on GT. This characteristic also results in elements to define FMCs that will feedback design. In this way, a great variety of product options can be obtained from limited design and manufacturing resources.

Concerning the manufacturing processes, unfortunately there is no way of preventing that production in small batches from raising the direct production cost. Strong integration of design with manufacture is necessary, including standards for processes, methods, and resources aiming at improvement in performance and reliability. The standardization of the parts to be manufactured become indispensable. The arraying of the entire system around families defined by the GT starts guiding not only design but also manufacturing and planning.

Many other implications occur. It is not difficult to observe what standardization means in term of economy in inventory, tools, molds, devices and so on. This perspective can be extended to the simplification of problems such as production routing and scheduling, material and supplier management, or to the potential role of standardization for building the manufacturing modules, assuring great evolutionary potential and low initial investments necessary to establish a system. The sum of these factors greatly compensates for the initial disadvantage of small batch production.

The contribution offered by the standardization to QAS is also important. The concept of design associated to GT offers rich potential in defining management and quality control standards, including the possibility of automatic generation of inspection programs (Krause, 1993). The feedback of design and production systems with data originated from QAS offers another important technological evolutive base to the existing patterns, as these data are the major primary sources for this feedback. This integrated conception is strongly dependent on a new distributed engineering database architecture, which is, in conception, at the same time robust (in selective access and retrieval of information meanings), real time access capable, able of showing different data versions and visions and transparent in relation to operational systems (Lepikson, 1990b).

3 ORGANIZATIONAL CULTURE: THE FLEXIBLE ORGANIZATION

Even nowadays, most organizational structures of companies follow a complex hierarchical models, whose articulation is very difficult. In this traditional structure, innovations are developed by the engineering department and commercialized by the sales people. This strategy, besides being a short term approach, overrates technical and economic aspects, relegating the customer to a secondary role. There are also distortions in the real understanding of the marketing concept. Too much emphasis is given to market surveys that contribute little to product innovation. After all, only what is already known can be surveyed. These surveys end up being worthwhile only to handle sales of products already well-known in the market, and this certainly is not the objective of a company who seeks differentiation.

The approach proposed is based on a structural and functional reformation. The tripod composed of marketing, industrial design, and engineering functions redefines each of the roles and its importance to the formulation of new strategic bases. These professionals start to work in a cooperative and integrated way seeking, therefore, an efficient strategic model aimed at efficiently attending market desires. Figure 2 shows how this is done, a concept that goes a little farther than those of concurrent or simultaneous engineering.

The marketing function has two primary purposes: (1) to detect market opportunities or inadequately satisfied demands by existing products and (2) to fulfill these spaces in an efficient way. This also involves the designers and engineers who will help to adjust the characteristics of the company's products to the ruling forces of the market.

The industrial design works between the technology (product development and manufacturing), and the market. It establishes a bridge between the customer and the company by converting the information obtained into desirable and useful products. Rather than dressing up the product, design helps to generate new concepts into new products in order to answer customers needs.

Engineering is connected to innovation, turning technology viable. Its role is well established in companies. Both design and engineering study the relationship between man and environment. While industrial design studies this relationship from the human point of view (physiological, psychological, and social conditions), engineering studies the physical, chemical, organic, and economical roles. Hence, they analyze the problems with different focusing. In other words, while engineering focuses on how to make product work, industrial design concentrates on making it readily usable (Harkins, 1994).

The engineering function must face the need for a deep structural change in order to prepare its professional for the market requests. Industrial design turns out to be a key element in the dialectic communication between marketing and engineering. Core competence acts as the embedded culture upon which the formal and informal relations are built and clear objectives are settled for the whole team. QAS interprets this culture and puts it on an operational basis, working with the performance factors (figure 2), which are made of quantitative and qualitative indices that measure and control the interfaces between functions These indices are built by the deployment of each element in the product and process modeling, using QFD (Quality Function Deployment) concepts (Krause, 1993).

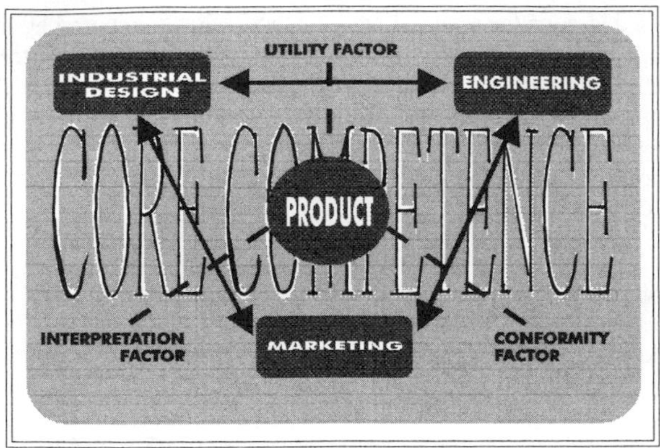

Figure 2 Relation of industrial design, engineering and marketing on product development.

The interpretation factor is involved in the product conception and is hard to deploy. Marketing must "feel" clients needs and be able to translate them to the industrial design which, in turn, must deploy desires into conception of products. Furthermore, this conception must be capable of being tested by potential clients. This means the capacity to bring together cultural context, human factors (even emotions) into one product. The conformity factor, on the other hand, brings rationality to the conception. It puts technical and economic questions into conceptions in order to make them viable as product It means design for manufacturability, pride in manufacturing. The utility factor translates conceptions into usable products as it introduces the human-product interface (pride of ownership, ease of use and maintenance, significant appearance, attractive aesthetic, good ergonomics, easy disposal at the end of life cycle).

These three performance factors induce the integration process and, in turn, build the company's core competence in a closed loop way which strengthens the technological base and the organization's culture.

4 CONCLUSION

The concept described represents the framework for the integration of the different functions within the company to accomplish its strategic objectives. This approach serves as the key concept for various projects dealing with the introduction and implementation of a differentiation strategy supported on a technological and organizational basis. This framework is not a collection of prescriptions made available to solve the operational or strategic problems of companies. It must be understood as an evolutive and educational matter, that can help companies interested in profound, progressive and consistent transformation on a long term basis.

The core competency plays an important role, as catalyst of the developed culture, that is beyond individual or even groups. It sets up the language, understanding, budgets, aims and so on. QAS is a global function which helps to support and develop the strategy and plays an important role in translating the core competency into operational basis. The computer-aided technologies are used to cope the required precision and velocity imposed by the environmental conditions. The organizational culture based on core competency causes the standardization process to be creative and evolutive, which is, by itself, an important conquest.

The impact of the consequent standardization process echoes throughout the whole company even to the clients and suppliers, developing a constructive relationship that really guarantees a competitive advantage that can hardly be surpassed by the competitors that use classic strategies for competition.

5 REFERENCES

Chen, I. Calantone R.J. and Chung, C.-H. (1992) The Marketing-Manufacturing Interface and Manufacturing Flexibility. *Omega- International Journal of Management Science*, **20**/4, 432-3.

Clark, K.B. and Wheelwright, S.C. (1993) *Managing New Product and Process Development.* Macmillan - The Free Press, New York.

Day, G.S. (1989) Deciding How to Compete. *Planning Review,* sept./oct., 18-23.

Harkins, J.R. (1994) Is Design Doing its Job?, *Machine Design*, feb. 7, 53-8.

Juran, J.M and Gryna, F.M. *Controle da Qualidade (Handbook).* McGraw-Hill, São Paulo.

Krause, F.-L., Ulbrich A. and Woll, R. (1993) Methods for Quality-Driven Product Development, in *Annals of CIRP*, .**42**/1, 151-4.

Lepikson, H.A. (1990a) *Padronização e Interação das Unidades de Fabricação, Manipulação e Inspeção de uma Célula Flexível de Fabricação.* UFSC Dissertations, Florianópolis.

Lepikson, H.A. (1990b) Requisitos para Sistemas de Bases de Dados para Manufatura Automatizada, in *Annals of* 4º *CONAI- Congresso Nacional de Automação Industrial*, São Paulo.

Logar B. and Peklenik, J. (1991) Feature-Based Database Design and Automatic Forming of Part Families for GT, in *Annals of the CIRP* **40**/1, 153-6.

Pahl G.and Beitz, W. (1988) *Engineering Design: Systematic Approach.* Springer Verlag, London.

Porter, M.E. (1986) *Estratégia Competitiva.* Campus, Rio de Janeiro.

Prahalad C.K. and Hamel, G. (1990) The Core Competence of the Corporations. *Harvard Business Review,* may-aug. 79-91.

Biography

Herman Augusto Lepikson, M.Sc.: Assistant Professor and Coordinator of the Industrial Automation and Metroloy Laboratory in the Department of Mechanical Engineering at the Federal University at Bahia State, Brazil. At the present, joining the GRUCON- Group on Industrial Automation and Numerical Controls in the Federal University at Santa Catarina State for a post-graduation program.

37

Balanced automatization levels in manufacturing systems

A. Adlemo
Dept. of Computer Engineering, Chalmers University of Technology
S-412 96 Göteborg, Sweden
tel. +46-31-772 1000, fax. +46-31-772 3663
adlemo@ce.chalmers.se

S.-A. Andréasson
Dept. of Computing Science, Chalmers University of Technology
S-412 96 Göteborg, Sweden
tel. +46-31-772 1043, fax. +46-31-16 56 55
andreasson@cs.chalmers.se

Abstract

The role of the human operator in manufacturing systems has changed in recent decades as computers, data networks and intelligent machines have been introduced. One effect is that the number of operators has been reduced as the cost of operators is higher than the cost of computers, intelligent machines and intelligent transportation systems. However, to reach a balanced automation system, it is necessary to consider the relative cost of labor as compared with advanced machinery. At the same time, it is essential that these considerations do not affect product quality.

The way of reaching a balanced automation system presented in this paper is through appropriate levels of control and automatization, considering cost and product quality. It is demonstrated how a modern machining cell can be stripped, according to the described levels of automatization, to save money and thus reach a balanced automation system without any loss of product quality. The bottom line of the discussion is that product supervision should not be left to human operators solely, but rather be performed automatically by the system. Apart from the machining cell, an assembly cell in a car manufacturing installation is also demonstrated to explain the full range of aspects of automatization when designing a balanced manufacturing system.

Keywords

Balanced automation systems, computer integrated manufacturing, quality control

1 INTRODUCTION

The role of the operator in a manufacturing system has changed with time. In early days of manufacturing, the tasks of an operator were to manually load, operate and unload machines. Later on, more advanced machines were introduced that could work autonomously, although without intercommunication. As progress continued towards more intelligent machines and with the introduction of computers, the manufacturing system was expected to take over more and more of the operator's work and finally become the ultimate, unmanned factory.This view has been moderated slightly in recent years and the role of the operator in a manufacturing system has become more appreciated. The knowledge and experience of a skilled operator is very difficult and expensive to achieve and build into a system in other ways.

Various papers have been written on the role of the human operator in manufacturing systems, e.g:

- Operator control activities, Adlemo and Andréasson (1995a), Sheridan *et al.* (1988).
- Operator roles and conflicts, Badham and Schallock (1991), Drury *et al.* (1986), Mårtensson and Stahre (1992), Stahre and Mårtensson (1992).
- Operator interfaces, Geary *et al.* (1992), Olsson and Lee (1994), Schneiderman (1987), Sylla (1992).

The tendency in modern manufacturing systems has been to reduce human interactions with systems to a minimum. One important reason for this, albeit not the only one, has been to reduce costs caused by increases in the salaries of human operators. At the same time, the cost of introducing intelligent machines and automatic control systems has decreased in relative terms as compared with operators' salaries. Reductions in human interactions with a manufacturing system, and thus indirect reduction in the number of human operators involved in production, are understandable adjustments made by chief executives. Humans are notoriously inconsistent, both inter-individually and over time. Humans tire easily. They often reject products they regard with suspicion more on the basis of fixed quotas than on actual defect levels. Computers and machines, on the other hand, are very patient supervisors, over long periods of time, and neither complains nor become bored. An important aspect, however, is how to obtain a balanced manufacturing system, meaning a system that has the proper mix of operators and machines to obtain the highest profit possible, taking into account the country in which the system is located, and without suffering any losses of product quality.

This paper discusses the levels of control and automatization in a manufacturing system appropriate to obtain a balanced system. This is achieved by studying a system and identifying conditions under which it is best to rely on an operator or on an automatic system. This adjustment is made considering various aspects, such as time (i.e. productivity), money (i.e. savings), safety (i.e. human safety), flexibility and quality. The objective of the research presented in this paper is to demonstrate situations in which an operator can perform a work task rather than have the system do it automatically. To illustrate this, a machining cell for the production of truck and bus axles is used. The machining cell is stripped in a number of steps, from a totally automatic system to a totally manual system. The main theme of the discussions in this paper is that, even when human operators perform activities in a manufacturing system that would normally be considered appropriate for automatic control and automatic production, the supervisory portion of the control should not be left to the operators solely if the quality goals of production are to be met. One reason for this is that the equipment needed for supervision (computers, data network and sensors) is relatively cheap as compared with advanced machines, such as a computerized milling machine. To obtain 100% inspection using humans typically requires a considerable amount of redundancy, often as much as three reinspections (Dreyfuss, 1989; Freeman, 1988). At the end of the paper is given another example of a balanced manufacturing system. Here, however, an assembly cell in a car manufacturing installation is used rather than a machining cell.

The work presented in this paper is the result of work by a cross departmental research group at Chalmers University of Technology, Göteborg, Sweden. The group consists of individuals from the Department of Computer Engineering, Department of Computing Science, Department of Production Engineering and the Control Engineering Laboratory. Other work performed by the group relates to high level operational lists (Andréasson *et al.*, 1995), the control system in a machining cell (Fabian *et al.*, 1995) and generic resource models (Gullander *et al.*, 1995). The research aims at demonstrating how a truly flexible manufacturing system can be achieved (see Adlemo *et al.*, 1995b, for some preliminary results).

2 CASE STUDY OF A MACHINING CELL

A machining cell for rear-axles has recently been installed at one site at Saab Scania Trucks and Buses in Sweden. The cell serves as an example of a highly automatic manufacturing system and of how a balanced system can be achieved in a country in which salaries are comparatively high. However, the following sections demonstrate how this system can be stripped of different parts to obtain balance in a country in which salaries are comparatively low.

As can be observed in Figure 1, the cell consists of seven resources; a lathe and a multi operational milling device, together with a quality control (QC) station; two exit buffers and one entry buffer; and a gantry crane (GC) for loading and unloading the devices. A local area network interconnects the resources with each other and with the cell controller. An initial premise was that all communication should take place by means of MMS messages (MMS, 1990). Finally, however, only the Read- and WriteVariable messages were chosen to be implemented.

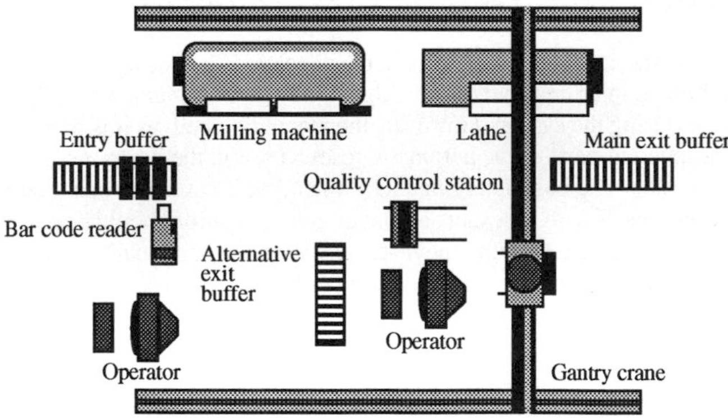

Figure 1 The machining cell in the case study.

Rear axles are manually entered by the operator at the entry buffer. Here, the bar code reader registers the incoming axles by identifying their article number. The operator can manually enter rework codes for those axles that have already been through the system but have been rejected by the QC. The normal flow through the system for each axle is first to visit the milling machine and then the lathe, and finally to exit through the main exit. However, the operator (or operators) can at any time request a specific axle to the QC station, where it will be tested for adherence to the tolerance specifications. This testing is done manually.

The GC is a special type equipped with twin grippers. Under normal operation, one of the grippers is always empty, while the other holds an axle. Loading a device with a new axle means first fetching the old axle with the empty gripper, rotating 180 degrees and then loading the device with the new axle. Thus, loading a device is actually an unload/load sequence under normal operation. When leaving an axle at one of the exit buffers, the GC becomes empty, and then always moves to the entry buffer to fetch a new axle. Special work cycles must naturally exist for the starting up and emptying of the cell. The fact that the GC has twin grippers, with one always empty, is equivalent to having one global buffer place within the system. Thus, the system can never deadlock. This fact significantly simplifies the implementation of the control functions.

With production times of about 10 minutes for each device, normal operation means that the GC will spend most of its time waiting for the axles to finish their work cycles. When the GC has fetched a new axle from the entry buffer, it will wait for the milling machine to finish its current task. When the milling machine has been served, the GC stands idle waiting for the lathe to finish its current job. Finally, when the lathe has been unloaded/loaded, the processed axle will be left at the main exit, whereafter the GC fetches a new axle from the entry buffer.

The result of normal operation is that incoming axles push the outgoing material in front of them. Axles will only be unloaded when new axles are to be loaded. This has several important implications. For instance, if the input flow of axles stops, axles will remain in the devices until a new axle forces the other axles to exit the devices. When the system is not in the start-up or emptying phase, there will usually be an axle at the QC station waiting to be moved to some other unit, so that axles may have to wait a good deal of time to be moved from the QC station.

3 CONTROL AND AUTOMATIZATION CONCEPTS

The production in the machining cell described in section 2 is designed to be more or less automatic, i.e. human interference in production should be minimum. The only interference is the operator supervising the system. However, the machining cell, as it is presented in the case study, can be stripped of parts of the automatic resources, e.g. the gantry crane. The result of a totally stripped system is a completely manual system with manually operated machines and manual transportations. Where the exact amount of automatization should be placed is governed by a number of concepts, e.g. time, money, safety, flexibility and quality. These concepts and their effect when making decisions are described below.

3.1 Time

The first concept, time, is normally associated with the level of automatization such that greater automatization leads to faster production and thus a reduction in the time spent. However, in some special situations, this may not be true. Another aspect is that it is difficult to manually optimize production, which leads to a longer overall production time.

3.2 Money

In second concept, money, it is not as easy to identify the relation between the degree of automatization and the costs or savings. Normally, the replacement of an operator with an automatic equivalent renders some reductions in cost, as the operator is one of the more expensive parts. However, in some countries, the cost of an operator is a relatively small expenditure. This leads to a situation in which it might be more favorable to use humans instead of unmanned, automatic machines.

3.3 Safety

When it comes to safety, the advantages of an automatic system are more obvious. The removal of humans in a manufacturing system is almost always a way to reduce the risk of injuries during normal production. However, in terms of service, it is usually safer for an operator to work in a manually operated system.

3.4 Flexibility

The advantages of an automatic system over a human operator as concerns flexibility is not as tangible. When it comes to flexibility, who can beat man? For instance, if the space encountered in a manufacturing cell is very limited, it might be more flexible to rely on a human rather than a big, bulky robot.

3.5 Quality

The upholding or improvement of production quality, finally, is more directly correlated to the automatization of production. By introducing computers, data networks and intelligent machines into a manufacturing system, quality can be increased. As mentioned initially, humans tire easily and are notoriously inconsistent.

4 STEPS OF AUTOMATIZATION IN A MANUFACTURING SYSTEM

The following section describes a number of possible steps of automatic production in a manufacturing system, using the manufacturing cell in section 2 as an example.

Six different steps of automatization have been identified:

1. Cell control system.
2. Automatic asynchronous material transportation.
3. Automatic synchronous material transportation.
4. Automatically operated machines.
5. Automatic supervision.
6. Data network.

These six different steps of automatization can be combined in a number of possible ways into what we call levels of automatization.

A system that is automatized with respect to cell control system, material transportation, machine operation, data transportation and control supervision represents a fully automated system. The following subsections describe possible ways of reducing the amount of automatization.

4.1 Cell control system

One way to reduce the cost of a fully automatized system is to remove the cell control system. As the cell control system takes care of such things as scheduling and dispatching, this means that the worker who takes the control systems place must be highly skilled and, thus, the control system cannot be replaced with any worker.

In the case study, this step of automatization can be obtained by removing the cell controller.

4.2 Automatic asynchronous material transportation

The transportation of material between different machines in a machining cell is asynchronous, i.e. there are no requirements concerning synchronization with other flows of transportation.

Another way to reduce the cost of a fully automatized system, apart from the cell control system mentioned above, is to remove the automatic asynchronous material transportation system. Instead, one relies on a manually operated material transportation system between different machines. One hazard when removing the automatic material transportation system is the increased risk of errors introduced by humans. Another hazard is the risk of human accidents when the automatic material transportation system is omitted.

In the case study, this step of automatization can be obtained by replacing the expensive gantry crane with a manually operated truck. Instead, here, printed truck orders are needed. If the quality of the products is to be maintained in this case, it is necessary to use sensors at the machines to verify that the correct product is delivered at the correct machine.

4.3 Automatic synchronous material transportation

In an assembly cell, as compared with a pure machining cell, there is also a synchronous transportation of material between different stations. The synchronous transportation is special, as the arrival of a piece of material at a station should be coordinated with the arrival of other pieces of material. This implies that synchronous transportation needs a better, and possibly more expensive, supervision than asynchronous transportation. By replacing an automatic synchronous transportation system with a manual equivalent, it is possible to reduce costs. As in asynchronous material transportation, the risk of errors introduced by humans increases.

Another aspect of the synchronous material transportation is that it is sometimes used to hold a piece of material in position while some other parts are joined to it. This type of work is usually performed by a robot that has the capability for both material transportation and material fixation. If synchronous material transportation that should perform both tasks is necessary, it is more difficult to replace this by a worker.

Thus, as synchronous material transportation is sometimes more complicated than asynchronous material transportation, it is advisable to try to omit asynchronous material transportation sooner than any synchronous material transportation.

4.4 Automatically operated machines

Another method for reducing costs is to exchange automatically operated machines with semi-automatic or manually operated machines. This can be done for all of the machines or only some of them.

In the case study, this step in reducing automatization can be obtained by replacing the expensive milling machine and the expensive lathe by manually operated equivalents. Instead, here, printed working orders for the machines are needed. If the quality of the products is to be maintained at the same time that automatically operated machines are omitted, it is necessary to use sensors that verify the quality of the products.

4.5 Automatic supervision

To further reduce costs it is possible to remove automatic production supervision. Hence, as there are no supervisory activities, it is impossible to verify the quality of the products and, in some cases, impossible to validate the correctness of the products. Instead, these activities must be performed by a human, with the increased risk of erroneous measurements. In spite of this, some information can still be passed through the manufacturing system via the data network.

In the case study, this level is obtained if the automatic production supervision activities are removed, e.g. by removing the bar code reader. The sensors that were introduced earlier to verify the delivery of correct products and the quality of the products are also omitted. Manual routines to verify the quality are then needed.

4.6 Data network

Yet another step to reduce costs is to remove the data network. This leaves us with a system with stand-alone machines, where the instructions for the machines must be entered manually as no data network exists to transport instructions.

In the case study, this level can be obtained by removing the local area network that interconnects the machines and the cell controller.

5 APPROPRIATE LEVELS OF AUTOMATIZATION IN A MANUFACTURING SYSTEM

The six different steps of automatization can be combined into different levels of automatization. Table I describes some examples of possible levels of automatization by combining the six steps of automatization in different ways. For example, one level is the fully automated production, i.e. level I. Another level is to introduce manual asynchronous material transportation but to allow the rest to be automatic, i.e. level II. Still another level would be to have everything manually controlled and have no data network, i.e. level VI. A level that actually exists in many installations is automatically operated machines and no further automatization, i.e. level III, although this level inflexible as well as unpractical. Furthermore, verification of the quality of the products is very difficult.

Table 1 Steps of automatization combined to achieve different levels of automatization

1. Cell control system
2. Automatic asynchronous material transportation
3. Automatic synchronous material transportation
4. Automatically operated machines
5. Automatic supervision
6. Data network

Y Automatic or present
N Manual or not present
M Mixed automatic and manual
— Not applicable

		Steps of automatization					
No.	*Examples of levels of automatization*	*1.*	*2.*	*3.*	*4.*	*5.*	*6.*
I	Fully automated manufacturing cell	Y	Y	Y	Y	Y	Y
II	Manual asynchronous material transportation	Y	N	Y	Y	Y	Y
III	Automatically operated machines only	N	N	N	Y	N	N
IV	Lowest acceptable level of automatization	N	N	N	N	Y	Y
V	Not acceptable level of automatization	N	N	N	N	N	Y
VI	Non automatic production	N	N	N	N	N	N
VII	The machining cell in the case study (section 2)	Y	Y	—	Y	M	Y
VIII	The assembly cell in the example (section 6)	Y	N	Y	Y	M	Y

The conclusion drawn from observing the different levels of automatization is that the lowest acceptable level of automatization for a manufacturer, without having to consider the country or type of production, is the case in which production supervision is left to the automatic system. This would be a system described by level IV. If the automatic supervision is removed, i.e. level V, the quality of the products would be affected.

To achieve a balanced manufacturing system, one should omit expensive automatic parts in favor of manual counterparts if the manual counterparts are cheaper in the long run and if the product quality can be maintained.

Taking into consideration the aspects of safety, quality etc., we believe that the following steps should be taken in the following order to achieve a balanced manufacturing system:

- Omit automatic asynchronous material transportation.
- Omit automatic synchronous material transportation.
- Omit automatically operated machines.

We believe that it is most cost effective to have manual material transportation, as described earlier, and then, if a greater reduction in automatization is required, to continue by omitting manually operated machines. The reason for this is that it is easier to maintain a certain level of product quality using manual transportation than it is to introduce manually operated machines, where there is a greater risk of reduced product quality. The next step toward reducing costs is to remove the automatic supervision of production. However, we strongly believe that this is an unwise step, as the verification of the quality of the products is left to a human inspector who is more liable to commit errors than a correctly installed automatic supervision system. Furthermore, the cost of automatic supervision with computers and sensors is far less expensive than highly complex machines and automatic transporting systems. In order to have automatic supervision, it is necessary to have a data network and, hence, this component should not be omitted in a balanced manufacturing system.

6 EXAMPLE OF A BALANCED ASSEMBLY CELL

The example describes a non-traditional assembly cell. Two of the primary differences, as compared with traditional cells, are the number of parallel assembly teams resulting in very long cycle times and the material feeding system, which is a pure kitting system. The cell has been studied extensively and the layout and function of the cell is described in Johansson and Johansson (1990). The cell is also illustrated in Figure 2.

This cell produces kits, which are plastic bags containing small size parts (e.g. screws, nuts, and plugs) to be used in the assembly department of a company. Each kit contains the small parts needed for a specific portion of the assembly of one specific product. There is a large variety of kits (several hundred), each containing between two and 20 different parts, i.e. parts with different part numbers, and between 10 to 40 single parts. A typical value is 8 part numbers corresponding to 25 single parts.

The production of kits is divided into two major activities corresponding to physical areas in the system: automatic counting and packaging (consisting of bagging and printing). Counting is done into sectioned trays, where each tray has 18 sections. Automatic counting is done by four identical vibratory bowl feeders that work in a so called counter-group, P_1. The feeder can handle only one part number at a time, which means that four part numbers per kit is the maximum number that can be counted at one time. For this reason, many of the kits must pass the counter-group more than once. Bagging is done by a bagging machine, P_{21}. In a special machine, the tray is turned and the kits are placed into a bucket elevator that leads to the bagging machine. A printer that to the bagging machine prints the internal part number of the kit on the bag, P_{22}. The quality of the kits with regard to part numbers and quantities of each part number is controlled by a check weigher connected to the bagging machine, P_{23}. The downloading of production information is retrieved from the VAXCluster, P_{-1}, via the personal computer, P_0.

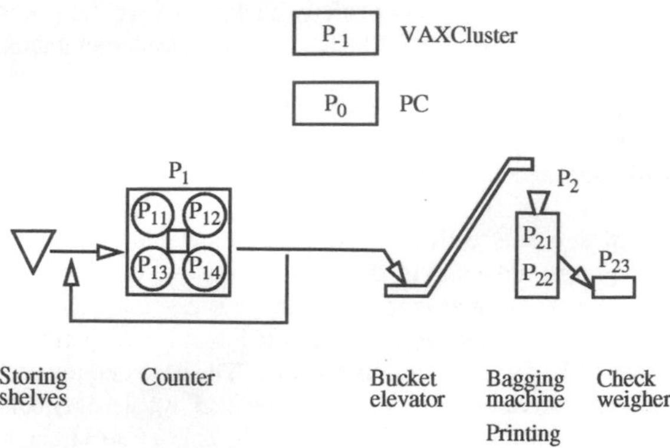

Figure 2 Layout of the assembly cell.

The assembly cell is balanced in the sense that, when the cell was planned, it was decided that the asynchronous material transportation be performed manually, e.g. the transportation between P_1 and P_2, while the synchronous material transportation be performed automatically, e.g. the positioning of the tray in the P_1 area. This measurement was taken in a car producing facility in an industrialized country, i.e. Sweden, even though the salaries were high. The reason for this was simply that the asynchronous material transportation was too expensive to perform automatically. The cell control system is automatic and performed by the personal computer P_0. The machines, i.e counting, P_1, bagging, P_{21}, printing, P_{22} and weighing, P_{23}, are automatically operated. Most of the supervision is done automatically but an operator manually enters the identification of the tray into the packaging system. Finally, a data network is connected to all the machines in the cell.

The weak part of the assembly cell is the manual supervision. There is an obvious risk for the operator to enter an erroneous identification. We therefore suggest that bar code readers are introduced, both at the counter, P_1, and at the packaging system, P_2. At P_1, the identification of the tray is read and saved in a database where it can be retrieved by P_2 afterwards. In this way, the quality of the products is not endangered, as it is the case of manual supervision.

7 CONCLUSIONS

A number of levels of automatic control and automatic production have been presented in order to identify activities that may be performed by a human operator rather than an automatic system to achieve a balanced automation system with respect to such relevant aspects as cost, flexibility and quality. To do so it is essential for a manufacturing system that the supervisory portion exists and is performed automatically by the system. It is not possible to leave this task to a human operator solely, as it is common for humans to commit errors that might reduce the quality of the products. In addition, the cost of the equipment to perform the automatic supervision is now relatively low, compared with certain highly specialized machines. Nonetheless, certain parts of an automatic system can be omitted to reduce costs without reducing the production quality.

Thus, by removing expensive machines, such as the gantry crane described in section 2, in exchange for humans and simple, cheap machines, it is possible to reduce costs, maintain the quality of the final product and uphold a higher degree of employment in countries in which salaries are low. However, it is essential in such situations that the manufacturing system be able to rely on some kind of automatic supervision to verify product quality. We believe that the cost of having a reasonable amount of computer power together with a data network that connects the computers with the sensors placed on the machines is reasonable and affordable. Even more important, this equipment is necessary if a manufacturer is to be able to provide products of good quality.

8 REFERENCES

Adlemo A. and S.-A. Andréasson (1995a). Operator control activities: a case study of a machining cell. To appear in *Proceedings of the 8th IFAC/IFIP/IFORS/IMACS/ISPE Symposium on Information Control Problems in Manufacturing Technology, INCOM'95*. Beijing, China.

Adlemo A., S.-A. Andréasson, M. Fabian, P. Gullander and B. Lennartsson (1995b). Towards a truly flexible manufacturing system. Will appear in *Control Engineering Practice*, **3**.

Andréasson S.A., A. Adlemo, P. Gullander, M. Fabian and B. Lennartsson (1995). A machining cell level language for product specification. To appear in *Proceedings of the 8th IFAC/IFIP/IFORS/IMACS/ISPE Symposium on Information Control Problems in Manufacturing Technology, INCOM'95*. Beijing, China.

Badham R. and B. Schallock (1991). Human factors in CIM: a human centred perspective from Europe. *International Journal of Human Factors in Manufacturing*, **1**, 121-41.

Dreyfuss D. D. (1989). Is industry ready for machine vision? - a panel discussion, in *Machine Vision for Inspection and Measurement* (ed. H. Freeman), Academic Press Inc., New York, U.S.A., 223-36.

Drury C. G., M. H. Karwan and D. R. Vanderwarker (1986). Two inspector problem. *IEEE Transactions*, **14**, 174-81.

Fabian M., P. Gullander, B. Lennartsson, S.-A. Andréasson and A. Adlemo (1995). Dynamic products in control of an FMS cell. To appear in *Proceedings of the 8th IFAC/IFIP/IFORS/IMACS/ISPE Symposium on Information Control Problems in Manufacturing Technology, INCOM'95*. Beijing, China.

Freeman H. (1988). *Machine Vision*. Academic Press Inc., Boston, U.S.A.

Geary G. M., R. Walker, H. Mehdi and W. Parks (1992). Improving the yield of a manufacturing process using an on-line product information and display system. *Proceedings of the 7th IFAC/IFIP/IFORS/IMACS/ISPE Symposium on Information Control Problems in Manufacturing Technology, INCOM'92*. Toronto, Canada, 177-81.

Gullander P., M. Fabian, S.-A. Andréasson, B. Lennartsson and A. Adlemo (1995). Generic resource models and a message-passing structure in an FMS controller. To appear in *Proceedings of the 1995 IEEE International Conference on Robotics and Automation, ICRA'95*. Nagoya, Japan.

Johansson M. I. and B. Johansson (1990). High automated kitting system for small parts - a case study from the Volvo Uddevalla plant. *Proceedings of the 23rd International Symposium on Automotive Technology and Automation*. Vienna, Austria, **1**, 75-82.

MMS (1990). *Industrial Automation Systems - Manufacturing Message Specification —Part 1: Service Definition*, International Standard, ISO/IEC 9506-1, First edition.

Mårtensson L. and J. Stahre (1992). Operator roles in advanced manufacturing, in *Ergonomics of Hybrid Automated Systems III* (eds. P. Brödner and W. Karwowski), Elsevier, Amsterdam, The Netherlands, 155-62.

Olsson G., P. L. Lee (1994). Effective interfaces for process operators - a prototype. *Journal of Process Control*, **4**, 99-107.

Schneiderman B. (1987). *Designing the User-Interface: Strategies for Effective Human-Computer Interaction*, Addison-Wesley, Reading, U.S.A.

Sheridan T. B., L. Charny, M. B. Mendel and J. B. Rosenborough (1988). Supervisory control, mental models and decision aids. *Proceedings of the 3rd IFAC Conference on Analysis, Design and Evaluation of Man-Machine Systems*. Oulu, Finland.

Stahre J. and L. Mårtensson (1992). Supervising the CIM system - new roles for the operator. *Proceedings of the Industrial Automation'92 Conference*. Singapore, 199-211.

Sylla C. (1992). Modelling the pairing of human and machine-vision in industrial inspection tasks. *Proceedings of the 7th IFAC/IFIP/IFORS/IMACS/ISPE Symposium on Information Control Problems in Manufacturing Technology, INCOM'92*. Toronto, Canada, 33-8.

9 BIOGRAPHY

Anders Adlemo received his M.Sc. in Electrical Engineering at Lund University of Technology, Lund, Sweden, in 1981, and his Ph. D in Computer Engineering at Chalmers University of Technology, Göteborg, Sweden, in 1993.

Dr. Adlemo is currently a researcher at the Department of Computer Engineering at Chalmers University of Technology. His current research interests include distributed computing systems, fault tolerant systems, and complex flexible manufacturing systems.

Sven-Arne Andréasson received his B.A. in Physics and Mathematics at the University of Göteborg, Sweden, in 1971, M.Sc. in Electrical Engineering at Chalmers University of Technology, Göteborg, Sweden, in 1976, and Ph. D. in Computer Science at Chalmers University of Technology, Göteborg, Sweden, in 1986.

Dr. Andréasson is currently an associate professor at the Department of Computing Science at Chalmers University of Technology. His current research interests include distributed computing systems, fault-tolerant systems, databases and complex flexible manufacturing systems.

Dr. Andréasson is a member of the Association for Computing Machinery (ACM), and the IEEE Computer Society.

PART FOURTEEN

CAE/CAD/CAM Integration

ARCHITECTURE SOLUTIONS FOR INTEGRATING CAD, CAM AND MACHINING IN SMALL COMPANIES

R. Boissier, R. Ponsonnet, M. Raddadi, D. Razafindramary
Groupe de Recherche sur la Production Intégrée (GRPI)
IUT de St Denis, Université Paris 13.
F-93206 St DENIS CEDEX 1, France.
Fax.: +33-1-49 40 61 96. E-mail : rboissier@d.univ-paris13.fr

Abstract.

The functions and flows in the process of machining complex geometry part are considered. The computer solutions as well as some of their effective implementation in a demonstration platform are described. The overall efficiency greatly relies on communicative numerical controllers and availability of local and wide area networking facilities.

Keywords.

CAD, CAM, CIM, CNC, quality control, industrial network.

1 THE NEED FOR NEW PERFORMANCES IN SMALL CNC SPECIALISED COMPANIES

In the field of mechanical industries, large industrial groups (automotive, aerospace,...) urge their subcontracting partners to adopt CAD/CAM tools and quality control procedures to reach the shortest development times. This means difficult choices for small subcontracting companies.

For several years GRPI has been devoting efforts to the design and implementation of performance inductive techniques for a particular category of subcontracting companies : those specialised in the machining of complex geometry parts. A part is said to have a complex geometry whenever the cutting path cannot reduce to a mere set of macroscopic linear or circular moves. In which case NC (Numerically Controlled) machining must compulsorily be associated to CAM (Computer Aided Manufacturing) tools. In a few years, NC machining of complex parts has been taking over traditional hand modelling as a result of the spreading of CAD/CAM techniques. It concerns :

- many consumer products the shape of which is mainly governed by aesthetics considerations,
- shapes resulting from mathematical optimisation criteria such as from the Finite Element Methods.

Complex shaped parts are seldom machined in large batches because it is a relatively slow and very expensive process. On the other hand this technique is very appropriate in the following cases :

- part **prototyping** (in the course of the product design phase)
- manufacturing of **tools** such as moulds, electrodes for electro-erosion, etc.

After considering the main technical activities to be performed in a CNC plant, the solutions studied and implemented will be presented.. The various achievements deal with the **preparation**, **machining** and **inspection** processes, as well as the **integration** of these processes.

2. TECHNICAL FUNCTIONS ANALYSIS

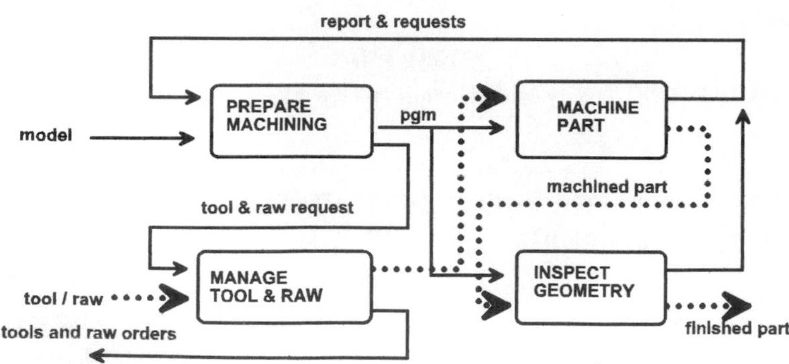

Figure 1 Technical functions in CNC plant, level 1.

Figure 1 is a SADT-like, level 1, display of the interconnection of the main technical functions to be performed in a typical subcontracting CNC company.

2.1 Function PREPARE MANUFACTURING can be split into :

- project file reception : transfer, unpack, store
- pre-analysis of received CAD data : choose and request necessary tools and machines
- geometry analysis and tool path generation
- choice of machining parameters
- machining simulation for error tracking
- post-processing of tool path data into specific machine-tool programs
- selecting elements and surfaces to be inspected
- generate surface sensing sequences

2.2 Function PERFORM MACHINING receives a PART FILE which consists of :

- NC machining programs, specific to the considered machine (Tape Files). The operator may be led to **chain** together several sub-programs, for example several raw and finishing cuts.
- Operator directives : the program or the appended directives may trigger tool search, either in the machine local store (automated or not) or in the workshop central tool store.

The machining data for a part consist of several computer files. The management of this Part File is performed at the machining station by the operator. The problems to be solved at this level are related with the fact that complex shape machining programs consist of an extensive number of very short linearly interpolated moves : most of the Numerical Controllers, except for very recent ones, are unable to memorise data amounting to more than 1 megabyte. On the other hand their computing power is limited to a few tens interpolations per second where a few hundred segments per second would be more appropriate, particularly for making prototypes out of soft polymer material.

2.3 Function INSPECT PART ultimately warrants conformance of the finished product:

The Coordinate Measuring station gets a Part Inspection File which consists of : (i) sensing sequences for the surfaces and elements to be inspected, (ii) tolerance margins in the form of associated software gauges. According to the sensing directives, the Coordinate Measuring Machine (MMT) senses the part, determines the part own referential and registers the coordinates of the specified points. These data are then collected and processed in order to show the operator whether the part is within the tolerances or not. More accurate and visual interpretation of the results can be performed on the preparation site and decision can then be taken about a bad part to modify any of the CAD/CAM processes.

2.4 Function MANAGE TOOLS AND MATTER :

This function is invoked right at the start by the project leader to specify the needs and trigger orders for raw material, equipment, tool preparation or just tools and machine-tool reservation. The Tool Data Base (TDB) is updated by the tool operator in the course of his tool management activity. The TDB is relatively complex because tools are compound objects whose characteristics are time changing due to wear, re-sharpening, re-assembly. The TDB must also manage tool location within the workshop and their allocability to the machines. The data in the TDB should be accessible from any work station. Also the NC machine-tool controllers may memorise tool compensation tables which must be closely linked with the TDB.

2.5 Needs for COMMUNICATION :

Links in Figure 1 emphasise the necessity of communication between the previously mentioned functions. In view of relieving the operators from cumbersome administration tasks and avoiding errors meanwhile any information transmission or transcription, the company must be equipped with :

- means for easy local and remote data transfer,
- software interfaces for inter-software communication
- man-machine interfaces for the management of data exchanges and retrieval.

3. IMPLEMENTATION OF A LIGHT INTEGRATED CNC PLANT.

In the following, the guiding principle for the choices of architecture has been one of standardisation and simplicity. This principle is well adapted to the limited means of a small company, furthermore it does not block the plant in a one time configuration (Figure 2).

3.1 Communication.

3.1.1 Remote communication.

Traditional means of communication in which humans interact at both ends remain of utmost importance and new technologies like telefax or electronic mail increase their efficiency.

A national or international automated system for information diffusion or retrieval is also a important factor of productivity. In that respect the French intensively use the TELETEL videotex service which offers :

- access to professional directories, particularly useful when subcontracting is intended;
- access to technical data bases like CETIM's (Centre d'Etude des Industries Mécaniques);
- easy server set up, very useful for a company with many clients who wish to automate information diffusion and order taking.

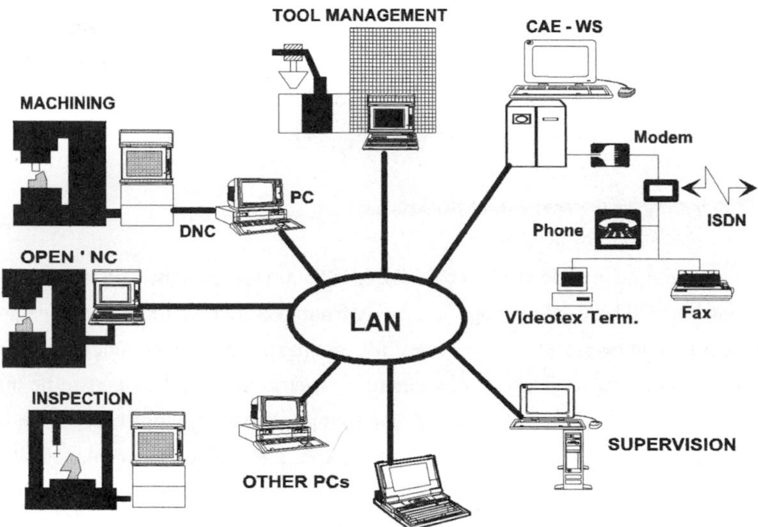

Figure 2 Schematic overview of CNC plant

Beside interactive exchanges, the company needs Digital Data Exchange with their customers for the transfer of manufacturing files and reports, and various solutions can be considered. Due to the variety and evolution of CAD systems and neutral format standards, it may be safer to call for the services of CAD file transfer nodes were expertise in format translation is concentrated.

As to the physical link, the Integrated Service Data Network (ISDN), in service in many countries, presently seems at short term the best compromise between cost and throughput. ISDN can forward Pulse Code Modulated voice and digital data at the rate of (2 x 64 + 16)Kbits/s. Transmission time for 1 megabyte is then reduced to some 2 mn/Mbyte which is an improvement by a factor of ten over usual asynchronous Modems. It seems judicious to connect at least the technical office to ISDN, one ISDN socket is sufficient to serve all telecommunication usage up to 8 various terminals.

3.1.2 Local communication.

Any workstation in the workshop in the plant consists in, integrates or is under control of a standard computer. These computers are CAD/CAM/CAE UNIX workstations, X-Windows terminals and industrial PC/PS microcomputers under DOS or OS/2.

Inter-computer communication is through a standard office Local Area Network such as Ethernet or Token Ring since traffic is intensive only during short bursts at file transfer time. Actual transmission time is some 10s/Mbyte due to protocol overhead and PC's limited throughput). Due to plant heterogeneity, TCP-IP has been adopted as the base protocol for the exchanges. TCP-IP has the advantage of being part of the UNIX/AIX operating system and available on MS-DOS or MS-WINDOWS systems.

3.2 Preparation of the machining process

Machining preparation is performed on IBM AIX-RS6000 workstations with high performance graphic controllers. The CAD package is CATIA from Dassault Systems, a surface and solid modeler. CATIA integrates CAM modules for multiaxis lathes and milling centres. The CAD/CAM operator gets delivery of the remote geometric model of the part to be machined and defines on which surface elements and along which curves the tool end will be moving, actually he must be a specialist in both CAD and CAM. The CATIA CAM module allows for introducing the surface tolerance and tool geometry, and simulating tool path. This visual check is a protection against the most obvious errors.

Machining data are produced in an intermediate APT-like format. These data must then be adapted to the selected machine-tool characteristics through a so called Post-Processor interface which produces machine specific Tape Files. Modular post-processors for multiaxis machining centres have been developed. The ISO Tape Files pertaining to a given part are grouped in a Part File and become available on request from the machining station operator.

The CAD/CAM workstations also support a Computer Aided Inspection software, VALISYS, which is interfaced with the CATIA generated models. On the graphic display, the CAD/CAM operator defines the tolerancing type with graphic ISO/ANSI symbols which are translated into Virtual Gauges. For any component surface to be inspected, a grid of test points is generated. Both inspection sequence and tolerancing data are available from the Inspection Station through the local network. Conversely the CAD/CAM operator can retrieve data issued during an inspection session and visually compare the specified and measured surfaces.

3.3 Machining stations

3.3.1 Computer as a Data Store and a Machine-tool Supervisor.

At the beginning of this project development, one urging need was present : how to carry out complex prototypes machining on a modern milling centre, yet equipped with a poor 60 Kbytes program memory. The first answer was the use of a Direct Numerical Control (DNC) mode in which the PC computer standing by the machine-tool manages data transfer along a 9600 bits/s asynchronous line according to a safe block mode protocol.. This was tested by executing machining programs over 1 megabyte long. Extension to PC control of several machines, using an intelligent communication card was also considered. When later on CAD/CAM workstations and a LAN were installed on the site, it was easy to include calls to TCP-IP application commands in the DNC software so as to let the PC-DNC operator ask for the downloading of a Part File from any CAD/CAM workstation (Boissier, 1988), (Razafindramary, 1991).

3.3.2 PC Computer as the kernel of an Open Numerical Controller.

The preceding tests made clear a serious limitation of the microprocessor based CNC controllers in the 80's : a low interpolation speed (5-10 blocks/s), affordable for macroscopic moves but speed limiting on short segment surface following.

Searching at once for high interpolation speed, communicativity, evolutivity and low cost, we developed a concept of OPEN NC using a standard PC computer component for managing dialogue with the operator, and the workshop, and a specialised module for rapid interpolation.

The machine functions and dynamics have been specified in Figure 3, using the Ward & Mellor hierarchical formalism(Ward, 1986). This specification has been mapped into the architecture in Figure 4 (Raddadi, 1991,1993).

The OPEN NC is composed of an industrial PC working as operator interface, program store with mass memory and workshop communication unit with a LAN adapter and the TCP-IP protocol. Machine-tool axes control (3 axes, extensible to 5) is performed by an add-on card from the Delta Tau Company whose specialised Digital Signal Processor (Motorola DSP 56000) achieves in a very small volume interesting interpolation performances.

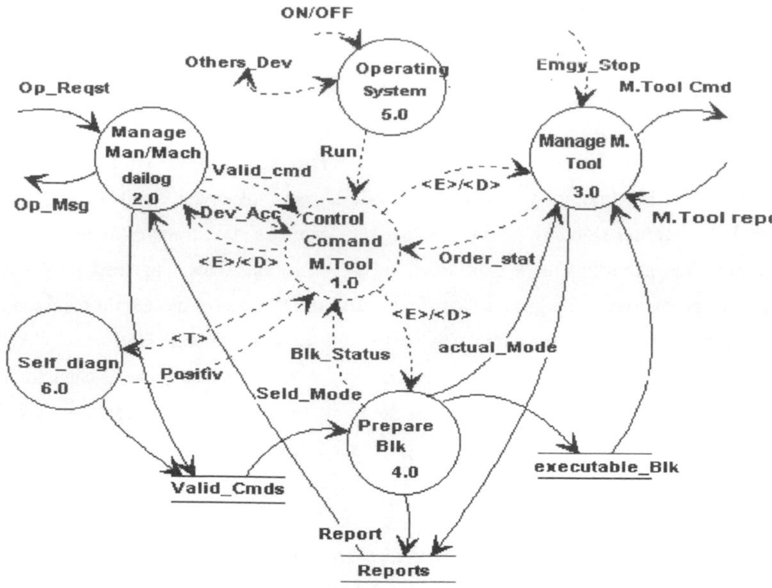

Figure 3 Functional and dynamic specification of OPEN NC

Control of a machine tool and its environment implies some logic and sequential functions, these are specified with Sequential Function Charts or Grafcet (IEC, 1992) and implemented as cyclic I/O tasks within the Real Time IBM-ACS environment (IBM, 1993). I/O data are conveyed to the operative part through a 1 Mbit/s field bus, such a link to remote I/O reduces cabling time and errors.

The achieved OPEN NC is very compact, very modular and extensible, it allows easy transformation of a mechanically sound machine tool with obsolete control into a very performing and evolutive industrial system.

The IBM-France Company have seen, in our solution, the possibility to directly integrate machining stations into their overall industrial communication architecture (IBM, 1993). Actually the control architecture is adaptable to a great variety of contexts.

The need and interest for openness in a machine-tool controller is real and has given way to several studies and realisations. A critic of traditional controllers has been presented in (Grégoire, 1989) together with the design of an experimental modular system, while recently NC manufacturer NUM SA expressed their interest for such systems and announced an architecture somewhat different than ours, in which a PC-OS/2 compatible card is plugged into the VME bus based existing numerical controller, allowing the local installation of DOS or OS/2 applications (Pritschow, 1993).

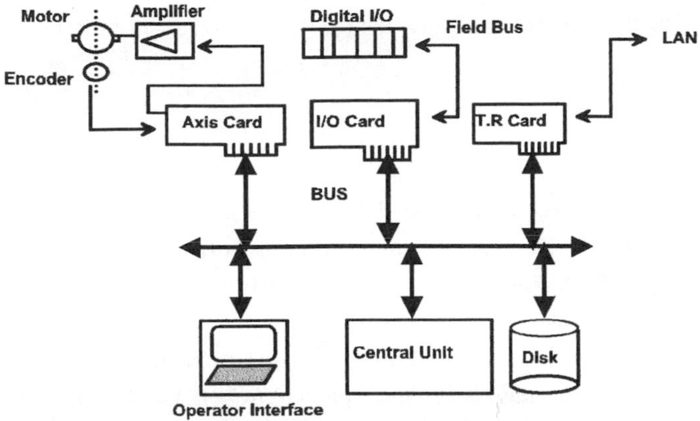

Figure 4 Architecture of OPEN NC

3.4 Coordinate Inspection Station

Integration of a coordinate inspection stand in the plant profits from the above mentioned developments. The VALISYS modules needed at inspection stand are installed in an AIX operated IBM-PS/2, just by the DEA Coordinate Measuring Machine (CMM). This micro-computer serves as a relay between the technical office and the CMM, it translates the VALISYS sensing instructions into specific DEA instructions which are downloaded into the CMM controller through an asynchronous line. Conversely, coordinate data retrieved from the CMM are locally formatted for the purpose of VALISYS processing.

4 CONCLUSION.

In this paper some particular aspects of CIM have been highlighted. We wish to make clear, particularly to industrials that in small plants integration needs a relatively modest investment compared to machine-tools ones. The light communication system, simple gateways and versatile Open Numerical Control we have set-up should contribute to CNC plant performance and leave more time for the operators to practise their job skills. Work in progress includes a methodological object oriented approach to tackle overall manufacturing management, particularly tool management, with inclusion of MMS primitives (Lefebvre, 1994).

5 ACKNOWLEDGEMENTS

The authors wish to thank Mr MOURIER, IBM consulting engineer, and his colleagues as well as Mr MONTES, IUT workshop manager, and the IUT technicians for their fruitful help.

6 REFERENCES

Boissier R. et Al. (1988).La chaîne CAO-Usinage dans un îlot de fabrication de pièces à géomètrie complexe. *Rev. Aut. & Prod. Appl. Paris.* **1(4)**, 38 - 49

Grégoire J-Ch (1989). Architecture de commandes numériques des machines-outils. *PhD Thesis,* EPFL-Lausanne.

IBM (1993). *Dossiers de l'Informatique Industrielle.* **4411-01/93**. IBM Paris.

IEC (1992). *Sequential Function Charts. International IEC-848 standard.* IEC,Geneva.

Lefebvre M. (1994). MMS sur TCP-IP, une solution qui utilise l'appel de procédure à distance. *D. Ing. dissertation.*, CNAM-Paris.

Pritschow G. (1993).Il faut ouvrir les CN. *NUM Information* **(18)**,13. Argenteuil.

Raddadi (1992). Elaboration d'un DCN multiaxes ouvert. *PhD Thesis* , Université de Nancy 1.

Raddadi M. et Al. (1993).Spécification temps-réel et réalisation d'une commande numérique ouverte. *Rev. Aut. & Prod. Appl.* **6(3)**, .303 - 322. Paris.

Razafindramary D. (1991). Conduite de Systèmes de Fabrication. *PhD Thesis,* Université de Paris-Nord.

Ward S. and Mellor (1986). *Structured development of Real Time Systems.* Yourdon Press, Englewood Cliffs.

7 BIOGRAPHY

Raymond Boissier graduated from Ecole Supérieure d'Electricité and is a Doctor in Physics, he has specialised in industrial automation and computing. Raymond Ponsonnet graduated from Ecole Normale Supérieure de Cachan and is a Doctor in Science He is the head of GRPI and has specialised in robotics and complex surface inspection. Mohammed Raddadi graduated from Centre d'Etudes Supérieures des Techniques Industrielles and is a Doctor from Université de Nancy 1. He has specialised in machine-tool control.Donné Razafindramary got his Engineering Diploma in telecommunication and his Doctorate from Université Paris-Nord. He has specialised in industrial computing and workshop communication.

A feature-based concurrent engineering environment

N. P. Moreira, A. C. Ferreira
GRUCON - Mechanical Engineering Department/
Universidade Federal de Santa Catarina
PBox. 476 - GRUCON/EMC/UFSC - 88040-900 -
Florianópolis/SC - Brazil
Tel. 55-482-319387 Fax. 55-482-341519
email: npm@grucon.ufsc.br acf@grucon.ufsc.br

Abstract

This paper presents an approach to a Concurrent Engineering Environment (CEE) development based on Feature Technology. This proposal is specially useful in the Design for Manufacturing (DFM). In this sense, the information, related to different areas of the product life-cycle development, must have a common meaning for everyone. This may be achieved by using a constructive element called "feature" where the information is modeled. In order to support the information management some Information System requirements are defined. Some aspects of the model have been implemented in a real industrial case study.

Keywords

Concurrent engineering, product information modeling, feature technology

1 INTRODUCTION

Concurrent engineering is more than a new word to old concepts. It is a natural human-centered way to organize the product life-cycle development. By this way most of the usual errors and delays, during the product development are avoidable. It is a mult-effort issue where many different professionals are involved in the product implementation, each one bringing his/her own experience and needs. Design for Manufacturing (DFM) and Design for Assembly (DFA) are two of the efforts to organize the information exchange among different areas.

The GRUCON, in close cooperation with a Brazilian enterprise and the University of Amsterdam is developing a computational environment to help in the product life-cycle development. It is structured in a heterogeneous and distributed database where the information is modeled based on feature technology. Currently the development is concentrated in the Design for Manufacturing methodologies. In this area, a feature-based interface (developed at UFSC) for the CAD system and the information modeling are under development.

The information is represented using feature technology once the product is the link between several completely different information along the product life-cycle development. In these sense, a feature can be used as an "information key" to users from any design, manufacturing, assembly and others activities.

This paper is organized as follows: section 2 describes the concurrent engineering (CE) concepts and approaches presented in the references as well as some not well-defined aspects of the CE definition; section 3 presents a feature technology revision; section 4 describes the CEE based on an Information System; section 5 describes the CEE architecture and the feature technology role in the information modeling; section 6 presents some useful aspects of the environment in the Design for Manufacturing area; finally section 7 presents some conclusions and next steps.

2 CONCURRENT ENGINEERING

Concurrent Engineering for Chen et al. (1992) is a systematic approach to the concurrent design of products and their related processes in a computer integrated environment. In another point-of-view, Terpenny et al. (1992) say that CE brings to the design phase considerations related to many factors affecting total cost and performance throughout the life-cycle of a product. For Kusiak et al. (1993) CE is defined as a systematic approach to the integrated simultaneous design of products and processes, including manufacturing and support. In short, CE is generally recognized as a practice of incorporating various life-cycle values into the early stages of design. These values include, not only the products primary functions, but also its esthetics, manufaturability, assembly, serviceability and recyclability (Jo, 1991, Ishii, 1993).

Terpenny (1992) identified three basic components to concurrent engineering systems. These components include: improved design strategies, decision methods and supporting information and knowledge. Design strategies include general principles for design that enable or improve the function of other life-cycle processes. Decision methods described include: multifunctional teams, classical optimization methods, and artificial intelligence techniques. Supporting information and knowledge includes a discussion on geometric modeling, design by features and CIM requirements to achieve an integrated system.

In few words concurrent engineering, in the scope of this work, is **a life-cycle development** methodology **that aims the activities integration through the information decentralization and cooperative work**. This means that, every product life-cycle activity may be much better supported by concurrent engineering.

This approach primarily suggests the use of teamwork, where the team is formed by engineers and experts from all activities related to the product life-cycle (Huthwaite, 1994). One main function of the multifunctional team is to negotiate the best solution for the product development. The team is created in the earliest life-cycle phases and is responsible for all decisions made regarding the product, until the product is out of the market. Hence, teamwork involves intense collaboration and exchange of information.

In spite of CE could be implemented without computer based support, in the industry, it is quite hard to do it once the information needs to be widely distributed. Then a computer based environment have been designed to help each activity development as well as the multifunctional team work. The CEE is designed as a heterogeneous and distributed database where each activity (an agent in this context) has its own information model and exports the relevant information to other agents (activities).

2.1 Concurrent Engineering: doubtful aspects

Concurrent engineering is, probably, the newest fashion in the manufacturing environment. So, it has lots of non well-defined aspects. In order to encourage the discussion some of them are briefly presented here.

Design-centered approach

Once the design is the life-cycle activity responsible for the major part of the cost definition in the product development (Bedworth, 1991), the concurrent engineering focus is in the design activity and its relationship with the remainder steps. Usually these proposals try to develop the design simultaneously with the process planning, assembly, etc.

This approach does not examine the fact that each life-cycle activity has its own problems, related with other activities. Then the same attention that has been dispensed to the design must be considered to any life-cycle activity.

Concurrent or Simultaneous Engineering

The terms concurrent engineering and simultaneous engineering are being used synonymously to refer the same methodology (Lindenberg, 1993). Specially in Brazil it is difficult to know which name is more intensely used. Generally the term is associated with its definition scope.

Analyzing the words semantics, in English or in Portuguese, is quite clear that both "concurrent" and "simultaneous" have the meaning "happen at the same time", but "concurrent" has the additional idea cooperating, agreeing Holanda (1988) and Longman (1983).

Looking at the common manufacturing process (complex parts and products), it is possible to assert that is quite uncommon to produce a part while the part has not yet been entirely designed. Besides, the CE concept means information exchange and cooperation in order to detect problems in advance and discuss solutions. Therefore the word "concurrent" seems to agree more with the CE main idea.

Is CE a new name for CIM?

In early 80's Computer Integrated Manufacturing (CIM) was considered the solution to any problem in an industry. During these CIM "golden" years several definitions, always increasing the scope, had been created. Unfortunately, due to the complexity of the practical aspects, no useful CIM implementations have been presented and the idea has been slowly discredited (Bedworth, 1991).

Concurrent engineering is a new name to an old idea, but not the CIM idea. CE has been used since Henry Ford and Ransom Olds automobile industries had created groups in the design phase with experts in design and production (Huthwaite, 1994). But CE is quite different from the CIM idea. CE can be implemented in a people-to-people approach and uses technologies (e.g., DFM-Design for Manufacturing and DFA-Design for Assembly) completely computer independent. In every case where it is possible to simplify the implementation and use non intensive computer support it should be the easiest way.

Currently CIM can be considered a computer tool to the information integration (Moreira, 1993). In this sense CIM can be a support to the information exchange in a complex Computer Engineering Environment.

3 THE FEATURE TECHNOLOGY

Features concept is not a unanimity in the scientific community and several works have been realized using the main idea (Shah, 1988a, b, 1991,Yeo, 1991, Hernandez, 1991, Moreira, 1992). The paper presented by Shah (1988b) shows the concept evolution. It can be very restrained like "Elements usage in generating, analyzing, or evaluating designs" - or generic - "Groups of associated or related data elements." In order to aggregate the information of the whole product life-cycle development some rules had been defined about features:

- A feature must be classified in such way that seems usual for any life-cycle development area.
- There is a finite field of parts. It should be represented by a finite domain of features.
- The set of features is enough to represent a part in the domain. That is, the design process uses a features library. Once some times it is necessary to create a new feature the environment must provide support to feature specification
- The feature library is a knowledge base used by every tool (CAD, CAPP, etc.) in the environment.

A feature is an entity composite by production system's data and knowledge about itself. The main idea of this approach is the possibility to represent, under the same name, or key, all information about design, process planning, cutter path generation and so on.

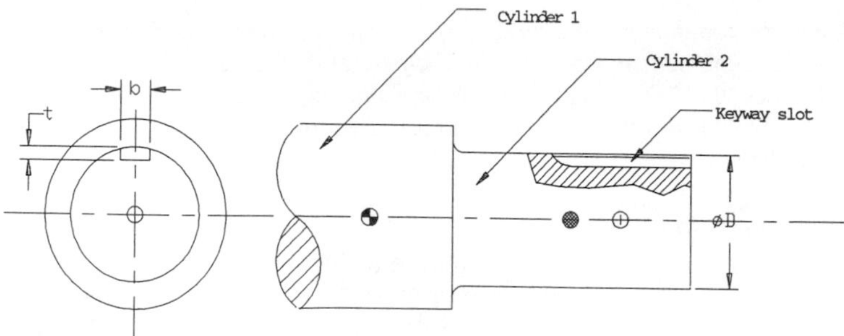

Figure 1 A shaft with keyway slot represented using features.

In Figure 1 there is a part that aggregate three features. The relationship among them is carried out by the reference system of each one. The features attributes are:

- geometrical description
- finishing surface
- form tolerances
- dimensional tolerances
- surfaces
- knowledge about the manufacturing process.

The feature domain identification and classification generate feature families that are organized in hierarchies. In order to model features in such way it represents what it really means, an Object-Oriented approach (Tonshoff, 1994) is used structuring features like class and instances. This approach is described in Moreira (1992).

4 THE INFORMATION SYSTEM SUPPORT TO CONCURRENT ENGINEERING

In Camarinha (1994), a very interesting approach is proposed to implement concurrent engineering. In that paper a concurrent engineering environment, able to support the communication whitin engineering team, is based in the solution of three main problems: the definition of a common model, engineering information management and, process supervision.

A common model and an information management infrastructure are interrelated issues concerning the information sharing and handle. The process supervision is necessary to model the engineering process and to implement a federated loose-control process interpreter or supervisor. In a few words, it is necessary to schedule and coordinate activities during the manufacturing development. Of course this is mainly necessary in a non human-centered environment, but concurrent engineering is mainly people exchanging information, and most efforts must point to the teamwork coordination.

Following the concepts presented in sections 2 and 3, the main requirement to the concurrent engineering implementation is the information sharing and handle. By this approach a natural framework., to support the management and sharing of information among different manufacturing phases or activities, is a network (federation) of heterogeneous and autonomous agents that are either loosely or some tightly-coupled. In this federation, an agent is involved in one activity (e.g. design) related to the product life-cycle, while several agents may take part in the same activity. On one hand production related activities are independent, (heterogeneous and autonomous) to serve different purposes. On the other hand, trivially these activities are interrelated (coupled) and need to cooperate and exchange information among themselves.

The Information System must do the following tasks:

- modeling information from different manufacturing areas;
- allow queries in any manufacturing activities;
- use the existent models;
- allow a secure information maintenance.

5 CEE ARCHITECTURE

The CEE architecture is described in (Afsarmanesh, 1994). In spite of concurrent engineering could be implemented without computer support it is necessary depending on the amount of information to be processed, and its complexity.

The main requirement of the concurrent engineering approach is the availability of all the information related to any life-cycle activity, to all team members who need to access it. The information representation is heterogeneous and distributed among different autonomous agents. Depending on the automation degree some of the following tools should be used in the environment:

- CAx tools
- information servers
- analysis tools.

Currently most of product life-cycle activities have its own Computer Integrated Manufacturing tools (CAx) providing a useful aid to the development. Those tools should provide mechanisms to integrate information servers and analysis tools access. Otherwise the information servers must maintain a user interface to provide information directly to the users.

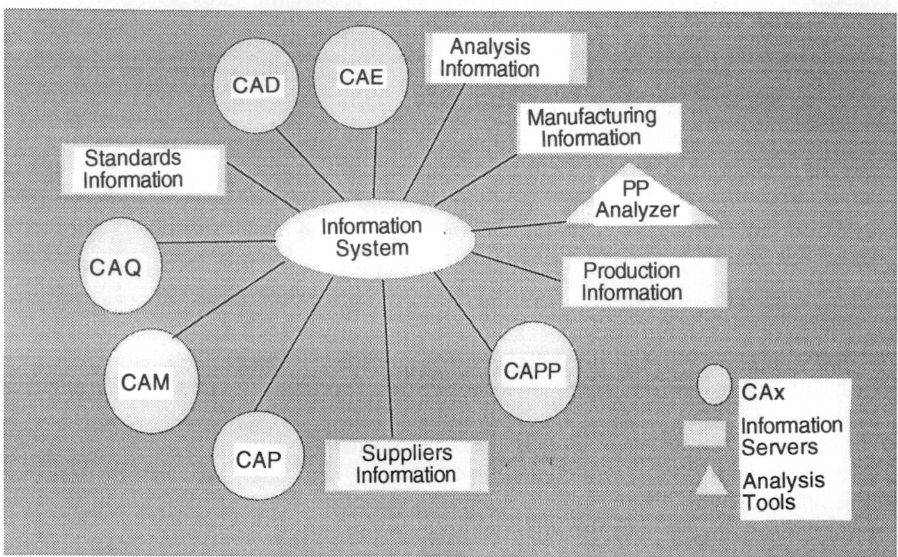

Figure 2 Information servers working together.

Each information server is related with one activity and provides answers to user questions. Each server must be modeled as an autonomous agent able to cooperate and shares information with others to solve problems. The exchange of information has to be done through an heterogeneous and distributed information system (Figure 2). By this way the requirements described in section 4 are preserved.

For instance, if a CAD user must know if a part can be produced in-house with minimal cost, he/she can submit the design to a manufaturability analysis agent. Then, the CAPP, CAQ and CAP agents will work together to analyze the part design and to deduce the best option, in terms of the production process costs and parts/machines availability. Besides, information about third part suppliers should be analyzed to decide if it is more efficient to get the part from other manufacturers. In the example, it is possible to see the need of several agents to access the process planning and scheduling information.

The analysis tools are expert systems developed as an agent with heterogeneous database management system support. The tool intelligence degree depends on the analysis difficulty. In most cases, the user's help is necessary for a good analysis. In this case, a friendly user interface and proper communication, among the tools, must be additionally developed.

6 CEE INFORMATION MODELING

The information required in a concurrent engineering environment has to be analyzed specifically to each case, once it depends on the activities involved in the implementation. In Afsarmanesh et al. (1993 and 1994), a description of the information flow study, using IDEF0, in an aerospace enterprise is presented. Here it is described briefly the product model.

During the product life-cycle many different kinds of information are handled by each activity, using its own representation approach and storage strategies. Marketing, design, manufacturing and assembly phases develop activities where the product information is defined and used to generate the final product. During this process, several other pieces of information are defined and aggregated to the product definition, so that it can be used as an integration element in a concurrent engineering environment. Then, it is possible to say that the product definition is the main concept to which other manufacturing information is related.

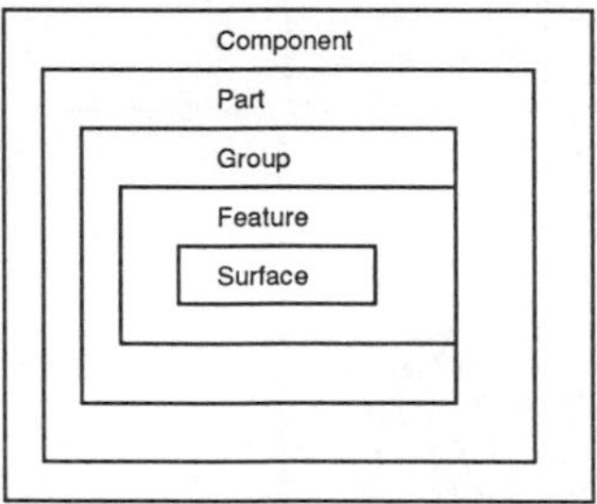

Figure 3 Product model hierarchy.

As described in Afsarmanesh et al. (1994), the product is defined hierarchically (Figure 3). The component level is user defined and can be reused in other product definitions. In this hierarchic definition, the component complexity is simplified in its sub-components definitions, while the functionality and characteristics of sub-components are more detailed. A simple component is described by one or more parts. A part is defined as a non-assembled element produced by one piece of raw material. Parts constitute the last level that can be defined by the user. One part can be composed of as many different preexistent groups and features as necessary to represent its functionality. Groups are composed of features and generally describe a standard constructive element. Both groups and features are available in libraries. Each library supports the manufacturing needs of one specific product domain.

The features are the principal element in the product representation (Moreira, 1991, 1992, 1993, Afsarmanesh, 1993, 1994 and Tonshoff, 1994). By those constructive components it is possible to link information from different product-life-cycle activities maintaining the link among them. This is possible because the features provide a functional representation valid for every activity. Each feature represents a design function used by the designer in the product conception phase. With this constructive element the part is built in a function assembly process. One feature can be generic defining a shape and its dimensions. A Form Feature can be used for many different purposes. In this basic feature a function can be aggregated, in order to represent more information. This new feature is called Functional Feature. Surface finishing and tolerances, machining process and thermochemical process are examples of such functions. In Figure 4 there are many features put together to build a part. Some of them, like the two cylinders and the hole, are just shapes where some other features are overlapped. The retention ring groove, for example is a shape with function. About the cylinder it is possible to say just dimensions, but concerning a specific retention ring groove it is possible to say tolerances, finishing surface, dimension range as well as, the machining process.

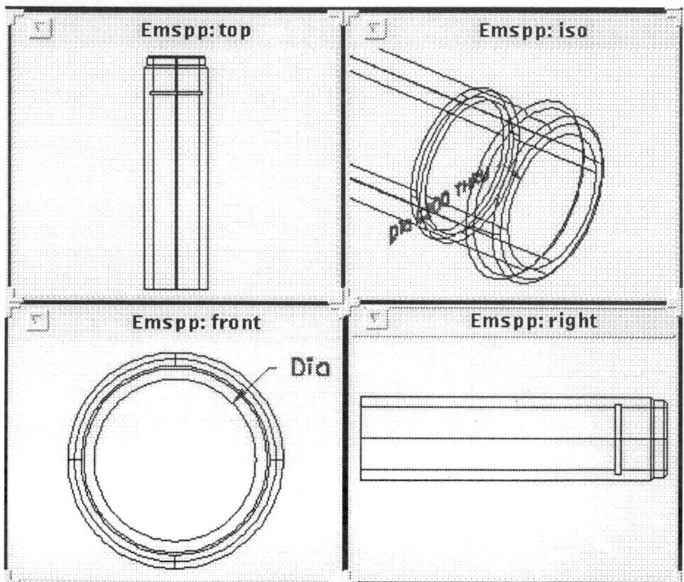

Figure 4 Feature-based product representation.

The product model is the important unifying piece of information among different activities within the Concurrent Engineering environment. As described in earlier section, it is represented hierarchically. Features are structured in subtypes that represents a specific part domain, for instance, rotational parts, prismatic parts, sheets, etc. As described earlier a feature should aggregate generic information (form-features) and information about functionality what can improve its representation. In the model presented in Figure 5 it is represented by a multiple inheritance between FORM_FEATTURE and FUNCTIONAL_FEATURE. The type RETENTION_GROOVE is a functional feature derived from a generic feature GROOVE and is described by additional attributes. In the example of the Figure 5 just rotational features are represented like a subtype of form_features, but any other kind of feature should be represented by the same way.

The methodology used to implement a Concurrent Engineering Environment consist, basically, of determining part families based on the geometry of the part and its features. A company may already have their parts classified into families. However, the families already existent in a company may not be appropriate for a manufacturing database based on features. Therefore, it is advised that a thorough study on all products of the company should be carried out, even if the company already has parts grouped into families.

In a company being studied, products such as valves and landing gears are produced (Moreira, 1994). In this case it was decided to focus attention on the hydraulic equipment such as valves and actuators. These products have:

- similar functions;
- similar geometric complexity;
- the same manufacturing technology;
- manufacturing planning for the following 12 months, i.e., these products are on order for delivery during the following 12 months;
- a higher frequency of new designs compared to other products.

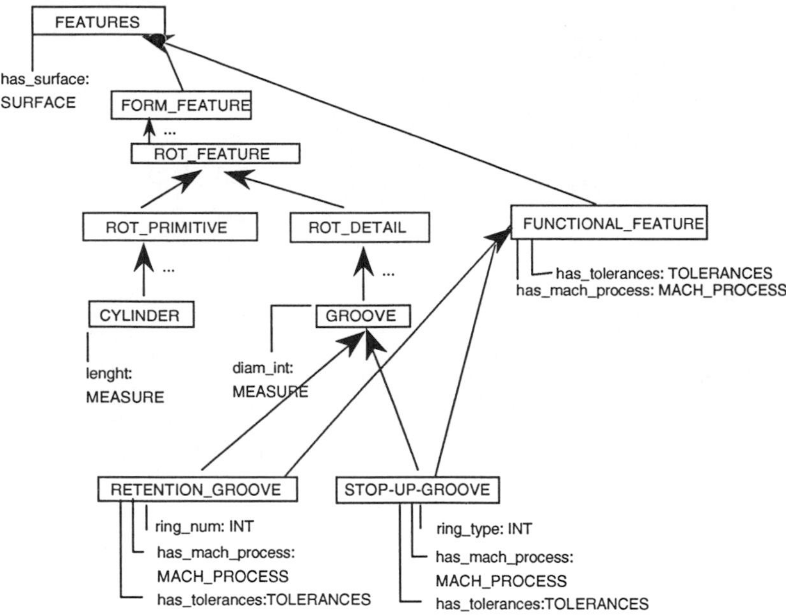

Figure 5 Feature model.

Among the hydraulic equipment, it was decided to gather information on the rotational parts belonging to these products. Attention was given to the following groups of parts (see Figure 6):

- shafts with constant diameter, without holes or with stepped holes;
- stepped shafts, without holes or with holes having constant diameter.

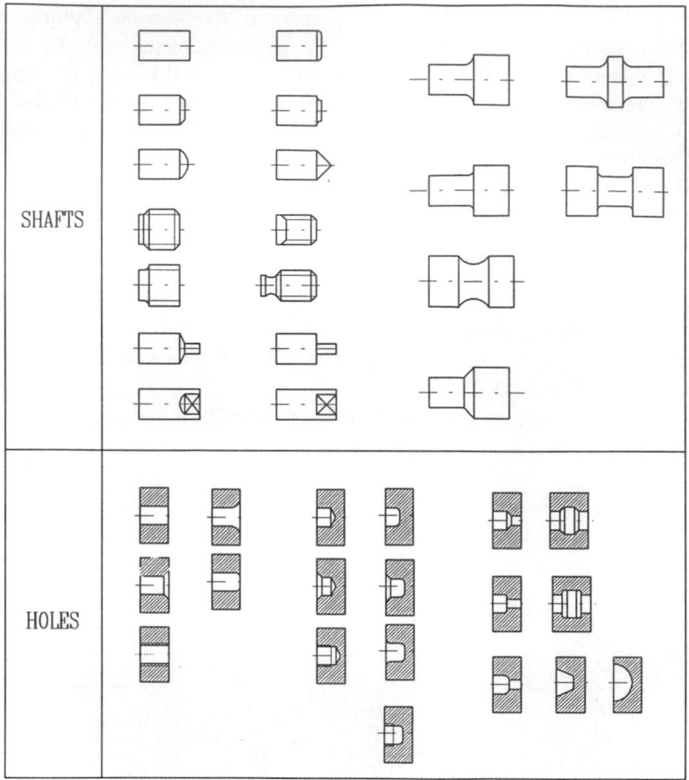

Figure 6 Features found in rotational parts.

7 CONCLUSIONS

In the old times the craftsman was, certainly, the first person to use the concept of CE, because the whole life-cycle development was completely integrated in his mind. Today, with the increasing of the products complexity, the whole life-cycle development is divided into several steps. Each one involve several sub-divisions, where each professional, using his own skills and experience, add value to the product. To do that each professional, usually, knows nothing about the subsequent steps.

In order to achieve the best results, aiming to come up with the most competitive product of the market, in a short period of time, all the steps of the whole life-cycle development must be fully integrated (as in the craftsman mind). This means human and information integration.

In order to advise this integration it is necessary multifunctional teamwork and Information System to provide the adequate support. The results achieved in a real industrial case study, have shown the Feature Technology potential to model and link the information coming from different areas of the product life-cycle development

The main contributions of the authors described in this paper could be considered as the feature-based approach to the part modeling in order to create a concurrent engineering environment. It is really important once product information central information in those environments.

We would like to acknowledge the European Commission for the possibility to develop work together.

8 REFERENCES

Afsarmanesh, H.; Wiedijk, M.; Ferreira, A.C.; Moreira, N.P. (november, 1993) An Approach to the Design of a Distributed CIM Database for a Brazilian Aerospace Industry. *Proceedings of the European-Community-Latin America Workshop on Computer Integrated Manufacturing 93.* Monte de Caparica, Portugal.

Afsarmanesh, H.; Moreira, N.P.; Wiedijk, M.; Ferreira, A.C. (june, 1994) Design of a Distributed Database for a Concurrent Engineering Environment. *Proceedings of the European-Community-Latin America Workshop on Computer Integrated Manufacturing 94.* Florianópolis, Brazil.

Bedworth, D.; Henderson, M.R.; Wolfe, P.M. (1991) *Computer-Integrated Design and Manufacturing.* McGraw-Hill, ISBN 0-07-004204-7.

Camarinha, L.M.; Osório, A.L. (1994) *An Integrated Plataform for Concurrent Engineering.* III CIMIS.net Workshop. Florianópolis, Brazil.

Chen, C-S.; Swift, F.; Lee, S.; Ege, R.; Shen, Q. (1992) Development of a Feature-Based and Object-oriented Concurrent Engineering System. *Proceedings of the International Conference on Object-Oriented Manufacturing-systems.* University of Calgary, Alberta, Canada.

Hernandez, J.; Rodrigues, M.; Ferreira, A.; Zendron, P. (august, 1991) *IAGE: A Feature Oriented CAPP System.* GRUCON/UFSC. Internal Report.

Holanda, A.B. (1988) *Pequeno Dicionário da Língua Portuguesa.* São Paulo, Brazil.

Huthwaite, B. Scheneberger, D. (1994) *Design for Competitiveness.* Institute for Competitive Design. USA.

Ishii, K. (1993) "Modeling of Concurrent Engineering Design" in *Concurrent Engineering: Automation, Tools, and Techniques.* Edited by Andrew Kusiak. John Wiley & Sons. Inc.

Jo, H.H.; Parsael, R.; Wong, J.P. (1991) Concurrent Engineering: The Manufacturing Philosophy for the 90's. *Computers in Industrial Engineering.* Vol. 21. Nos 1-4.

Kusiak, A. and Wang, J. (1993) "Decomposition in Concurrent Design" in *Concurrent Engineering: Automation, Tools, and Techniques*. Edited by Andrew Kusiak. John Wiley & Sons. Inc.

Lindenberg, L. (1993) Notes on Concurrent Engineering. *Annals of the CIRP*. Vol 42/1/1993.

Longman (1983) *Dictionary of American English*. Edited by Longman. New York, USA.

Moreira, N.P. and Ferreira, A.C. (november, 1991) *Feature Technology in CIM: tools and application*. Simpósio sobre CAE/CAD/CAM. Sobracon. Brazil (in Portuguese).

Moreira, N.P. and Ferreira, A.C. (july, 1992) *Open environment aiming CIM: a basic proposal*. Congresso Internacional de Computação Gráfica. Sobracon. Brazil (in Portuguese).

Moreira N.P. (1993) *An Information Modeling Proposal to the Manufacturing Integration and Concurrent Engineering*. Master Thesis. Universidade Federal de Santa Catarina (in Portuguese).

Moreira, N.P.; Cabral, P.H.; Oliva, S. (1994*) An Approach to Concurrent Engineering Implementation in EMBRAER-EDE*. Annals of SAE Brasil 94. São Paulo, Brazil.

Shah, J.; Rogers, M. (february, 1988a) *Functional Requirements and Conceptual Design of the Feature-Based Modeling System*. Computer-Aided Engineering Journal.

Shah, J.; Sreevalsan, P.; Rogers, M.; Billo, R.; Mathew, A. (november, 1988b) *Current Status of Features Thechnology*. CAM-I Report R-88-GM-04.1.

Shah, J. (june, 1991) Assessment of Feature Technology. *Computer Aided Design*. Vol 23. Nro 5.

Terpenny, J.P. and Deisenroth, M.P. (july, 1992*) A Concurrent Engineering framework: three basic components*. Flexible Automation and Information Management - FAIM.

Tonshoff, H.A.; Aurich, J.C.; Baum, T. (1994) Configurable feature-based CAD/CAPP System. *Proccedings of IFIP international conference*. France.

Yeo, S.H.; Wong, Y.S.; Rhaman, M. (1991) Integrated knowledge-based machining system for rotational parts. *International Journal of Production Research*. Vol 29. No 7.

40

A framework for feature based CAD/CAM integration

C.F. Zhu, N.N.Z. Gindy and Y. Yue
Department of Manufacturing Engineering and Operations Management
University of Nottingham, Nottingham NG7 2RD, UK
Tel +44-115-9514031, Fax +44-115-9514000
Email epxcfz@unicorn.nott.ac.uk

Abstract

This paper proposes a framework for a CAD/CAPP/CNC integration using the latest solid modelling and CNC development tools. The central concept for the integration is to use a generalised component data model throughout all the activities from design to manufacture. A feature-based component data model (FBCDM) is introduced. The creation and enhancement of FBCDM are also described.

Keywords

CAD/CAM integration, CAPP, component model, features, geometric reasoning

1 INTRODUCTION

The goal of integrating computer aided design and manufacturing (CAD and CAM) into computer integrated manufacturing (CIM) has been long desired. It is understood that it would be impossible to achieve real integration between CAD and CAM without automating computer aided process planning (CAPP). One of the problems encountered here is that it is difficult to interpret component design automatically by a computer for use in downstream applications in the production process (Anting and Zhang 1990, ElMaraghy 1993).

This paper proposes a framework for a CAD/CAPP/CNC integration using the latest solid modelling and computer numerical control (CNC) development tools. The central concept for the integration is to use a generalised form of component data throughout all the activities from design to manufacture. In the following sections, definitions for features including form features, design features and machining features are discussed, and a feature-based component data model (FBCDM) suitable for design, process planning and CNC programming is introduced. An intelligent feature based design system (IFBDS) is used to creates a generic FBCDM containing basic component information which is enriched through geometric reasoning. A feature based process planning system (FBPPS) extends and refines the FBCDM and automatically generates a process plan. The refined FBCDM is then directly used by a feature based numerical control programming system (FBNCPS) to generate the cutter path. A preliminary investigation of the framework has indicated a high potential of the methodology employed to achieve a fully integrated CAD/CAPP/CNC system.

2 ARCHITECTURE OF THE CAD/CAPP/CNC INTEGRATION SYSTEM

The architecture of the CAD/CAPP/CNC integration system is depicted in Figure 1. It consists of three subsystems: IFBDS, FBPPS and FBNCPS.

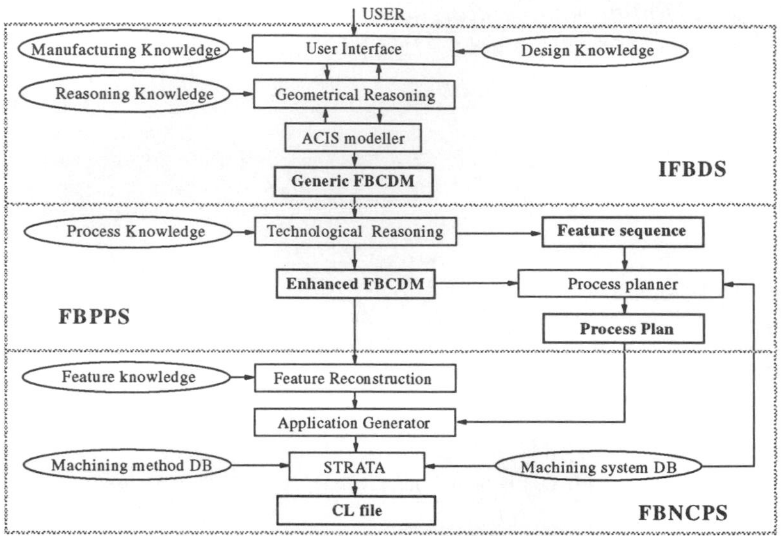

Figure 1 System architecture.

IFBDS uses ACIS (Spatial Technology Inc.1994) as its underlying modeller. It has a user interface with design and manufacturing knowledge, an inference engine with reasoning capabilities, and an attribute tagging facility. The output is a generic FBCDM. The user interface contains a user-system interaction window in which command menus are provided. It allows the user to make parametric designs using the feature library, or modify/edit existing designs. A design knowledge database contains the feature library, standards and other necessary information; and a manufacturing knowledge database stores information on raw materials, machines, tools and fixtures.

In front of ACIS, there is an inference engine, the core of IFBDS, that serves three purposes: finding information of constituent entities of features, checking feature validity and identifying feature interactions. These tasks are mainly performed through geometric reasoning during the design stage.

FBPPS comprises a design interface and a process planner. The function of the design interface is to take the generic FBCDM as input, enhance it through technological reasoning and refines the interacting features so that they can be treated individually for machining. Feature refinement aspects include the feature geometry, technological constraints and potential technological solutions. The feature sequence is also determined.

Based on the enhanced FBCDM, the process planner automatically determines the technological solutions for each feature, taking into account its technological constraints and the processing system capabilities. FBPPS output is an enhanced FBCDM and a process plan for the component.

FBNCPS uses Strata development environment (Spatial Technology Inc. 1994). It consists of a system interface, an application program generator and Strata CNC code generator. The system interface takes the enhanced FBCDM as input, converts features into Strata geometric entities with manufacturing information and reconstructs some of the features if necessary.

Based on the enhanced FBCDM and process plan, the application program generator automatically creates program in Strata command format. Strata executes the application program and creates a CNC toolpath for each feature. The toolpaths may be grouped into operations, while operations may be clustered into setups and setups into sub-programs which are combined into a full CNC program for the component. FBNCPS output is a cutter location (CL) file.

The key issues of integration between CAD/CAPP/CNC are the creation and enhancement of FBCDM as described in Sections 3 and 4.

3 FEATURE-BASED COMPONENT DATA MODEL

Conventional geometric models are created from geometric primitive objects (e.g. cylinders, blocks) by Boolean operations or from lines and curves by other geometric construction techniques (e.g. sweeping profiles to get volumes, or fitting a skin to a series of cross sections arranged in the space). They have proved deficient for applications such as automated process planning. Feature based models which provide higher conceptual meaning of component characteristics are considered as a suitable communication medium between design and manufacturing.

3.1 Feature definition

Many definitions appear in the literature (Case and Gao, 1993) since features originate from various design, analysis and manufacturing activities, and are often associated with particular application domains. In essence, features represent the engineering meaning of the geometry of a component (Shah 1992). Form features, simply defined as portions of nominal geometry or recurring shapes, have been widely used.

Since features are often application-specific, they can be defined and classified from the viewpoint of different processes occurring in the product life-cycle (Pratt 1993, Young and Bell 1993). Design and manufacturing features are the two most common types: design features are geometry and specifications which fulfil certain functional requirements while manufacturing features are geometry that can be related to specific manufacturing methods.

It can be seen from the above discussion that all application-specific features are related back to form features. Therefore design features can be defined as form features with technological constraints while manufacturing features as form features with technological solutions.

3.2 Feature taxonomy

While it is difficult to enumerate all features, it is useful to try to classify them. This research uses Gindy's (1989) feature taxonomy in which form features are treated as volumes enveloped by entry/exit and depth boundaries. A feature may consist of any number of real surfaces and a fixed number (0 - 6) of imaginary surfaces. Each surface has a direction represented by a normal vector (NV). The NV of each imaginary surface of a feature is termed an external access direction (EAD) which represents a possible machining direction for the feature.

The features are divided into three categories: protrusion, depression and surface based on the number of their imaginary surfaces (i.e. EADs). These categories are further divided into nine classes based upon the feature boundary type (open or closed) and exit boundary status (through or not through). Each class of feature has a perimeter shape. This feature taxonomy is proving suitable to represent machined components for process planning and CNC programming. It is a hierarchical structure which can be implemented using object oriented techniques.

3.3 A feature-based component data model

FBCDM is used to represent the set of features included in component definition, their attributes and the relationships between features.

The feature list contains high-level information about each feature, such as its type, location, orientation, dimensions, accuracy and surface finish, some of which may be regarded as technological constraints for process planning and CNC programming.

From process planning point of view, there are a number of relationships between the features on a component, which are kept in the feature relationship list:

- Feature connectivity: Features with common EADs can be considered together for machining. The common EADs are called the potential approach directions (PADs) representing the possible machining directions for the features.
- Parent-child relationship: If features A and B are adjacent and A has an imaginary surface contained within the boundary of any real surface of B, then A is the child feature of B and B is the parent feature of the A. The parent-child relationship is useful in the decision-making (e.g. determining feature sequence).
- Compound feature: A compound feature is a set of primitive features treated as a single entity because they may perform a single function and/or need to be machined by the same procedure. In some cases, compound features have a parent-child relationship (e.g. a T-slot). Information of compound features can be used for determining setups.
- Tolerance relationship: When the designer specifies positional tolerances between features to guarantee some functional requirements, some tolerance relationships between the features are established. One feature can be considered the reference feature and the others tolerancing features. A tolerancing feature may have more than one reference feature and vice visa. The tolerance relationship is necessary for making setup and machining strategies.

The ACIS data structure is extended to include the feature list and feature relationships. The advantage of FBCDM is that it provides sufficient and easily-accessible information for the applications while the model is kept in a generic and compact format.

4 GENERATION AND ENHANCEMENT OF FBCDM

A generic FBCDM is generated by the IFBDS through interactions between the user and the system. Geometric reasoning is performed where necessary; it minimises the demand on designers, and ensures that the model is valid and contains sufficient information for the downstream applications. The generic model is enhanced by technological reasoning to facilitate the FBPPS which then creates a process plan. The features in the enhanced FBCDM are reconstructed and used by the FBNCPS to generate a CL file.

4.1 Geometric reasoning

Based on component geometric information during the design stage, the tasks of geometric reasoning include the followings:

- Feature entity identification: Before a feature is added to the feature list, some of its constituent entities are identified and tagged with necessary information (e.g. tolerance and surface finish on a real face). The constituent entities include faces, edges and vertices among which faces are the most significant. The imaginary faces are identified using a predefined feature convention while other constituent entities are identified as necessary (e.g. when tolerance and surface finish attributes are to be tagged).
- Feature validity: This is a common problem in feature-oriented research. e.g. a designed slot may become a step due to feature interaction or manipulation. It is necessary to check that a feature is valid before it is added to the feature list. Thus ensuring the validity of the attributes tagged on the feature constituent entities.
- Parent-child feature identification: A valid feature's imaginary faces are examined against the real faces of other features. If an imaginary face lies within a real face of another feature, there is a parent-child relationship and this information is added to the model data.

4.2 Technological reasoning

The tasks of technological reasoning include feature sequencing, feature refinement and feature conversion based on both component geometric information and process knowledge.

Feature sequencing

Factors considered here include feature relationships such as parent-child relationships and tolerance relationships. The basic principles for feature sequence are:

- Reference features before other features;
- Parent features before child features if there are no other constraints;
- Main functional features before other features;

- Large-sized features before small-sized features; and
- Orthogonal features before non-orthogonal features.

Feature sequence influences both operation sequence and features themselves which need to be refined based on the feature sequence. It may thus be necessary to change a feature sequence based on the feedback from feature refinement. Hence an iterative procedure is used.

Feature refinement

Because of feature interactions, it can be ambiguous to match design feature with manufacturing feature. This is solved by feature refinement, examples include:

- Feature splitting: Due to feature interaction, it is necessary to split a design feature into two or more manufacturing features to be machined separately (i.e. depending on the machining sequence). Figure 2a shows a design feature "hole" broken into hole1 and hole2 by the slot which is machined first. If the length to diameter ratio (*l/d*) of the hole is too large for the hole to be drilled in one direction, it must be split into two holes for machining.
- Feature merging: On the other hand, it may be necessary to merge two or more design features into one manufacturing feature. For example, it is reasonable to treat hole1 and hole2 as a single hole for machining before the slot as illustrated in Figure 2b.
- Feature extension: A design feature may have to be extended owing to feature interactions and/or the allowance of raw material. Feature extension lengthens the feature volume as much as possible, e.g. the depth *h* of the hole shown in Figure 2c can be extended to *h1* if the hole is machined prior to the slot.
- Feature degeneration: Due to feature interactions and/or technological constraints, some EADs of a feature can not be used as feasible cutter access. Thus a design feature may become a manufacturing feature with fewer EADs by degeneration. Figure 2d provides an example where a hole (through hole) is degenerated to a pocket (blind hole).

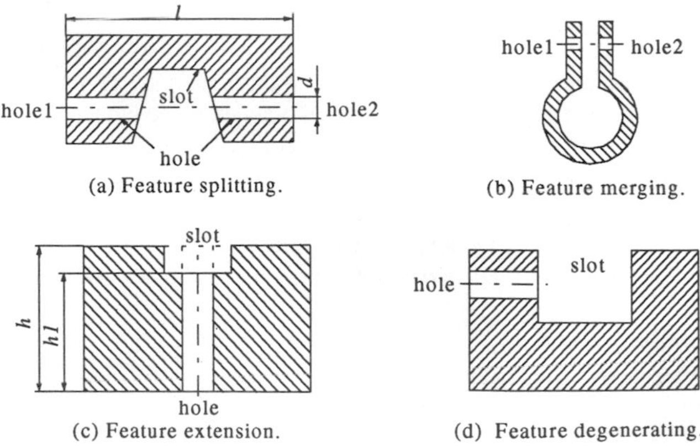

Figure 2 Feature refinement.

Feature conversion

The design features after refinement are converted to manufacturing features by attaching manufacturing information called the technological solutions at the feature level (TSFs). A TSF is an ordered set of operations that satisfy the feature geometry and technological requirements (Gindy et al. 1993). Each operation is described as a form generation schema, i.e. a set of motions and tools for performing the operation, which facilitate machine tool and cutting tool selection in the process planning stage.

TSFs are stored in the manufacturing database according to feature transition diagrams (FTD) which represent the available processing methods for producing the feature. Based on the individual feature technological constraints, the technological solutions for producing a feature can be obtained by interrogating the manufacturing database.

4.3 Feature reconstruction

After feature sequencing, refinement and conversion, the generic FBCDM is enhanced. Due to the consistent use of manufacturing features defined in FBPPS and FBNCPS, most manufacturing features in the enhanced FBCDM can be used directly by FBNCPS. However, feature "hole" in Strata is different and has a special data structure. So it is necessary to reconstruct it from some manufacturing features in the enhanced FBCDM.

In Strata, feature "hole" is defined as a round pocket (blind hole) or round hole (through hole) machined by drilling (drilling, boring and reaming) operations, which can be identified using expert system rules. The rules for identifying Strata "hole" from the FBCDM can be written as follows:

IF a feature class is pocket or hole
AND its subclass is circular
AND its TSFs only include drilling, boring or reaming operations
THEN the feature is "hole"

Having been identified, feature "hole" can be recreated by Strata commands. Other features in the enhanced FBCDM are machined by milling and can be used directly by Strata.

5 IMPLEMENTATION AND CONCLUSIONS

The work discussed above is being implemented on a Sun SPARC station10 with ACIS version 1.5.2 and Strata version 1.7.3. The major advantages of ACIS are its complete and open data structure and functionality which have made it possible to use the ACIS model as the core of the FBCDM. The ACIS attribute function has been used to tag attributes onto feature entities in generating a generic FBCDM. Strata is a next-generation manufacturing system based on ACIS models. It addresses operations planning and the automatic creation, validation, and verification of CNC toolpaths. The Strata command language provides a toolkit for developing CNC programming systems. These features of Strata have proved useful in the CAD/CAPP/CNC integration through the use of a generalised form of component data based on ACIS models.

The FBCDM is created by human-computer interaction in design stage. It contains sufficient information regarding the component which can be used by downstream applications such as process planning and CNC programming through automatic processing. This saves time and costs by avoiding repeat of input and reducing human errors. Furthermore, the current implementation of the framework has indicated a high potential of the methodology employed to achieve fully functional integration between CAD and CAM tasks.

6 REFERENCES

Alting, L. and Zhang, H.C. (1989) Computer-aided process planning: the state-of-the-art survey, *Int. J. of Production Research,* **27**(4), 553-85.

Case, K. and Gao, J. 1993, Feature technology: an overview, *Int. J. of Computer Integrated Manufacturing*, **6**(1 & 2), 137-51.

ElMaraghy, H. (1993) Evaluation and future perspectives of CAPP, *Annals of the CIRP*, **42**(2), 739-51.

Gindy, N.N.Z. (1989) A hierarchical structure for features, *Int. J. of Production Research,* **27**(12), 2089-103.

Gindy, N.N.Z., Huang, X. and Ratchev, T.M. (1993) Feature based component model for computer aided process planning systems, *Int. J. of Computer Integrated Manufacturing,* **6**(1&2), 20-6.

Pratt, M.J. (1993) Applications of feature recognition in the product life-cycle, *Int. J. of Computer Integrated Manufacturing*, **6**(1&2), 13-9.

Shah, J.J. (1992) Features in design and manufacturing, *Intelligent Design and Manufacturing,* (ed. A. Kusiak), John Wiley & Sons Inc.

Spatial Technology, Inc. (1994) *ACIS Modeller Interface Guide.*

Spatial Technology, Inc. (1994) *Strata Command Reference.*

Young, R.M. and Bell, R. (1993) Design by features: advantage and limitations in machine planning integration, *Int. J. of Computer Integrated Manufacturing*, **6**(1 & 2), 105-12.

7 BIOGRAPHIES

Cheng-Feng Zhu, B.Sc., M.Sc., is currently a Ph.D student at the Department of Manufacturing Engineering at the University of Nottingham, UK. He worked as a lecturer at Shanghai Jiao Tong University China until 1992. His major interests include CAD/CAM integration and Group Technology.

Nabil Gindy, B.Sc., M.Sc., Ph.D, is professor of Advanced Manufacturing Technology and leads the responsive manufacturing research group at the University of Nottingham, UK. His research interests include machining and tooling technology, feature-based CAD/CAM systems, generative process planning, integrated planning and scheduling systems, CNC part programming and the design and configuration of manufacturing systems.

Yong Yue, B.Sc., Ph.D, is currently a research assistant at the University of Nottingham, UK. He earned a Ph.D in 1994 from Heriot-Watt University, UK. He worked as a design and research engineer in Shenyang Aluminium and Magnesium Institute China from 1982 to 1990. His major interests include CAD/CAM and integrated manufacturing.

PART FIFTEEN

Monitoring and Sensors

CNC Machines Monitoring Using Sensorial Stochastic Models

Manuel M. Barata, Ricardo R. Jardim Gonçalves, Adolfo S. Steiger Garção, *José Álvaro Assis Lopes
Universidade Nova de Lisboa - FCT
Departamento deEngenharia Electrotécnica
Quinta da Torre, 2825 Monte de Caparica
Portugal
E-Mail mmb@uninova.pt
**Instituto Superior Técnico, Secção Autónoma de Economia e Gestão*

Abstract

This paper presents the work that is being done by the Intelligent Robotics Group on a CNC Monitoring and Prognosis System. It is based on an integrated hardware/software environment including CNC lathe and mill machines. The machining process is monitored in real time using sensors for vibrations, sound and power consumption. This paper presents our approach for sensor signal modelling using auto regressive stochastic models. Models for machining process characterisation have already been investigated for different situations. The applicability of the *ARMA(p,q)* model for process monitoring and evolution tracking, are discussed as well as the results obtained with real data are presented.

Keywords

Monitoring, Prognosis, CNC, Sensor Integration, Expert Supervision, Stochastic Process.

1 INTRODUCTION

CNC machines play an important role in Flexible Manufacturing Systems: They produce parts to be assembled. Solving the problem of making products featuring the ability of defects forecasting will considerably contribute to improve the system's productivity and product quality. This goal leads to an exhaustive analysis of related tasks, in order to identify and enumerate the relevant aspects suitable to be improved. In our case numerically controlled machines work as a test bed for experimentation and demonstration.

System's degradation, opposite to random accidents, has to be adequately monitored and evaluated if we are committed to corrective action or intervention. As almost in all the defective situations, a broad band of possibilities can occur: from well identified symptoms to very fuzzy and unclear scenarios. Dynamics in those cases will play a major role, because decision will be based not only on statistical measured values, but also on their temporal evolution. Advanced monitoring systems, depending on the particular design ambition, can address or not, automatic refinement of decision making mechanisms. That stresses the fact that statistical accumulation and assessment are important functions to be incorporated in that kind of systems.

Faults can occur randomly or as a consequence of structural internal degradation. Little hope exists for the first case. However, the second one opens the field for prediction of deviation of quality, enlarging the state-of-the-art methodologies. Prediction of faults can represent an important economic issue as an effective technique aimed to operate in run time and not based on a posterior analysis of samples.

An experimental setup specially developed by our group for experimental research on monitoring the machining process was developed and it has been used for the assessment of our developed models. Additional information describing the system's architecture and processing facilities are described in (Steiger-Garção 1989), (Barata 1990a), (Barata 1990b) and (Barata 1990c). In this paper we present the work done in sensorial data characterisation using autoregressive stochastic process and, in our point of view, is a refinement of the approach presented in (Barata 1994) because it contributes with an additional technique for the machining process monitoring and failure prognosis.

2 APPROACH STRATEGY

2.1 Machining process characterisation

Monitoring and Supervisioned action based on sensors is considered. Sensing, characterisation of the different participating processes and learning their behaviour, are the base for the implementation of decision making mechanisms. Our models are mainly supported by learning. They are built taking into consideration the process physics, and are tuned and trained under real time sensor data acquisition.

To build a suitable model of the machining process we need to identify a set of low level building blocks (process steps). These blocks are commonly called geometric machining features. The machining of a specific piece is thus understood as the execution, in a proper order, of a sequence of these blocks. The corresponding Numerical Control (NC) programs can be understood as a way of implementation of such building blocks. Inspecting several designs of different pieces and the corresponding NC programs, a great level of similarity between the segments of NC instructions corresponding to the machining of the same geometric feature in different pieces can be found. This particularity and considering that the machine's behaviour is mainly determined by the NC instructions it executes, it is plausible to expect a characterisation of the machining process based on machining features, independently of a specific piece. This characterisation is done by learning the associated sensorial patterns in terms of *good* and *bad* situations or when this is not possible in terms of the good situations already found. This approach was previously presented in (Barata 94) and we shortly describe it here in order to give the reader a better understanding of the problem we are addressing.

2.2 Machining process model

At physical level the machine, machining process and installed sensors can be modelled as a system having one input (the sequence of the NC instructions) and several outputs (one for each sensor). Such a model is presented in figure 1. During the machining process, while executing the NC instructions, each installed sensor presents a signal that can be used as a signature for the characterisation of the process. For each status (execution of one NC instruction) the installed sensors output a proper signal pattern. We understand such patterns as the system response to the given stimulus. In part b) of figure 1 it is sketched how the sensor data can be related to the CN commands during the execution of a complete NC program. Thus the time series of the data must be labelled with the CN command that was in execution during data capture.

According to figure 1 the machining process is characterised by the NC instruction and the sensor patterns produced by the sensors while the NC instruction is being executed. Accordingly, the first step is to get knowledge about the sensors' behaviour for the set of all situations. Using Pattern recognition terminology, we denote by *class* the execution of each NC instruction, and the sensorial signal characterisation by *features*.

After getting knowledge about the sensor data behaviour, the sensor model can be implemented. That model includes the representation of the nature of the low level data processing and the feature extraction method. Also the correlation of the resulting features and the process status related to the collected data must be evaluated. After this evaluation decisions can be made about which features are good or not for process characterisation. That leads us to the selection of a correct feature extraction method. The reader can find more detailed information about this subject on (Barata 1994). In (Rauber 1993) a tool developed for statistical characteristics evaluation of the feature vector is presented. For, example using the FFT for low level processing, it can be determined which subset of harmonics represents the best correlation with the variable class values (Guinea 1991). Some results of applying this procedure are presented in (Barata 1992), (Rauber 1993), (Barata 1994).

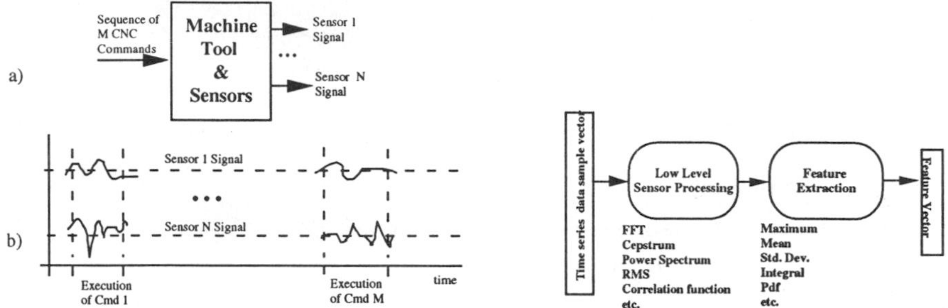

Figure 1 Machine model and sensor data capturing. **Figure 2** P.R. feature extraction.

Figure 2 summarises the block diagram of low level sensor data processing and feature extraction. Besides the mentioned techniques, the work presented in this paper concentrates in the application of the theory of *Stochastic Time Series Analysis* for doing low level sensor processing and feature extraction. This approach is motivated by the nature of the data produced by some kind of sensors that present some stationary random characteristics, e.g. the sound or vibrations of a cutting process. For several repetitions of the same machining process, these sensors never reproduce the same signal pattern but similar patterns evidencing some randomness.

3. DISCRETE SYSTEMS MODELLING USING TRANSFER FUNCTION

3.1 Discrete systems

A discrete system is a rule for assigning to a sequence $f[n]$ another sequence $g[n]$. Thus, a discrete system is a mapping (transformation) of the sequence $f[n]$ into the sequence $g[n]$. We shall use the notation $g[n] = L\{f[n]\}$ for this mapping. The sequence $f[n]$ will be called de *input*, and the sequence $g[n]$ the *output*, or response (Papoulis 1984).

The response of a discrete system to the delta sequence is usually called the ***system transfer function***:

$$h[n] = L\{\delta[n]\}. \tag{1}$$

and it defines the response of the system to an input sequence $f[n]$. It is possible to determine the response of the system doing the discrete convolution between the input sequence $f[n]$ and the system function $h[n]$:

$$g[n] = f[n] * h[n] = \sum_{k=-\infty}^{\infty} f[k]h[n-k] = \sum_{k=-\infty}^{\infty} f[n-k]h[k]. \tag{2}$$

For our aims, the system transfer function must have the property of linearity, time invariance and causality given respectively by the following equations:

$$L\{a_1 f_1[n] + a_2 f_2[n]\} = L\{a_1 f_1[n]\} + L\{a_2 f_2[n]\} \text{ for any } a_1, a_2, f_1[n] \text{ and } f_2[n]. \tag{3}$$

$$L\{f[n-k]\} = g[n-k] \text{ and } h[n] = 0 \quad \text{for} \quad n < 0. \tag{4}$$

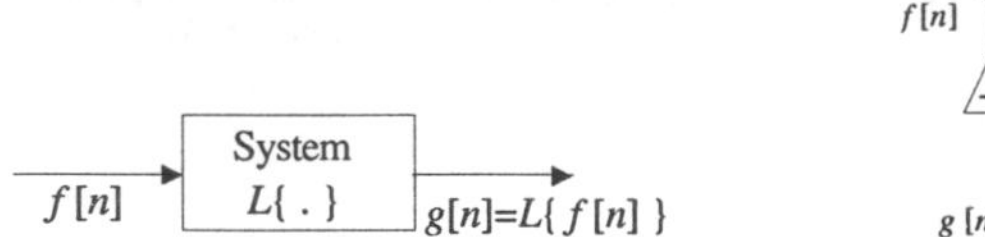

Figure 3 Discrete system model.

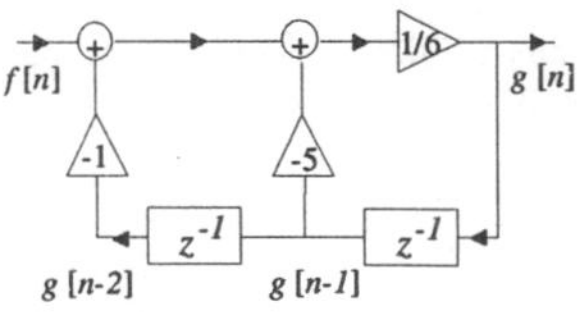

Figure 4 Discrete realisation of a system.

3.2 System Function

Computing the Z transform of the system transfer function $h[n]$, we obtain the usually called *System Function* in the complex Z domain:

$$H(z) = \sum_{n=-\infty}^{\infty} h[n]z^{-n} \ ; \ L\{z^n\} = H(z)z^n. \tag{5}$$

in words: If the input sequence is z^n the corresponding output is the same sequence multiplied by the System Function $H(z)$. Discrete systems are useful for simulating analogue systems in a computer. For example if we program the system $6g[n] + 5g[n-1] + g[n-2] = f[n]$ the corresponding system function $H(z)$ is :

$$H(z) = \frac{1}{6 + 5z^{-1} + z^{-2}} \tag{6}$$

The block diagram of the corresponding filter is presented in figure 4. The z^{-1} boxes correspond to one sample time delay and the triangles are amplifiers.

4. AUTOREGRESSIVE MODELS

Sensorial data can be modelled using the statistical models and techniques developed for characterisation of stochastic time series. They assume that most of the stochastic process realisations can be modelled by filtering white noise. This approach was first presented by (Box.Jenkins 1970). In this sense white noise is a special stochastic process, which realisations' $\varepsilon[n]$ form the input of a system and the corresponding response $g[n]$ is another stochastic process realisation. These two stochastic process, and of course the corresponding realisations, are related by the system transfer function $H(z)$. The white noise process is stationary and has: constant expected value $E\{\varepsilon[n]\} = \eta_\varepsilon$, normally $\eta_\varepsilon = 0$, constant variance $V\{\varepsilon[n]\} = \sigma_\varepsilon^2$ and correlation:

$$\rho_k = \begin{cases} 1 & \text{for } k = 0 \\ 0 & \text{for } k \neq 0 \end{cases} \tag{7}$$

In general if the system input sequence $f[n]$ has the autocorrelation function $\phi_{ff}(n) = E\{f[k]f[k+n]\}$ the correlation between the input and output is:

$$\phi_{gf}(n) = E\{g[k]f[k+n]\} = E\{g[k-n]f[k] = \phi_{fg}(-n) \tag{8}$$

If $\Phi_{gf}(z)$ and $\Phi_{ff}(z)$ are the Z transforms of $\phi_{gf}(n)$ and $\phi_{ff}(n)$ the following relation holds:

$$\Phi_{gf}(z) = H(z)\Phi_{ff}(z) \tag{9}$$

It describes the effect of the system transfer function on the input sequence. Detailed presentation of this subject can be found in (Box-Jenkins 1970), (Abraham 1983), (Harvey 1992).

4.1 *ARMA(p,q)* Models

A stochastic process $g[n]$ is *ARMA(p,q)* when it satisfies the following difference stochastic equation:

$$g[n] - \phi_1 g[n-1] - \cdots - \phi_p g[n-p] = \varepsilon[n] - \theta_1 \varepsilon[n-1] - \cdots - \theta_q \varepsilon[n-q] \tag{10}$$

and the corresponding system transfer function is:

$$H(z) = \frac{1 - \theta_1 z^{-1} - \cdots - \theta_q z^{-q}}{1 - \phi_1 z^{-1} - \cdots \phi_p z^{-p}} \tag{11}$$

For sensorial data modelling, given a sensor times series sample $g[n]$, the goal is to find a system transfer function, such that when feed by white noise results in a similar process realisation. It is necessary to determine the values of p and q, and the coefficient values $\phi_1 \, \mathrm{L} \, \phi_p$ and $\theta_1 \mathrm{L} \, \theta_q$. This process is iterative and can be done using a statistical package, e.g. SPSS (SPSS 1993).

5. APPLICATION TO MONITORING THE MACHINING PROCESS

Stochastic autoregressive models *ARMA(p,q)* described in paragraph four can be used for monitoring the machining process in three ways:

- *Doing low level sensor signal processing and feature extraction.* Using the sensor data, the parameters of the *ARMA(p, q)* model are estimated. The feature vector corresponds to the vector built with the p, q parameters estimated. In this case the feature vector describes the transfer function of a filter that transforms white noise in a signal similar to the sensor output. The clustering of the vector of p, q parameters is used by the upper level for identifying the class of the signal, i.e. identify the status of the process.

- *As identifiers of the sensor data signal.* In this situation they work as signal predictors, and match the signal to the class model that presents the lowest error in prediction. A monitor can decide about the situation by comparing the identified class with the expected one.
- *Tracking the evolution of some situation*, e.g. the lubrication status. In this case the *ARMA(p,q)* model is used for doing continuous parameter estimation and the feature vector is used for tracking a typical known path. If critical situations are marked in the path, it is possible to predict how far is the current vector from that zone. In this case tracking the parameters is equivalent to track the system's transfer function. If the transfer function is changing that means a change in the system.

Figure 5 a) presents the block diagram corresponding to the low level sensor processing and feature extraction. An identifier uses the feature vector for identifying the class (i.e. process status) by comparing it with a set of feature clusters learned during a training phase.

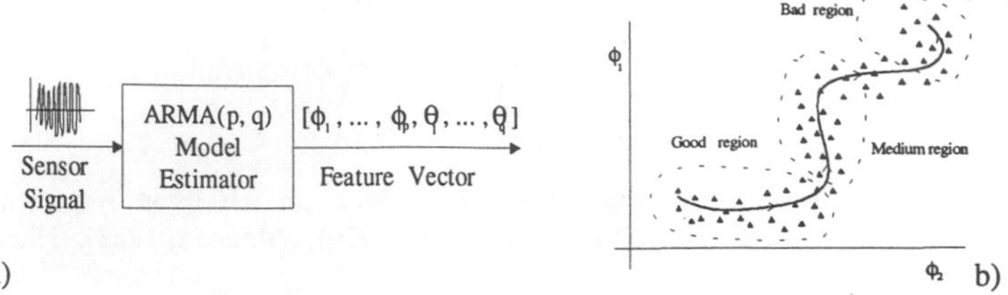

Figure 5 a) Application of the *ARMA(p, q)* model for low level signal processing and b) Application of *ARMA(2,0)* for process quality tracking.

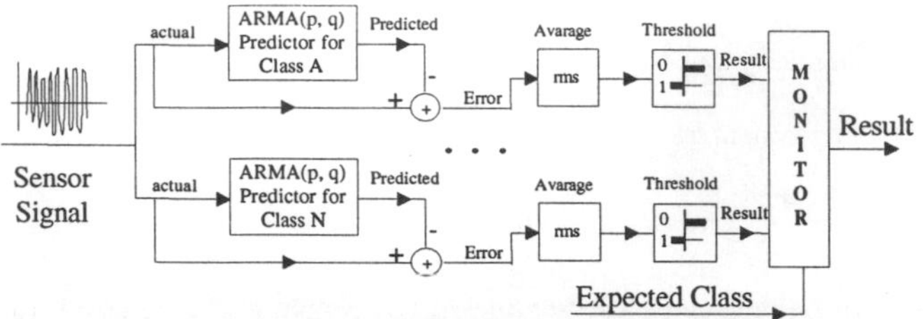

Figure 6 Application of the *ARMA(p, q)* model for signal class identification

The second case is illustrated in figure 6. Here we have a fixed *ARMA(p,q)* model for each class of the process status. Each model works in parallel as a signal predictor, and compares the error between the predicted and incoming signal. For a given situation the model presenting the lowest error level is selected for signal identification. Each threshold unit is programmed with a threshold value computed during the estimation of the corresponding class model. The monitor decides about the situation by comparing the expected class value with the class identified. The average units are necessary, because during estimation, the error estimated is based on the average prediction error. Moreover it does some filtration avoiding false errors due to noise.

An example of parameter tracking is illustrated in figure 5 b). In this case, for simplicity of explanation, it was assumed a model with two parameters (higher dimensions are easily extrapolated). The construction of the parameters' path is done by a previous correlation with the model's parameters and known situations. For instance, suppose that the example of figure 5 b) explains the lubrication status of some component. To have that characterisation, it was necessary to do a first track of the process during a complete phase, starting with a good level of lubrication, passing by intermediate levels until a low level was reached. By this way we have learned the process characterisation and recorded the parameters' clustering for all situations.

6. EXPERIMENTS DONE

In order to evaluate the applicability of the monitoring approaches presented in the previous paragraph, it was decided to do preliminary experiments. Four sensors for machining process sensing were installed in a lathe STARTURN 4. This machine is a small laboratorial CNC lathe with 1/2 Hp spindle power and the axes are driven by step motors. One sensor, senses the mains current consumed by the machine. Besides some constant consumption done by the machine's electronics, the main consumption is related with the spindle motor. A second sensor is a microphone that captures the machine working sound. The third and four are two accelerometers installed on the tool handler carriage. One accelerometer detects the vibrations on the X axis direction and the other for the Z axis. Figure 7 illustrates the installation of the sensors on the machine.

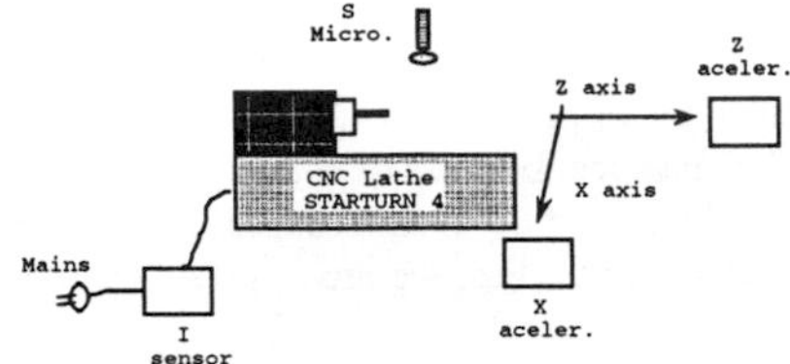

Figure 7 Locations of lathe sensors.

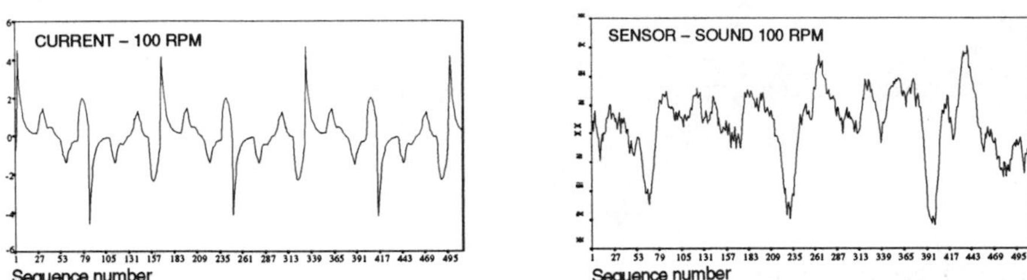

Figure 8 Plotting of current and sound sensors' data collected with spindle at 100 RPM.

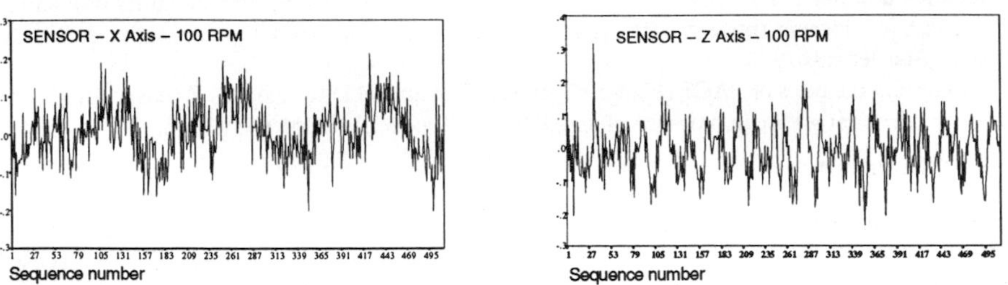

Figure 9 Plotting of axis X and Z sensors' data collected with spindle at 100R RPM.

The experiments performed correspond to a small subset of all machining situations. It was decided to analyse only the possibility of identification of the spindle speed by means of the described sensors. Sensor data was captured for the following spindle rotations: 100RPM, 300RPM, 500RPM, 700RPM and 900RPM. The sampling rate was 8192Hz, corresponding to 122,07 μs sampling intervals.

Figures 8 and 9 depict the series got from sensors while the spindle was operated at 100 RPM.

In order to obtain the correspondent *ARMA(p,q)* models, blocks of 512 data sample points were considered. The procedure described below for determining the *ARMA(p,q)* model using the axis Z sensor data was performed similarly for the data of the other sensors.

There is a model building procedure proposed by (Box-Jenkins 1970). This procedure was followed, and consists of three steps:

1. Identification of a suitable model, i.e. the identification of the *p*, and *q* order for the *ARMA(p,q)* model.
2. Estimation of the coeficients ϕ and θ.
3. Diagnosis, i.e. the evaluation of the adequacy of the model.

6.1 Identification

The first step is subjective, because it is necessary to identify the underlying processes. First it must be determined the values of the integers *p* and *q* of the supposed *ARMA(p,q)* process.

Parameter *d* is the first to be achieved, because to obtain parameters *p* and *q* the series has to be stationary. As it can be seen in the plot of the Z axis sensor series (Figure 9) there is no trend and seasonally, evidencing good stationary. Thus, we are in presence of an *ARMA (p,q)* model, with zero mean.

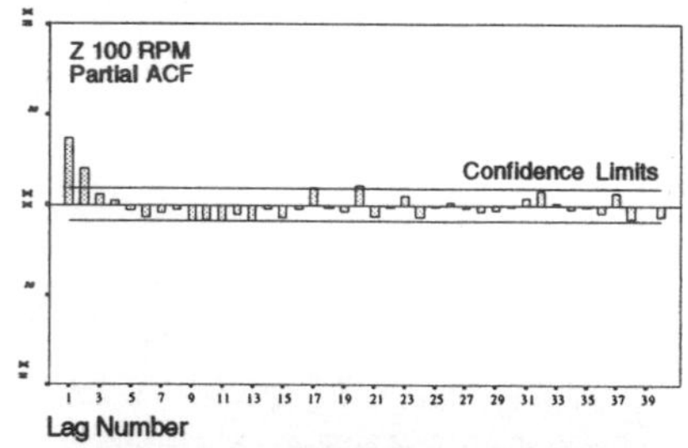

Figure 10 Plotting of PACF axis X sensor

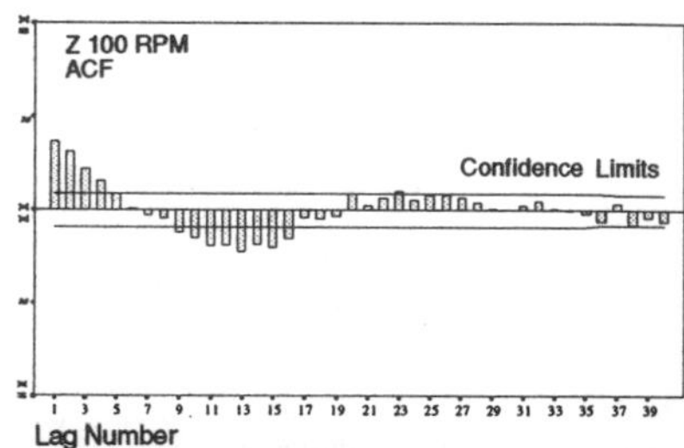

Figure 11 Plotting of ACF axis Z sensor data

To identify the values for *p* and *q*, the analysis of autocorrelation function (ACF) and partial autocorrelation function (PACF), calculated at lags 1, 2, ..., reveals the correct values of *p* and *q*. There is a relationship between *ARMA(p,q)* models and the corresponding data series ACF and PACF (Box.Jenkins 1970).

Analysing the plots of PACF (Figure 10) and ACF (Figure 11) for the axis Z sensor data, it can be concluded that we are in presence of an ARMA(2,0) model, with $\phi 1$ and $\phi 2 > 0$.

6.2 Estimation

After knowing the order of the model, the next step is the coeficients' estimation for an *ARMA(2,0)* model: $\phi 1$ and $\phi 2$. To perform this task, that requires a lot of computation, it was used the SPSS statistical package (SPSS 1993). The results obatained are presented in table 1.

Several statistics describ how well the model fits the data. The standard error, the Log likelihood, the Akaike Information Criterion (AIC) and the Schwartz Bayesian Criterion (SBC). They measure how well the model fits the series, taking into account the fact that a more elaborate model is expected to fit better. The AIC is for *ARMA(p,0)* models while the SBC is a more general criterium. Using these statistics, we can choose from different models for a given serie. The model with lowest AIC, SBC, standard error and highest Log likelihood is the best.

Inspecting the estimated standard error of the coeficientes it can be verified that the coefficientes are statistically significant, and they are not correlated between them.

Table 1 Result of $\phi1$ and $\phi2$ estimation

```
Number of residuals  512                        Correlation Matrix:
Standard error       .06106404                                    AR1          AR2
Log likelihood       705.81381                  φ1         1.0000000    -.4276129
                                                φ2         -.4276129    1.0000000

Variables in the Model:
PROB.          B           SEB          T-RATIO        APPROX.
φ1      .32809913    .04311531    7.6098048       .000000          AIC   -1407.6276
φ2      .23221007    .04312045    5.3851500       .000000          SBC   -1399.151
```

Table 2 - ACF error data in low-resolution

```
Auto- Stand.
Lag  Corr.   Err. -1   -.75   -.5 -.25   0   .25  .5   .75   1   Box-Ljung  Prob.
                  +----+----+----+----+----+----+----+----+
 1    .054   .070                   .  |* .                       .593    .441
 2   -.018   .070                   .  *  .                       .661    .719
 3    .100   .070                   .  |**.                      2.729    .435
 4   -.002   .069                   .  *  .                      2.730    .604
 5    .020   .069                   .  *  .                      2.815    .728
 6    .019   .069                   .  *  .                      2.893    .822
 7   -.009   .069                   .  *  .                      2.908    .893
 8   -.023   .069                   .  *  .                      3.020    .933
```

6.3 Diagnosis

To validate the achieved model, the following checks are essential: The ACF and PACF of the error series should not be significantly different from 0. And the residuals should be without pattern (i.e. should be white noise). A common test for this is the Box-Ljung Q statistic. To accept it, this statistic should not be significant. Inspecting table 2, we found that the values of the Box-Ljung statistic for the ACF function is not statisticaly significant at any lag.

6.4 The best *ARMA(p,q)* model for axis Z sensor data (100 RPM)

Considering the analysis already performed, we can state that we are in presence of an *ARMA(2,0)* model with coeficientes $\phi1$=0.328 and $\phi2$ = 0.232.

The stochastic process $g[n]$ for axis Z sensor data (100 RPM) satisfies the following difference stochastic equation:

$$g[n]-0.328g[n-1]-0.232g[n-2]=\varepsilon[n] \tag{12}$$

where $\varepsilon[n]$ is a white noise process. The corresponding system transfer function is:

$$H(z)=\frac{1}{1-0.328z^{-1}-0.232z^{-2}} \tag{13}$$

Figure 12 depicts the axis Z sensor data (100RPM) and the fitted *ARMA(2,0)* model. Figure 13 shows the error data obtained from the axis Z sensor data (100RPM) and the fitted *ARMA(2,0)* model. As it can be seen it is white noise.

6.5 Results obtained

The models obtained for the current, machining sound and axes' vibrations sensors, are presented in table 3. Because of space limitations we only present the coeficients for the sound sensor in table 4.

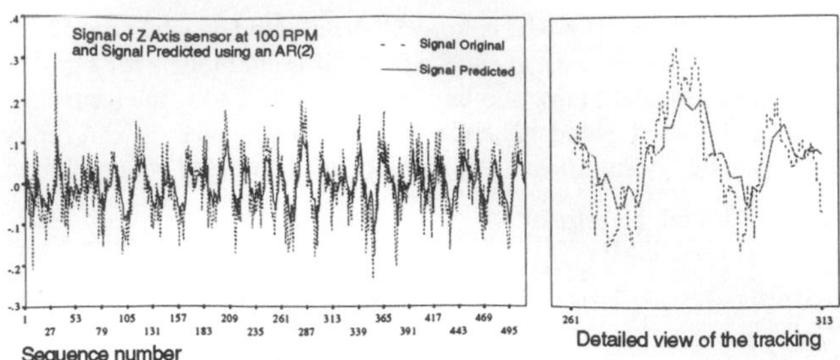

Figure 12 Axis Z sensor and fitted *ARMA(2,0)* model data.

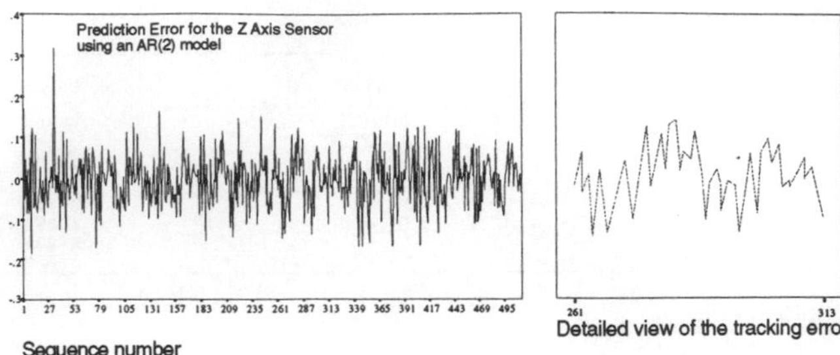

Figure 13 Plotting of error data.

The applicability of these models for building feature vectors as previouslly described in paragraph 5 and figure 5 a), was also confirmed experimentally. The extracted features have good clustering regions well separated, meaning their adequacy for a process status pattern recognition identifier.

Table 3 - ARIMA models achieved for diferent sensor data

	Idle					Cuting	Cuting
ARIMA Model	**100 RPM**	**300 RPM**	**500 RPM**	**700 RMP**	**900 RPM**	**300 RPM**	**500 RPM**
Current	AR(1)	AR(1)	AR(1)	AR(1)	AR(1)		AR(1)
Sound	AR(4)	AR(3)	AR(3)	AR(3)	AR(3)		AR(3)
Axis X	ARI(2,1)	ARI(2,1)	ARI(2,1)	ARI(2,1)	ARI(2,1)	ARI(3,1)	
Axis Z	AR(2)	AR(4)	ARMA(1,2)	ARMA(2,3)	ARMA(2,3)	ARMA(2,1)	

Table 4 - Estimated coefficientes for the machining sound model

Idle					Cuting
100 RPM	**300 RPM**	**500 RPM**	**700 RMP**	**900 RPM**	**500 RPM**
$\phi_1, \phi_2, \phi_3, \phi_4$	ϕ_1, ϕ_2, ϕ_3	ϕ_1, ϕ_2, ϕ_3	ϕ_1, ϕ_2, ϕ_3	ϕ_1, ϕ_2, ϕ_3	ϕ_1, ϕ_2, ϕ_3
1.20, -.37, .59, -.45	.71, -.36, .60	.58, .40, .72	.44, -.37, .81	.33, -.26, .76	.85, -.53, .49
1.20, -.46, .67, -.43	.80, -.39, .53	.58, .39, .71	.39, -.27, .74	.37, -27, .76	.99, -.64, .45
1.19, -.35, .50, -.40	.81, -.45, .60	.57, .45, .74	.36, -.31, .81	.36, -.27, .74	.72, -.40, .61

7. CONCLUSIONS AND FUTURE WORK

A methodology for characterising the machining process is proposed. It is based on sensorial data modelling using autoregressive *ARMA(p,q)* models for stochastic signal characterisation. The relationship between the *ARMA(p, q)* models obtained and system transfer function theory is presented to explain the meaning of the models obtained and their relationship with the real system under observation. Because of the complexity of the machining process, it is also proposed a method for modelling the complete process as a sequence of elementary and more simple sub-processes. Each one of these simple sub-process corresponds to the NC instructions under execution.

The applicability of the proposed methodology was assessed by doing preliminary experiments on a laboratorial lathe. The results obtained, are very encouraging for pursuit the proposed approach in future works.

Future work will be done in two directions: first, pursuit the research for characterisation of more situations of the machining process and second, develop real-time monitors based on our approach. The integration of the proposed approach in our Prognostic and Monitoring System for CNC Machines, (for detailed description see (Barata 94)) will be done. Tansputer tasks will carry out all the real time processing necessary for doing the estimation of the *ARMA(p, q)* parameters. The applicability of adaptive systems described in (Widrow 1985) seems to be an interesting approach for real time parameters estimation.

8. REFERENCES

Abraham, Bovas and Johannes Ledolter(1983). *Statistical Methods for Forecasting.* Jhon Wiley & Sons, New York, 1983.

Barata, Manuel M. (1990a). Maintenance and Prognostic Systems: State of the Art. *Internal report UNL DI NT-2-90 GR RP-PS-90.*

Barata, Manuel M., José C. Cunha and A. Steiger-Garção (1990b). Transputer Environment to Support Heterogeneous Systems in Robotics. *Transputer Applications 90 Conference,* Southampton, 11-13 July, 1990.

Barata, Manuel M. and A. Steiger-Garção (1990c). Sensor Environment for Prognosis and Monitoring Systems Support. *3rd ISRAM International Symposium on Robotics and Manufacturing,* British Columbia, Canada, 20-22 July, 1990.

Barata, Manuel M., Thomas W. Rauber, A. Steiger-Garção (1992). Sensor Integration for Expert CNC Machines Supervision. *Etfa'92 Emerging Technologies in Factory Automation.* Melbourne, Autralia, 11-14 August, 1992.

Barata, Manuel M., Thomas W. Rauber, A. Steiger-Garção (1994). Prognostic and Monitoring System for CNC machines. *Revue Européenne Diagnostic et Sûreté de Fonctionnement.* Editions HERMES, Vol. 4 - N° 2/1994, 1994.

Box- Jenkins, George E. P. and Gwilyn M. Jenkins (1970). *Time Series Analysis Forecasting and control.* Holden-Day, San Francisco, 1970.

Steiger-Garção, A., M. M. Barata and L. F. S. Gomes (1989). "Integrated Environment for Prognosis and Monitoring System Support". *1st UNIDO workshop on Robotics and Computer Integrated Manufacturing.* Lisbon, Portugal, 11-15 September, 1989.

Guinea, D. et al.(1991). Multi-sensor integration - An automatic feature selection and state identification methodology for tool wear estimation. *Computers In Industry,* Vol. 17 , no. 2&3, pp 121-130, Nov. 91.

Harvey, Andrew C. (1992). *Forecasting, Structural Time Series and the Kalman Filter.* Cambridge University Press, 1982.

Papoulis, Athanasios (1965). *Probability, Random Variables and Stochastic Processes.* McGraw-Hill International Students Editions, 1965.

Papoulis, Athanasios (1984). *Signal Analysis.* McGraw-Hill International Editions, 1982.

Rauber, Thomas W., Manuel M. Barata e A. Steiger-Garção (1993). A Toolbox for Analysis and Visualization of Sensor Data in Supervision. *TOOLDIAG'93*, Toulouse-France, April 5-7, 1993.

SPSS, Inc. (1993), SPSS Statistical Package for Windows Release 6.0. *SPSS Inc.*, 444 N. Michigan Avenue, Chicago, Illinois 60611, 1993.

Widrow, B. and S.D.Stearns (1985). *Adaptive signal Processing.* Prentice-Hall, Englewood cliffs, N.J., 1985.

9. BIOGRAPHY

Manuel M. Barata is adjunct professor at Instituto Superior de Engenharia de Lisboa of the Polytechnic Institute of Lisbon where he teaches computer science subjects. Since 1987 he has joined the Intelligent Robotics Centre of the UNINOVA Institute, where he is finishing his Ph.D. thesis on Monitoring and Prognostic Systems for CNC Machines. His main research interests are: Monitoring Systems, Real-time architectures, Transputer based systems and Artificial Intelligence systems for real-time operation. He has participated in several international European projects and has been the group leader of the UNINOVA participation in the CIMTOFI BRITE/EURAM project.

Adolfo Steiger-Garção is full professor at Department of Electrical Engineering of the New University of Lisbon. He is also the president of UNINOVA Institute and the director of the Intelligent Robotics Centre. His main research interests are: Robotics / Perception / Monitoring. He is responsible for several international projects and has more than 100 publications on national and international conferences.

Ricardo R. Jardim Gonçalves is graduated in Computer Science, and now is coursing his master in Operational Research and Systems Engineering. He has a large experience in the areas of computer science research specially in integration of industrial tools and standards, STEP in particular. He his the working group leader of the UNINOVA participation in the ESPRIT III RoadRobot project and has participated has work coordinator in the CIMTOFI BRITE/EURAM project .

José Álvaro Assis Lopes is associated professor at the Autonomous Section of Economy and Management in the Instituto Superior Técnico of Technical University of Lisbon. He has been graduated in Chemical Engineering, got a master degree on Operational Research, and made a Ph.D. on Operational Research and Systems Engineering. His main research interests are: Forecasting, Operational Research and Management Science.

42

No Knowledge to Waste – Learning more about Processes

Ir. K. Loossens & Ir. G. Van Houtte
Expert Systems Applications Development Group
Department of Chemical Engineering
Katholieke Universiteit Leuven
de Croylaan 46, B-3001 Heverlee, Belgium
Koen.Loossens@cit.kuleuven.ac.be
Guy.VanHoutte@cit.kuleuven.ac.be

Abstract

Automating your process is one thing, using your investment in an optimal way is another. Automation systems provide you with a lot of process data, which should lead to a better understanding of the process. This is seldom the case, the culprit being an overdose of information. As a result, a lot of useful information remains hidden in these data. In this article we present a system that is capable of distilling knowledge out of process data. Based on the current process situation and using information about previous operations, knowledge is generated on the fly. As a result existing knowledge can be validated and hidden process characteristics can be revealed. Without expensive expert system technology, the engineer can maximally use the information that is being logged in the plant and gradually expand his know-how of the process. He can then put this knowledge at the disposal of the operators who can use this to better manage the process. The proposed system is generally applicable throughout the process industry. Two industrial examples are referenced to: a waste paper plant (continuous operation), and a pulp factory (batch operation of digesters).

1. INTRODUCTION

Today's automation systems are capable of measuring and collecting every desired detail about the production process. This massive amount of information should lead to a better understanding and management of the process. However, the problem is how to cash in on this information as to get the most out of your investments in an automation system. An engineer can attempt to uncover some knowledge with statistical tools and modelling approaches but often they are confronted with a process that is too complicated and too less known to gain insights in the process [Leitch, 1992], or to have a clue where to start. Still the key to a better understanding of the process resides within these data. In a next stage, the expert systems were hailed as the miracle solution for process management [Mark, 1991]. It turned out differently due to various causes. As impressive as an expert system can be, it still lacks the flexibility of a human. The knowledge in such a system is static [Rich, 1987] and doesn't adapt to different process conditions or evolutions in a plant. To keep it up to date and useful for an operator will require a lot of maintenance efforts and the expensive intervention of a knowledge engineer [Mengshoel, 1993]. Firstly one has to gather knowledge to put it into the system. This acquisition of knowledge is the acknowledged bottleneck in developing an intelligent system [Steier, 1993]. It takes up a lot of resources and the result is often poor because the knowledge engineer usually doesn't know the process [Muraret, 1993]. The presented system offers an elegant solution to these problems.

2. GENERAL PICTURE

The system consists of different modules. Each of them has a distinct task and can be adapted separately to specific processes and requirements.

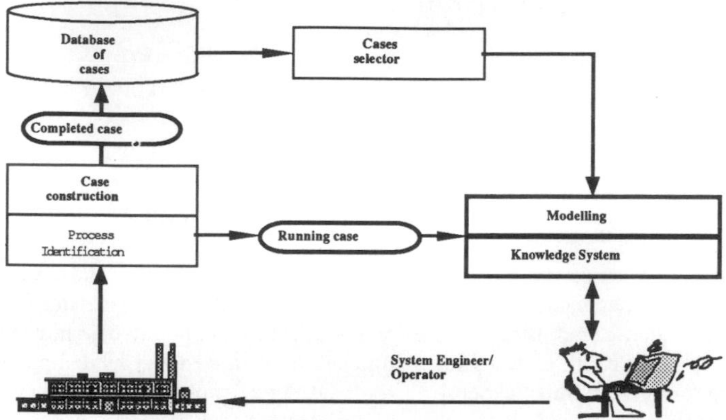

Figure 1. General architecture.

The data come from the automation system into the process identification module. This module filters out the relevant information and performs mathematical and statistical manipulations [Cheung, 1990] to come to a predefined structure: a set of process parameters that are representative for the state of the production process. We call this the *running case:* this is a snapshot of the current situation in the plant. Once a logical stage in the production process is reached - this of course is very process dependent -, the running case is considered complete and sent through to the database of cases. This database contains the information about all the previous plant operations structured into a standardised case form. One can search and retrieve cases out of the database through the cases selector module.

The knowledge system works with two parts: a static and a dynamic one. The static part is knowledge about your process that is already present and is considered true: it is not supposed to be changing over time. The dynamic part is knowledge that is generated on the fly. This dynamic knowledge is distilled out of a model. This model is build with the running case (the actual process situation) as a basis and using cases out of the database (information about previous operations). The dynamic knowledge is then confronted with the static part of the knowledge system. As a consequence the engineer working with the system can discover new knowledge and update and refine his perception of the process.

3. CONCEPT OF A CASE

Process information is represented in the form of cases. The purpose is to create a reference framework to organise the data in logical units. This facilitates comparing the information coming from the process with information about the previous operations in the plant. Typically a case will consist of a series of process parameters and quality parameters. Process parameters are data relating to the process under consideration. Quality parameters are constraints you want to see fulfilled at the end of your case. How you define process and quality parameters and hence your cases at hand greatly depends upon the process you are working with. Let us illustrate this with the two examples where a prototype has been installed.

The first example is a chemical pulping plant. The system is installed in the digester house. There are four digesters, each pulping wood chips with acid in a batch way. Logically you take a batch to be a case. Process parameters are everything you are logging during batch operation, including the input conditions. Quality parameters will be anything you want to achieve with the batch. The main quality parameter is here the P Number (which is an indication of the amount of lignin left in the wood chips).

The second example is a plant that manufactures paper board out of waste paper. The process is a continuous one. However the paper is winded on rolls and every 20 minutes a roll is finished and a new one is started without interruption of the machinery. The production of a roll was taken as the definition of a case: this was even more obvious since every roll was controlled on several qualities like weight and strength.

The purpose of creating cases is to be able to compare them to each other: so compare one batch to another, compare the making of one roll against another.

4. DATA HANDLING MODULES

The data handling modules are those modules that were designed to structure and handle the data from the automation system.

The process identification and case construction module have the task to shape the data collected about the process into the case form as seen previously. The process parameters you want to see in the case might not be readily available from the automation system. To represent your process in a case you might need averages of measured values, evolutions of set points, historical data, combinations of parameters, ...[Cheung, 1990]. Also the issues of missing values and faulty measurements can be addressed at this level [Qin, 1993]. The process identification module takes care of the necessary manipulations, calculations and statistical processing of the data from the automation system to build a case. As soon as all process data are filled in and the quality parameters are available (this may take a while when laboratory analyses are involved) the case is considered complete and shipped to the database of cases. The process identification produces also the running case during the processing of the case. The easiest way to imagine this is as an empty skeleton of a complete case where gradually as the process continues values of process parameters are filled in. This running case must give the knowledge system an idea of the situation in the plant and will form the basis for the modelling.

The database of cases contains all the cases collected during operations in the plant. You might consider this as the memory the system has about what occurred in the plant. It is the information that implicitly lies in these past experiences that will be used by the modelling and the knowledge system to generate and validate knowledge concerning the present situation [Barletta, 1993].

The case selector module acts as a support to the modelling module in that it provides a set of cases from the database requested by the modelling. Basically the modelling asks for cases that are in some way close to the running case in order to construct a reliable model. Hence this module is able to scale and quickly retrieve cases from the database that comply to a certain criterion of nearestness to the running case. We defined the degree of nearestness as the sum of least squares of differences between a set of selected parameters, complemented with weight factors to put emphasis on more important parameters and with a correction to get a spatially uniform distribution [Loossens, 1995].

5. MODELLING

The modelling part of the system provides the knowledge system with a process model that can be used to derive process behaviour (the so-called knowledge rules) in a quantitative manner [Stephanopoulos, 1990]. Two modelling approaches were chosen: a linear approach (using the Partial Least Squares technique), and a non-linear approach using a backpropagation neural network [Rumelhart, 1986]. However, any modelling technique can be used (physical models about the process as well as mathematical models), since the modelling can be considered as a black-box as far as the knowledge system is concerned. Purpose of the modelling essentially is to produce relations between process and quality variables.

With the Partial Least Squares modelling approach [Geladi, 1986], we use a set of so-called nearest cases to develop a linearised system behaviour around the current point of operation of the process, hoping that the effects of the process parameters towards the qualities will be rather linear. The first tests at the industrial sites however showed that real processes rarely behave in a linear manner, so results may be completely out of scope. And even so, the behaviour is only known in the exact point of operation, where the linearisation took place so no boundary influence limits (parameter limits within which the obtained model is valid) can be calculated.

Using the Neural Network as the modelling tool requires two steps. The first step is building the model: the so-called training of the neural network with historical data. The second step is using the network to derive the process's behaviour around the current running case. In order to train the network to identify general behaviour, one can use all data logged in the database in a recent history. When one wants to look at specific behaviour (different input, start-up conditions, special production series,...), one can select only those cases that comply to these conditions to train the network to model this specific behaviour. The neural network we developed makes training easy for the novice user, as well as allows the experienced user to play extensively with the internal variables of the network. The result of the training contains the internal configuration of the neural network, as well as the prediction error for each case in the used dataset. The user can easily use these results to determine if neural network training was successful of not.

Initially we use the neural network to give an idea of the process evolution when changing a process parameter [Catfolis, 1994]. And this is exactly the kind of information both the system engineer and the operator is interested in. The network determines the direction of change, the range of within which the direction of change remains the same, and the order of magnitude of this change from the current point of operation onwards. These relations will be used in a qualitative manner by the knowledge system. To get an idea of the reliability of the network we used it also to predict the outcome of quality parameters. Our experiences with the prototypes installed at the industrial partners show that the neural network follows the system behaviour over a broad range (Figure 2). This strengthens our confidence in this modelling technique and motivates further investigation into this quantitative way of data mining [Loossens, 1995].

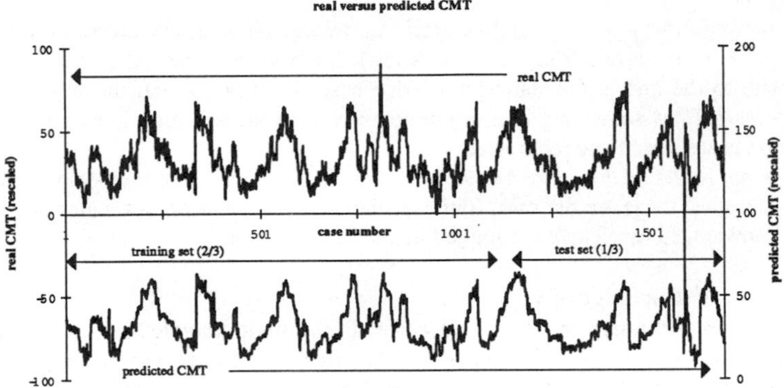

Figure 2: NN result for quality parameter CMT on the paper machine.

6. KNOWLEDGE PROCESSING

Dynamic rule construction

First the knowledge system converts the relation between the selected process parameters and quality parameters from the modelling into a knowledge form that is comprehensible for humans yet easy to handle by a computer. For this purpose we developed a general rule-like structure. For example, the neural net approach results in rules of the following structure:

IF <parameter_Pi> changes with <increment_Pi>, THEN
<quality_Qi> will change with <increment_Qi>

Validity of direction of change from <lower_limit_Pi> to <upper_limit_Pi>.

These are essentially one-to-one relations, but if a user decides to produce multiple relations (by adapting the neural network modalities) or incorporate a totally different type of modelling the rule construction is able to cope with this. Each time the modelling is triggered, the rule construction produces a set of rules. As a result we have generated knowledge about our process, not having to go through the tedious process of knowledge acquisition, but simply by taking the current situation in the plant and using previous data to derive knowledge on the fly.

Dynamic rule classification

After generating the dynamic knowledge we have to filter it through the static knowledge base (static because it is supposed to be well known and true in all circumstances). The rules that are constructed from the modelling are catalogued in three ways:

- *Rules found in the static knowledge base:* these are rules from the modelling that qualitatively match those from the static knowledge base. Qualitatively means that the direction of change is the same irrespectively of the absolute value of the change. We also keep track of how many times a rule in the static knowledge base has been validated and rejected. The user can consult this to quickly have a grasp as what the evolution is of his static knowledge base.
- *Rules in contradiction with a rule from the static knowledge base.* These are rules found by the modelling that match a rule from the static knowledge base but where the direction of change is opposite to the one in the static knowledge base. This means that the model we built of the process tells us something that is the contrary of what we think is true for the process. These rules are stored and referenced.
- *Rules that are not found in the static knowledge base.* This comprises in the beginning of the majority of the rules: these are the rules found by the modelling that do not match any of the rules in the knowledge base. These are logged in a separate rule base.

Validation of static knowledge

The engineer can browse through the three rule bases. In the static knowledge base he can check how many times the rules have been validated. An often validated rule means that what the modelling found is what our initial idea of the process was. The engineer can then augment the priority of this rule. This priority will be of influence on the way the operator at the control panels is presented with process knowledge from the static knowledge base.

The rule base of contradictory rules contains rules which have been found to be opposite of what the engineer's idea of the process was. He can use the knowledge system to easily backtrack to the conditions when this occurred. Different scenario's are plausible.

- An abnormal process condition occurred; a condition of which we did not have any knowledge in the static database. It is quite possible to find then that rules for normal process conditions are no longer valid. So this learns the engineer how the process behaves in these circumstances.
- The rule in the knowledge base is too general: it seems that the rule as it is stated is only valid given that certain parameters remain within certain limits. For example the rule:

IF <parameter_P1> changes with <increment_P1> THEN
<quality_Q1> will change with <increment_Q1>

appears to be correct when parameter P1 lies between 0 and 50 whereas when P1 is between 50 and 100, the opposite behaviour is observed. This indicates that this rule in the static knowledge base should be adapted and expanded into two rules.

• The process is evolving, due to machine wear, replacement of machinery, different inputs, change in process management, ... As a result the knowledge about the system will be evolving and gradually you will see that some rules that were valid before are becoming more and more invalid.

• The rules in the static knowledge base were wrong in the first place. Perhaps the idea about the process was wrong.

With the knowledge system, the engineer has an intuitive tool to refine his perception of the process under his supervision and keep track of changes in the plant. The system has a user friendly interface so no extensive training is required.

Other functionalities

The knowledge system constantly generates new knowledge in the rule base of new rules. The engineer who is familiar with the process can easily distinguish between rules that don't make sense and the ones that are sensible. To test his newly gained knowledge he can put them in the static knowledge base (with a low priority) to see if they are validated in time. With the database of cases (past experiences) at his disposal, he can focus his research on a particular point of interest of his process (some correlation he suspects to exist, distinct process parts or conditions) by adapting the modelling. Hence the system skips any forgoing acquisition of knowledge.

Also the operator can benefit from this system. What the operator has at his disposal is the proofed knowledge (with a priority that is high enough) which he can consult in two ways. First he might be confronted with a deterioration of a certain quality parameter. This could be that the parameter is slowly shifting or that he gets a warning from the laboratory that the quality has turned bad. He can then query the knowledge system to see what he can do to adjust the quality. Secondly he might wonder what the effect would be if he was to change some parameter setting. The system will reason through the knowledge base to give him the implications of that change in parameter settings on the quality parameters.

Future development includes putting the system on-line: we then use the neural network model to predict the outcome (quality) of the process and in case of a bad prospect, the knowledge system uses the information from the network to advise the operator to the best corrective action(s).

9. CONCLUSION

We have presented a modular architecture that is able to extract and validate knowledge out of data collected from the process in the past and based on the current situation in the plant. This system provides the process engineer with a comprehensive tool to maintain and keep up to date with the knowledge about his process. Besides this it offers support to the operator to better manage his task. At the paper manufacturer, all known correlations have been confirmed and already one unknown relation has been revealed. At the pulp mill, there were at the time of writing not enough cases collected to produce reliable results (a batch takes +/- 10 hrs.), but the first tests already confirm the known relations.

Acknowledgement

This architecture has been developed within the CLEAN P5285 project. This is an industrial project with 5 partners sponsored by the EC that resorts under the BRITE-EURAM program.

The authors are researchers at the Katholieke Universiteit Leuven (Belgium), in the Expert Systems Applications Development Group (part of the department of Chemical Engineering), headed by Prof. dr. ir. M. Rijckaert.

References

Barletta R. (1993) Case-based reasoning and information retrieval: opportunities for technology sharing. *IEEE Expert,* **October**, 2-4.

Catfolis, T. (1994) Generating Adaptive Models of Dynamic Systems with Recurrent Neural Networks. *Proceedings of the IEEE International Conference on Neural Networks '94*, **vol. 5**, 3238-3243, Orlando, Florida.

Cheung J. T.-Y.and Stephanopoulos G. (1990) Representation of process trends - part II: the problem of scale and qualitative scaling. *Computers and Chemical Engineering*, **Vol. 14, No. 4/5**, 511-539.

Geladi P. and Kowalski B.R. (1986) Partial least squares regression: a tutorial. *Analytica Chimica Acta*, **185**, 1-17.

Leitch R. (1992) Artificial intelligence in control: some myths, some fears but plenty of prospects. *Computers & Control Engineering Journal*, **July**, 153-163.

Loossens K. and Van Houtte G. (1995) Predicting Process Evolution with a Neural Network. A window on relevant data. *Proceedings of the IEEE Industrial Automation and Control Conference '95, Taipei, Taiwan.*

Mark W.S. and Simpson R.L.Jr. (1991) Knowledge-based systems: an overview. *IEEE Expert,* **June**, 12-17.

Mengshoel O. J. and Delab S. (1993) Knowledge validation: principles and practice. *IEEE Expert*, **June**, 62-68.

Muratet G.and Bourseau P. (1993) Artificial intelligence for process engineering - state of the art. *Computers and Chemical Engineernig - special issue: European symposium on computer aided process engineering,* **February**, S381-S385.

Qin, S. J. and Rajagopal, B. (1993) Combining statistics and expert systems with neural networks for empirical process modelling. *Advances in Instrumentation and Control*, **Vol. 48, part 3**, 1711-1720.

Rich S.H. and Venkatasubramanian V. (1987) Model-based reasoning in diagnostic expert systems for chemical process plants; *Computers and Chemical Engineering*, **Vol. 11, No. 2**, 111-122.

Rumelhart, D.E., Hinton, G.E. and Williams, R.J. (1986) Learning internal representations by error propagation. In *D.E. Rumelhart , J.L. McClelland and the PDP Research group (eds.) Parallel Distributed Processing: Explorations in the Microstructure of Cognition, Volume 1: Foundations.*

Steier, D. M. et al. (1993) Combining multiple knowledge sources in an integrated intelligent system. *IEEE Expert,* **June**, 35-44.

Stephanopoulos G. (1990) Artificial intelligence in process engineering - current state and future trends. *Computers and Chemical Engineering.*, **Vol. 14, No. 11**, 1259-1270.

43

3-D Vascular reconstruction on a standard X-ray angiography machine

R. Gupta
General Electric Corporate R&D
1 River Rd, Schenectady, NY 12309, USA

Y. Trousset, C. Picard, R. Rom`eas
GE Medical Systems --- Europe
283 rue de la Miniere, 78530 -- Buc, France

Abstract

Angiography gantries are used for a diverse set of procedures such as catheterization, embolization of aneurysms, angiography, digital subtraction angiography, and balloon angioplasty. Traditionally, these gantries have been used as a source of 2-D images of the pathology. Special machines have been designed to allow 3-D imaging. However, these new machines represent additional cost and require a separate angio suite to be maintained. This paper outlines a methodology for doing 3-D reconstruction on a standard angiography gantry. The complete procedure is described and the first in-vivo results are presented.

Keywords

X-Ray angiography, digital subtraction angiography, 3-D reconstruction

1.0 Introduction

Fig. 1 schematically shows a typical X-ray angiography gantry. It consists of an X-ray tube and an image intensifier mounted on a large C-shaped structure called the C-arm which, in turn, is mounted on a large L-shaped structure. This assembly, which is sometimes referred to as an LC gantry, allows the imager (i.e. the X-ray tube and the image intensifier) to be oriented at any desired angular position around the patient. These gantries are very versatile and are used for such procedure as, to name a few, catheterization, embolization of aneurysms, angiography, digital subtraction angiography, and balloon angioplasty.

This paper primarily deals with digital subtraction angiography (DSA), a routine X-ray procedure for observing vasculature (Macovski, 1983). In DSA, a catheter is inserted in the blood vessel to be studied and a set of so called *mask images* are acquired from various orientations in space. An X-ray opaque dye, typically an iodine solution, is then injected into the vessel through the catheter and another set of *opacified images* are acquired at the same orientations in space. The only difference between the mask and the opacified images is that in the latter the blood vessels are visible. When the mask image is subtracted from the opacified image, everything but the blood vessel is subtracted out resulting in a digital subtraction angiogram of only the blood vessels.

One can pivot the center of the C-arm and rotate it around the patient while acquiring 2-D images. Many new gantries can rotate quite fast (e.g. 30 or even 60 degrees/second). Since the dye washout time for humans is about 3 seconds, one can acquire a set of DSAs covering 180 degrees. Thus, at the video frame rate of 25 frames/sec, this setup yields about 75 images with an angular separation of about 2.4 degrees. We shall refer to the process of acquiring images while the imager is continuously rotating at 30 degrees/sec or more as a *fast spin acquisition.* In this paper we outline a methodology for constructing a 3-D volume from a fast spin acquisition of mask and opacified images.

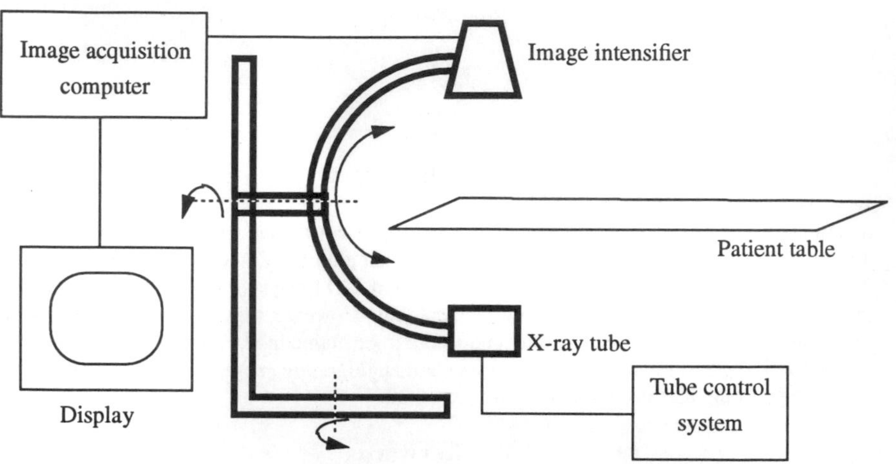

FIGURE 1. Schematic of a typical angiography gantry.

2.0 Clinical Usefulness of 3-D Angiography

In order to evaluate the clinical usefulness of 3-D images of the vasculature, GE Medical Systems --- Europe designed a prototype machine called 3-D Morphometer specifically for this application (Didier, 1993). Two prototypes of the 3-D Morphometer have been installed in hospitals and several exams have been performed on them. This initial clinical evaluation phase has demonstrated that 3D X-Ray angiography is highly valuable both for diagnosis and for therapy planning (e.g., surgery, interventional radiology, and radiotherapy).

Using a machine such as the Morphometer, one can examine, in 3-D, the intra-cranial arteries with selective arterial injections. The 3-D images allow precise analysis of the pathologies such as aneurysms and arterio-venous malformations (AVMs). One can also study the carotid arteries in the neck region with non-selective injection in aortic arch. It has been demonstrated that even with a non-selective injection, the 3D images have a quality which is sufficient to provide a safe and accurate diagnosis (e.g., pre- or post-stenosis surgery control). Our experience has shown that 3D exams are faster and much less invasive than a standard 2D exam as only one non-selective injection, comparable to Spiral CT Angio studies), is done. At the same time, there is no loss in the diagnostic quality of the imagery.

In both protocols, the quality of the 3D images is good and stable. Further, it has been demonstrated that the 3D image generated contains more readily visible diagnostic information as compared to other 2D or 3D techniques. This realization has led to "mainstreaming" of Morphometer examinations in the hospitals it was installed in.

Bone studies have also been performed on the Morphometer, and the results are very encouraging. The spatial resolution, especially in the vertical direction, is very good. At the same time, the X-Ray dose and exam duration in dramatically decreased compared to a CT exam.

3.0 The Market Reality

The clinical usefulness of 3-D angiography is unquestionable. However, a Morphometer-like machine requires a separate vascular room to be maintained just for 3-D angiography. For developing countries, and even for industrialized countries undergoing the current health-care crisis, this additional expense may not be justifiable.

This paper shows that by using a more anthropocentric and somewhat more complex protocol, one can realize 3-D X-Ray angiography on a "standard" vascular system based on an LC gantry. This approach brings together the best of both worlds as it allows specialized 3-D procedures to be done on the existing, installed base of machines. The reconstruction done on an LC requires greater human intervention, and is not as automatic as that on the Morphometer. However, this anthropocentric approach is worth the extra effort as it alleviates the need for a separate machine and a separate angio suite.

3.1 Main Challenge

At first glance, the task of assembling 2-D DSAs into a 3-D volume may appear to be no different from that encountered in conventional computed tomography (CT). However, the following differences between the two make it necessary to devise a completely new protocol for generating 3-D vascular images.

Calibration: In CT, the gantry is very stable and precisely calibrated. Such precise calibration is not available with angiography gantries.

Gantry Deformation: Because of the weight of the image intensifier and the X-ray tube, typical angiographic gantries deform slightly when they rotate. Thus, even if a precise *prior* calibration was available, it cannot be used over and over again because of the variability in the deformation from one run to the next.

Dynamic Range: The dynamic range of image intensifiers used in angiography is much less than that for the detectors used for CT: only about 8 bits/pixel are available in angiography while as many as 20 bits/pixel may be available in CT. As a consequence, only high contrast anatomical structures such as bones and opacified vasculature can be reconstructed from angiographic images. From a set of mask images alone, one can reconstruct the bones in 3-D, while the same information with CT-type detectors would enable reconstruction of bones as well as the soft tissue.

Speed of Rotation: The speed of rotation for angiographic gantries, though much faster than what it used to be only a few years ago, is still considerably slower that of a typical CT gantry. Thus, during the washout period of the X-ray opaque dye injected for DSA, only a limited number of images can be acquired. As a result, 3-D reconstruction has to be done with incomplete information (Trousset, 1994).

1-D vs. 2-D Detector: Traditionally, a 1-D detector array is used for CT. This results in an image of a 2-D slice through the body. Several such slices are stacked together to form a 3-D volume. Because each DSA is a 2-D image, the approach outlined in this paper results in ``true volumetric CT''--- though only for high contrast structures.

Each image acquired by the imager, after distortion correction, is well approximated by a perspective projection geometry. In fact, one can think of the imager as no different from an ordinary pinhole camera with the difference that the object being imaged is inside the camera (i.e., between the optical center and the image plane). Knowing the parameters of the perspective projection associated with each image is essential if they are to be collated into a 3-D stack. This task, in classical computer vision, is referred to as *camera calibration.*

Typically, the gantry has on-board sensors to monitor its position and orientation parameters. However, for images acquired at 25 frames/sec during a continuous rotation, the sensed parameters need to be refined if high accuracy is desired. Because of vibrations, gantry deformation, and slight mis-registrations in the imager position, each image is taken by a camera with slightly different parameters. Not only do the acquisition parameters change from one image to the next, they vary from one run to the next as well. Thus camera calibration for images acquired on standard X-ray angiography gantry is a key task that must be accomplished before 3-D reconstruction can be done.

This paper outlines a new methodology for calibrating deformable gantries. It also presents the first in-vivo results of 3-D reconstruction using cross-calibration.

4.0 Image Acquisition

For digital subtraction angiography, an X-ray image of normal anatomy is taken before and after an X-ray opaque contrast agent is injected into the blood vessels (Macovski, 1983). Logarithmic subtraction of the before image from the after image subtracts out all but the opacified blood vessels. In log-subtraction, from the logarithm of each pixel value in the opacified image, the log of the corresponding pixel value in the mask image is subtracted out.

Fig. 2 shows a digital subtraction angiogram of a dog's carotid. The catheter inserted into the carotid for injecting the dye is clearly visible. Instead of direct subtraction, log subtraction was used. For 3-D angiography, several such 2-D DSA images around a patient are acquired and then combined to form a 3-D description of the blood vessels.

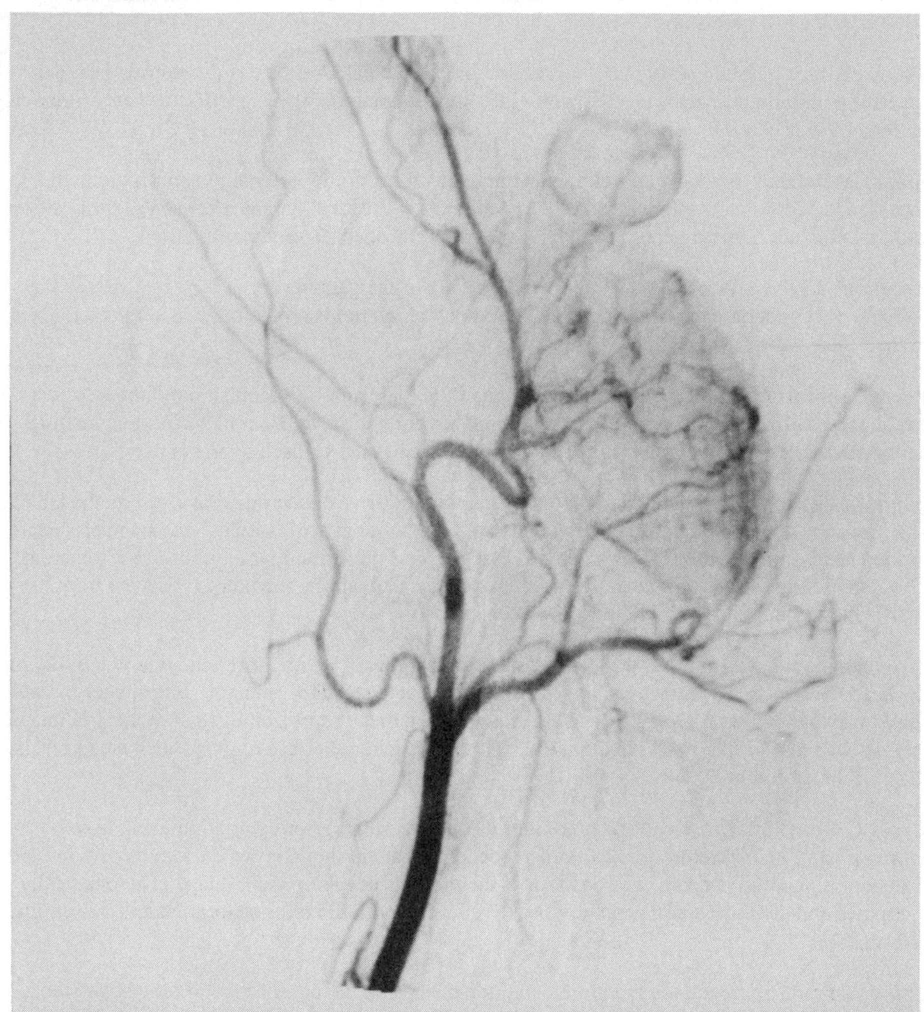

FIGURE 2. Digital subtraction angiogram of a dog's carotid and other blood vessels.

5.0 Calibration and 3-D Reconstruction

Once all the 2-D images have been acquired, we need to know, with high accuracy, all the geometrical parameters that describe the conic projection for each image used for reconstruction. The gantry, which obviously deforms during rotation, makes this task more difficult. Gantry deformation introduces artifacts both in the fast spin studies (subtraction artifacts) and in 3-D reconstruction (the reconstructed morphology is blurred if the geometry of the acquisition is known with a low precision).

There are three possible ways to do the calibration for each DSA image (i.e. determine the perspective transform that maps the 3-D world into the 2-D image). These possibilities are discussed below.

Ideal Calibration. In ideal calibration, the nominal values of the internal parameters (e.g. the focal length) associated with the gantry are used. For the external parameters such as the position and the orientation of the gantry, the values measured by on-board sensors are used.

Experiments have shown that the internal parameters of the gantry vary as it rotates around the patient. This phenomenon renders ideal calibration, which holds all the internal parameters fixed to their ideal value, a crude approximation.

Prior Calibration. If the deformations were reproducible from one acquisition to the next, we could use a calibration procedure prior to the image acquisition, in order to perform geometric calibration of each projection. This *prior calibration* can then be used over and over.

In prior-calibration, a calibration phantom containing known fiducial markers is put on the table. Calibration is done without the patient on the table. A fast-spin acquisition --- the same procedure as the one that will be used with the patient later on --- is performed on the calibration phantom. The X-Ray images are automatically analyzed and the acquisition geometry is derived from the localization of the fiducial markers in each projection.

To determine the feasibility of this prior-calibration approach, we measured the reproducibility of the deformations from one acquisition run to the next on a standard gantry. Unfortunately, it was found that the reproducibility is poor. For example, the mean reprojection error in the 2-D image is of about 3 to 4 pixels, in 512x512 matrix. This approach, nonetheless gives much better results than ideal calibration.

Cross Calibration. Cross-calibration refines the results of the prior-calibration using in-vivo information. Some markers are positioned around the patient in unknown but static positions and fast spin acquisition is done. The position of the markers in each patient image is automatically computed and the acquisition geometry is derived using both the prior calibration and this on-line informations.

This calibration approach requires some additional steps on the part of the radiologist and the X-ray technicians: they have to position the fiducial markers around the patient, guide their extraction and supervise the calibration. However, there are several advantages inherent in this anthropocentric approach. Cross-calibration places no restrictions on the internal parameters of the camera (e.g., unlike auto-calibration, it does not assume that each image is taken by the same camera). In addition, only those parameters that change from one run to the next are refined. Finally, image to image match points rather than 3-D control points are used. Thus, the fiducial markers can be places very close to the patient and we do not need to know their 3-D location.

In experiments conducted on a biplane system, cross-calibration gave very good results compared to the prior calibration. These results will be presented in later section.

5.1 3-D Reconstruction

The classical algebraic reconstruction technique (ART) was used for reconstructing the 3-D volume. The details of this tomographic reconstruction technique are beyond the scope of the paper (see Rrouge'e (1988) for details).

6.0 In-Vivo Results

In order to test the overall approach, we acquired a sequence of images on a rabbit. Only the mask images were acquired and a 3-D reconstruction of the bones was done. As far as the overall protocol is concerned, there is no difference between 3-D reconstruction of bones and that of the vasculature. So one can infer the accuracy of the blood vessel reconstruction from these results. In this experiment, a set of pins were put around the head of the rabbit to provide for the fiducial marks used for cross-calibration.

Fig. 3 shows a 3-D reconstruction of the bones. The pins used for cross-calibration are clearly visible. Reconstructed volumes obtained using the different calibrations procedures discussed in this paper were compared. It was found that the results of ideal calibration are quite poor. Those of prior-calibration are much better but still the reconstruction has many visible defects (e.g. the cross-section of the pins is not circular). Finally, cross-calibration results in good definition for both the pins as well as the prominent bones.

7.0 Future of 3-D Angiography

What will a radiologist do with a 3D image? In the following, we outline the kinds of medical applications that will benefit from 3D.

3D is used today to analyze the morphology of a pathology. When the reconstruction is finished, the system will automatically send the 3D image to a display monitor in the angio suit, in a rotating mode. The radiologist can then use a dedicated tools to analyze the vessels and the pathology. One can also envision specialized software tools, for example, for virtual endovascular view, and automatic aneurysm and neck sizing.

By manipulating the 3D database on the workstation, a radiologist can also choose the right viewing incidence to better display and visualize the pathology. The next step then is to automatically instruct the to gantry reach this viewing incidence.

One can also superimpose the fluoroscopic images on the 3D database. This will result in a ``3D road-map'' for catheter planning and intervention. The uses outlined here are probably the tip of the iceberg. We anticipate a host of new applications and procedures in the field of interventional radiology, therapy planning, and surgery once safe and accurate 3D imagery becomes available on routine basis on as standard gantry.

FIGURE 3. 3-D reconstruction of bones from mask images using cross-calibration.

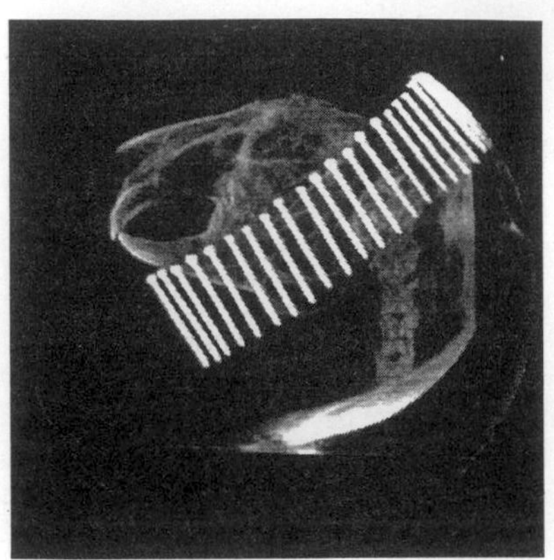

8.0 References

Macovski, A. (1983) Medical Imaging System}. Prentice-Hall, Englewood Cliffs, NJ.

Rouge'e, A., Hanson, K.H., and Saint-Fe'lix, D. (1980) Comparison of 3-d tomographic algorithms for vascular reconstruction. In SPIE Medical Imaging II, volume 914.

Saint-Fe'lix, D., Picard, C., Ponchut, C., Rome'as, R., Rouge'e, A. and Trousset, Y., Campagnolo, R., LeMasson, P., Schermesser, P., Crocci, S., Gandon, Y., Rolland, Y., Scarabin, J.M., Amiel, M., Finet, G., and Moll, T. (1993) A new system for 3D computerized X-ray angiography: First in vivo results. In SPIE Medical Imaging VII, volume 1897.

Trousset, Y., Desecures, H., and Grimaud, M.. (1994) On the interest of a region of support in 3-D vascular reconstruction. In SPIE Medical Imaging, Newport Beach, CA, February 1994.

INDEX OF CONTRIBUTORS

KEYWORD INDEX